Biomathematics

Volume 7

Edwin R. Lewis

Network Models in Population Biology

With 187 Figures

Springer-Verlag Berlin Heidelberg New York 1977

Edwin R. Lewis
College of Engineering, University of California, Berkeley, CA 94720, USA

AMS Subject Classifications (1970): 92-02, 92 A 15

ISBN 3-540-08214-X Springer-Verlag Berlin Heidelberg New York
ISBN 0-387-08214-X Springer-Verlag New York Heidelberg Berlin

Library of Congress Cataloging in Publication Data. Lewis, Edwin R. 1934– . Network models in population biology. (Biomathematics; v. 7). 1. Population biology – Mathematical models. I. Title. [DNLM: 1. Genetics, Population. 2. Models, Biological. W1 BI852T v. 7/QH455 L673n] QH352.L48. 574.5′24. 77-5873

Typesetting, printing, and binding:
Universitätsdruckerei H. Stürtz AG, Würzburg
2145/3140-543210

To
Betsy, Ned and Sarah

Preface

This book is an outgrowth of one phase of an upper-division course on quantitative ecology, given each year for the past eight at Berkeley. I am most grateful to the students in that course and to many graduate students in the Berkeley Department of Zoology and Colleges of Engineering and Natural Resources whose spirited discussions inspired much of the book's content. I also am deeply grateful to those faculty colleagues with whom, at one time or another, I have shared courses or seminars in ecology or population biology, D.M. Auslander, L. Demetrius, G. Oster, O.H. Paris, F.A. Pitelka, A.M. Schultz, Y. Takahashi, D.B. Tyler, and P. Vogelhut, all of whom contributed substantially to the development of my thinking in those fields, to my Departmental colleagues E. Polak and A.J. Thomasian, who guided me into the literature on numerical methods and stochastic processes, and to the graduate students who at one time or another have worked with me on population-biology projects, L.M. Brodnax, S-P. Chan, A. Elterman, G.C. Ferrell, D. Green, C. Hayashi, K-L. Lee, W.F. Martin Jr., D. May, J. Stamnes, G.E. Swanson, and I. Weeks, who, together, undoubtedly provided me with the greatest inspiration.

I am indebted to the copy-editing and production staff of Springer-Verlag, especially to Ms. M. Muzeniek, for their diligence and skill, and to Mrs. Alice Peters, biomathematics editor, for her patience.

A very large part of this book was written while I was on leave from the University of California with my family and living at Kamaole, Maui. I am grateful to Prof. H. Morowitz for pioneering the locale and for encouraging us to follow, and especially to my family for their constant support.

September 9, 1977

Table of Contents

1. Foundations of Modeling Dynamic Systems

2. General Concepts of Population Modeling

3. A Network Approach to Population Modeling

4. Analysis of Network Models

Introduction

The chief purpose of this book is to introduce the biological scientist to some of the basic notions and methods of dynamical systems and network theories, especially to the powerful and heuristic yet beautifully simple topological aspects of those theories. Although the emphasis here is on the application of these tools to population biology, I hope that the reader will see beyond this limited scope to the much more general applicability of the methods to biology at large. The notions of particulate processes and of particle state spaces, for example, to which chapter 1 is devoted almost entirely, provide a sound theoretical substrate for biophysical models at all organizational levels, from the macromolecular through the organismal to the biospheric. Add to these notions the principle of detailed balance and the rather simple elements of lumped-circuit theory and one has a powerful set of intellectual tools for forming models of the irreversible thermodynamic processes taking place at those various levels. Add further the simple topological methods described in chapter 4, and one has the means of deducing the consequences of such models.

The book is written for the biologist with a workable knowledge of elementary algebra. The most simple tools of calculus are applied in one or two instances, but these applications are neither complicated nor crucial. Knowledge of elementary probability theory would facilitate understanding of much of the text; but the book was written under the assumption that the reader lacked such knowledge, and the basic concepts are introduced as they are needed.

The text describes dynamics in terms of discrete time rather than continuous time. This is a natural mode for dealing with population models; and it has the advantage of allowing the description and analysis of dynamics to be carried out without the calculus. On the other hand, the text just as easily could have been couched in terms of continuous time, employing differential equations and calculus (in fact it originally was intended to be so). However, virtually all of the methods employed in the text are equally applicable to continuous-time models, as any reader familiar with differential equations and Laplace transforms clearly will see. Therefore, there is very little loss of generality in the restriction to discrete-time models.

Because the book is intended to introduce a modeling methodology rather than to be an exposition of existing population models, it does not include descriptions of many of the elegant, classical models of population biology, particularly those that are based on continuous time. Expositions of these models now are abundant in ecology and population-biology texts, and their inclusion here would be of little or no value. The text opens with a discussion of dynamic processes and epistemological considerations related to such processes and our models of them. It moves logically from the dynamics of individuals to the dynamics of ensembles, treating explicitly the special aspects of ensembles of organisms. Then it moves on to an extensive discussion of the problem of actually constructing models for the purpose of deducing dynamics. This is followed by an equally extensive discussion of the methods presently available for carrying out the deductive process (i.e., analyzing the dynamic behavior of a model).

Along the way, the reader is introduced to the notions of *states, state spaces,* and *state equations,* to the *Markov process* and its dynamics, to the logical transition from *probabilistic models* to *large-numbers* ("deterministic") models, to the notions of *linearity* and *nonlinearity*, to *uncertainty* and its impact on models, to the logic of *state aggregation* (or "lumping"), to the elementary principles of lumped network modeling, to the elementary theory of probabilistic interaction (e.g., search and encounter) as a source of nonlinearity in population systems, to the notion of the *critical point* and the deduction of dynamical behavior in the neighborhood of such a point, to the methods of *linear transforms,* to the topological methods of *signal flow graphs,* to the *numerical methods* of system characterization, and to the basic concepts of *systems control.*

Why Model?

Each year, the International Whaling Commission and its scientific advisors sit down and review the present status of each of the exploited whale populations. Unfortunately, since accurate census of pelagic whale populations is impractical, the Commission must depend to a large extent upon inference rather than direct observation. Noting that the average size of whales taken during the past year is less than that for the previous year, while the average effort expended to catch a whale is greater, some Commission members conclude that the whale populations have been diminished markedly by whaling and propose reduced catch limits. Other members argue that the data are misleading and that the whale populations actually have not been diminished significantly. Compromise catch limits are established, and the Commission adjourns for the year.

In a remote village, young men and women are plagued by a recently established endemic population of blood flukes (schistosomes), which reside in the human circulatory system and whose eggs migrate through human tissues, causing considerable irritation. Working within the constraints of a limited public-health budget, field workers attempt to eradicate the schistosome population by semi-annual extermination of the schistosome's intermediate hosts (water snails), occasional chemotherapy treatments of the infected villagers, education of the villagers with respect to hygiene, and development of improved sanitation facilities. Periodically, the field workers check the young men and women of the village to determine whether, by any chance, their methods have been successful yet in eliminating the endemic.

Each of these situations represents an experiment directly on real populations and indirectly on an entire ecosystem. Experiments such as these, carried out directly on populations in order to devise (or in lieu of devising) management plans for those populations, clearly have significant costs and risks. The great whale experiment has had enormous costs, leading in succession to the commercial extinction of the Blue Whale and the Finback Whale, and undoubtedly to the same status eventually for the Sei Whale. The concomitant costs with respect to whale's food chain and the ecosystem associated with it have not been counted yet. In the case of the schistosome experiment, the costs have included considerable time and effort on the part of public-health field workers, large quantities of molluscicide and chemotherapeutic materials, and considerable human discomfort from chemotherapy and misery from unabated schistosomal infection.

More and more, where population management is involved and effective management strategies must be devised, workers have come to realize that experiments performed on population models rather than on the populations themselves generally involve considerably less risk and considerably less cost. Thus, in fisheries management, game management, pest control, epidemiology, and other population related fields, one finds increasingly greater use of models in the development and *quantitative* testing of tentative strategies and tactics.

It has been said that the immense complexities of the biological world are better suited to description in natural human language (e.g., German, Mandarin, Russian) than they are to the language of mathematics. We certainly should be aware of those complexities when we attempt to manipulate any component of an ecosystem; and natural human language presently may provide the best vehicle for qualitatively deducing a conservative approach. On the other hand, the language of mathematics also is a "natural human language", one devised for the special purpose of providing a vehicle for *systematic quantitative deduction*. Occasionally the resulting deductions are quite "counter-intuitive", i.e., contrary to the qualitative conclusions reached through the other natural human language. Thus, for example, Macdonald and others have concluded, through modeling carried out in the language of mathematics, that certain qualitatively very promising schistosome control strategies were doomed quantitatively to failure. Among these doomed strategies were several employed by field workers in their "experiments". The point is not that quantitative models and reasoning should displace qualitative thought in biology, but that each mode should have a place in the biologist's intellectual repertoire.

1. Foundations of Modeling Dynamic Systems

1.1. Time

We are faced with a dynamic universe. Every object we perceive is undergoing change of one sort or another, rapidly or slowly. To deal with such a universe, we need a scale by which to gauge the sequencing or flow of events that are important to our survival and wellbeing; and that scale, of sourse, is time.

"Time", as we perceive it, may be a rather abstract, strictly human concept. However, it is quite clear that all organisms face a dynamic universe and somehow must deal with it; and it is equally clear that many (probably all) organisms have built-in gauges of what we call "time".

Our individual, egocentric conceptions of time undoubtedly are quite different from one another. On introspection, I find that I tend to perceive observable epochs, such as one day, as passing at different rates, depending on my own activities. I tend to value present time much more highly than time past, begrudging any interruption from my chosen task, then turning and without a qualm discarding weeks of work from the past. I tend to perceive of remote times, future and past, in "compressed" form and to perceive present time in expanded form. As a result of my own cultural background, certain epochs in the past are not nearly so compressed as others. The dawn of man is extremely compressed; the epochs of Greek and Roman civilization are not as compressed as the Dark Ages that followed; the reigns of Henry VIII and Elizabeth I are far less compressed than those of the Seventeenth Century combined, yet the Seventeenth Century expands for me when I think of events in the New World. I suspect that most other human beings have compressions and expansions in their own perceptions of time.

To set their thinking straight with respect to the relative sequencing of events and to aid in their communication with one another concerning sequences of events, human beings have attempted to develop standard measures of time and standard clocks with which to make the measurements. Probably the simplest and most obvious clocks are natural sequences of discrete events that we perceive to be periodic. The standard measure of time often is a running count of such events, in which case the resolution of time is strictly limited to the interval between events. On the other hand, we seem to perceive the passage of time not as saltatory, from one discrete interval to another, but as a continuous flow. The simplest and most obvious clocks consistent with this perception of time are those comprising objects moving smoothly through physical space with velocities that we perceive to be constant. Thus, we often conceive of the passage of time in terms of one of the coordinates of physical space, with the *present* being represented as a point moving along

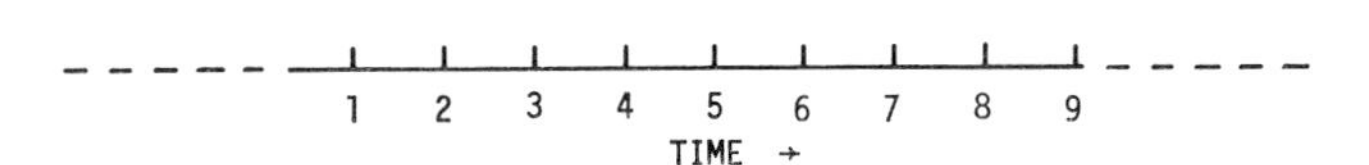

Fig. 1.1. The standard time line. The concept of time represented by a linear mapping onto a line

that coordinate with a *constant* velocity. For practical reasons, most clocks of this type involve points that sweep out circles or spirals as they move along. Of course, the passage of time can be conceived as the movement of a point along a straight line, whose extent is a matter for philosophers to ponder, but whose segments are extremely useful for picturing the ordering of events.

By convention, the standard mapping of time onto a line in physical space has one interesting and important property, namely it is taken to be a linear mapping. Loosely stated, this means that a centimeter of the resulting time line, regardless of where it is taken along the line, always represents the same temporal duration. Thus, if we were plotting discrete, periodic events along the standard time line, the distance between successive events would be constant.

1.2. Dynamics

With our standard time line, we now are in a position to have standard pictures of the dynamics in the universe about us. In order to discuss those dynamics on a rational basis, however, we also need descriptions of the things that are undergoing change. If we focus our attention on a single object[1], then the things that will be changing are the various attributes of that object, such as its position or distribution in space, its motion, its mass, its energy content, its color, its age or the age distribution of its components, and so forth. These attributes may be interdependent to a greater or lesser extent; but, ultimately, their changes all are gauged with respect to time. Time is the one strictly independent variable of dynamics.

1.3. State

The notion of "state" is extremely useful in the discussion of dynamics and is central to virtually all quantitative modeling of dynamic objects. As we shall see later on, the practical, precise definition of "state" depends on the type of object being modeled, taking one form for models of large ensembles of individuals (such as populations of organisms), and a fundamentally different form for models of the individuals themselves. Strictly speaking, the "state" of an object at any given moment is the collection of values attained at that moment by all of its attributes. From a practical point of view, however, not all of its attributes are observable, nor do all of them change perceptibly during

[1] "Object", in this case, is used in a very general sense and can refer to any identifiable entity, such as an individual whale, a pod of whales, a herd of whales, or the entire population of whales in the Southern Hemisphere.

normal periods of observation; so many attributes often are left out of state descriptions. The various attributes included in the state description of an object are called the "state variables" of that object.

Concisely put, *the "dynamics of an object" comprise the changes of its state as time progresses.*

1.4. Discrete and Continuous Representations of Time

To depict graphically the dynamics of an object, one might observe each of its state variables and plot its values over the time line (i.e., on a graph with time represented along the horizontal axis and the value of the attribute represented along the vertical axis). Occasionally, state variables are monitored continuously and their values plotted automatically (e.g., by an oscillograph or pen recorder) as a continuous curve. On the other hand, data points often are taken and plotted at discrete intervals of time, in which case one has several choices for the final graphical representation. He can fill in between the data points by any of the various interpolation schemes, or he simply can acknowledge the incompleteness of his data and leave the points as they are or convert them into a histogram (bar graph).

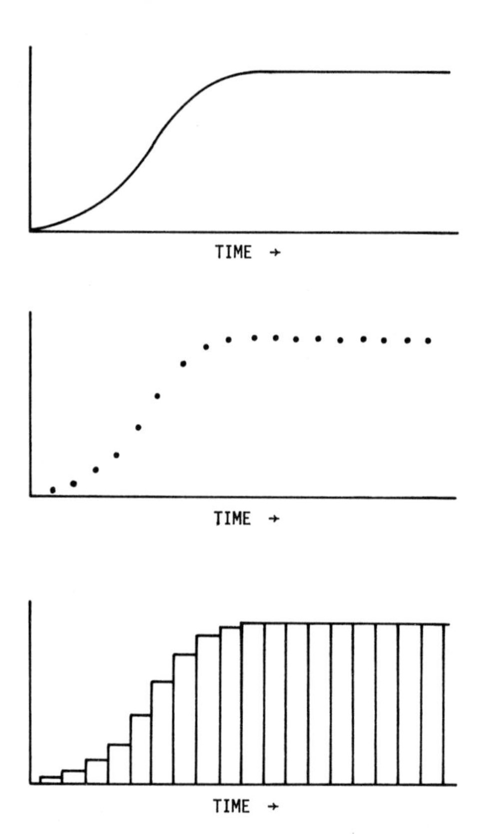

Fig. 1.2. Three ways to display data along the standard time line

From Figure 1.2 it is clear that time can be represented in two fundamentally different ways. It can be represented point by point as a continuous independent variable, with the dependent variable assigned a separate value for each point; or it can be broken up into a sequence of distinct epochs, forming a "discrete" independent variable, with the dependent variable assigned a single value for each entire epoch. Normally, when the latter approach is taken, the epochs are assigned equal durations. This need not always be the case, however.

Since the standard time line is a linear representation of time, one can picture an object moving along it with a constant velocity, the point of its present position representing the moment of present time. It is both interesting and useful to realize that because points of time (i.e., "instants" or "moments") are infinitesimal in extent, the object does not dwell in them, but merely passes by them. The object does dwell in discrete epochs of time, however, moving from one to the next in saltatory fashion.

1.5. The Discrete Nature of Observed Time and Observed States

According to currently accepted physical principles, quite a large number of state variables are inherently discrete. Thus electrical charge presumably comprises discrete protonic or electronic charges; matter comprises discrete particles; energy comprises discrete quanta; and so on. On the other hand, we generally seem to perceive physical space as a continuum. Furthermore, perhaps because we are such visually oriented organisms, we perceive a very wide variety of physical variables by first translating them into position in physical space; and therefore we might perceive in those variables the same continuous nature we attribute to physical space itself. By the same token, our resolution of variables that are translated in this manner clearly will be limited by our resolution of physical space. Whether or not physical space indeed is a continuum, there always has been a limit on the resolution with which we can perceive it; and I suspect that there always will be such a limit. One important variable affected by this limit is time. Even if we use the second type of clock described in section 1.1, our resolution of time nonetheless will be strictly limited.

A separate but equally important consideration with respect to resolution of a state variable is the time required to observe and record it. Here we run headlong into a fundamental limitation. Precision in our plots of an object's dynamics requires not only precision in our determinations of the values of its state variables, but precision in our determination of the corresponding time. According to the Heisenberg uncertainty principle, if a measurement is to be at all precise, it will require some finite span of time to make. As we approach the limits specified by the principle, increased temporal resolution will require increased energy delivered to the object being observed and will result in increased uncertainty with respect to the amount of that energy absorbed by the object and the amount of energy that the object would have had on its own during the interval of observation. Even without this fundamental limitation, we are faced with the very practical fact that observations require time, and increased precision almost always is purchased at the expense of increased time of measurement and a concomitant decrease in the resolution of the value of time corresponding to the value obtained by the measurement. This problem is compounded, of course, when the state of an object is described by several variables, all of which must be measured.

Thus, it can be argued quite logically that observed time and observed state variables are discrete not only in practice but also in principle. Any measurement of time will be recorded and communicated as an integral number of discrete elements whose magnitudes are equal to each other and to the minimum discernible interval of time, which we shall call the unit of temporal resolution. Similarly, a measurement of any other variable will be recorded and communicated as an integral number of appropriate, discrete units of resolution.

1.6. State Spaces

An individual object, of course, may have a very large number of attributes. Attributes of importance for a nonliving object might be position in space, velocity, mass, temperature, and the like. For living objects, these same attributes may be important, but so are other attributes, such as physiological, anatomical and behavioral variables. If the value of each attribute is represented as a location (of finite extent, not a point) along a line, as is often the standard practice, then one can imagine (rather vaguely) a space in which each attribute line is a dimension. The object itself would be located in a finite hypervolume of this space, representing its present state. As time progresses, the object would jump from finite hypervolume to finite hypervolume. The finite hypervolumes would be the fundamental, discrete *elements* of the space; and their sizes (which might vary from place to place in the space) would be determined by the precision limits with which the value of each attribute was measured. Thus would the *state variables* of the object be brought together in the construction of a multidimensional *state space* for that object.

One can think of the discrete, elementary hypervolumes as the grains of the space, with their size determining the resolution within the space. Presumably, increased resolution can be obtained with increased effort and expense in attribute measurement; but, according to the discussion in the previous section, there very likely will be unavoidable trade-offs and ultimate resolution limits. Whether in principle, or merely in practice, the grain size will be finite.

Now for those readers who, like the author, have great difficulty imagining hypervolumes in multidimensional spaces, there is another way to view the state space of an object, a way that is less foreign to us dwellers in three-dimensional physical space. Since there is a finite number of elementary hypervolumes or elements in the space, each of them can be represented as a small area on a plane. Following an often-used convention, we shall bound these areas by small circles when we draw the state space in this manner. Each circle (or the area within it), then, represents one possible state of the object. We now can view the dynamics of the object as its saltatory progression from circle to circle as time passes.

Continuing our spatial and graphical analogy, we can speak of a state (i.e., the hypervolume or circle representing the state) as being *occupied* by the object

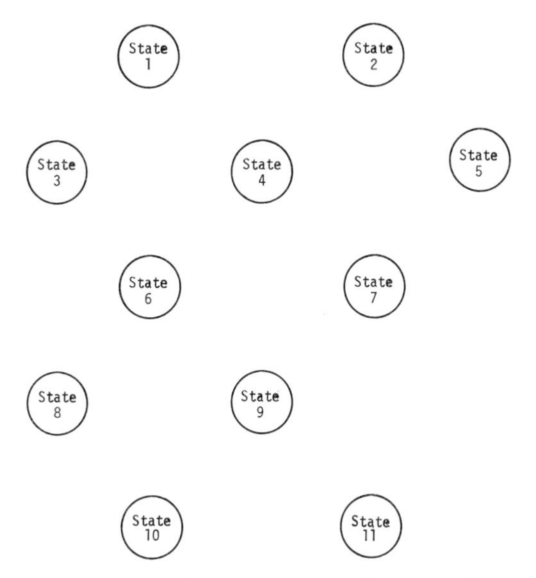

Fig. 1.3. Grains of resolution in state hyperspace, mapped onto a plane and representing the discrete elements of a finite state space

during an element of time. What we mean is that the values of the various attributes of the object at that particular time fall within the finite ranges defining that particular state. We can speak equally well of a state as *occurring* during a given element of time. The notion of a state being occupied will be especially useful later on, when we consider populations of objects, many of which can occupy the same state (i.e., exhibit the same attribute values, within the limits of precision) during the same element of time.

The state space is the foundation of any model of dynamic processes. Its formulation is quite arbitrary, both in number of attributes represented and in its grain size or resolution. During any period in the development of technology, there will be a maximum precision imposed on the measurement of each attribute and trade-offs imposed on the precision of measurements when several attributes are observed simultaneously. Thus will the resolution of the state space be limited. When a state space has the presently achievable maximum resolution, we shall call it a *primitive state space,* and we shall call its elements *primitive states*. Clearly, if trade-offs are involved in the measurement of attribute values, then an object will have several or even many primitive state spaces, a different one for each different combination of specific trade-offs.

Surely, one should expect the number of discrete elements of a primitive state space to be very large, in fact enormous. With each element being represented by a circle in our graphical depictions of the space, it is quite obvious that we generally shall be able to show only an extremely small part of the complete primitive state space. Such partial depictions, nonetheless, can be useful in aiding our visualization of the dynamics of the object. In later sections we shall reduce greatly the number of elements in the space by one method or another, and our depictions will become complete, but less precise.

Although time is *the* independent variable of dynamics, it is not included as a dimension of the state space (however, an important concomitant of time, namely the age of an object, often is included). Instead, time either is treated as a parameter or is ignored altogether in state-space representations. Often, on the other hand, the dynamics of an object are represented very informatively by a plot of the progression of the object as a time function of one or two important variables of the state space. Since time is not a state variable, however, such plots are not state-space representations in the strict sense.

Example 1.1. Figure 1.4a shows a plot of the mass of a Finback whale as a function of time. Figure 1.4b shows a plot of age as a function of time for the same whale. These two plots can be combined to yield mass plotted against age, as shown in Figure 1.4c, with time as a parameter. The resulting age-mass plane is a projection of the state space of the whale onto a single plane; such projections are called "state planes." The reader can imagine the beginnings of the construction of the entire state space by adding one more dimension (e.g., length of the whale) to the graph.

In spite of the fact that time is the independent variable of dynamics and is not included as a dimension of state space, its treatment is not independent of the structure of the state space. Limits of precision always will impose a minimum interval of time that can be discerned. Ideally, one might wish to use this minimum discernible time interval as the basis for following the progres-

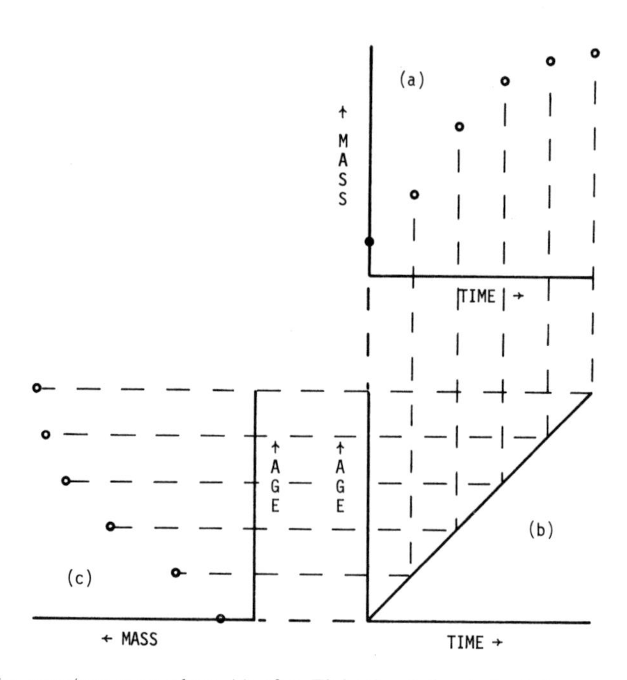

Fig. 1.4a–c. The mass/age state plane (c) of a Finback whale, derived from mass (a) and age (b) plotted as functions of time

sion of an object through its state space (i.e., find the state occupied by the object during each minimum discernible time interval). However, since measurement of traits cannot be accomplished instantaneously, one generally must invoke a compromise trade-off between resolution in time and resolution in state space. Such a compromise will lead to a basic temporal unit of resolution that is greater than the minimum discernible interval and is as important to the primitive state space as any of the resolution limits contained in it explicitly.

1.7. Progress Through State Space

It is not at all unreasonable to propose that the entire history of an object's progression through its state space is embodied in its present state. In other words, if we knew *precisely* the present state of an object and if we knew precisely all of the extrinsic forces that had acted on the object in the past, then in principle we should be able to determine precisely the entire history of the object's wanderings through state space. Furthermore, if we also knew precisely all of the extrinsic forces that would act on the object in the future, then we also should be able to predict precisely its future wanderings through state space. This is the *deterministic ideal.*

However, whether in practice or in principle, we are not able to locate an object precisely in state space. The best we can do is locate the element of the primitive state space that it occupies. We cannot determine where it is within that element. In fact, if the state description comprises state variables that cannot be observed simultaneously (i.e., in the same minimum discernible interval of time), then location of an object in a single element of its primitive state space becomes problematical; and perhaps the best one can do is assign a probability to each of several states it might have occupied during the observation period. In either case, the question of whether or not the statement in the previous paragraph is true or philosophically acceptable is important only because of its converse: Since we cannot know precisely the present state of an object and we cannot know precisely the extrinsic forces that will act on it in the future or that have acted on it in the past, we cannot deduce precisely its past wanderings in state space or predict precisely its future wanderings. On the face of it, this last statement is quite trivial. However, it can be strengthened to form the following statement which we shall hold as a basic tenet, without attempting to prove its truth: Because we cannot know precisely the present state of an object, but merely can locate it in an element of its primitive state space, we cannot predict with certainty even the element of that primitive state space to which the object will progress next.

In other words, as a basic tenet, we hold that because of the inherent uncertainty in our measurements and the resulting less-than-perfect resolution in the primitive state spaces of the interacting objects we observe, the transitions of those objects from state to state are inherently stochastic (i.e., governed by probabilistic laws).

Therefore, if we observe the present primitive state of an object, our predictions, based on that state, of future primitive states will be less than certain. As a result of the uncertainty of each and every transition, our uncertainty concerning the location of the object in its primitive state space usually will increase as the time since our last observation of the actual primitive state increases. The word *usually* is included for two reasons. First, we must exclude a very special subset of those objects whose probabilities of transition from state to state vary with time. Second, although for most objects the uncertainty does grow with time since observation, that growth is not always purely monotonic (see section 2.2). In a very real sense, it is the dynamics of the object that produce this *amplification of uncertainty*.

The amplification of uncertainty is demonstrated very dramatically by a common exercise given to students in elementary physics courses, which shows that even the relatively minute uncertainty postulated by Heisenberg is amplified enormously by just a few elastic or nearly elastic collisions, such as one expects between particles in a perfect gas. To understand this amplification, consider the following planar analogy. A ray of light, restricted to the plane, is bouncing back and forth between two perfectly cylindrical mirrors, as depicted in Figure 1.5. If the ray is aligned perfectly along the line passing through the centers of the two cylinders, it will bounce back and forth forever between them along that line. On the other hand, if it is aligned an iota off of that line, it very quickly will escape from the two mirrors. The direction of the final escape route is an extremely sensitive function of the initial path; so errors in our estimate of that path are amplified enormously in the escape route, and errors in our prediction of the direction of the escape route are converted in time to ever-increasing errors in our subsequent estimates of the position of the beam. Because of the addition of the third spatial dimension, the growth of uncertainty in the paths of perfectly spherical molecules making perfectly elastic collisions would be even greater, in spite of the fact that the same rules of collision would apply (the angle of incidence equals the angle of reflection).

The amplification of uncertainty is even greater if more interacting state variables are involved. For example, if the particles in the gas were not spherical, then the effective collision angles would depend on the orientations of the particles as well as on their relative positions and velocities. Uncertainties in estimation of orientation and rate of change of orientation would be amplified

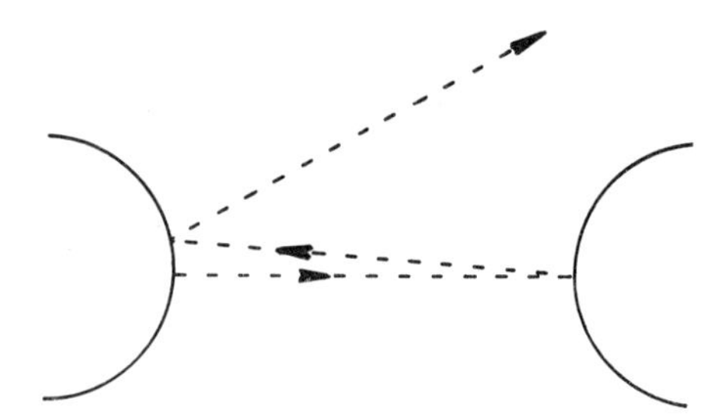

Fig. 1.5. Imagined path (dashed line) of a light ray bouncing back and forth between two cylindrical mirrors. If the alignment of the initial segment of the path is one iota off of the center-to-center line of the mirrors, the ray very quickly will escape

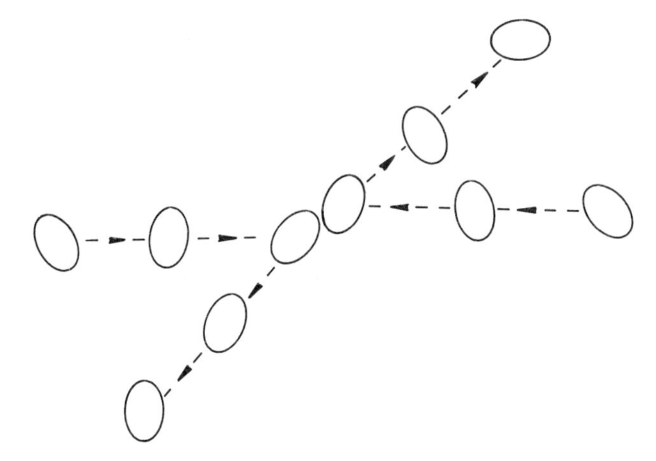

Fig. 1.6. Imagined collision of a pair of rotating, nonspherical particles, illustrating that the effective collision angle depends upon the rotational orientations of the particles as well as upon their relative positions and velocities

through time and would compound the uncertainties of future positions and velocities.

Other complexities might be added. For example, if the particles collide with appropriate velocities and orientations, they might stick to one another, only to be knocked apart in subsequent collisions. As the number of interacting variables increases, the amplification of uncertainty with each collision becomes greater and greater. Interacting, or "colliding" organisms very likely have a huge number of such variables, leading one to expect extremely rapid amplification of uncertainty with respect to the state of any individual organism.

Exercise. In analogy with the gas discussed in the previous paragraphs, consider a population of idealized paramecia in a drop of water, with simple rules of motion and rules of collision such that appropriate new paths of motion are selected after each collision. Formulate the rules yourself, then convince yourself that the uncertainty concerning the position of a given paramecium increases with the time since it last was observed.

Exercise. Adding slightly to the complexity of the previous exercise, imagine that the drop of water contains two populations of paramecia, with the members of one preying upon the members of the other. Convince yourself that the uncertainty concerning the time of demise of an individual prey increases with the time since the last observation.

The very strong amplification of uncertainty is the result of an inherent, strongly divergent nature of the paths or *trajectories* available to an object through state space when it is interacting with other objects or with boundaries. If the object were free to move about deterministically and without such interactions, then one very reasonably might expect its trajectories not to diverge,

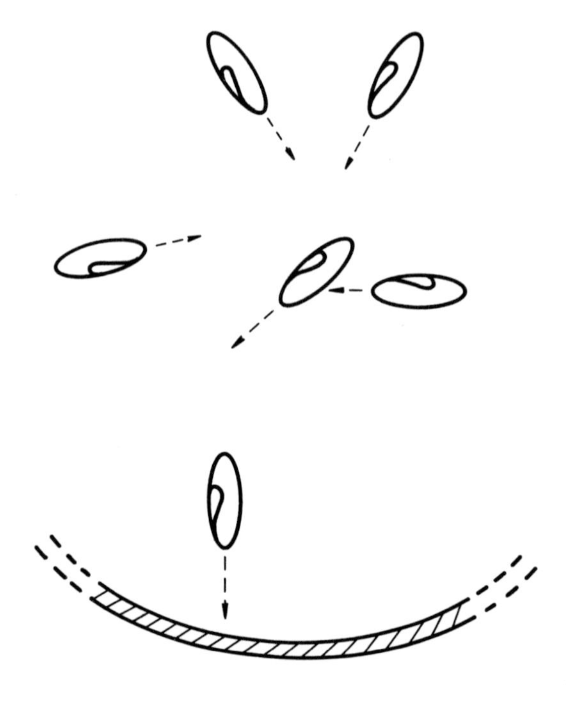

Fig. 1.7. Imagine the chaotic behavior of paramecia swimming about in a container of water, colliding with each other and with the walls of the container

in which case the transitions from primitive state to primitive state also could be deterministic, as long as the boundaries of the states were properly aligned with one another. In fact, the degree of determinancy very likely could be increased systematically as the hypervolumes enclosed by the boundaries were increased, making the error in alignment less and less significant. However, in this text we shall be dealing with the state spaces of interacting objects, whose available trajectories diverge markedly, even though they might be deterministic. It is this divergence that provides the most compelling logical basis of our tenet that transitions from discrete primitive state to discrete primitive state are stochastic.

1.8. The Conditional Probability of Transition from State to State

Once we have constructed a state space for an object, the next step in the formulation of a model is the specification of the rules of transition from state to state in that space. Up to now we have alluded to extrinsic and intrinsic forces that drive the object from one state to another. The notion of *force* is natural and heuristic for many situations. Its application to the model, however, requires that the forces and their effects be specified or hypothesized. In dealing with populations of interacting objects, such as organisms, the notion of *transitional probability,* meaning the conditional probability of transition from state to state, usually is natural and much more heuristic than the notion of force. From this point on, therefore, we shall replace the notion of *force* with that of *transitional probability*. As we shall see, this does not eliminate the necessity of specification or hypothesis formation.

The probability of transition is labelled *conditional* because it depends on certain conditions being met before it applies to the object. The most important aspect of specification of rules of transition in terms of transitional probabilities is the specification of those conditions. It is here that one must formulate the basic hypothesis of his model by answering the question, "How much hysteresis is exhibited in the transition from state to state?" In other words, is the probability of transition from any one state (which we shall label U_i for convenience) to any other state (U_j) dependent, or conditional, not only on the presence of the object in state U_i but also on the states that it occupied prior to moving into U_i? If so, how far back into history must one go in order to specify adequately the conditions on the probability?

If we held to the tenet proposed in the first paragraph of the previous section and if we were able to know precisely the state of the object, our question would be answered. The nature of the transition would depend on the present state, but not on any previously occupied states. However, the lack of perfect resolution in our primitive state spaces precludes precise knowledge of an object's present state; and just as this lack of precise knowledge of state dictates that the transition be stochastic (owing to limits of precision), it also very well could dictate that the stochastic nature of the transition depend on the history of states occupied by the object. One might even go so far as to propose that knowledge of the complete history of an object's travels about its primitive state space could compensate completely for the imperfect resolution of that space and once again return us to a deterministic situation. In other words, by present observation we may not be able to determine an object's location *within* an element of its primitive state space; but, *perhaps,* we can come closer and closer to pinpointing its location within that element as we know more and more about the route it took to reach the element.

Since we must begin our observation at a finite time, thus being unable to observe the object's history prior to that time, there always will be an uncertainty that precludes our determining precisely the laws of historical dependence. Because of this uncertainty, the hypothesis of complete historical dependence can never be proven incorrect. This essential intestability makes the hypothesis scientifically valueless. It is scientifically valueless for another reason as well; it completely precludes further generalization of the rules of transition. Empirical observations of transitions will forever become part of the rules themselves, never tests of those rules. Models based on the hypothesis of complete historical dependence thus never can be completed; and as they grow, they are nothing more than complete records of the object's observed travels through its state space (i.e., unreduced data).

For these reasons, we shall discard the hypothesis of total historical dependence. We nevertheless still must decide whether or not to replace it with some hypothesis of partial historical dependence (i.e., dependence of the transitional probability on finite segments of the history of the object's travels). In principle, such hypotheses are testable statistically (i.e., by observations of correlations). The plausible alternatives can be sufficiently numerous, however, to preclude their exclusion in reasonable observation time. For example, one very reasonably might propose that the history of an object's travels through state space fades gradually in its reflection in present transitions. Thus, the occurrences of previous states would be weighted according to their times. They also might be given different weights according to the states themselves, some states having longer-lived effects than others. The combinations and variations of such factors in hypotheses of partial historical dependence obviously could be extremely numerous. This is not sufficient reason to reject such hypotheses out of hand. In fact, one always must bear in mind the possibility of partial historical dependence.

On the other hand, returning to the arguments at the end of section 1.7, one might note that if the divergence of precise trajectories takes place sufficiently rapidly (i.e., in a hypervolume that is very small compared to the hypervolume of the elements of the primitive state space), then it would be very reasonable to propose that all historical information is lost by repeated divergences within the individual element. In a subsequent section, we introduce the notion of maximum (or asymptotic) uncertainty, and one of the various theorems (called *ergodic theorems*) related to that maximum. In the abstract, we can think of each primitive state (each discrete element of the primitive state space) as a collection of a large number (less than infinity) of states that are undifferentiable from one another by observation. In a sense, what we can do here, then, is propose that a form of ergodic theorem can be applied to this collection of undifferentiable states, that as the object progresses stochastically from one of these finite states to another, the uncertainty with respect to its location among them grows and very rapidly approaches an asymptotic value that is independent of the route by which the object entered the collection (i.e., the route by which it came to its present primitive state). If this growth of uncertainty is sufficiently rapid compared with the time that the object normally dwells in the primitive state, then the route by which it entered its present state will provide negligible information concerning the routes by which it might leave.

What we are approaching here is a situation which very well could be called the *Markovian ideal*. It is the counterpart of the *deterministic ideal* and applies to situations in which transitions are stochastic. An object that conforms to the Markovian ideal is one whose transitional probabilities depend on the state from which transition is to take place (i.e., upon the present state) but *not* on the route by which it reached that state (i.e., not on the history of states prior to the present one). *Note that we have not excluded the possibility that the transitional probabilities depend on time-varying extrinsic factors; we have excluded only historical dependence.*

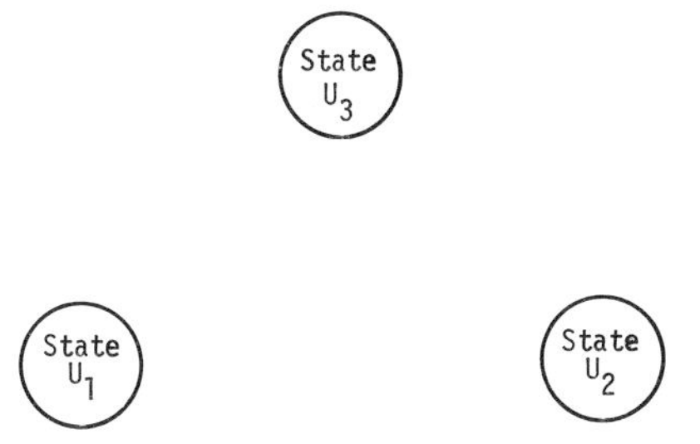

Fig. 1.8. Three non-Markovian states. Transition probabilities depend on the present state and the immediately preceding state occupied by the particle

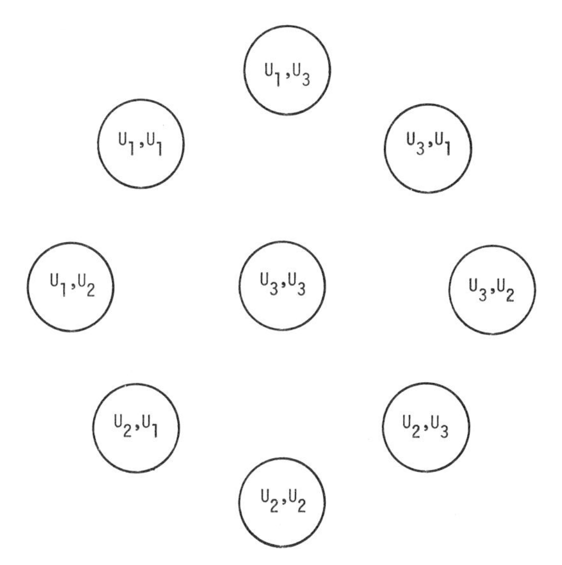

Fig. 1.9. Nine Markovian states constructed from the three non-Markovian states of Fig. 1.8. Each state embodies the two-interval history required to specify the transition probabilities; the label U_i, U_j represents a particle in non-Markovian state U_j having arrived there from non-Markovian state U_i. Clearly the transitions *from* state U_i, U_j will carry the particle into a state of the form U_j, U_x (i.e., a state whose stated history is consistent with arrival from U_i, U_j)

Whereas, in its primitive state space, an object conforms to the Markovian ideal only under rather special circumstances, having discarded the hypothesis of total historical dependence and accepted in its place the hypothesis of dependence of the transitional probabilities on finite segments of history, then in principle we always will be able to construct a new state space in which the Markovian ideal is met. In order to construct this new state space, we first must identify or specify the segments of history upon which the transitional probabilities depend. Then we simply make each segment an element of our new space, thus embodying the history in the new states. This process is illustrated in Figures 1.8 and 1.9 for a very simple case.

The generality of the process depicted in Figure 1.9 should be self-evident. As a result of that generality, throughout the remainder of this text we shall accept as a matter of principle the tenet that any primitive state space can be converted to a state space in which the states and state transitions conform to the Markovian ideal. This new state space we shall identify as the *primitive Markov state space,* and its elements will be called *primitive Markov states.* Our hypothesis concerning the historical dependence of state transitions will be embodied in the conversion of the primitive state space into the primitive Markov state space. If we hypothesize no historical dependence, then the two state spaces will be identical.

1.9. Network Representations of Primitive Markovian State Spaces

At this point, we can begin to establish a pattern that will be followed throughout the remainder of the text: the use of network diagrams as the bases for construction, discussion and analysis of our models. Such diagrams often provide invaluable heuristic aids to our intuitions concerning the dynamics of objects and groups of objects.

In Figure 1.3 we introduced a graphical representation of a multidimensional state space made up of a finite number of discrete elements. Each circle on a plane represented one of the elements, or states. From this point on, we shall continue to draw the circles with thin lines and we shall limit the use of these symbols (the circles with thin lines) to primitive Markovian states. Thus, they will represent either primitive states or sets of primitive states. Now, we can extend this graphical representation of the primitive Markov state space to include paths of possible transitions between the states, simply by drawing arrows connecting all pairs of states between which transitions are possible, as illustrated in Figure 1.10. In each case, the arrow points in the direction of the transition. Having drawn the arrows, we now have a *network* made up of primitive Markov states and transition paths connecting them; and this network is a *general model* of the object that moves about in the primitive state space we have depicted. The model will become specific when we specify an initial state of the object or when we specify the initial probabilities of its occupying the various states (which we shall call its initial probability distribution). In fact, specification of a single initial state simply is a special case of specification of an initial probability distribution. In this special case, the probability is one that the object is in a certain state and zero that it is in any other state.

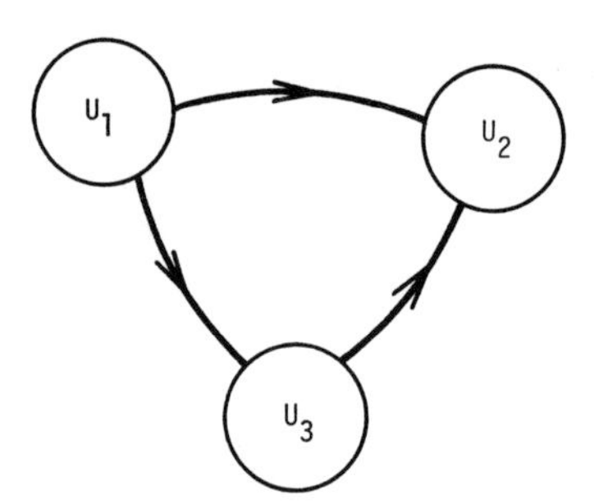

Fig. 1.10. Three primitive Markovian states with available transitions represented by directed paths

So far, the rules of transition have been specified only to the extent that the various transition paths have been specified. Even with this limited amount of information, however, we can deduce certain key properties of the network. For example, certain regions of the network may have arrows leading into them, but no arrows leading out; clearly if the object enters these regions, it never can leave. Similarly, a region of the network may have arrows leading out, but none leading in. One would conclude that if the object were in such a region, it eventually would leave, never to return. A network or a part of a network is called *irreducible* if every state in that network can be reached directly or indirectly (in several steps) from every other state. Irreducible networks have interesting properties that are discussed in subsequent sections (e.g., see section 2.1).

To complete our network model, it would be useful to label each transition path with the corresponding conditional probability of transition. In the last two sections, we have ignored the time parameter in our state spaces. Now we must return to it explicitly. At the end of section 1.6 we stated that there will be a basic unit of temporal resolution that is an essential part of the primitive state space, even though time is not one of the dimensions of that space. Having chosen the basic resolution unit of time on the basis of a trade-off with resolution in state space, we should include that unit in the primitive state description, which should read as follows:

> "Sometime during the basic resolution-interval of time bounded by t_1 and t_2, the state of the object under consideration fell somewhere within the hypervolume designated U_i in primitive state space."

The basic resolution-interval of time in this case is the (integral) number of minimum discernible time units between t_1 and t_2.

We now are in a position to define precisely our transition probabilities. $p_{i,j}(t_k)$ is the conditional probability that an object will occupy state U_i *sometime* in the interval between t_k and t_{k+1}, provided that it occupied state U_j *sometime* during the immediately previous interval (i.e., between t_{k-1} and t_k); where the time intervals involved (i.e., the interval from t_k to t_{k+1}) are equal to the basic temporal unit of resolution for the state space at hand. From this point onward in the text, we shall assume that one discernible transition has occurred during each such interval; and we shall call the interval simply the *transition interval.* Of course, after a single transition interval we may find that the object has not proceeded to a perceptibly different state. For that reason, a complete network model for the object must include self-transition paths (i.e., paths leading directly from a state back to itself), with associated transition probabilities. In order to distinguish sharply the self-transitions from the transitions from one state to a different state (interstate transitions), we shall label the self-transition probabilities q. Thus $q_i(t_k)$ is the probability that the object will occupy state U_i sometime during the interval between t_k and t_{k+1}, given the fact that it occupied state U_i in the immediately preceding interval.

Example 1.2. Consider the state diagrams of Figures 1.8 and 1.9. Assuming that each state in 1.8 is Markovian and is directly coupled to every other state,

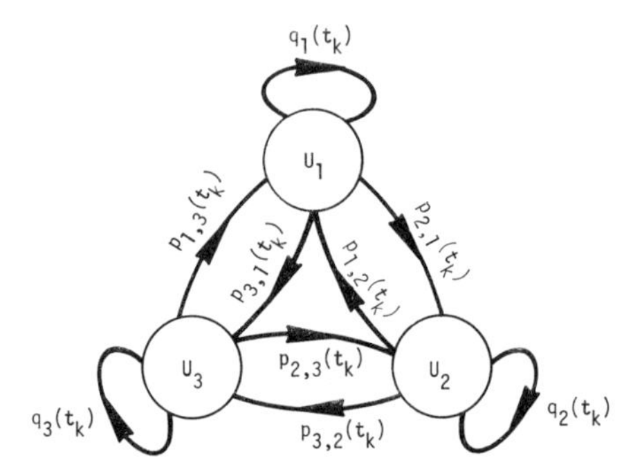

Fig. 1.11. A three-state system with available transitions represented by directed paths each labeled with the associated conditional probability of transition

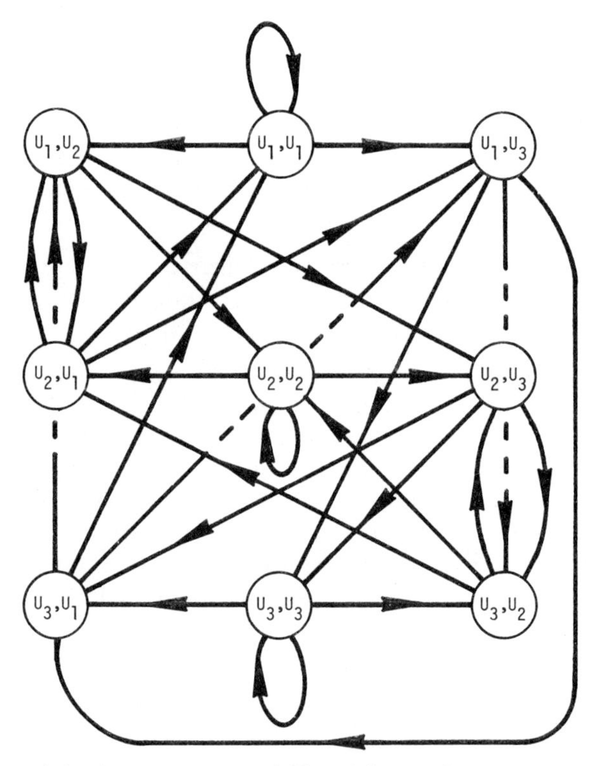

Fig. 1.12. Expansion of the three-state system of Fig. 1.9 into a nine-state system, with available transitions represented by directed paths

draw and label arrows to show transitions and transition probabilities in both diagrams.

Answer: Clearly, the process is very straightforward for Figure 1.8. One simply connects a pair of arrows between each pair of states, plus a self-transition arrow from each state to itself, and labels each arrow appropriately, as shown in Figure 1.11. The process for Figure 1.9 is complicated slightly by the fact that not every state can be reached directly from every other state. However, the connections can be determined rather easily by the rule stated in the caption of Figure 1.9. The corresponding network diagram is shown in Figure 1.12.

1.10. Conservation

One very important and useful constraint that can be imposed on our networks is that of conservation. The details of our conservation principle may vary from network to network when more than one object is being considered; and a statement of the conservation principle is an important part of the final model. When we are considering a single object in its state space, however, the principle of conservation usually is not flexible. It can be stated quite succinctly: "One and only one object is in the state space and it remains there always."

The embodiment of this principle in the network is mediated by the transition probabilities. An individual probability can range from zero (impossible) to one (certainty), taking on any value in between. Values outside of this range are without interpretation. One way to develop an intuitive feeling for the meaning of a probability is to think of it in terms of an empirically observed frequency of occurrence. For example, $p_{i,j}$ can be interpreted as the proportion of the times in which the object was observed in state U_j that its next transition carried it directly to state U_i. Zero would mean none of the times; one-quarter would mean one-quarter of the times; one would mean all of the times. Clearly, if when the object is in state U_j it goes to state U_i one-quarter of the times and to state U_k one-half of the times, then three-quarters of the times it will go to one or the other of those two states (U_i or U_k). Thus, if we add up all of the transition probabilities associated with the paths leading from a given state (including both self-transitions and interstate transitions), the sum will be the proportion of times (or probability) that after a transition from the given state the object still will be somewhere in the state space. Clearly, that sum *must* be one in order to guarantee that our conservation principle be met.

We shall return to the matter of conservation often again in later sections of the text. For the present purposes, however, we can state simply that *in order to represent a single object remaining forever in its state space, the sum of all of the probabilities of transition from each state in the model must be one.*

Notice that we have placed absolutely no constraint on the sum of the probabilities of transition *into* a given state.

Example 1.3. Apply the conservation principle stated above to the networks of Figures 1.11 and 1.12.

Answer:
$$\sum_i p_{i,j} + q_j = \mathbb{1} \qquad \text{for all } j.$$

Conservation also applies to the specification of the initial distribution. Clearly, the sum of the initial probabilities must be one in order that the object's initial presence somewhere in the space be guaranteed.

1.11. State Variables Associated with Individual Organisms

In section 1.5 we mentioned some of the state variables associated with individual organisms. Obviously there are many variables that should be included in the primitive state space of an organism, and construction of state spaces for certain types of organisms (i.e., those that reproduce by fission) may provide some interesting conceptual difficulties. Ignoring, for the moment, the vastness of the state space as well as these conceptual difficulties, we nevertheless can make a few useful observations.

In the first place, the state space of an individual organism is not irreducible. There are in fact certain irreversible changes that can occur in the physical, physiological, anatomical and behavioral variables associated with an organism. The irreversible changes that often are considered most important from a population-dynamics point of view are those associated with death. Although the location of the boundary is the subject of considerable debate, there clearly are two distinct regions of an organism's state space, one associated with life, the other associated with death. Clearly, each region contains a huge number of primitive states. In the usual sense of the word *death* (but by no means the only sense) the states associated with life are unreachable from the states associated with death, whereas each of the states associated with death can be reached eventually from one or more of the states associated with life. There are many irreversible changes that can occur other than those associated with death (their partial enumeration is left to the reader's imagination). Each of these results in isolation of two regions of state space, the states of the second region being reachable from states of the first, but no state in the first being reachable from any state in the second.

Another variable that very often is included in the state-spaces of individual organisms is *age*. Once again, we might have some interesting difficulties in applying this variable to certain organisms, but for many it can be defined in a useful manner. Strictly speaking, age can be considered as an individual's private segment of the time line; and, like time, it is not a variable in the primitive state space of an individual. By rather standard convention in the study of population dynamics, however, age is employed in the state space to replace a very large number of physiological, anatomical and behavioral variables that are strongly correlated with age. Being an artificial state variable, and therefore of arbitrary design, age is the one state variable in which transitions are deterministic. Attempting to treat all state variables in the same manner, some modelers have dealt with the process of aging as if it were probabilistic (e.g., a one-year old rabbit would be given a finite probability of becoming ten years old in one year's time). This logical absurdity has led to public ridicule of some models, but has gone totally unnoticed in many others. Considered in the light of the probabilistically changing variables that age is replacing, however, the probabilistic progression of age may not be totally absurd. It certainly is not very neat, on the other hand; and it definitely tends to make a mess out of bookkeeping in any model. We shall postpone further elaboration of these points to later sections, where they are discussed amply. The major point here is that, if age is included as a state variable in a neat and logical manner (i.e., with unidirectional, deterministic progression), then, being an irreversible process, aging automatically isolates a very large number of regions of the state space (one region for each unit of resolution of age).

The replacement of a whole myriad of state variables by a single variable such as age, is a universal process in modeling, known as *aggregation or lumping.* The result is a new type of state space, much reduced in size from that of the primitive state space. The logical basis for lumping is discussed in sections 2.1 and 2.3. The process itself is introduced here, however, in order that some examples and exercises of interest to the reader can be presented in the intervening sections. By replacing many variables with one, we have taken very large numbers of primitive states and combined them into single elements of our *reduced state space.* We shall call these elements *lumped states;* and we shall use the term *reduced state space* to describe any state space in which lumping has taken place (i.e., in general, any state space that is not a primitive state space). It is a basic fact of life (based on sound scientific methods) that when they are conceived originally, state spaces are reduced and states are lumped. The state spaces then are expanded to include more variables only when the modeler finds it necessary or amusing to do so. The primitive state space clearly is an ideal, useful perhaps for developing a modeling philosophy, but toward which very few practical modelers do or should strive. Nevertheless, the degree and nature of lumping is a crucial aspect in the specification of any model.

Throughout the remainder of this chapter and the next, reduced state spaces of individual objects will be presented occasionally. When they are illustrated, the lumped states will be represented by a circle bordered by a thick line, as shown in Figure 1.13.

From a population dynamics point of view, the two processes associated with the individual that are most important are survival and reproduction. State variables often are considered important only to the extent that they affect those two processes. This is the basic rationale behind the process of lumping in organism-population models; and, clearly, age is a natural variable in which to lump many others, since most of those primitive variables that affect survival and reproduction are very strongly correlated with age.

Occasionally, a variable that affects these processes is sufficiently independent of age that a modeler decides to include it separately. Mass is such a

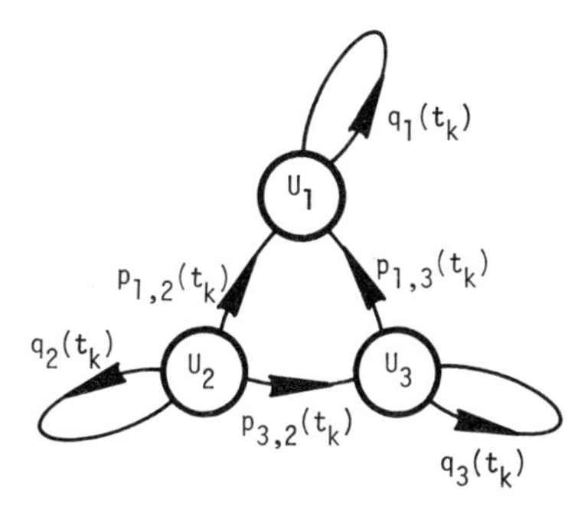

Fig. 1.13. Lumped Markovian states with associated transitions

variable. In some organisms, the ability to reproduce is markedly dependent on mass; and in many, the ability to survive is extremely dependent on mass. Thus, occasionally, modelers include mass along with age as a state variable of the individual organism.

Position in space is another variable that can be crucially important to an individual organism. It can make the difference between finding food or shelter and not finding food or shelter, finding a mate and not finding a mate, being preyed upon or infected by a parasite and not being preyed upon or infected. Furthermore, position in space often is not correlated very much at all with age or mass, so it very well might be included as a separate state variable.

The physiological and behavioral variables most closely associated with reproduction (e.g., the physiological condition of the gamete-producing organs, the behavioral state with respect to mating, and the like) often are lumped into a single, age-dependent variable called fecundity or fertility. At other times, *fecundity* is taken to be the rate of production of eggs and *fertility* to be the proportion of eggs that develop into viable offspring. In this text, we simply shall avoid the issue by *describing* the reproductive variable under consideration, rather than *naming* it.

1.12. Basic Analysis of Markov Chains

In order to relate our discussion of models to the dynamics of the objects involved, it is useful at this point to introduce some very basic and very simple analytical methods. Up to this point, we have constructed (in our imaginations) primitive state spaces for individual objects; we have converted those state spaces into primitive Markovian state spaces; and for the transitions from state to state we have assigned rules (transition probabilities) that are consistent with our specified principle of conservation. In this manner, we have constructed a model on the foundation provided by the primitive state space. Then, in section 1.11, we disclaimed the entire process by stating that models always are developed on reduced state spaces. For the time being, however, it will be useful to continue our contemplation of models on primitive state spaces; but we shall choose examples on reduced state spaces.

The question at this point is, "Now that we have a model, what do we do with it?" Since it is a model on a primitive state space, we can assume that it is a very good model, so we can use it to deduce good estimates of future behavior of the object being modeled. In other words, we can observe the object's present state, within the limits of state-space resolution, and from that observation deduce estimates of the object's states at future times. When we construct models on reduced state spaces, we may very well have other objectives, such as attempting to determine the mechanisms underlying a certain observed mode of behavior. For the present, however, we shall be content with the deductive aspects of models with respect to future behavior.

Since we have constructed our model on a Markovian state space (i.e., one in which the entire relevant history of the object's dynamics is embodied in its present state), when we observe the present state or the present probability distribution of the object we are establishing an entity known as a *Markov chain.*

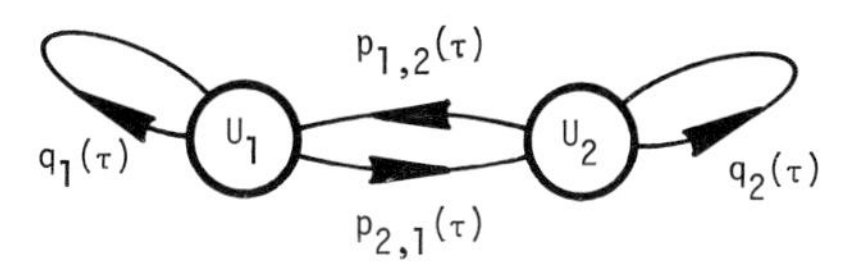

Fig. 1.14. Two lumped Markovian states with associated transitions

A *Markov chain* is defined simply as a set of Markovian states of a single object with their transition probabilities, taken together with an estimate of an initial distribution for the object. In other words, a Markov chain is specified by our model constructed on the primitive Markovian state space taken together with an initial distribution of the object occupying that space. To deduce from the initial (or present) distribution of the object estimates of its future states, (future distributions), we simply apply the very simple, basic analytical methods for Markov chains.

The actual analytical procedure can be visualized quite easily with a network diagram; the one provided in Figure 1.14 is very simple, but illustrative. Notice that the subscript has been dropped from the symbol for time. On the basis of our previous discussion, we shall assume that every Markov chain of interest to us has associated with it a fixed transition interval length, which we shall designate Δt. For consistency, the interval during which the state of the object is observed (i.e., that during which the initial distribution is established) will be labelled $\mathbb{0}$; and all previous intervals and subsequent intervals will be labelled sequentially with sequentially increasing integers, beginning at the interval of observation (the initial interval). Following the usual convention, intervals prior to the initial interval will be indicated with a minus sign, intervals that follow the initial interval will carry no sign. The symbol for the interval label will be τ. Thus, τ is a variable very closely related to time, but with a slightly different intepretation, applying to our discrete representation of time. It also can be interpreted as the number of transitions that have taken place since the observation of the initial state or, if it carries the negative sign, the number of transitions prior to the observation. Thus, $\tau=\mathbb{0}$ represents the initial interval; $\tau=\mathbb{5}$ represents the fifth interval after the initial interval (and occurs $\mathbb{5}\Delta t$ later, or $\mathbb{5}$ transitions later); and $\tau=-\mathbb{5}$ represents the fifth interval before the initial interval.

Suppose, now, that during $\tau=0$, the object under consideration is in state U_1 of the network of Figure 1.14. From this fact, we wish to deduce estimates of the state it was in during each interval after $\tau=\mathbb{0}$. This can be done interval by interval, working forward from $\tau=\mathbb{0}$. *According to our definitions of p and q*, the probability that the object still occupies state U_1 after one transition is $q_1(\mathbb{0})$; and the probability that it has moved on to state U_2 is $p_{2,1}(\mathbb{0})$. This is the extent of the deductions available to us for the first transition. It can be abbreviated as follows:

$$\Pr[\mathsf{U}_1(\mathbb{1})]=q_1(\mathbb{0}) \tag{1.1}$$

$$\Pr[\mathsf{U}_2(\mathbb{1})]=p_{2,1}(\mathbb{0}) \tag{1.2}$$

where $\Pr[\mathsf{U}_i(\tau)]$ is the probability that the object occupies state U_i during the τth time interval. We could complicate this notation by including in the argument of $\Pr[\,]$ the fact that the probability is based on the observation (or assumption) that the object occupied state U_1 initially. However, as long as the initial conditions and the dependence upon them always are understood and borne in mind, the additional notation is unnecessary and will be omitted for the time being.

The next question is, "What are the values of $\Pr[\mathsf{U}_1(\mathbb{2})]$ and $\Pr[\mathsf{U}_2(\mathbb{2})]$?" This question can be answered by remembering two simple rules from probability theory, the first is that the probability of a specific sequence of independent events (and Markov transitions qualify as independent events) is simply the product of the probabilities of the individual events. For example, the probability that the object in Figure 1.14 begins in state U_1, then moves to U_2 in the first transition, then back to U_1 in the second transition is simply the product $p_{2,1}(\mathbb{0})\,p_{1,2}(\mathbb{1})$. Notice that this product represents a specific route from U_1 back to U_1 in two transitions. The second rule that we need to invoke is that the probability of a given outcome is the sum of the probabilities of the various paths by which that outcome could be achieved. For example, there are two routes by which the object could begin in U_1 and be there again after two transitions: (1) it could remain in U_1 during two successive transitions, or (2) it could move to U_2 during the first transition and back to U_1 during the second. The probability that the object takes one or the other of these two paths is simply the sum of the probabilities associated with each of them:

$$\Pr[\mathsf{U}_1(\mathbb{2})] = p_{2,1}(\mathbb{0})\,p_{1,2}(\mathbb{1}) + q_1(\mathbb{0})\,q_1(\mathbb{1}). \tag{1.3}$$

There also are two routes by which the object could begin in U_1 and end up in U_2 after two transitions: (1) it could remain in U_1 during the first transition, them move to U_2 during the second, or (2) it could move to U_2 during the first transition and remain there during the second. Therefore

$$\Pr[\mathsf{U}_2(\mathbb{2})] = q_1(\mathbb{0})\,p_{2,1}(\mathbb{1}) + p_{2,1}(\mathbb{0})\,q_2(\mathbb{1}). \tag{1.4}$$

Obviously, calculation of $\Pr[\mathsf{U}_1(\mathbb{3})]$ and $\Pr[\mathsf{U}_2(\mathbb{3})]$ will be more complicated, since there are many more routes available to each outcome. However, in the previous paragraph we have established an algorithm whereby these values and those of $\Pr[\mathsf{U}_1(\tau)]$ and $\Pr[\mathsf{U}_2(\tau)]$ in general can be found. One simply traces out each possible route to the outcome, forming as he moves over it the product of all transition probabilities along the route; then he takes the sum of the resulting products over all of the routes to that outcome. Clearly, as τ increases, the number of routes to a given outcome and the length of each route both will increase. Thus if we set out to calculate $\Pr[\mathsf{U}_1(\tau)]$ for large τ without having made any calculations for smaller values of τ, the task might seem rather difficult. For example, just how does one go about finding all of the routes to a given outcome; and how does he know that he has included them all? The task becomes very simple, however, if the calculations are made for one value of τ at a time, beginning with $\tau = \mathbb{0}$ and progressing sequentially to higher values. This is simply because of the distributive property of multiplication over addition.

$\Pr[\mathsf{U}_1(\tau)]$ is the sum of the products of all transition probabilities in all routes leading to U_1 in τ transitions. Therefore $q_1(\tau)\Pr[\mathsf{U}_1(\tau)]$ is the sum of the products of all transition probabilities in all routes leading to U_1 in interval τ and thence to U_1 in interval $\tau+\mathbb{1}$. Similarly, $p_{1,2}(\tau)\Pr[\mathsf{U}_2;\tau]$ is the sum of the products of all transition probabilities in all routes leading to U_2 in interval τ and thence to U_1 in interval $\tau+\mathbb{1}$. Therefore, for the network of Figure 1.14,

$$\Pr[\mathsf{U}_1(\tau+\mathbb{1})]=q_1(\tau)\Pr[\mathsf{U}_1(\tau)]+p_{1,2}(\tau)\Pr[\mathsf{U}_2(\tau)] \tag{1.5}$$

and, similarly,

$$\Pr[\mathsf{U}_2(\tau+\mathbb{1})]=p_{2,1}(\tau)\Pr[\mathsf{U}_1(\tau)]+q_2(\tau)\Pr[\mathsf{U}_2(\tau)]. \tag{1.6}$$

These relationships provide a sequential algorithm (known as the Chapman-Kolmogorov equation) that automatically takes care of bookkeeping the routes. Such an algorithm would be extremely easy to realize with a digital-computer program.

It would be very interesting to be able to work backward in time as well as forward, to deduce from the state at $\tau=\mathbb{0}$ estimates of the object's states for negative values of τ. In certain models (including that of Figure 1.14), such estimates can be made. In other models, they cannot be made without supportive assumptions. When they can be made, their calculation is more complicated (see the discussion of Bayes's Rule in section 3.5).

1.13. Vector Notation, State Projection Matrices

We can very easily extend the logic behind equations 1.5 and 1.6 to cover situations with more than two states. For example, if the state space has v states, every pair of which is connected by a transition, then we can write

$$\begin{aligned}
\Pr[\mathsf{U}_1(\tau+\mathbb{1})]&=q_1(\tau)\Pr[\mathsf{U}_1(\tau)] +p_{1,2}(\tau)\Pr[\mathsf{U}_2(\tau)]\\
&\quad +p_{1,3}(\tau)\Pr[\mathsf{U}_3(\tau)]+\cdots+p_{1,v}(\tau)\Pr[\mathsf{U}_v(\tau)]\\
\Pr[\mathsf{U}_2(\tau+\mathbb{1})]&=p_{2,1}(\tau)\Pr[\mathsf{U}_1(\tau)]+q_2(\tau)\Pr[\mathsf{U}_2(\tau)]\\
&\quad +p_{2,3}(\tau)\Pr[\mathsf{U}_3(\tau)]+\cdots+p_{2,v}(\tau)\Pr[\mathsf{U}_v(\tau)]\\
\Pr[\mathsf{U}_3(\tau+\mathbb{1})]&=p_{3,1}(\tau)\Pr[\mathsf{U}_1(\tau)]+p_{3,2}(\tau)\Pr[\mathsf{U}_2(\tau)]\\
&\quad +q_3(\tau)\Pr[\mathsf{U}_3(\tau)] +\cdots+p_{3,v}(\tau)\Pr[\mathsf{U}_v(\tau)]\\
&\cdots\cdots\cdots\cdots\cdots\cdots\\
\Pr[\mathsf{U}_v(\tau+\mathbb{1})]&=p_{v,1}(\tau)\Pr[\mathsf{U}_1(\tau)]+p_{v,2}(\tau)\Pr[\mathsf{U}_2(\tau)]\\
&\quad +p_{v,3}(\tau)\Pr[\mathsf{U}_3(\tau)]+\cdots+q_v(\tau)\Pr[\mathsf{U}_v(\tau)].
\end{aligned} \tag{1.7}$$

Even if direct transitions do not occur between every pair of states, this formulation still applies perfectly well. One simply must substitute a zero for each p or q representing a connection that does not exist.

Clearly, as the number of states in our models increases, this description of our deductive process becomes extremely cumbersome. It can be simplified very much, however, through the use of conventional algebraic shorthand, namely vector and matrix notation. As the first step in the development of this shorthand, let us find a convenient abbreviation for the set of states on which our model is constructed. We could display that set horizontally $[U_1, U_2, U_3, \ldots, U_v]$, as is often done in textbooks on probability theory; or we could display it vertically

$$\begin{bmatrix} U_1 \\ U_2 \\ U_3 \\ \vdots \\ U_v \end{bmatrix}$$

as is often done in textbooks on network and systems theory. The two methods are compared and discussed in Appendix A. For the remainder of this text, however, we shall employ the vertical display, since it is network and systems theory that we are applying to population models. Displayed in this manner, the states form a *column vector;* and each state is an *element* of that vector. Displayed horizontally, the states form a *row vector*, each state being an *element.* It is very convenient to be able to represent an entire vector by a single symbol. There are several conventions that could be followed for this representation; but for consistency we shall employ just one of them throughout this text. Vectors will be represented by letters in bold-face type. The elements of vectors will be represented by letters in normal type. In the case of the column vector of states, we shall use the symbol **U.**

$$\mathbf{U} \text{ will represent } \begin{bmatrix} U_1 \\ U_2 \\ U_3 \\ \vdots \\ U_v \end{bmatrix}.$$

Thus, in a very real sense, **U** represents the state space of our model; the elements of **U** are the elements of the state space, and every element of the state space is represented in **U.**

Now, as the next step, we can find a very convenient description for our deductions concerning the state of the object of our model during interval τ:

$$\mathbf{Pr}\,[\mathbf{U}(\tau)] \text{ will represent } \begin{bmatrix} \Pr[U_1(\tau)] \\ \Pr[U_2(\tau)] \\ \Pr[U_3(\tau)] \\ \vdots \\ \Pr[U_v(\tau)] \end{bmatrix}.$$

Thus $\mathbf{Pr}\,[\mathbf{U}; \tau]$ is a new vector, defined in terms of our state-space vector, **U.** $\mathbf{Pr}\,[\mathbf{U}(\tau)]$ is known as the *probability distribution vector* for the object being modeled.

As the next step, we somehow must represent the expanded version of $\mathbf{Pr}\,[\mathbf{U}(\tau + \mathbb{1})]$ that appears on the right-hand sides of equations 1.7 and 1.8:

$$\mathbf{Pr}[\mathbf{U}(\tau+\mathbb{1})]=\begin{bmatrix} q_1(\tau)\Pr[\mathsf{U}_1(\tau)] \quad +p_{1,2}(\tau)\Pr[\mathsf{U}_2(\tau)] \\ \qquad +p_{1,3}(\tau)\Pr[\mathsf{U}_3(\tau)]+\cdots+p_{1,v}(\tau)\Pr[\mathsf{U}_v(\tau)] \\ p_{2,1}(\tau)\Pr[\mathsf{U}_1(\tau)]+q_2(\tau)\Pr[\mathsf{U}_2(\tau)] \\ \qquad +p_{2,3}(\tau)\Pr[\mathsf{U}_3(\tau)]+\cdots+p_{2,v}(\tau)\Pr[\mathsf{U}_v(\tau)] \\ p_{3,1}(\tau)\Pr[\mathsf{U}_1(\tau)]+p_{3,2}(\tau)\Pr[\mathsf{U}_2(\tau)] \\ \qquad +q_3(\tau)\Pr[\mathsf{U}_3(\tau)] \quad +\cdots+p_{3,v}(\tau)\Pr[\mathsf{U}_v(\tau)] \\ \cdots\cdots\cdots\cdots\cdots \\ p_{v,1}(\tau)\Pr[\mathsf{U}_1(\tau)]+p_{v,2}(\tau)\Pr[\mathsf{U}_2(\tau)] \\ \qquad +p_{v,3}(\tau)\Pr[\mathsf{U}_3(\tau)]+\cdots+q_v(\tau)\Pr[\mathsf{U}_v(\tau)] \end{bmatrix} \tag{1.8}$$

This can be done very conveniently by representing the array of transition probabilities with a single symbol and then properly defining a process of combining two arrays (such as the array of transition probabilities and the probability distribution vector). First, for the array of transition probabilities, we shall let

$$\mathbf{P}(\tau)=\begin{bmatrix} q_1(\tau) & p_{1,2}(\tau) & p_{1,3}(\tau) & \dots & p_{1,v}(\tau) \\ p_{2,1}(\tau) & q_2(\tau) & p_{2,3}(\tau) & \dots & p_{2,v}(\tau) \\ p_{3,1}(\tau) & p_{3,2}(\tau) & q_3(\tau) & \dots & p_{3,v}(\tau) \\ \vdots & \vdots & \vdots & & \vdots \\ p_{v,1}(\tau) & p_{v,2}(\tau) & p_{v,3}(\tau) & \dots & q_v(\tau) \end{bmatrix}.$$

Then we simply shall *define* the *inner product* of $\mathbf{P}(\tau)$ and $\mathbf{Pr}[\mathbf{U}(\tau)]$ to be given by the vector on the right-hand side of equation 1.8. The step-by-step procedure for carrying out the formation of inner products of arrays is outlined in Appendix A. At this point, however, the details are not nearly as important as the fact that the process can be defined, which allows us to represent equations 1.7 and 1.8 in the following, very concise form:

$$\mathbf{Pr}[\mathbf{U}(\tau+\mathbb{1})]=\mathbf{P}(\tau)\cdot\mathbf{Pr}[\mathbf{U}(\tau)]. \tag{1.9}$$

The array that $\mathbf{P}(\tau)$ represents is a *matrix*, and each p and q entered in the array is an *element* of the matrix. Since the matrix $\mathbf{P}(\tau)$ converts our deductions concerning the state of the object during interval τ into deductions concerning its state during the next interval, $\tau+\mathbb{1}$, $\mathbf{P}(\tau)$ is called the *state projection matrix.*

Notice that the inner product $\mathbf{P}(\tau)\cdot\mathbf{Pr}[\mathbf{U}(\tau)]$ is itself a column vector, namely, $\mathbf{Pr}[\mathbf{U}(\tau+\mathbb{1})]$. Clearly, both equation 1.9 and our process of inner-product formation apply generally for all values of τ; so we can write

$$\mathbf{Pr}[\mathbf{U}(\tau+\mathbb{2})]=\mathbf{P}(\tau+\mathbb{1})\cdot\mathbf{Pr}[\mathbf{U}(\tau+\mathbb{1})]$$

or,

$$\mathbf{Pr}[\mathbf{U}(\tau+\mathbb{2})]=\mathbf{P}(\tau+\mathbb{1})\cdot\{\mathbf{P}(\tau)\cdot\mathbf{Pr}[\mathbf{U}(\tau)]\}.$$

Now, it so happens that by virtue of the way in which it is defined conventionally, the formation of inner product obeys the commutative law but not the associative law. In other words, given three arrays, **A**, **B** and **C**, $\mathbf{A}\cdot(\mathbf{B}\cdot\mathbf{C})$ is not equal to $(\mathbf{A}\cdot\mathbf{B})\cdot\mathbf{C}$, although $\mathbf{A}\cdot\mathbf{B}$ is equal to $\mathbf{B}\cdot\mathbf{A}$. Therefore, when a series of inner products is to be carried out, it must be done in the correct sequence. By simple extension, it is clear that

$$\mathbf{Pr}[\mathbf{U}(\tau)] = \mathbf{P}(\tau-\mathbb{1}) \cdot [\mathbf{P}(\tau-\mathbb{2}) \cdot [\mathbf{P}(\tau-\mathbb{3}) \cdot [\ldots \mathbf{P}(\mathbb{0}) \cdot \mathbf{Pr}[\mathbf{U}(\mathbb{0})] \ldots]]]. \quad (1.10)$$

An alternative to the sequential formation of inner products is the process of *array multiplication* (also defined in Appendix A). This process does not obey the commutative law, but it does obey the associative law. Thus, if **A**, **B** and **C** are three arrays, $\mathbf{A}(\mathbf{BC})$ is equal to $(\mathbf{AB})\mathbf{C}$, but $\mathbf{AB}$ is not equal to $\mathbf{BA}$. All but the very first in a sequence of inner-product operations can be replaced by array multiplications. Thus, for example, the sequence $\mathbf{A} \cdot (\mathbf{B} \cdot (\mathbf{C} \cdot \mathbf{D}))$ can be replaced by $(\mathbf{ABC}) \cdot \mathbf{D}$. Using array multiplication, therefore, we can rewrite equation 1.10 as follows:

$$\mathbf{Pr}[\mathbf{U}(\tau)] = [\mathbf{P}(\tau-\mathbb{1})\,\mathbf{P}(\tau-\mathbb{2})\,\mathbf{P}(\tau-\mathbb{3}) \ldots \mathbf{P}(\mathbb{0})] \cdot \mathbf{Pr}[\mathbf{U}(\mathbb{0})]. \quad (1.11)$$

The product of arrays $[\mathbf{P}(\tau-\mathbb{1})\,\mathbf{P}(\tau-\mathbb{2})\,\mathbf{P}(\tau-\mathbb{3}) \ldots \mathbf{P}(\mathbb{0})]$ is a matrix with the same number of elements as $\mathbf{P}(\tau)$. However, it represents a τ-step transition rather than a one-step transition; and it is conventionally called the *state transition matrix*.

From equations 1.10 and 1.11 it is clear that in order to draw deductions about the present state of an object, one needs an estimate of the initial probability distribution over the states, which he or she then can express as the initial probability distribution vector $\mathbf{Pr}[\mathbf{U}(\mathbb{0})]$. Of course, in order to make deductions, one also must have estimates of the various p's and q's for all of the intervals of interest. These, in general, will be much more difficult to obtain than the observation of the initial state, for their determination requires repeated observation not only of the present state of the object, but also of the state to which it progresses next.

As long as we remember what they stand for, the abbreviations outlined in this section can be extremely useful, and they can simplify to a great extent our conceptions of the analytical procedures involved in forming deductions. The carrying out of these procedures then can be postponed or left to a digital computer.

1.14. Elementary Dynamics of Homogeneous Markov Chains

So far, we have considered Markov chains in which the transition probabilities can be dependent on extrinsic factors that vary with time. The deductions concerning the future states of the object in such a chain will depend markedly on the variations with time of the various p's and q's in the state projection matrix. We can state this more concisely by saying that the dynamics will depend markedly on the function $\mathbf{P}(\tau)$, which reflects the dependence of **P** on extrinsic factors which, in turn, depend on time.

The most thoroughly studied form of the Markov chain is that in which **P** does not depend on time at all, but remains constant (i.e., every p and q remains constant). Such a chain is called a *homogeneous Markov chain*. The Markov chain in which **P** varies with time is called *nonhomogeneous*. Clearly, with the

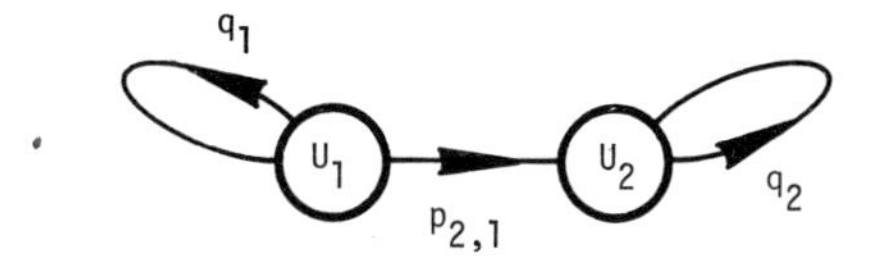

Fig. 1.15

algorithms presented in sections 1.12 and 1.13 of this chapter, one can employ a digital computer to do the relatively routine bookkeeping involved in forming deductions about nonhomogeneous chains. With many homogeneous chains, on the other hand, the deductions are sufficiently stereotypic and the necessary bookkeeping sufficiently limited that one can form the deductions rather quickly even without a digital computer. In some cases, in fact, deductions can be formed in one or two steps; and use of a digital computer would be extremely wasteful of both time and money.

In this section, we shall discuss three very simple structures that occur frequently in homogeneous Markov chains. First of all, consider the situation depicted in Figure 1.15, a Markov chain comprising states U_1 and U_2 coupled *unidirectionally* (i.e., there is a path from U_1 to U_2, but no return path from U_2 to U_1), with the initial state specified as U_1 (i.e., $\Pr[\mathsf{U}_1(\mathbb{0})]=\mathbb{1}$; $\Pr[\mathsf{U}_2(\mathbb{0})]=\mathbb{0}$).

In this very simple situation, one does not need to invoke the vector notation of the previous section. It is quite a simple matter to keep track of all of the possible routes to a given outcome and sum the probabilities associated with them. The calculations for state U_1 are especially easy. The object occupies state U_1 during the interval $\tau=\mathbb{0}$. In order to be in U_1 after τ transitions, the object clearly must make τ passes around the self-transition of U_1; so there is only one route to the outcome that the object occupies U_1 during interval τ. Furthermore, the probability associated with that outcome clearly is simply

$$\Pr[\mathsf{U}_1(\tau)]=q_1^\tau. \tag{1.12}$$

Now, we can calculate $\Pr[\mathsf{U}_2(\tau)]$ immediately, simply by noting that the object at all times must be either in U_1 or U_2:

$$\Pr[\mathsf{U}_1(\tau)]+\Pr[\mathsf{U}_2(\tau)]=\mathbb{1}. \tag{1.13}$$

This is another statement of our conservation principle. It automatically will be true for all times if it is true for $\tau=\mathbb{0}$ and if the condition stated at the end of section 1.10 is met (i.e., the sum of the probabilities of transition from each state is one). This type of conservation statement very often leads to simplification in probability calculations. In this case it leads directly to

$$\Pr[\mathsf{U}_2(\tau)]=\mathbb{1}-(q_1)^\tau. \tag{1.14}$$

Of course, we could have considered each route from U_1 to U_2 in precisely τ transitions. It is a little more difficult, but not much. Clearly, there are τ different routes, one for each place that the transition $\mathsf{U}_1\rightarrow\mathsf{U}_2$ could take among the τ

transitions (the transition $U_1 \rightarrow U_2$ only occurs once on any given route). Specifically, the object could pass directly to U_2, with no passes around the self-transition at U_1; or it could pass to U_2 after one pass around the self-transition at U_1, or after two self-transitions, or three, etc., all the way to $\tau - 1$ self-transitions followed by the passage to U_2. Summing the probabilities of all of these routes, we find

$$\Pr[U_2;\tau] = p_{2,1} + q_1 p_{2,1} + q_1^2 p_{2,1} + q_1^3 p_{2,1} + \cdots + q_1^{\tau-1} p_{2,1}. \tag{1.15}$$

Noting that conservation requires that

$$p_{2,1} + q_1 = 1 \quad \text{or} \quad p_{2,1} = 1 - q_1$$

we can simplify equation 1.15 to yield

$$\Pr[U_2(\tau)] = (1 - q_1) + q_1(1 - q_1) + q_1^2(1 - q_1) + \cdots + q_1^{\tau-1}(1 - q_1) = 1 - q_1^{\tau}.$$

We now have a rather simply statement of our deductions concerning the state of the object during any interval after the initial one:

$$\mathbf{Pr}[\mathbf{U}(\tau)] = \begin{bmatrix} \Pr[U_1(\tau)] \\ \Pr[U_2(\tau)] \end{bmatrix} = \begin{bmatrix} q_1^{\tau} \\ 1 - q_1^{\tau} \end{bmatrix}.$$

Since the passage of time is explicitly represented in these deductions, they are not merely descriptions of static probability distributions. Instead, they are descriptions of the changes of the probability distribution; and as such they are dynamic descriptions of the object being modeled. From this point on, in fact, we often shall refer to $\mathbf{Pr}[\mathbf{U}(\tau)]$ simply as the *dynamics of the object*. On some occasions, we will be interested in the probability of occupancy of a given state and the changes in that probability with time. In that case, we shall refer to the *dynamics of the state*. Thus, $\Pr[U_i(\tau)]$ often will be referred to simply as the dynamics of state U_i.

Figure 1.16 shows the dynamics associated with the model of Figure 1.15, for a situation in which $p_{2,1} = 1/2$, $q_1 = 1/2$. The general shapes of the two components of the dynamics are typical for all two-state, homogeneous Markov chains, regardless of the values of the p's and q's.

Figure 1.17 depicts a slightly more complicated situation. Here the initial state again is specified as U_1, but the transitional coupling between U_1 and U_2 is *bidirectional*. The dynamics of the object in this case are a bit more difficult to calculate, but not much. Tracing the various routes to an outcome occurring in some arbitrary interval, τ, however, is not at all practical, since, as the reader can determine very easily for himself, the number and complexity of the routes even in this simple case is rather terrifying. Another approach might be successive applications of the state projection matrix, following the rules of inner-product formation or multiplication given in Appendix A. We very easily can represent this process in our vector/matrix notation:

$$\mathbf{Pr}[\mathbf{U}(\tau)] = \begin{bmatrix} q_1 & p_{1,2} \\ p_{2,1} & q_2 \end{bmatrix} \cdot \mathbf{Pr}[\mathbf{U}(\tau - 1)]$$

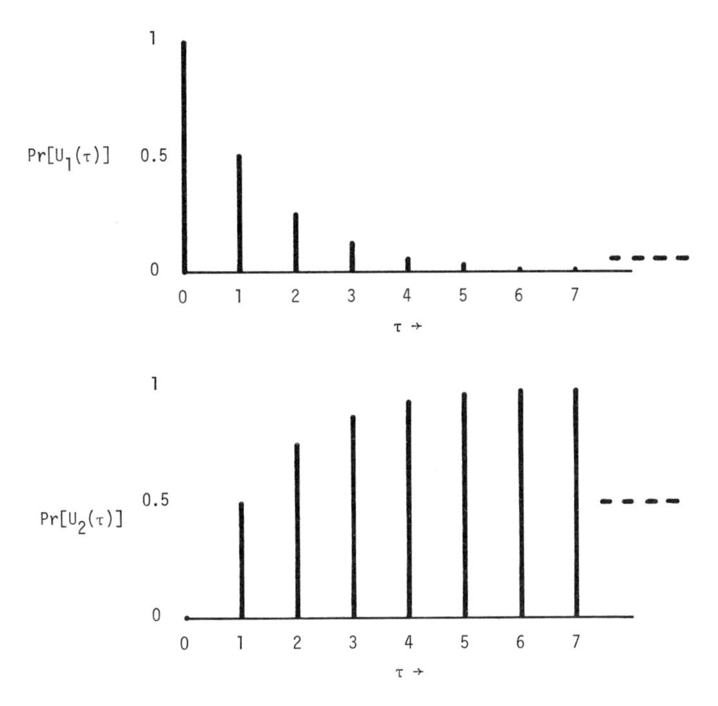

Fig. 1.16. Dynamics of states U_1 and U_2 in the model of Fig. 1.15 with $p_{2,1}=1/2$

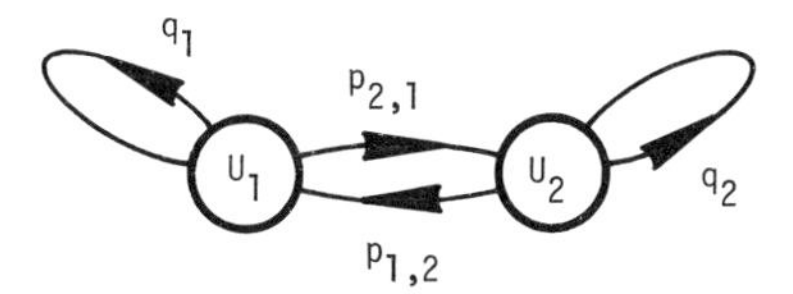

Fig. 1.17

or

$$\mathbf{Pr}\,[\mathbf{U}(\tau)]=\left(\begin{bmatrix} q_1 & p_{1,2} \\ p_{2,1} & q_2 \end{bmatrix}\begin{bmatrix} q_1 & p_{1,2} \\ p_{2,1} & q_2 \end{bmatrix}\begin{bmatrix} q_1 & p_{1,2} \\ p_{2,1} & q_2 \end{bmatrix}\cdots\begin{bmatrix} q_1 & p_{1,2} \\ p_{2,1} & q_2 \end{bmatrix}\right)\cdot\mathbf{Pr}\,[\mathbf{U}(\mathbb{0})]$$

where the multiplication within the parentheses is carried out τ times. In analogy with repeated multiplication of real numbers, we can abbreviate this process as follows:

$$\mathbf{Pr}\,[\mathbf{U}(\tau)]=\begin{bmatrix} q_1 & p_{1,2} \\ p_{2,1} & q_2 \end{bmatrix}^{\tau}\cdot\mathbf{Pr}\,[\mathbf{U}(\mathbb{0})].$$

It must be remembered, however, that the order of multiplication generally cannot be reversed in the case of matrices. In other words,

$$\begin{bmatrix} q_1 & p_{1,2} \\ p_{2,1} & q_2 \end{bmatrix}^{\tau}=\begin{bmatrix} q_1 & p_{1,2} \\ p_{2,1} & q_2 \end{bmatrix}\begin{bmatrix} q_1 & p_{1,2} \\ p_{2,1} & q_2 \end{bmatrix}^{\tau-\mathbb{1}}$$

but

$$\begin{bmatrix} q_1 & p_{1,2} \\ p_{2,1} & q_2 \end{bmatrix}^{\tau} \text{ does not equal } \begin{bmatrix} q_1 & p_{1,2} \\ p_{2,1} & q_2 \end{bmatrix}^{\tau-1} \begin{bmatrix} q_1 & p_{1,2} \\ p_{2,1} & q_2 \end{bmatrix}$$

(see Appendix A).

Now we very easily can represent the process of repeated application of the state projection matrix; but with a little experimentation, the reader soon will discover that the process itself quickly becomes very unwieldy. So much so, in fact, that she or he might be tempted to replace the generalized p's and q's with specific values and let a digital computer carry out the work. If this process were repeated several times, for several sets of values for the p's and q's, then one might be able to discern a pattern in the results and develop a generalization. On the other hand, if he were unable to do so, then his computer results will be very much *ad hoc* and not especially enlightening.

The generality of the situation can be maintained (i.e., the p's and q's can remain in literal form), while the calculations are carried out by a very simple method, well-known in linear algebra and discussed in considerable detail in chapter 4 (in the context of networks). The method involves development of a *characteristic* equation from the state projection matrix and the determination of the roots of that equation followed by application of the conditions imposed by the initial state of the object. The procedure for this particular case is outlined here, but not discussed in detail. The characteristic equation for the state projection matrix is

$$m^2-(q_1+q_2)m+q_1q_2-p_{1,2}p_{2,1}=0.$$

The roots of the equation are

$$m_1=\frac{q_1+q_2}{2}+\sqrt{\left(\frac{q_1-q_2}{2}\right)^2+p_{1,2}p_{2,1}}$$

and (1.16)

$$m_2=\frac{q_1+q_2}{2}-\sqrt{\left(\frac{q_1-q_2}{2}\right)^2+p_{1,2}p_{2,1}}\,.$$

The general expressions for $\Pr[\mathbf{U}(\tau)]$, without taking into account the initial state of the object are

$$\mathbf{Pr}[\mathbf{U}(\tau)]=\begin{bmatrix}\Pr[\mathsf{U}_1(\tau)]\\ \Pr[\mathsf{U}_2(\tau)]\end{bmatrix}=\begin{bmatrix}Am_1^{\tau}+Bm_2^{\tau}\\ Cm_1^{\tau}+Dm_2^{\tau}\end{bmatrix}. \tag{1.17}$$

One then can apply the initial conditions as follows:

$$\begin{aligned}&\Pr[\mathsf{U}_1(0)]=1=A+B\\ &\Pr[\mathsf{U}_2(0)]=0=C+D\end{aligned}\qquad (m^0=1)$$

$$\begin{aligned}&\Pr[\mathsf{U}_1(1)]=q_1=Am_1+Bm_2\\ &\Pr[\mathsf{U}_2(1)]=p_{2,1}=Cm_1+Dm_2\end{aligned}\qquad (m^1=m)$$

$$A=\frac{m_2-q_1}{m_2-m_1} \qquad B=\frac{m_1-q_1}{m_1-m_2}$$
$$C=-\frac{p_{2,1}}{m_2-m_1} \qquad D=\frac{p_{2,1}}{m_2-m_1}. \tag{1.18}$$

Combining equations 1.16, 1.17 and 1.18, we have a very simple description of the dynamics of the object. The object's behavior will depend on the magnitude and signs of the roots, m_1 and m_2. In this particular case, the roots must be positive and lie somewhere in the range from (and including) zero to (and including) one:

$$0 \leq m \leq 1.$$

When the root is one, we have $m^\tau = 1^\tau = 1$; when the root lies somewhere between zero and one, we have m^τ decreasing steadily as τ increases; and when the root is zero, we have $m^\tau = 0^\tau$ which is zero for all values of τ greater than zero and which we shall interpret to be one when τ equals zero. Therefore, regardless of the values of the p's and q's, the dynamics of the object of Figure 1.17 will comprise three types of terms, those that do not vary with τ, those that decrease steadily as τ increases, and those that decrease abruptly as τ goes from 0 to 1.

Example 1.4. Consider the situation in which $q_1 = q_2 = 0.9$, $p_{1,2} = p_{2,1} = 0.1$. Determine the dynamics of the object of the model in Figure 1.17, given that it initially occupies state U_1.

Answer: First, we can apply equations 1.16 to determine m_1 and m_2; then we can go on to find A, B, C, and D from equations 1.18; and, finally, we can insert these findings into vector equation 1.17 to obtain the description of the object's dynamics:

$$m_1 = 0.9 + \sqrt{0.01} = 1$$
$$m_2 = 0.9 - \sqrt{0.01} = 0.8$$
$$A = 0.5; \qquad B = 0.5$$
$$C = 0.5; \qquad D = -0.5$$

$$\mathbf{Pr}\,[\mathbf{U}(\tau)] = \begin{bmatrix} 0.5 + 0.5\,(0.8)^\tau \\ 0.5 - 0.5\,(0.8)^\tau \end{bmatrix}.$$

The dynamics of states U_1 and U_2 are plotted separately in Figure 1.18.

Example 1.5. Repeat the problem in Example 1.4 for the situation in which $q_1 = q_2 = 0.5$.

Answer: It is clear from conservation that $p_{1,2} = p_{2,1} = 0.5$. Following the same procedure that we did in example 1.4, we find

$$m_1 = 1; \qquad m_2 = 0$$
$$A = 0.5; \qquad B = 0.5; \qquad C = 0.5; \qquad D = -0.5$$

$$\mathbf{Pr}\,[\mathbf{U}(\tau)] = \begin{bmatrix} 0.5 + 0.5\,(0)^\tau \\ 0.5 - 0.5\,(0)^\tau \end{bmatrix}.$$

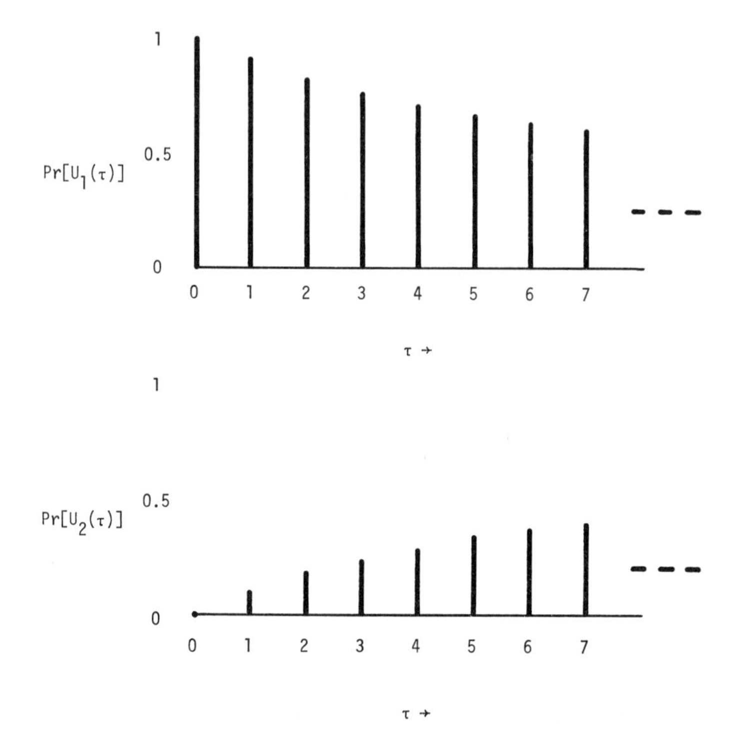

Fig. 1.18. Dynamics of states U_1 and U_2 in the model of Fig. 1.17 with $p_{2,1}=p_{1,2}=0.9$

In other words, for *all* values of τ greater than 0, $\Pr[\mathsf{U}_1(\tau)]$ and $\Pr[\mathsf{U}_2(\tau)]$ both are equal to 0.5. During the interval $\tau=0$, we have the form $(0)^0$, which we must interpret to be equal to one, giving us the desired description of the initial state of the object. The dynamics of states U_1 and U_2 are plotted separately in Figure 1.19. Notice the abrupt jump of $\mathbf{Pr}[\mathsf{U}(\tau)]$ to its final, steady distribution.

The final structure that we shall consider in this section is a concatenation of unidirectionally coupled states with identical transition probabilities, as illustrated in Figure 1.20. For convenience of notation, we shall assume that the initial state of the object is U_0 and we shall determine for each state in the chain the probability that the object occupies it after τ transitions. In this very special situation, we can consider the individual routes because, as we shall see, they are highly stereotyped.

For example, there is only one route by which the object can remain in state U_0 after τ transitions, and that is τ passes around the self-transition of U_0. The probability for that outcome clearly is q^τ. There are τ routes that lead to U_1 after τ transitions: $\mathsf{U}_0\to\mathsf{U}_1\to\mathsf{U}_1\to\mathsf{U}_1\to\cdots\to\mathsf{U}_1$; $\mathsf{U}_0\to\mathsf{U}_0\to\mathsf{U}_1\to\mathsf{U}_1\to\cdots\to\mathsf{U}_1$; $\mathsf{U}_0\to\mathsf{U}_0\to\mathsf{U}_0\to\mathsf{U}_1\to\cdots\to\mathsf{U}_1$; ...; $\mathsf{U}_0\to\mathsf{U}_0\to\mathsf{U}_0\to\mathsf{U}_0\to\cdots\to\mathsf{U}_0\to\mathsf{U}_1$; and all of these routes have the same probability of being taken, namely $(pq^{\tau-1})$. Therefore, the probability of occupation of U_1 during interval τ is $\tau pq^{\tau-1}$ (i.e., $\tau pq^{\tau-1}$ is the sum of $pq^{\tau-1}$, for each route, taken over all τ routes). Similarly, all routes to state U_i have exactly i interstate transitions and $\tau-i$ self-transitions,

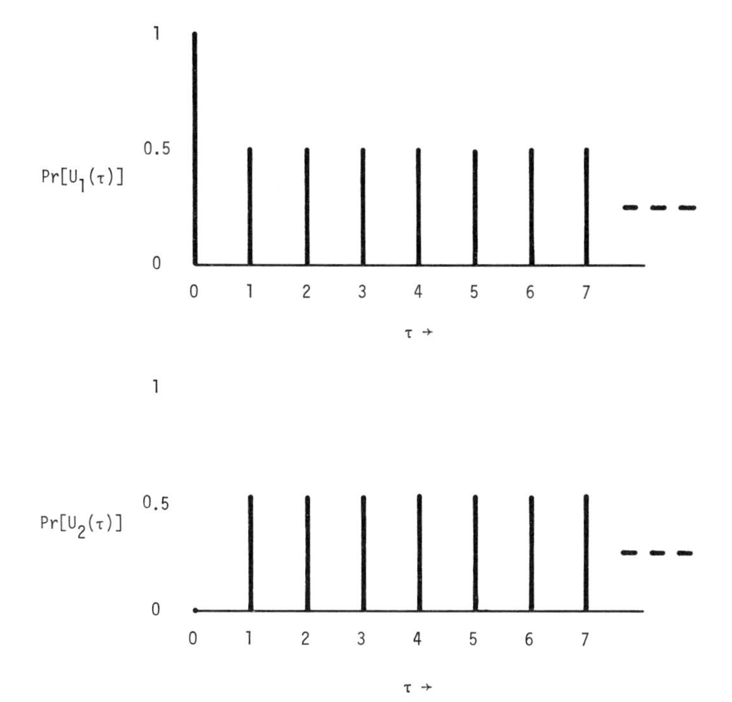

Fig. 1.19. Dynamics of states U_1 and U_2 in the model of Fig. 1.17 with $p_{1,2}=p_{2,1}=1/2$

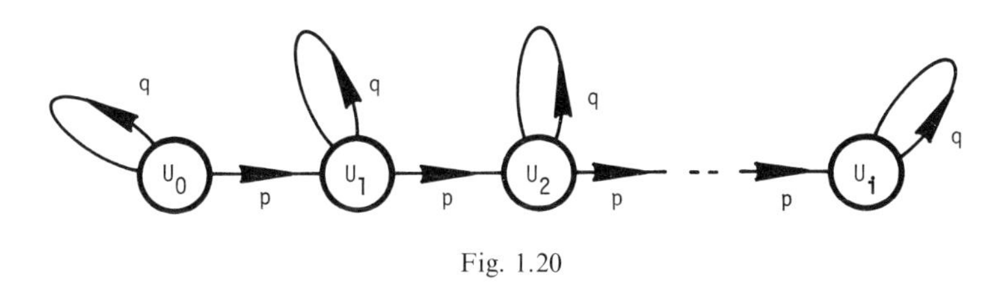

Fig. 1.20

and therefore have identical probabilities, namely $p^i q^{\tau-i}$. All that remains is to count the number of routes to U_i and multiply that number by $p^i q^{\tau-i}$.

The number of routes to state U_i is equal to the number of ways that i interstate transitions can be distributed among all τ transitions. This is a very well-known number, and is derived in Appendix B.

It usually is written in shorthand form as

$$\left\{\begin{matrix}\text{The number of ways to distribute } i \text{ indistinguishable}\\ \text{items among } \tau \text{ places, one to a place}\end{matrix}\right\}=\binom{\tau}{i}$$

and it is given by the following expression:

$$\binom{\tau}{i}=\frac{\tau!}{i!(\tau-i)!}$$

where $\tau!$ has its usual meaning (i.e., $\tau!$ is the factorial of τ):

$$\tau! = \tau(\tau-1)(\tau-2)(\tau-3)(\tau-4)\ldots(3)(2)(1)$$
$$0! \equiv 1.$$

This leads to the general expression for the dynamics of state U_i in Figure 1.20:

$$\Pr[U_i(\tau)] = 0 \qquad \text{when } \tau < i \text{ (not enough transitions to reach } U_i)$$
$$\Pr[U_i(\tau)] = \binom{\tau}{i} p^i q^{\tau-i} \qquad \text{when } \tau \geqq i \tag{1.19}$$

which can be recognized by the reader familiar with probability theory as the binomial distribution. It can be generated simply by expanding the binomial $(q+p)^\tau$ and, beginning with q^τ for state U_0, applying each successive term to its corresponding, successive state: q^τ for state U_0, $\tau p q^{\tau-1}$ for state U_1, $[\tau(\tau-1)/2!]\, p^2 q^{\tau-2}$ for U_2, and $[\tau(\tau-1)(\tau-2)\ldots(\tau-i+1)/i!]\, p^i q^{\tau-i}$ for U_i.

Example 1.6. Consider the model of Figure 1.20 under the situation in which $p=0.5$. Determine $\Pr[U_i(\tau)]$ in general, and plot the dynamics of states U_0, U_4 and U_9.

Answer: Directly from equations 1.19, we have $\Pr[U_i(\tau)] = \binom{\tau}{i}(0.5)^\tau$ for $\tau \geqq i$.

This is plotted in Figure 1.21 for

$$\Pr[U_0(\tau)] = (0.5)^\tau; \qquad \Pr[U_4;\tau] = \frac{\tau(\tau-1)(\tau-2)(\tau-3)}{4\times 3\times 2\times 1}(0.5)^\tau; \qquad \text{and}$$

$$\Pr[U_9(\tau)] = \frac{\tau(\tau-1)(\tau-2)(\tau-3)(\tau-4)(\tau-5)(\tau-6)(\tau-7)(\tau-8)}{9\times 8\times 7\times 6\times 5\times 4\times 3\times 2\times 1}(0.5)^\tau.$$

Example 1.7. Consider the model of Figure 1.20 with $p=0.1$. Determine $\Pr[U_i;\tau]$ and plot the dynamics of states U_0, U_4 and U_9.

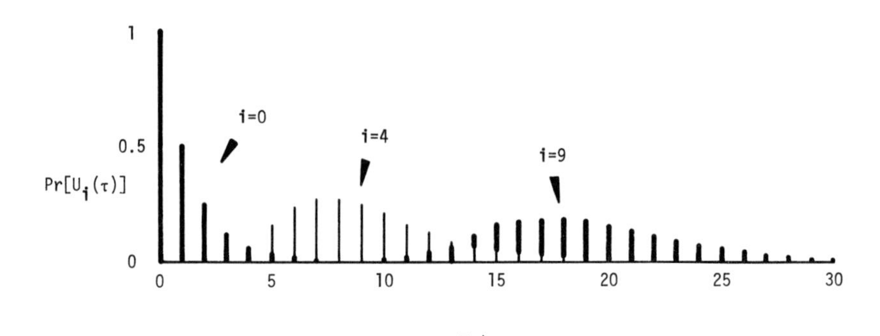

Fig. 1.21. Dynamics of states U_0, U_4, and U_9 in the model of Fig. 1.20 with $p=0.5$

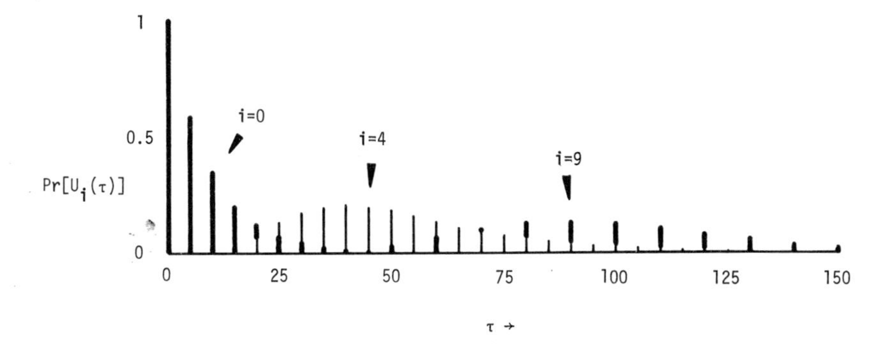

Fig. 1.22. Dynamics of states U_0, U_4, and U_9 in the model of Fig. 1.20 with $p=0.1$

Answer: From equation 1.19 we have

$$\Pr[\mathsf{U}_i(\tau)] = \binom{\tau}{i}(0.1)^i(0.9)^{\tau-i} \qquad \text{for } \tau \geqq i.$$

This is plotted in Figure 1.22 for

$$\Pr[\mathsf{U}_0(\tau)] = (0.9)^\tau; \qquad \Pr[\mathsf{U}_4(\tau)] = \binom{\tau}{4}(0.1)^4(0.9)^{\tau-4};$$

$$\Pr[\mathsf{U}_9(\tau)] = \binom{\tau}{9}(0.1)^9(0.9)^{\tau-9}.$$

Example 1.8. Consider the model of Figure 1.20 with $p=1.0$. How would you interpret $\Pr[\mathsf{U}_i(\tau)]$? Plot the dynamics of states U_0; U_4 and U_9.

$$\Pr[\mathsf{U}_i;\tau] = \binom{\tau}{i}(1)^i(0)^{\tau-i} = \binom{\tau}{i}(0)^{\tau-i} \qquad \text{for } \tau \geqq i. \tag{1.20}$$

Considering the Markov chain itself, it is clear that the transitions now are deterministic; each transition carries the object one state to the right in Figure 1.20. Therefore, the object will be in state U_i with certainty during the interval $\tau = i$, and it certainly will not be in state U_i during any other interval. Thus, $\Pr[\mathsf{U}_i(\tau)]$ must be one when $\tau = i$ and zero for all other values of τ. This requires that we interpret $(0)^{\tau-i}$ to be zero whenever τ is not equal to i and one whenever τ is equal to i, which is consistent with our earlier interpretation of $(0)^0$ as being equal to 1. The dynamics of states U_0, U_4 and U_9 are plotted in Figure 1.23.

The dynamics of the individual states have an attribute that we might call "sharpness." The sharpest dynamics occur when $p=1$. When $p=0.5$, the dynamics are less sharp (i.e., more spread out in time); and when $p=0.1$, they are even less sharp. Thus, sharpness in this case represents the degree of certainty with respect to the time an object will occupy a certain state (i.e., certainty about *when* it will occupy that state and *when* it will leave it). The sharpness of the dynamics is greatest when $p=1$ and decreases steadily as p decreases from 1

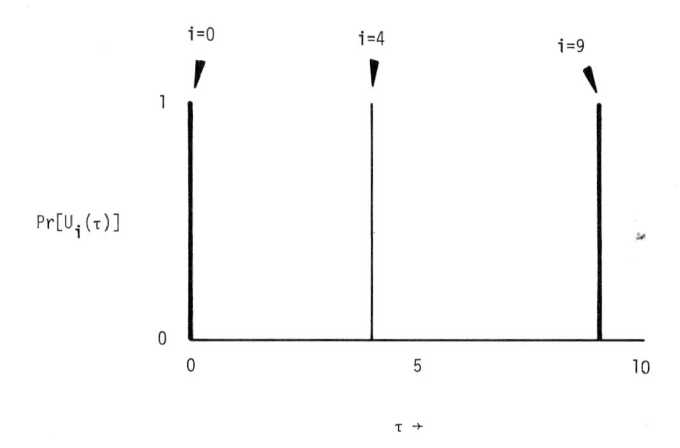

Fig. 1.23. Dynamics of states U_0, U_4, and U_9 in the model of Fig. 1.20 with $p=1$

toward 0. Here we have a rather interesting situation, however. If p were precisely 0, we would know with certainty that the object never would leave state U_0; yet as p approaches zero, we become increasingly uncertain about when the object will leave U_0. With some reflection, the reader should be able to convince himself that the latter is the workable concept.

As illustrated in Examples 1.6 and 1.7, when p is less than 1, the sharpness of the dynamics of an individual state decreases steadily as that state becomes more and more remote from the initial state. Of course, when p equals 1, the sharpness of the dynamics is independent of the position of the state.

The dynamics of the object (i.e., the dynamics of all the states) also can be displayed rather easily for the simple network of Figure 1.20. Figures 1.24, 1.25 and 1.26 show $\mathbf{Pr}[\mathsf{U}; \tau]$ plotted for the same three values of p that were used in the examples ($p=1.0$, 0.5, 0.1). The situation in Figure 1.24 is deterministic; interstate transition is a certainty, and the object moves through the states in a precisely predictable manner and with the greatest possible speed (i.e., one interstate transition per transition interval).

In Figures 1.25 and 1.26, progression through the states is probabilistic. In Figure 1.25, p is moderately large, and progression through the states is relatively rapid. As p decreases, the rate of progression decreases, as indicated by the distributions in Figure 1.26. As time passes, the state of the object (i.e., its precise location along the Markov chain) becomes increasingly uncertain. During any interval of time after $\tau=0$, the sharpness of the distribution for $p=0.1$ is noticeably greater than that for $p=0.5$, indicating that the uncertainty concerning the state of the object is greater in the case of $p=0.5$. As p approaches zero, in fact, there will be less and less uncertainty that the object will be in or very close to U_0. Similarly, as p approaches 1, there will be less and less uncertainty that the object will be in state U_τ during interval τ. The greatest uncertainty and the most rapid increase in uncertainty will occur with intermediate values of p. Identification of the greatest uncertainty, however, requires that we first have a *measure* of uncertainty. In a section 2.2, we shall

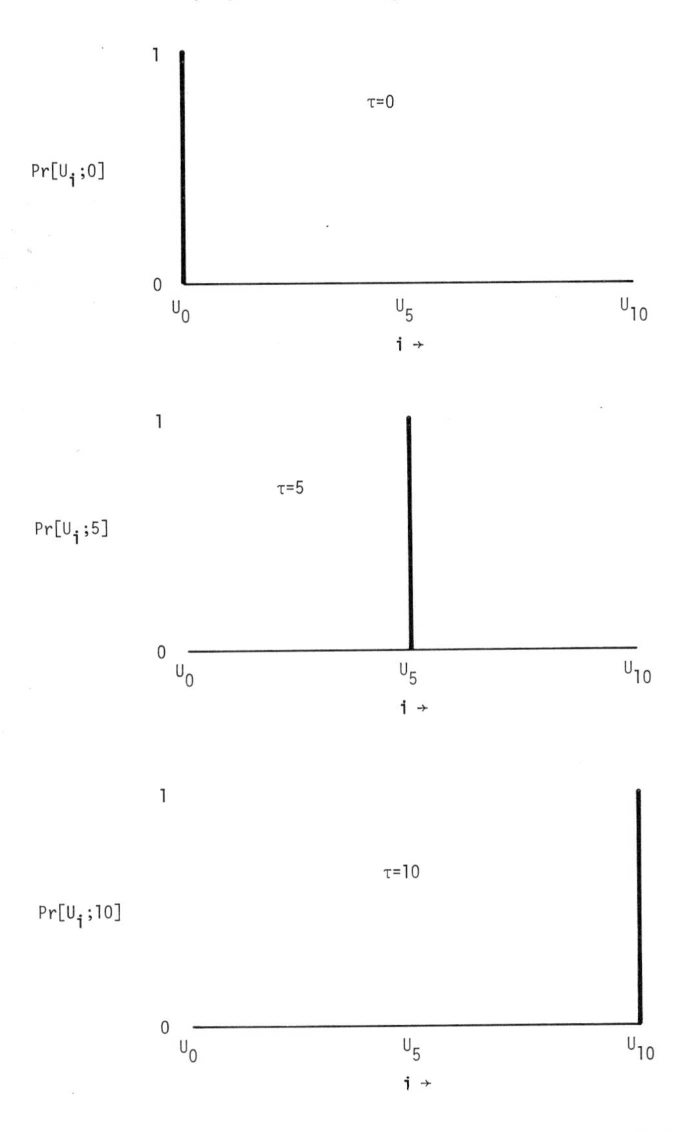

Fig. 1.24. Dynamics of the object of the model of Fig. 1.20, sampled at $\tau=0$, 5 and 10. In this case $p=1$, and the progress of the object through the states is deterministic

introduce the measure that was established by Shannon. Until then, our notion of uncertainty must remain somewhat vague.

Even though that notion is vague, we nonetheless now have a specific, graphical picture of the phenomenon discussed in section 1.7. Except in the very special case of deterministic transitions (p equal to one or zero), the uncertainty concerning the state of an object in a Markov chain is amplified by the dynamics of the object. Our initial uncertainty is expressed by the fact that

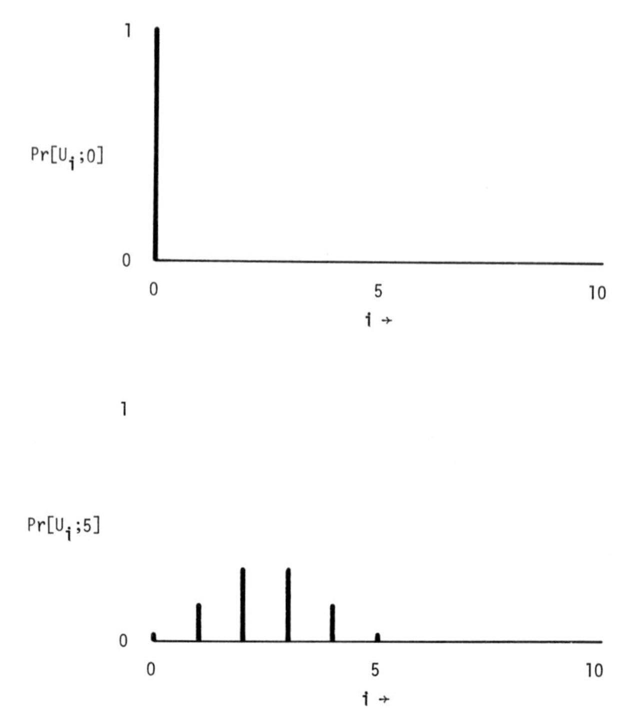

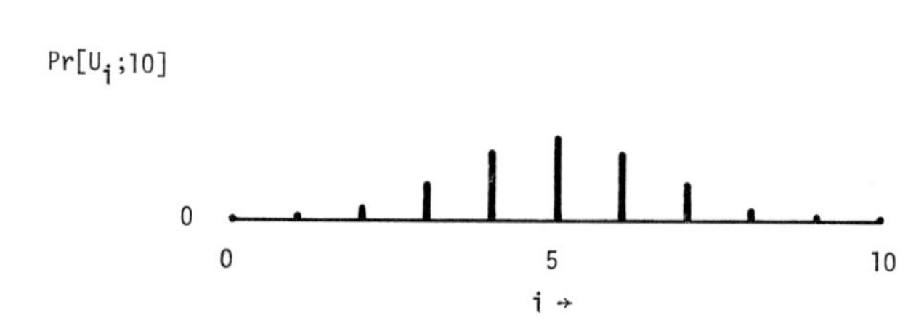

Fig. 1.25. Dynamics of the object of the model of Fig. 1.20, sampled at $\tau=0$, 5 and 10. In this case $p=0.5$, and the progress of the object through the states is probabilistic

during $\tau=0$ we have placed the object in a discrete state, the bounds of which represent the bounds of our uncertainty in state space, and by the fact that we have selected a discrete transition interval, the length of which represents the bounds of our uncertainty in time. The subsequent distributions in Figures 1.25 and 1.26 represent our best predictions, based on that initial, uncertain state, of where the object will be as time progresses into the future.

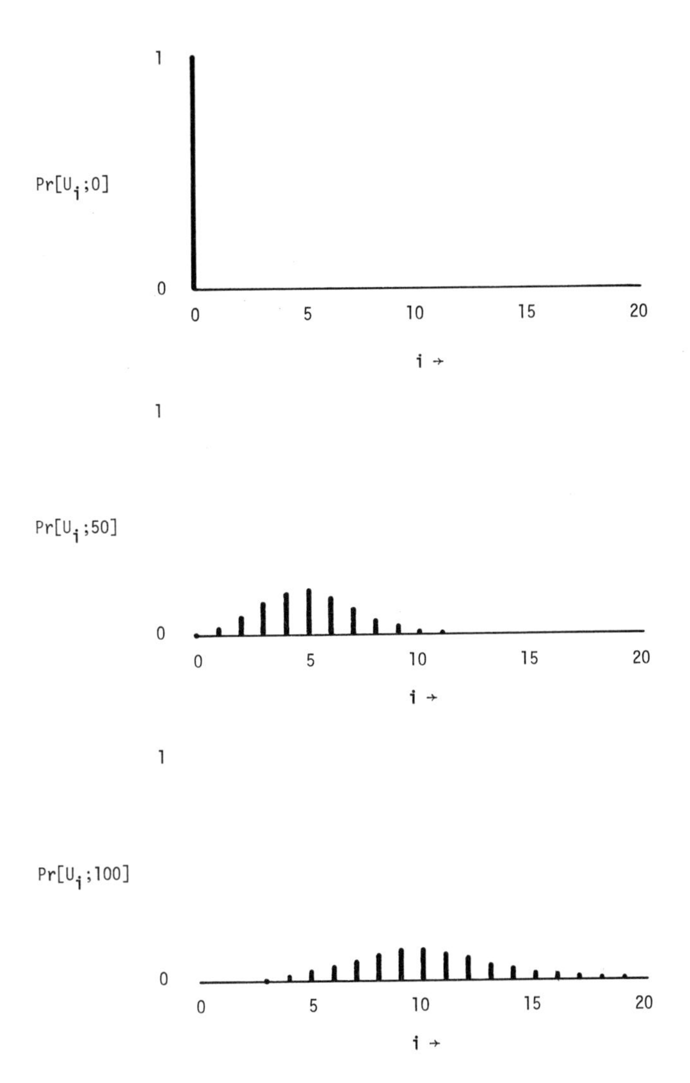

Fig. 1.26. Dynamics of the object of the model of Fig. 1.20, sampled at $\tau=0$, 50 and 100. In this case $p=0.1$, and the progress of the object is relatively slow as well as probabilistic

1.15. Observation of Transition Probabilities

Up to this point, we have assumed that transition probabilities exist for Markovian states, primitive or reduced; and, tacitly, we have assumed that somehow those probabilities can be specified. However, the actual determination of a transition probability can be extraordinarily difficult, and often forces the observer to make important assumptions about the objects being observed. To understand some of these difficulties, the reader is invited to consider a few situations involving idealized objects.

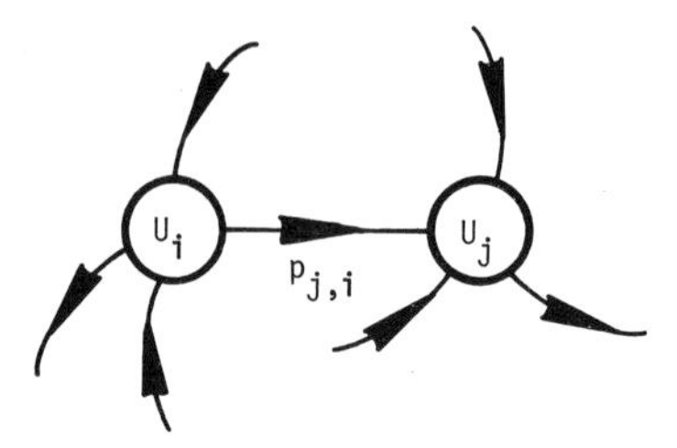

Fig. 1.27

First, consider an idealized situation in which two Markovian states are coupled by a path with a constant transition probability, as depicted in Figure 1.27. Other paths lead from each of these states, but for the moment we are interested only in the path connecting them. How can we determine the transition probability, $p_{j,i}$, associated with that path?

Evidently, one answer was provided in section 1.10. We might observe the object repeatedly in state U_i and count the number of times that its next transition carries it into state U_j. Dividing that number by the total number of transitions observed from U_i (counting one transition for each transition interval and thus including both self-transitions and interstate transitions from U_i), we would have the proportion of observed transitions from U_i that carry the object to U_j. At first, when we have made only a few observations, that proportion probably will change rather drastically from observation to observation. Gradually, as the number of observations increases, the changes may be limited to a narrower and narrower range; and with increasingly high probability, the actual value of $p_{j,i}$ will lie somewhere within that range. Eventually, we may be satisfied that we have reduced the range and increased the probability sufficiently to provide adequate specification of $p_{j,i}$.

That was a fairly straightforward process, depending only on our ability to make repeated observations of the object in a single state (state U_i). Of course, if we are dealing with a primitive state space, its grain size presumably will be extremely small, making the number of states very large. Therefore, we might have to wait a very long time between visits of the object to state U_i. In fact, we might wish to accelerate the process by repeatedly placing the object in, or close to that state ourselves, if we somehow are able to do so.

Suppose, however, that we cannot observe the object repeatedly in the same primitive state. This could happen for any of several reasons. A nonliving object may return to a given state too infrequently and be impossible to manipulate with sufficient accuracy for repeated return by experimental intervention; or it may be part of a population of objects that are identical in appearance and thus empirically indistinguishable, in which case it may be impossible to identify and follow a particular individual for a sufficient length of time. With living objects, there is another, even stronger reason: *the primitive state space of an organism inherently is not irreducible.* Instead, it is divided into a vast number of semi-isolated regions, connected by unidirectional transitions (corresponding to irreversible changes in physiological, anatomical and behavioral variables). Consider, for example, the transition from a primitive state associated with life to a primitive state associated with death. For a given organism it can occur only once, and no amount of experimental intervention can make it occur again.

What can we do in such a situation? Suppose, for example, that the transition from state U_i to U_j in Figure 1.27 can be made only once. During interval τ we observe the object in U_i and during $\tau + \mathbb{1}$ we observe it in U_j. What can we say about $p_{j,i}$? Certainly it is greater than zero (i.e., we have proved that the transition is possible). We can say virtually nothing more. Thus, in a very real sense of the word, the transition probabilities associated with the primitive states of an individual organism are, in principle, *unobservable*.

The consequences of this unobservability reach beyond the transition probabilities themselves. In general, we would delineate the primitive state space of an object by empirical observation, deducing which states are available to it by observing which states it actually occupies. In the case of an organism, vast numbers of states that it might occupy never even are approached, simply because, by pure chance, it took a route that circumvented them. Therefore, if we depended on a single organism for our information, the primitive state space of that organism forever would be undefined. We would have no state space; we would have no transition probabilities, we would have no model with which to deduce the future dynamics of the organism. In fact, all we would have is *ad hoc* history.

In order to develop a model, then, we are forced to make assumptions of one sort or another. For example, we simply might assume the entire state space and all of the transition probabilities (i.e., invent them with our own imaginations). On the other hand, we might take the conventional (almost universal) approach and assume that we have available to us numerous individual objects that are identical in the following senses:

1. Every object occupies the same primitive state space as every other object (i.e., all objects are fully described by the same set of primitive state variables, such as color, location in space, size, etc.)
2. The transition probabilities from state to state are the same for all of the objects.

Notice that we have not said that in order to be identical all of the objects must have the same values of all state variables (e.g., all be red, at the same location, the same size, etc.). In fact, such a situation not only is extremely unlikely, but it would provide us with no advantage with respect to model building. The collection of "identical" objects would be a trivial extension of the single object and, by definition, would move as one through state space.

Condition 1 in our definition of *identical objects* really is not at all restrictive. In fact, it always can be met. Take any two objects, each with its primitive state space; combine the two spaces, and we immediately have a space occupied by both objects. In fact, carrying this logic to its extreme, one can imagine a *universal primitive state space*, occupied by all objects. Condition 2, on the other hand, is a very strong assumption. Essentially, we shall assume that we can deduce the transition probabilies of a particular object by observing the transitions of other objects that we deem to be identical to it. This proposition is utterly untestable, but without it, we can do very little in the way of rational modeling.

Once we have made the assumption, on the other hand, the determination of $p_{j,i}$ is as straightforward as it was when we could observe a single object repeatedly in U_i. We simply count the number of all transitions made from state U_i (including self-transitions), and count the number of those that carried the objects to U_j. Divide the latter by the former to keep a running total of the proportion leading to U_j; and, eventually, we may have a reasonably accurate estimate of $p_{j,i}$, *based on our premise that all of the objects included in the count were identical in accordance with conditions 1 and 2.* Of course, in order to carry out this procedure we somehow must know when *individual* objects leave and enter U_i, which requires that we label the individual objects in some way in order to distinguish them from one another.

Before moving on to another topic, let us consider briefly the complications that might arise if $p_{j,i}$ varied with time (i.e., if it were different from one transition interval to another). If the variation in time follows a pattern, so that $p_{j,i}(\tau)$ is correlated some how with $p_{j,i}(\tau - \mathbb{1})$, then, in principle, one should be able to discern the time variation of $p_{j,i}$. On the other hand, if there is no discernible pattern ($p_{j,i}(\tau)$ is not correlated in any way with $p_{j,i}(\tau - \mathbb{1})$), then only the mean value of $p_{j,i}(\tau)$ is discernible, and the fact that $p_{j,i}$ varies with time is discernible only if the mean varies with time or the occupancy of U_i is very large.

Even if $p_{j,i}(\tau)$ does exhibit a pattern that is discernible in principle, the actual detection of that pattern will depend on the availability of a sufficient number of observed transitions in a given span of time. If we think of $p_{j,i}(\tau)$ as being made up of periodic components, then the shortest period we could detect is equal to twice the transition interval (i.e., twice our sampling interval). Furthermore, if any of the periods of the components of $p_{i,j}(\tau)$ are not integral multiples of the transition interval, then our sampling procedure will produce a stroboscopic effect (often called *frequency aliasing*) that can lead to significant errors in our models. This effect is discussed in some detail in section 2.4.

1.16. The Primitive State Space for an Entire Population of Identical Objects

In previous sections, we have examined the progress of an individual object through parts of its primitive state space (and, occasionally, though reduced state spaces). In light of the discussion in section 1.15 it is natural now to extend our considerations slightly, but importantly, to the progress of an entire population of objects through the primitive state space of the population. We shall assume that all of the objects in the population are identical according to conditions 1 and 2 of the previous section (i.e., they occupy the same primitive state space and have the same transition probabilities). We could make this assumption as a theoretical idealization; but we need not do so, since it is forced upon us by empirical constraints.

If the objects making up a population are identical in accordance with conditions 1 and 2, then the state of the entire population evidently is given by the distribution of its individual objects, or *members*, over the primitive state space of the individual member (which is the same for all members). Now we face a problem of nomenclature. On the one hand we have the primitive state space and primitive states of the individual members; and on the other hand, we have the primitive state space and primitive states of the population as a whole. To simplify the discussion, we shall use the terms *member state space* and *member state* for the former, and *population state space* and *population states* for the latter. In later sections, we shall deal almost exclusively with member states, and we shall revert to calling them simply *states*.

In constructing a primitive population state space, we must begin by making a choice. Do we want to treat the individual members as though they were distinguishable, or do we want to treat them as though they were indistinguishable? If we take the former approach, we would name and label each individual, and describe the state of the population in terms of the member state of each individual (e.g., Mary is in member state U_i; Bill is in member state U_j; ... ; etc.). On the other hand, if we take the latter approach, we would not name or label individuals, but would describe the population state in terms of the number of members occupying each member state (e.g., so many people are in state U_i; so many are in U_j; ... ; etc.).

Here we run into a conceptual difficulty, however. Presumably two or more individuals can be the same color, the same size, in the same physiological state, and the like; but if indeed we are dealing with a primitive state space, how can two or more indistinguishable[2] objects occupy the same element of physical space? We can deal with this difficulty in either of two ways. We can separate physical space from the rest of the primitive state space, and treat it separately; or we simply can recognize the fact that if indeed we are dealing with a primitive state space, the number of individuals undoubtedly will be very much less than the number of states, and the number of individuals occupying any given state will be either zero or one. Generally, however, we shall be dealing with reduced state spaces, in which case the difficulty is not at all likely to arise. Our prescribed resolution in physical space will be sufficiently coarse that many individuals will be able to occupy a given state simultaneously. In the discussion that follows in this and subsequent sections, we shall ignore the difficulty and treat primitive states as though they also could be occupied simultaneously by several individuals.

The number of population states in the population state space depends very much on whether we choose to treat individuals as distinguishable or indistinguishable, being much larger if the members are treated as being distinguishable. If there are v states in the primitive member state space and N *distinguishable* members in the population, then the number (D_1) of primitive population states is simply the number of ways to distribute N distinguishable objects over v states (with no restriction on the number that can occupy any given state):

$$D_1(N, v) = v^N. \tag{1.21}$$

[2] In the case of distinguishable objects, the finite span of our transition interval makes it possible that two or more distinguishable objects could be observed, one at a time, during one interval. Thus, of course, the concept of simultaneity in discrete time is significantly different from that of simultaneity in continuous time.

When the members are treated as being *indistinguishable*, the expression for the number (D_2) of primitive population states is more complicated:

$$D_2(N, \nu) = \binom{N+\nu-1}{N}. \tag{1.22}$$

However, when we are dealing with primitive state spaces, the number of member states is extremely likely to be very much larger than the number of members (i.e., $\nu \gg N$), in which case the expression simplifies to

$$D_2(N, \nu) \simeq \frac{\nu^N}{N!} \qquad (\nu \gg N). \tag{1.23}$$

In the case of a reduced state space (with lumped states), the situation very likely will be reversed, with a large number of members being distributed over a small number of states. In that case

$$D_2(N, \nu) \simeq \frac{N^{\nu-1}}{(\nu-1)!} \qquad (N \gg \nu). \tag{1.24}$$

Now, in models of many physical situations involving populations of nonliving objects, the sizes of the populations are assumed to be fixed. In biological populations, on the other hand, this usually is not the case. We could treat a varying population by allowing our primitive population state space to vary. This is extremely awkward, however. A better approach would be to construct the population state space in such a manner that it includes all possible population states over the full range of population sizes. In that case, the number (D') of primitive population states is equal to the sum of $D_1(N, \nu)$ or $D_2(N, \nu)$ taken over all values of N in the range of population sizes. In the special case in which the size can range from zero to some maximum value, N_{max}, we have

$$D_1'(N_{max}, \nu) = \sum_{N=0}^{N_{max}} D_1(N, \nu) = \frac{\nu^{(N_{max}+1)} - 1}{\nu - 1} \tag{1.25}$$

for distinguishable members, and

$$D_2'(N_{max}, \nu) = \sum_{N=0}^{N_{max}} D_2(N, \nu) = D_2(N_{max}, \nu+1) \tag{1.26}$$

or

$$D_2'(N_{max}, \nu) \simeq \frac{(\nu+1)^N}{N!} \qquad (\nu \gg N)$$

$$D_2'(N_{max}, \nu) \simeq \frac{N^\nu}{\nu!} \qquad (N \gg \nu) \tag{1.27}$$

for indistinguishable members.

Example 1.9. Consider a population comprising two identical members distributed over 50 primitive member states. Determine the number of primitive population states under the assumptions (1) that the members are indistinguishable and (2) that they are distinguishable.

Answer: (1) Since $\nu \gg N$ in this case, we can apply equation 1.23 to obtain an approximate answer:

$$D_2(2,50) \simeq \frac{50^2}{2!} = 1{,}250.$$

(Note: the actual value of $D_2(2,50)$ is 1,275, which is within 2 % of our approximation.)

(2) In this case, the relative magnitudes of N and ν do not affect our calculation.

$$D_1(2,50) = 50^2 = 2500.$$

Example 1.10. Consider a population comprising identical members with 100 primitive member states. Compute the numbers of primitive population states for each of the following cases: (1) the population comprises 5 distinguishable members; (2) the population ranges from zero to 5 distinguishable members; (3) the population comprises 5 indistinguishable members; and (4) the population ranges from zero to 5 indistinguishable members.

Answer: (1) In this case, we can apply equation 1.21:

$$D_1(5,100) = 100^5 = 10^{10}.$$

(2) In this case, we apply equation 1.25:

$$D_1'(5,100) = \frac{100^6 - 1}{100 - 1} \simeq 1.01 \times 10^{10}.$$

(3) In this case, we can use the approximation of equation 1.23:

$$D_2(5,100) \simeq \frac{100^5}{5!} \simeq 0.833 \times 10^8.$$

(4) And here we can use the approximation of equation 1.27:

$$D_2'(5,100) \simeq \frac{(101)^5}{5!} \simeq 0.876 \times 10^8.$$

Exercise. Consider a population of 200,000,000 people, each occupying one of two states (e.g., U_1 = male; U_2 = female). Determine the number of population states if each individual is treated as being distinguishable, and the number if each individual is treated as being indistinguishable.

1.17. Dynamics of Populations Comprising Indistinguishable Members

Once we have made the assumption that all members are identical, then as far as our deductions with respect to the population as a whole are concerned, it makes no difference *which* individuals are occupying a particular member state; the only thing that matters is *how many* are occupying that state. Since the population state space is largest by far when one labels individuals, deductions based on models on that space will be by far the most expensive to come by. Therefore, if we are interested in the dynamics of the population as a whole, the most reasonable approach we can take is to treat the members as if they were indistinguishable (except, of course, when we are attempting to observe the $p_{i,j}$'s).

With indistinguishable members, the primitive state of the population is given by the actual number of members occupying each of the primitive member states. The number of members occupying a particular member state is a pointwise discrete variable, whose value is a real integer. Thus, the primitive population state is given by a set of integers, one for each primitive member state. For convenience, the description of each population state can be written as a vector; and, following the convention employed in this text, we shall use column-vector notation. For example, if there are v member states in the primitive state space of each member (i.e., $\mathsf{U}_1, \mathsf{U}_2, \mathsf{U}_3, \ldots, \mathsf{U}_v$), then we shall represent the population state as follows:

$$\mathbf{n} = \begin{bmatrix} n_1 \\ n_2 \\ n_3 \\ \vdots \\ n_v \end{bmatrix}$$

where the vector $\mathbf{n}$ represents the primitive population state; and n_i is the number of members occupying primitive member state U_i. Each set of values, $n_1, n_2, \ldots, n_v$ that can be taken on by this vector represents one state in the primitive state space of the population.

Example 1.11. Consider a population of three indistinguishable members distributed over two primitive member states (U_1 and U_2). Diagram the primitive member state space and the primitive population state space.

Answer: Using a circle for each primitive state, we can depict the member state space with two circles, as illustrated in Figure 1.28a. Evidently, the population has four primitive states:

$$\mathbf{n} = \begin{pmatrix} n_1 \\ n_2 \end{pmatrix} = \begin{pmatrix} 3 \\ 0 \end{pmatrix}, \begin{pmatrix} 2 \\ 1 \end{pmatrix}, \begin{pmatrix} 1 \\ 2 \end{pmatrix}, \text{ and } \begin{pmatrix} 0 \\ 3 \end{pmatrix}$$

and can be depicted with four circles, as illustrated in Figure 1.28b.

In general, every transition in the population state space represents a set of simultaneous transitions of several members in the member state space. In fact, there generally will be *several* sets of simultaneous transitions in the member state space that correspond to the same transition in the population state space. Consider the transition from $\mathsf{U}'_2 = \begin{pmatrix} 2 \\ 1 \end{pmatrix}$ to $\mathsf{U}'_3 = \begin{pmatrix} 1 \\ 2 \end{pmatrix}$ in the population state space

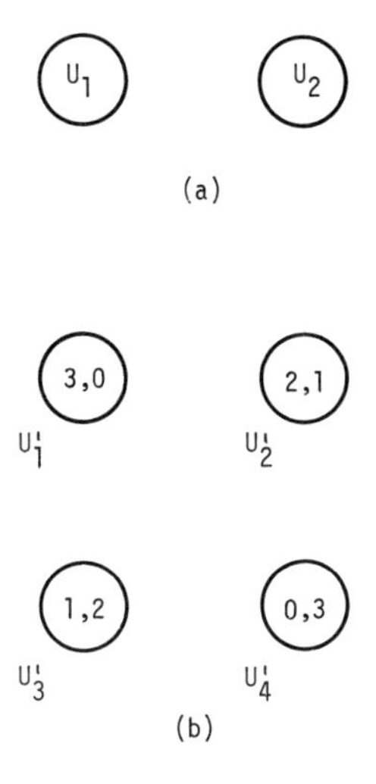

Fig. 1.28a and b. Member state space (a) and corresponding population state space (b) when three indistinguishable members are present

of Example 1.11. It can be accomplished by either of the following sets of transitions by individual members:

(a) one member undergoes the transition $U_1 \rightarrow U_2$
one member undergoes the transition $U_1 \rightarrow U_1$
one member undergoes the transition $U_2 \rightarrow U_2$

or (b) $U_1 \rightarrow U_2$; $U_1 \rightarrow U_2$; $U_2 \rightarrow U_1$.

In principle, given the transition probabilities associated with the member states, one can calculate the transition probabilities for the population states. He identifies all of the sets of simultaneous transitions in member state space that correspond to a given transition in population state space. Then he computes the probability of occurrence of each of the sets, then sums those probabilities to find the desired transition probability in population state space. Although it can be stated simply, this procedure can be extremely lengthy and difficult when the number of members and number of member states are at all large. For a very small population distributed over a very small number of member states, on the other hand, the calculations are not difficult.

Example 1.12. In terms of the general p's and q's of the member state space of Example 1.11, calculate the p's and q's of the population state space.

Answer: With four population states, there are four self-transition paths and twelve interstate transition paths, or a total of sixteen paths for which transition probabilities must be calculated. Let the transition probabilities between members states be designated p and q, those for population states be p' and q'. Thus $p_{2,1}$ is the probability that a given member in state U_1 undergoes the transition from U_1 to U_2, while $p'_{2,1}$ is the probability that when the population is in state

$U'_1 = \binom{3}{0}$ it will undergo the transition to $U'_2 = \binom{2}{1}$. The calculations proceed as follows: The transition $U'_1 \rightarrow U'_1$ corresponds to only one set of simultaneous transitions in member state space: $U_1 \rightarrow U_1$; $U_1 \rightarrow U_1$; $U_1 \rightarrow U_1$. Since each of these transitions in member state space is independent, the probability of their simultaneous occurrence is the product of the probabilities of their individual occurrences:

$$q'_1 = q_1 x q_1 x q_1 = q_1^3.$$

Similarly,

$$q'_4 = q_2^3.$$

The transition $U'_2 \rightarrow U'_2$ corresponds to two sets of simultaneous transitions in member state space: $U_1 \rightarrow U_1$; $U_1 \rightarrow U_1$; $U_2 \rightarrow U_2$ and $U_1 \rightarrow U_2$; $U_1 \rightarrow U_1$; $U_2 \rightarrow U_1$, the probabilities for which are $q_1 x q_1 x q_2$ and $2 x p_{2,1} x q_1 x p_{1,2}$ (there being 2 ways to achieve the second set):

$$q'_2 = q_1^2 q_2 + 2 p_{2,1} q_1 p_{1,2}.$$

Similarly

$$q'_3 = q_1 q_2^2 + 2 p_{1,2} q_2 p_{2,1}.$$

The remaining transition probabilities can be calculated in the same basic manner.

Once the transition probabilities have been calculated for the population state space, and the initial state of the population has been specified, we have established a Markov chain, from which we can deduce the dynamics of the population. Of course, with a very large population state space, the calculation of the dynamics will be rather formidable; and the resulting descriptions of the dynamics will be very lengthy. For very small populations distributed over a very small number of member states, however, the calculations are reasonably easy.

Example 1.13. Consider the population of Examples 1.11 and 1.12 being initially in state $U'_1 = \binom{3}{0}$, with the following probabilities of transitions for the member states: $q_1 = 1/2$; $p_{2,1} = 1/2$; $p_{1,2} = 0$; $q_2 = 1$. Calculate the description of the population dynamics for the first seven transitions.

Answer: The transition probabilities for the population states now take rather simple forms:

$$q'_1 = q_1^3 = 1/8 \qquad q'_2 = q_1^2 q_2 = 1/4 \qquad q'_3 = q_1 q_2^2 = 1/2 \qquad q'_4 = q_2^3 = 1$$

$$p'_{1,2} = p'_{1,3} = p'_{1,4} = p'_{2,3} = p'_{2,4} = p'_{3,4} = 0$$

$$p'_{2,1} = 3 q_1^2 p_{2,1} = 3/8 \qquad p'_{3,1} = 3 q_1 p_{2,1}^2 = 3/8 \qquad p'_{3,2} = 2 q_1 p_{2,1} q_2 = 1/2$$

$$p'_{4,1} = p_{2,1}^3 = 1/8; \qquad p'_{4,2} = p_{2,1}^2 q_2 = 1/4; \qquad p'_{4,3} = p_{2,1} q_2^2 = 1/2$$

from which we can construct the network in Figure 1.29.

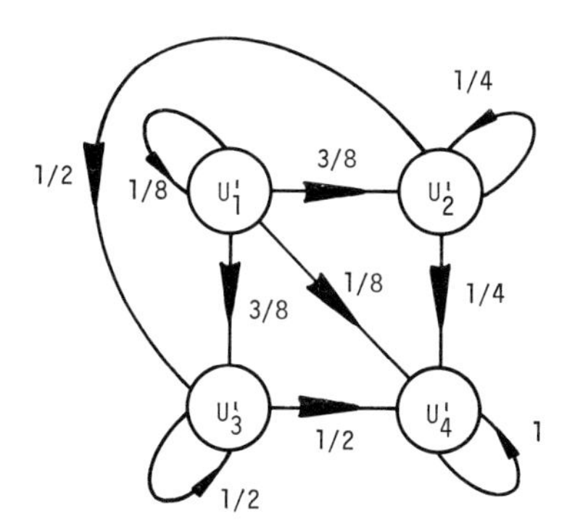

Fig. 1.29. Population states of Fig. 1.28b with transition probabilities calculated for Example 1.13

The state equations for this network are

$$\Pr[\mathrm{U}_1'(\tau+1)]=(1/8)\Pr[\mathrm{U}_1'(\tau)]$$
$$\Pr[\mathrm{U}_2'(\tau+1)]=(3/8)\Pr[\mathrm{U}_1'(\tau)]+(1/4)\Pr[\mathrm{U}_2'(\tau)]$$
$$\Pr[\mathrm{U}_3'(\tau+1)]=(3/8)\Pr[\mathrm{U}_1'(\tau)]+(1/2)\Pr[\mathrm{U}_2'(\tau)]+(1/2)\Pr[\mathrm{U}_3'(\tau)]$$
$$\Pr[\mathrm{U}_4'(\tau+1)]=(1/8)\Pr[\mathrm{U}_1'(\tau)]+(1/4)\Pr[\mathrm{U}_2'(\tau)]+(1/2)\Pr[\mathrm{U}_3'(\tau)]+\Pr[\mathrm{U}_4'(\tau)]$$

or

$$\mathbf{Pr}[\mathbf{n}'(\tau+1)]=\begin{bmatrix}1/8 & 0 & 0 & 0\\ 3/8 & 1/4 & 0 & 0\\ 3/8 & 1/2 & 1/2 & 0\\ 1/8 & 1/4 & 1/2 & 1\end{bmatrix}\mathbf{Pr}[\mathbf{n}'(\tau)].$$

Applying these interval by interval, we can generate the following description of the population dynamics for the first seven transitions:

$\tau \ldots 0$	1	2	3	4	5	6	7
$\Pr[\mathrm{U}_1'(\tau)] \ldots 1$	1/8	1/64	1/512	0.000244	0.000031	0.000004	0.0000005
$\Pr[\mathrm{U}_2'(\tau)] \ldots 0$	3/8	9/64	21/512	0.010986	0.002838	0.000721	0.0001818
$\Pr[\mathrm{U}_3'(\tau)] \ldots 0$	3/8	27/64	147/512	0.164795	0.087982	0.045422	0.0230715
$\Pr[\mathrm{U}_4'(\tau)] \ldots 0$	1/8	27/64	343/512	0.823975	0.909149	0.953853	0.9767462

Clearly, a description of population dynamics such as that in Example 1.13 is very detailed. From it, we can extract the probability of any particular distribution of the members over the member states. On the other hand, it is clear from the equations in section 1.16 that if either the number of members or the number of states is more than just a few, then the number of population states will be very large, and detailed descriptions such as that in Example 1.13 will be very, very lengthy. For example, if we increased the number of member states to three and the number of indistinguishable members to 100 (a fairly modest situation), the number of rows in the dynamic description would be 5,151.

Furthermore, the required, prior calculation of the transition probabilities for the population state space would be indeed a formidable task.

A dynamic description comprising 5,000 rows is virtually incomprehensible as it stands. In order to interpret it in a comprehensible fashion, one almost certainly will be required to consolidate and summarize it, throwing out much of the information contained in it. The most common summary used for population dynamics is the set of *expected* (or *mean*) *occupancies* of the individual member states during each transition interval. By using *expected occupancies*, one extracts a picture of the population dynamics that is much easier to display and interpret, even when the number of members and the number of member states are considerably larger than three and two. Furthermore, as we shall demonstrate in the following section, the calculation of the expected occupancies may not require previous calculation of the transition probabilities for the population state space. In fact, one may be able to deduce the expected occupancies directly from the member state space and the member state transition probabilities.

The *expected* or *mean* value of a variable is a very simple concept. To understand it in terms of a population, the reader may consider the following experiment: The indistinguishable members of a population repeatedly are returned to the same initial distribution over the member state space and then observed for a certain number of transition intervals. The number $n_i(\tau)$ of members occupying state U_i is counted during the τth interval after each reestablishment of the initial distribution. Every such count constitutes a trial. The counts are summed and the total is divided by the number of trials. The result is the mean or expected value of $n_i(\tau)$; which we shall call the *mean* or *expected occupancy* of state U_i. Thus, *expected* or *mean value* has its usual statistical meaning.

Empirically, one usually determines means in just this manner. In modeling, on the other hand, it is useful to be able to deduce means from prior knowledge of the probabilities (which themselves very likely were determined in the manner of the previous paragraph). When the probabilities are given for the various values of $n_i(\tau)$, then the expected or mean value of $n_i(\tau)$ can be deduced from the following relationship:

$$\mathrm{Ex}[n_i(\tau)] = \sum_{j=1}^{N} j \,\mathrm{Pr}[n_i(\tau)=j] \tag{1.28}$$

where N is the largest value assumed by $n_i(\tau)$, and $\mathrm{Pr}[n_i(\tau)=j]$ is the probability that $n_i(\tau)$ is equal to j (see Appendix A for probability notation). With this relationship, we can reduce detailed dynamic descriptions, such as that in Example 1.13, to much simpler forms.

Example 1.14. Using equation 1.28, reduce the detailed description of population dynamics in Example 1.13 to a set of expected occupancies of member states U_1 and U_2.

Answer: Recalling that U'_1 corresponds to $n_1=3$, $n_2=0$; U'_2 corresponds to $n_1=2$, $n_2=1$; U'_3 corresponds to $n_1=1$, $n_2=2$; and U'_4 corresponds to $n_1=0$,

$n_2 = 3$, we can apply equation 1.28 as follows:

$$\mathrm{Ex}[n_1(\tau)] = 3\,\mathrm{Pr}[\mathsf{U}_1'(\tau)] + 2\,\mathrm{Pr}[\mathsf{U}_2'(\tau)] + 1\,\mathrm{Pr}[\mathsf{U}_3'(\tau)] + 0\,\mathrm{Pr}[\mathsf{U}_4'(\tau)]$$
$$\mathrm{Ex}[n_2(\tau)] = 0\,\mathrm{Pr}[\mathsf{U}_1'(\tau)] + 1\,\mathrm{Pr}[\mathsf{U}_2'(\tau)] + 2\,\mathrm{Pr}[\mathsf{U}_3'(\tau)] + 3\,\mathrm{Pr}[\mathsf{U}_4'(\tau)]$$

from which we can generate the following sets of expected values:

τ	0	1	2	3	4	5	6	7
$\mathrm{Ex}[n_1(\tau)]$	3	1.5	0.75	0.375	0.1875	0.09375	0.046875	0.0234375
$\mathrm{Ex}[n_2(\tau)]$	0	1.5	2.25	2.625	2.8125	2.90625	2.953125	2.9765625

Even for this very simple situation, the use of expected occupancies provides a clearer, more concise picture of the essence of the population dynamics: the gradual increase in the occupancy of state U_2 at the expense of a decline in occupancy of state U_1.

1.18. Deduction of Population Dynamics Directly from the Member State Space

We closed the previous section by introducing the notion of expected occupancy of a member state during interval τ (i.e., the expected number of members simultaneously occupying a given member state during interval τ). If we ever are to observe an expected (mean) occupancy of a member state, we somehow must be able to make repeated observations on the population under essentially identical conditions. The greatest impediment to such observations very likely will be temporal variations of the transition probabilities associated with the member states. The problems inherent with such variations were discussed very briefly at the end of section 1.15.

In general, however, we can deal rather well with the situation if one of the following conditions prevails:

1. The transition probabilities are constant.
2. The transition probabilities vary with time in a random, uncorrelated manner (e.g., $p_{i,j}(\tau)$ is not correlated with $p_{i,j}(\tau+\gamma)$ for any γ) but with a constant mean, and state occupancies are not very large.
3. The transition probabilities vary with time with discernible, reliably repeated patterns.
4. The variations in transition probabilities can be controlled completely by experimental intervention.

Situations 1 and 2 are effectively identical. In either case, the transition probabilities automatically will be identical for all of the repeated observations. Situation 3, on the other hand, will require proper timing of the observations, so that each repetition is made at the same phase of the repeated temporal pattern of transition-probability variation (see section 2.4). In situation 4, the transition probabilities can be fixed by experimental intervention, allowing repetition under identical conditions.

Assuming that one of these four conditions prevails, consider the following experiment: Repeatedly, under the same conditions with respect to the transition probabilities, we place all of the N members of a population in the same initial member state, which we shall call U_0. Each time that we do so, we wait until the τth interval and then count the number of members occupying member state U_i.

Now, using the methods of section 1.15 we can calculate the probability (under our established conditions) that any given member that initially was in state U_0 will be in state U_i during interval τ. We simply would sum the observed occupancies of U_i over all k repetitions to determine the total number of members observed in U_i during τ. Then we would divide that total by the total number of members that might have occupied U_i during τ (i.e., N members per repetition, k repetitions). As k becomes large the quotient will approach $\Pr[\mathsf{U}_i(\tau); \mathsf{U}_0(\mathbb{O})]$, the *conditional probability* that a member initially in state U_0 will be in state U_i at the τth interval (i.e., will have proceeded to U_i in τ transitions).

$$\Pr[\mathsf{U}_i(\tau); \mathsf{U}_0(\mathbb{O})] = \lim_{k\to\infty} \sum_{\psi=1}^{k} n_{i,\psi}(\tau)/kN \tag{1.29}$$

where $n_{i,\psi}(\tau)$ is the observed occupancy of U_i during the τth interval of the ψth repetition.

On the basis of the discussion at the end of the previous section, and on the basis of elementary statistics, it is clear that the expected value of $n_i(\tau)$, under the established experimental conditions is given by

$$\mathrm{Ex}[n_i(\tau)] = \lim_{k\to\infty} \sum_{\psi=1}^{k} n_{i,\psi}(\tau)/k. \tag{1.30}$$

Comparing equations 1.29 and 1.30, we can see immediately that

$$\mathrm{Ex}[n_i(\tau)] = N \Pr[\mathsf{U}_i(\tau); \mathsf{U}_0(\mathbb{O})].$$

This will be an extremely useful relationship. With it, we often shall be able to calculate the expected occupancy of a member state directly from the Markov chain of the individual member, without first going through the very difficult process of expanding to the population state space, calculating the transition probabilities associated with the population states, and generating the detailed descriptions of population dynamics.

If the N members of the population are distributed over several or many initial member states, then we can invoke a simple rule for expectations to aid in our calculations:

> The expectation of a sum of random variables is the sum of the expectations of those variables.

If we initially have $n_0(\mathbb{O})$ members in state U_0, we expect $\Pr[\mathsf{U}_i(\tau); \mathsf{U}_0(\mathbb{O})]$ of them to be in state U_i during τ. If, in addition, we have $n_1(\mathbb{O})$ members in state U_1 initially, then we also expect $\Pr[\mathsf{U}_i(\tau); \mathsf{U}_1(\mathbb{O})]$ of *them* to be in U_i during τ. The total number of members that we expect to be in U_i is

$$\mathrm{Ex}[n_i(\tau)] = n_0(\mathbb{O}) \Pr[\mathsf{U}_i(\tau); \mathsf{U}_0(\mathbb{O})] + n_1(\mathbb{O}) \Pr[\mathsf{U}_i(\tau); \mathsf{U}_1(\mathbb{O})].$$

Extending this reasoning to a general initial distribution, $\mathbf{n}$, we have

$$\begin{aligned} \mathrm{Ex}[n_i(\tau)] &= n_0(\mathbb{0})\,\Pr[\mathsf{U}_i(\tau);\mathsf{U}_0(\mathbb{0})] + n_1(\mathbb{0})\,\Pr[\mathsf{U}_i(\tau);\mathsf{U}_1(\mathbb{0})] + \cdots \\ &\quad + n_v(\mathbb{0})\,\Pr[\mathsf{U}_i(\tau);\mathsf{U}_v(\mathbb{0})] \\ &= \sum_{\zeta=0}^{v} n_\zeta(\mathbb{0})\,\Pr[\mathsf{U}_i(\tau);\mathsf{U}_\zeta(\mathbb{0})]. \end{aligned} \tag{1.31}$$

Returning to the vector/matrix notation introduced in section 1.13, we can write the entire set of expected occupancies of all member states as a column vector

$$\mathbf{Ex}[\mathbf{n};\tau] = \begin{bmatrix} \mathrm{Ex}[n_1(\tau)] \\ \mathrm{Ex}[n_2(\tau)] \\ \mathrm{Ex}[n_3(\tau)] \\ \vdots \\ \mathrm{Ex}[n_v(\tau)] \end{bmatrix}.$$

Considering equation 1.31 in light of the rules for multiplication of arrays (Appendix A), the interested reader will see that this vector is the product of a matrix and the vector $\mathbf{n}$:

$$\mathbf{Ex}[\mathbf{n};\tau] = \begin{bmatrix} \Pr[\mathsf{U}_0(\tau);\mathsf{U}_0(\mathbb{0})] & \Pr[\mathsf{U}_0(\tau);\mathsf{U}_1(\mathbb{0})] \ldots & \Pr[\mathsf{U}_0(\tau);\mathsf{U}_v(\mathbb{0})] \\ \Pr[\mathsf{U}_1(\tau);\mathsf{U}_0(\mathbb{0})] & \Pr[\mathsf{U}_1(\tau);\mathsf{U}_1(\mathbb{0})] \ldots & \Pr[\mathsf{U}_1(\tau);\mathsf{U}_v(\mathbb{0})] \\ \Pr[\mathsf{U}_2(\tau);\mathsf{U}_0(\mathbb{0})] & \Pr[\mathsf{U}_2(\tau);\mathsf{U}_1(\mathbb{0})] \ldots & \Pr[\mathsf{U}_2(\tau);\mathsf{U}_v(\mathbb{0})] \\ \vdots & \vdots & \vdots \\ \Pr[\mathsf{U}_v(\tau);\mathsf{U}_0(\mathbb{0})] & \Pr[\mathsf{U}_v(\tau);\mathsf{U}_1(\mathbb{0})] \ldots & \Pr[\mathsf{U}_v(\tau);\mathsf{U}_v(\mathbb{0})] \end{bmatrix} \cdot \begin{bmatrix} n_0(\mathbb{0}) \\ n_1(\mathbb{0}) \\ n_2(\mathbb{0}) \\ \vdots \\ n_v(\mathbb{0}) \end{bmatrix} \tag{1.32}$$

which can be written

$$\mathbf{Ex}[\mathbf{n};\tau] = \mathbf{P}'(\tau)\cdot\mathbf{n}(\mathbb{0}) \tag{1.33}$$

where $\mathbf{P}'$ is the matrix of equation 1.32 and $\mathbf{n}(\mathbb{0})$ is the initial distribution of the members of the population.

The calculation of $\mathbf{Ex}[\mathbf{n};\tau]$ now depends simply on knowledge of the matrix $\mathbf{P}'(\tau)$. With a little reflection, considering equation 1.11 of section 1.13, the reader should be convinced rather easily that $\mathbf{P}'$ is simply the product of successive member-state projection matrices, beginning with that for the initial interval and extending to that for the interval $\tau-\mathbb{1}$:

$$\mathbf{P}'(\tau) = \{\mathbf{P}(\tau-\mathbb{1})\,\mathbf{P}(\tau-\mathbb{2})\,\mathbf{P}(\tau-\mathbb{3})\,\mathbf{P}(\tau-\mathbb{4})\ldots\mathbf{P}(\mathbb{0})\}. \tag{1.34}$$

This new matrix, $\mathbf{P}'(\tau)$, the *state transition matrix*, converts estimates or actual observations of the initial population state, $\mathbf{n}(\mathbb{0})$, into deductions concerning the expected state during interval τ.

From equations 1.33 and 1.34, it is clear that $\mathbf{Ex}[\mathbf{n}(\tau)]$ can be calculated in a stepwise manner with the relationship

$$\mathbf{Ex}[\mathbf{n}(\tau+\mathbb{1})] = \mathbf{P}(\tau)\cdot\mathbf{Ex}[\mathbf{n}(\tau)]. \tag{1.35}$$

Furthermore, if we identify

$$\mathbf{Ex}[\mathbf{n}(\mathbb{0})] = \mathbf{n}(\mathbb{0}) \tag{1.36}$$

then we can write

$$\mathbf{Ex}[\mathbf{n}(\tau)] = \mathbf{P}'(\tau) \cdot \mathbf{Ex}[\mathbf{n}(\mathbb{0})]. \tag{1.37}$$

Thus, we have all of the same relationships that we had for the primitive Markov chain of the individual member, but with $\mathbf{Pr}[\mathbf{U}(\tau)]$ replaced by $\mathbf{Ex}[\mathbf{n}(\tau)]$.

If any one of the four situations listed at the beginning of this section prevails, then in principle we should be able to determine the state transition matrix $\mathbf{P}'(\tau)$. If, in addition, we know the initial distribution $\mathbf{n}(\mathbb{0})$ of members over the member state space, then we can use equation 1.33 to calculate $\mathbf{Ex}[\mathbf{n}(\tau)]$ for future times on the basis of that initial distribution. If we know $\mathbf{P}'(\tau)$ and $\mathbf{n}(\mathbb{0})$, and we are willing to accept $\mathbf{Ex}[\mathbf{n};\tau]$ as a description of the dynamics of the population, then we effectively have eliminated the need to deal with the population state space, and can deal directly with the member state space instead. In the following section, we shall establish an important restriction in this simplification.

Before moving on, however, it is useful to consider the cost of the reduction of the size of population state space. Succinctly put, we have traded an entire probability distribution for its mean. We can consider this trade in terms of the member states. On the basis of the member state or states initially occupied by the members of a population, we shall be able to predict for all future times the expected or mean occupancies of all member states. Because of our trade, however, we shall not be able to predict the probability that the expected occupancies actually will occur, or the probability that any other, specific set of occupancies will occur. To do so, we must obtain the detailed description of population dynamics from the expanded population state space. Of course, this in turn has *its* cost, namely the formidable calculations involved in establishing a model on the expanded state space and in deducing dynamics from that model. Therefore, in order to have the entire probability distribution, we shall have to pay for it.

1.19. A Situation in Which Member State Space Cannot be Used to Deduce Population Dynamics

We closed the previous section by concluding that if we know the initial distribution $\mathbf{n}(\mathbb{0})$ of members over the member states, and if we know the state transition matrix $\mathbf{P}'(\tau)$ for the individual member, then we can deduce the dynamics of the population in terms of the expected occupancies $\mathbf{Ex}[\mathbf{n};\tau]$ of all member states during interval τ. The most serious difficulty in this process is the determination of $\mathbf{P}'(\tau)$. The elements of this matrix are sums of products of the transition probabilities associated with the member states (i.e., the p's and q's of the member states); and our ability to determine those elements depends upon the nature of the time variation of the p's and q's and our ability to discern it.

We have mentioned the possibility that the transition probabilities associated with member states might depend on extrinsic factors, which in turn vary with time. This would endow the transition probabilities themselves with time dependence. Many of the time-varying factors that affect individual organisms in the earth environment are well known. Among these are the periodic revolution of the earth about its axis (leading to the various physical concomitants of the diurnal cycle), the periodic rotation of the moon about the earth (periodically altering noctural illumination and combining with the earth's revolution to generate the complex tidal cycles), and the periodic rotation of the earth about the sun (leading to the various physical concomitants of the seasonal cycle). Another factor, which is sufficiently independent of these to require separate consideration, is the *presence of other organisms*. This often is the most important factor of all.

When we are dealing with an object of any sort in its own primitive Markovian state space (e.g., an entire population in population state space, a member in member state space), then there is only one variable intrinsic to the state space and that is the state of the object itself. According to the Markovian property of the states, if the object is in any state, U_i, there are certain transitions available to it and certain probabilities, $p_{i,j}(\tau)$, associated with each of those transitions. Since the object cannot occupy two states at once, it would be utter nonsense even to discuss the conditional probability of transition $p_{h,k}(\tau)$ from state U_k to state U_h during an interval, τ, that the object was known to occupy state U_i. In other words, it would be absurd to discuss the dependence of p's and q's on the state of the object when they come into play only when the object occupies the state they are associated with. Therefore, in a very real sense of the word, the only variations that are *possible* in the p's and q's are those depending on *extrinsic* factors. The only *intrinsic* factor, the state of the object, automatically is accounted for fully in the very nature of the Markovian states.

However, when we attempt to deduce the dynamics of an entire population on the basis of the member state space rather than the population state space, as we did in the previous section, we run headlong into a situation in which the intrinsic factors might not be accounted for fully. It is quite conceivable that the p's and q's associated with the member states could depend on the distribution, $\mathbf{n}$, of the population over those states. If this is true, then one can deduce $\mathbf{Ex}[\mathbf{n}(\tau+\mathbb{1})]$ from $\mathbf{Ex}[\mathbf{n}(\tau)]$ only if the total number of members of the population is sufficiently large that $\mathbf{n}(\tau)$ and $\mathbf{Ex}[\mathbf{n}(\tau)]$ are essentially the same (the *Strong Law of Large Numbers* from statistics proclaims that N can be sufficiently large to bring this about). On the other hand, if N is not especially large, we are faced with a situation in which none of the four conditions listed at the beginning of the previous section is met. Because of its dependence on $\mathbf{n}(\tau)$, $p_{i,j}$ will not be constant; because $\mathbf{n}(\tau)$ is correlated with $\mathbf{n}(\tau-\mathbb{1})$, $p_{i,j}(\tau)$ will not vary in an uncorrelated manner from interval to interval; because of the probabilistic nature of $\mathbf{n}(\tau)$ for any interval beyond the present, $p_{i,j}(\tau)$ cannot be counted upon to vary with a reliably repeated pattern; and for the same reason, $p_{i,j}(\tau)$ cannot be controlled unless $\mathbf{n}(\tau)$ is completely controlled, in which case modeling and deduction serve no purpose other than to aid in the control itself. As a result, we cannot determine the state transition matrix $\mathbf{P}'(\tau)$ for the individual member, and therefore cannot carry out the deductive process of equation 1.37.

To underscore this problem, we can extract the dependence of the p's and q's on $\mathbf{n}(\tau)$ from their general dependence on time. Abstractly, we can represent this by using two separate arguments in our representations of the state projection matrix:

$$\mathbf{P}(\mathbf{n},\tau)=\begin{bmatrix} q_1(\mathbf{n},\tau) & p_{1,2}(\mathbf{n},\tau) & \dots & p_{1,v}(\mathbf{n},\tau) \\ p_{2,1}(\mathbf{n},\tau) & q_2(\mathbf{n},\tau) & \dots & p_{2,v}(\mathbf{n},\tau) \\ \vdots & \vdots & & \vdots \\ p_{v,1}(\mathbf{n},\tau) & p_{v,2}(\mathbf{n},\tau) & \dots & q_v(\mathbf{n},\tau) \end{bmatrix}.$$

If we deal with the primitive Markovian state space of the member, then the present transition probabilities may depend on the present distribution, $\mathbf{n}$, of all members over that state space; but they cannot depend on the history of $\mathbf{n}$, otherwise the state space would not be Markovian even for the single member (i.e., the history of the member would be part of the history of $\mathbf{n}$ and thus would affect the p's and q's, making the states non-Markovian). Therefore, the argument $\mathbf{n}$ in $\mathbf{P}(\mathbf{n},\tau)$ always must be interpreted as $\mathbf{n}(\tau)$; in other words, we always must evaluate $\mathbf{n}$ for the interval τ when we apply it to the function $\mathbf{P}(\mathbf{n},\tau)$. The use of the second argument in the expansion of the state transition matrix requires more explicit notation:

$$\mathbf{P}'(\mathbf{n},\tau)=\{\mathbf{P}[\mathbf{n}(\tau-\mathbb{1}),\tau-\mathbb{1}]\,\mathbf{P}[\mathbf{n}(\tau-\mathbb{2}),\tau-\mathbb{2}]\,\mathbf{P}[\mathbf{n}(\tau-\mathbb{3}),\tau-\mathbb{3}]\dots\mathbf{P}[\mathbf{n}(\mathbb{0}),\mathbb{0}]\}. \tag{1.38}$$

Now, let us concede for the moment that we know precisely the function $\mathbf{P}(\mathbf{n},\tau)$. In other words, for any specific, precise values of $\mathbf{n}$ and τ that we observe, are given, or deduce, we can find a corresponding value of $\mathbf{P}$. Note that this implies that we know the dependence of $\mathbf{P}$ on $\mathbf{n}$ and that one of the four conditions at the beginning of the previous section be met for the remaining time dependence of $\mathbf{P}$. From the very construction of the primitive state space, the variable τ either is known or can be determined (by counting transition intervals); so the precise value of the argument τ is available for all future intervals. On the other hand, we already have determined that it is quite impossible to know the precise distribution $\mathbf{n}$ during any interval beyond the present. All that we are able to do is observe the present distribution and deduce from it the probability associated with each possible future distribution. Therefore, the two arguments of $\mathbf{P}(\mathbf{n},\tau)$ are fundamentally different, the precise value of one being inherently knowable for all intervals, the set of values of the other being inherently unknowable for all intervals beyond the present. Therefore, even though we know precisely the function $\mathbf{P}(\mathbf{n},\tau)$, we cannot evaluate it precisely for any interval beyond the present, so we cannot use it to deduce the future dynamics of the population.

This problem has arisen because in our attempts to simplify the calculations, we have employed a state space (the member state space) in which crucial intrinsic variables ($\mathbf{Pr}[\mathbf{n}(\tau)]$) are inextricably hidden. These variables are available only in the population state space; and to carry out our deductions when N is not *very* large and $\mathbf{P}$ depends on $\mathbf{n}$, we must revert to that state space, with its formidable constructions and calculations.

1.20. The Law of Large Numbers

In the previous section, we concluded that if the size, N, of the population is not *very* large and if the p's and q's associated with the member states depend on the distribution, $\mathbf{n}$, of members over those states, then the member state space cannot be used to form deductions concerning the dynamics of the population. Evidently, if neither of these constraints applies, and if any one of the four conditions at the beginning of section 1.18 is met, then the member state space can be used for the analysis of population dynamics. We discussed the dependence of $\mathbf{P}$ on $\mathbf{n}$ in reasonable detail, but we merely mentioned the size constraint; yet the assumption of the presence of a large population is perhaps the most common assumption made in the modeling of populations. This assumption is made effective by the *Strong Law of Large Numbers;* but, when $\mathbf{P}$ depends on $\mathbf{n}$, its effectiveness depends on the fulfillment of another condition as well.

The stipulations, constraints, and proofs of the Strong Law of Large Numbers can be found in most elementary texts on probability theory (e.g., see Feller, 1968, pp. 258–261). Its consequences for situations in which the population size is very large can be summarized as follows:

According to the Strong Law of Large Numbers

As N increases, the difference between the expected value of the occupancy of any member state deduced for any interval and the actual value observed during that interval will become increasingly negligible in comparison to the occupancy itself. In other words, as N becomes larger and larger, the ratio $\mathrm{Ex}[n_i(\tau)]/n_i(\tau)$ aproaches one; and we become more and more certain that the error contained by the approximation

$$n_i(\tau) \simeq \mathrm{Ex}[n_i(\tau)] \tag{1.39}$$

is completely negligible for all member states.

If the p's and q's associated with the member states *do not* depend on $\mathbf{n}$, then we can express this increasing certainty by rewriting equation 1.35 in the following form:

$$\mathbf{n}(\tau+1) \simeq \mathbf{P}(\tau) \cdot \mathbf{n}(\tau) \qquad \text{for large } N \tag{1.40}$$

or, if we are especially optimistic, we simply might write

$$\mathbf{n}(\tau+1) = \mathbf{P}(\tau) \cdot \mathbf{n}(\tau).$$

In order to draw a similar conclusion in cases in which the p's and q's do depend on $\mathbf{n}$, we first must be certain that the dependence contains *no disproportionately large jumps*. For the sake of discussion, consider a single element, $p_{j,k}$, of the projection matrix for member states. Suppose that it depends on the entire distribution of members over the member states, but that we somehow have fixed the occupancies of all states but one, U_i. In other words, we are interested in the dependence $p_{j,k}[n_i(\tau), \tau]$. Consider two consecutive values of the discrete argument $n_i(\tau)$:

$$n_i(\tau) = \chi$$

and

$$n_i(\tau) = \chi + \mathbb{1}.$$

If χ is small, we might expect the value of $p_{j,k}(\chi, \tau)$ to be distinctly different from $p_{j,k}(\chi + 1, \tau)$. On the other hand, as χ becomes larger and larger, we might expect the two values to become less and less distinguishable. For example, we might not expect $p_{j,k}(\mathbb{1{,}000{,}000{,}000}, \tau)$ to be noticeably different from $p_{j,k}(\mathbb{1{,}000{,}000{,}001}, \tau)$; whereas $p_{j,k}(\mathbb{1}, \tau)$ might be quite different from $p_{j,k}(\mathbb{2}, \tau)$. If these expectations are met, then the ratio $p_{j,k}(\chi, \tau)/p_{j,k}(\chi + \mathbb{1}, \tau)$ should approach one as χ becomes large. If this is true, then the Strong Law of Large Numbers also leads to

$$p_{j,k}[n_i(\tau), \tau] \simeq p_{j,k}\{\mathrm{Ex}[n_i(\tau)], \tau\} \tag{1.41}$$

with increasingly negligible error as N increases. In that case, we can use the approximation

$$\mathbf{n}(\tau + \mathbb{1}) \simeq \mathbf{P}(\mathbf{n}, \tau) \cdot \mathbf{n}(\tau) \qquad \text{for large } N \tag{1.42}$$

with increasing confidence as N increases.

On the other hand, if $p_{j,k}(\chi, \tau)/p_{j,k}(\chi + \mathbb{1}, \tau)$ does not approach one as χ becomes very large, then we can say that the function $p_{j,k}[n_i(\tau), \tau]$ contains *disproportionately large jumps*, in which case we can have no confidence in approximation 1.42, even when N is very large.

1.21. Summary

So far, we have stated that the dynamics of any object (e.g., a member of a population, or an entire population) can be described in terms of that object's travels through its primitive state space. Furthermore, according to the Markovian Ideal, the entire history of an object's travels through state space is embodied in its present state; and if we reject the untestable hypothesis of total historical dependence, then any primitive state space can be modified to form a new state space that conforms to this ideal. Thus, in principle, we can construct a primitive Markovian state space for a member of a population, and we can construct a primitive Markovian state space for a population as a whole. Because of the inherent uncertainty in our measurements, as well as the inherently discrete nature of many attributes, all state variables should be considered as being discrete. This leads to primitive state spaces with finite numbers of elements (states), each of which represents a unit of resolution within which differences in attribute values cannot be detected. Trade-offs among the resolutions of individual state variables make the selection of state-space elements arbitrary within the constraints of uncertainty.

Although time is not one of the dimensions of the primitive state space, the size of the chosen unit of resolution of time markedly affects the limits of resolution of the state variables. Therefore, the specification of temporal resolution is an important aspect of the specification of the primitive Markovian state space. The dynamics of the object will be deduced or observed from one unit of temporal resolution (one transition interval) to the next, and will be described as a sequence of states, one for each transition interval.

Deduction of future dynamics is carried out with a model, the foundation of which is provided by the primitive Markovian state space. The next step in the construction of the model is the determination of the transition probabilities associated with each of the primitive Markovian states. The final step is the observation of an initial state of the object. Once these two steps have been carried out, a model (in the form of a Markov chain) has been defined. Deductions can be carried out on the model by repeated application of the relationship

$$\mathbf{Pr}[\mathbf{U}(\tau+\mathbb{1})]=\mathbf{P}(\tau)\cdot\mathbf{Pr}[\mathbf{U}(\tau)] \tag{1.43}$$

where $\mathbf{P}(\tau)$ is the state projection matrix for the object; or they can be carried out by application of

$$\mathbf{Pr}[\mathbf{U}(\tau)]=\mathbf{P}'(\tau)\cdot\mathbf{Pr}[\mathbf{U}(\mathbb{0})] \tag{1.44}$$

where $\mathbf{P}'(\tau)$ is the state transition matrix for the object.

Generally, the dynamics of an object tend to amplify the initial uncertainty concerning its state (which is reflected in the resolution of the elements of the primitive state space). Thus, as time progresses after our last observation of the state of the object, we become increasingly uncertain as to its present state. This is reflected in $\mathbf{Pr}[\mathbf{U}(\tau)]$ becoming increasingly less sharp as τ increases. According to an ergodic theorem, which was stated but not proved, the uncertainty concerning the distribution often will approach an asymptotic limit, which would be represented by an asymptotic distribution $\mathbf{Pr}[\mathbf{U}(\tau)]$.

If the primitive Markovian state space is such that the transition from any given state to any given state cannot be observed repeatedly, then the transition probabilities are in principle unobservable and the model (the Markov chain) cannot be constructed without an enabling assumption. That assumption generally is the identity of a large number of objects, which allows the repetitive observation of a single object undergoing a given transition to be replaced by the observation of many, identical objects making that transition.

The primitive Markovian state spaces of populations of identical objects can be constructed by expansion from the primitive Markovian state space of the individual member. The construction of models on the primitive population state space is difficult, however; and the deductions from those models are essentially incomprehensible without extensive summarization. On the other hand, if the elements of the state transition matrix for the individual member do not depend on the distribution, $\mathbf{n}$, of members over the member states, then one can deduce the expected present state of the population directly from the initial state of the population and the state projection matrix of the individual member:

$$\mathbf{Ex}[\mathbf{n}(\tau+\mathbb{1})] = \mathbf{P}(\tau)\cdot\mathbf{Ex}[\mathbf{n}(\tau)] \tag{1.45}$$

or

$$\mathbf{Ex}[\mathbf{n}(\tau)] = \mathbf{P}'(\tau)\cdot\mathbf{Ex}[\mathbf{n}(\mathbb{0})] \tag{1.46}$$

where $\mathbf{P}(\tau)$ is the state projection matrix of a member; $\mathbf{P}'(\tau)$ is the state transition matrix of the member.

Even if the elements of **P** depend on **n**, as long as the total population is sufficiently large and the dependence does not have any disproportionately large jumps, we can invoke the Strong Law of Large Numbers (which was cited but not proved) to support the approximations

$$\mathbf{n}(\tau) \simeq \mathbf{Ex}[\mathbf{n}(\tau)] \tag{1.47}$$

$$\mathbf{n}(\tau+1) \simeq \mathbf{P}(\mathbf{n},\tau)\cdot\mathbf{n}(\tau) \tag{1.48}$$

$$\mathbf{n}(\tau) \simeq \mathbf{P}'(\mathbf{n},\tau)\cdot\mathbf{n}(\mathbb{0}). \tag{1.49}$$

On the other hand, if the size of the population is not sufficiently large to validate these approximations, we shall be forced to construct our models on the expanded, population state space.

Finally, returning to the situation in which the elements of **P** do not depend on **n**, if the size of the population is sufficiently large, the Strong Law of Large Numbers leads directly to the approximations

$$\mathbf{n}(\tau+\mathbb{1}) \simeq \mathbf{P}(\tau)\cdot\mathbf{n}(\tau) \tag{1.50}$$

$$\mathbf{n}(\tau) \simeq \mathbf{P}'(\tau)\cdot\mathbf{n}(\mathbb{0}). \tag{1.51}$$

These *deterministic approximations* (relations 1.48, 1.49, 1.50, and 1.51) often are written as equalities and are the most common by far of population models.

1.22. Some References for Chapter 1

The numbers in the following lists refer to entries in the Bibliography at the end of the text. The lists themselves are designed to guide the interested reader into some of the literature relevant to particular topic areas treated in Chapter 1.

On Fundamental Principles of States, State Spaces, Dynamics

13, 19, 32, 34, 70, 72, 83, 89, 101, 121, 122, 124, 141, 168, 174, 243, 257, 266, 270, 301.

On Markov Chains

10, 14, 20, 72, 85, 87, 98, 115, 125, 174, 216, 217, 229, 238, 269.

On Vectors and Projection Matrices

87, 141, 150, 157, 174, 180, 194, 205, 215, 229, 237, 238, 257, 277, 293, 301.

On Probabilities, Large Numbers

10, 13, 14, 20, 72, 125, 130, 216, 217, 238, 243.

2. General Concepts of Population Modeling

2.1. Lumped Markovian States from Irreducible Primitive Markovian State Spaces

In section 1.18 we concluded that we could achieve considerable simplification of our models if we were willing to sacrifice detailed probabilistic deductions concerning population dynamics, accepting in their place deductions of expected, or average behavior (i.e., expected state occupancies). By doing so, we would be able to base our models on the relatively simple primitive Markovian state space of the individual member of a population, rather than on the extremely extensive primitive Markovian state space of the entire population.

Even though it is quite simple relative to the primitive state space of the entire population, the primitive state space of the individual member nevertheless is far too extensive to suit most modelers, with their limited analytical tools and limited computer capacities. Therefore, we require further simplification. That which we shall introduce here underlies virtually every deductive model and logical modeling process. We shall presume ourselves willing to sacrifice resolution in any primitive Markovian state space, and we shall combine the primitive Markovian states to form many, many fewer *lumped* or *aggregated* states, thereby reducing considerably the size of the state space and the complexity of the models constructed upon it.

We actually might begin with the detailed, primitive Markovian state space of the individual (or our best estimate of that state space), and reduce its extent by systematically combining states in some manner. On the other hand, we might follow the usual modeling procedure and begin directly with the candidate lumped states themselves. In either case, the end result must be logically consistent if we are to carry out valid modeling efforts based on it. Clearly, a model cannot be divorced from the deductive manipulations that are applied to it; its state space foundations *must* be consistent with the deductive logic that is to be applied to it. The deductive logic that we shall apply in this text, indeed, the logic that almost universally is applied to deductive models, is that of the Markov chain. This logic is embodied in the equations of section 1.21. Since this is our logic, the state space foundations of our models must conform to it. Therefore, in order to provide valid bases for our models, *our state spaces must comprise only Markovian states*. This will be one of two key criteria of validity, the second criterion being consistency with whatever principle of conservation we have established. The latter relates most directly to the available transitions from state to state and to the transition probabilities (as discussed in section 1.10); but, indirectly, it also relates to the selection of lumped states. For example, if our established principle of conservation requires that any object represented as being in our state space must remain there forever, then clearly, every state of the object must be represented in the state space.

Developing consistency of the state space with respect to conservation (i.e., specifying an exhaustive set of states) apparently provides little challenge for most modelers; one finds few errors in the literature in this facet of model construction. Inconsistencies do appear from time to time, however, between the assigned transition probabilities and the stated or tacit principles of conservation (e.g., occasionally one finds models in which the sums of transition probabilities from individual states do not equal one, as they must according to the discussion in section 1.10). Such inconsistencies invalidate the models but not necessarily the state spaces upon which the models were constructed.

On the other hand, there exist numerous published examples of models to which the logic of the Markov chain has been applied, but which had been constructed on state spaces comprising clearly non-Markovian states. In each of these cases, the lumping of states had led to an invalid reduced state space. Thus it appears that the first criterion of validity is the more difficult to meet; so we shall focus our attention on it in this section.

Each state space that we construct will have associated with it a transition interval, which is the unit of temporal resolution. According to the logic of the Markov chain, an object occupying the state space is assumed to undergo exactly one transition for every transition interval. Clearly, one transition path likely to be required for any lumped state is the "self-transition," representing the possibility that an object remains in a given lumped state from one transition interval to the next. In order for a lumped state to be Markovian, the probability that any object occupying it takes any given transition path from it *must* be independent of the history of that object. In other words, as far as any object occupying the state is concerned, the transition probabilities associated with the paths leading out of the state *must not* depend upon how the object arrived at the state *nor upon how long it has occupied the state* (i.e., how many self-transitions the object has undergone since arriving at the state). It is this last constraint that seems to cause the trouble with some models.

Consider, for example, the situation depicted in Figure 2.1, where the states of a rather simple primitive Markovian state space arbitrarily have been segregated into two lumped states (U_A and U_B). Clearly, the probability that an object in lumped state U_A undergoes the transition to lumped state U_B depends upon how many transition intervals have passed since the object last entered state U_A from state U_B. If the object entered state U_A very recently, then it probably will be close to the boundary between U_A and U_B, and its likelihood of returning to U_B in a given transition will be relatively high. On the other hand, if the object in U_A did not enter that state recently, then it probably has drifted away from the border between the two states, and its likelihood of returning to U_B on a given transition will be relatively low. One can see that this argument does not depend upon the actual location of the boundary between states U_A and U_B. In fact, it applies equally well to any grouping of states that one might choose in Figure 2.1. It also applies to any other primitive Markovian state space that is irreducible (in the sense of section 1.9). Thus, in general, one would expect lumped states in such spaces to be inherently non-Markovian.

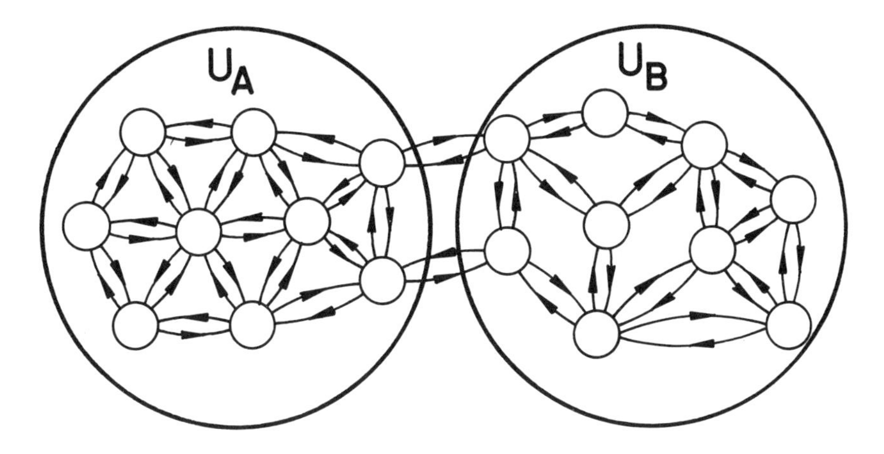

Fig. 2.1. Two sets of primitive Markovian states grouped to form two lumped states, U_A and U_B

On the other hand, under certain circumstances, an irreducible primitive Markovian state space can give rise to truly Markovian lumped states. As an example of such circumstances, consider the situation in which every primitive state is connected to every other primitive state by a path with the same probability. In this very special case, Markovian lumped states are produced by any arbitrary grouping of the primitive states.

Furthermore, certain primitive Markovian state spaces can, with judicious grouping of the primitive states, give rise to lumped states that closely approximate the Markovian criterion. For example, in the system of Figure 2.1 imagine that the transitions from U_A to U_B and from U_B to U_A were extremely unlikely compared with the self transitions (U_A to U_A, and U_B to U_B). In that case, it is possible that by the time an object leaves a given lumped state, the route by which it entered no longer is important. In other words, once an object has entered a lumped state, its probability distribution over the corresponding cluster of primitive states may, under certain conditions, approach a unique steady state that is independent of the route of entry. If that steady-state distribution is approached in times that are short compared to the time that the object is expected to remain in the lumped state, then the path of exit from the state and the probability associated with that path will be nearly independent of the original path of entry; and the lumped state will approximate a Markovian state. For convenience, we shall label such lumped states *apparently-Markovian*. The conditions that lead to the existence of a unique steady-state probability distribution are the following:

1. Every primitive state in the lumped state can be reached from every other primitive state by at least one route that does not pass outside the lumped state (a route consists of one or more sequential transition paths; a transition path is a direct connection from one primitive state to another).
2. The number of primitive states in the lumped state is finite.

3. The greatest common divisor of the number of transition paths within the lumped state from any primitive state back to itself is one (a condition automatically fulfilled if condition 1 is met and any of the primitive states in the lumped state has a self-transition path).

To show that these conditions are sufficient, we can invoke a well-known theorem from probability theory. In order to apply it, however, we first must stipulate that an object occupying the lumped state under consideration will remain there (i.e., we are interested in the object only so long as it remains there). In that case, we can ignore all paths leading from primitive states within the lumped state to primitive states outside of it, thus treating the lumped state as a closed system. Under this condition, and under the three conditions listed in the previous paragraph, each of the primitive Markovian states making up the lumped state will have the following properties:

A. Each primitive state will be *persistent* (in the sense that once an object has left that state, its probability of eventually returning to it is one).

B. Each primitive state has a *finite mean recurrence time* (in the sense that an object having left the state is expected to return to it in a finite time).

C. Each primitive state is *aperiodic* (in the sense that the greatest common divisor of possible recurrence times is a single transition interval).

Property A applies only as long as the object is stipulated to remain in the lumped state. Since, in fact, the object eventually will leave the lumped state, we should not label its component primitive states *persistent*, implying that they are absolutely persistent, but rather we might label them *apparently persistent*, indicating that they are indistinguishable from absolutely persistent states *as long as the object under consideration remains in the lumped state.* Property B also depends upon the object remaining in the lumped state; so again we should modify its label to read *apparently finite mean recurrence time*, indicating that the primitive states are indistinguishable from states with absolutely finite mean recurrence times *as long as the object under consideration remains in the lumped state.* Property C, on the other hand, does not depend on the object remaining in the lumped state, but follows directly from conditions 1 and 3 of the previous paragraph. Given that the object does remain in the lumped state, property A follows from condition 1, and property B follows from conditions 1 and 2. To see the relationships between these properties and the various conditions cited, the reader is referred to Feller (1968), chapter 15.

Any Markovian state with the three properties (A, B, and C) is called an *ergodic state*. Therefore, the primitive states making up our lumped state and conforming to conditions 1 through 3 will be *apparently ergodic* (indistinguishable from absolutely ergodic states *as long as the object under consideration remains in the lumped state*). The following theorem can be found (stated in various ways) in numerous texts on probability theory:

An Ergodic Theorem: *In an irreducible Markov chain (i.e., one conforming to condition 1) whose states are ergodic, the probability that an individual occupies any given state will approach a corresponding constant value as the number of transitions increases (i.e., as time progresses), and that constant value is indepen-*

dent of the initial state of the individual or the initial probability distribution for the individual over all of the states.

Thus we have the basis of our unique steady-state probability distribution. The steady state probability, $p_i(ss)$, for a given state, i, is the reciprocal of the mean recurrence time for that state and is given by the following equation:

$$p_i(ss) = \sum_j p_j(ss)\, p_{i,j} + p_i(ss) q_i \tag{2.1}$$

where the $p_{i,j}$'s and q_i's are those for the set of states under consideration when it is closed (i.e., when paths leading into and out of the set of states are eliminated).

Example 2.1. For the three-state system of Figure 1.11, find the steady-state probability distribution assuming that all transition probabilities are constant with respect to time.

Answer: Applying equation 2.1, we develop three equations for the three steady-state probabilities, $p_1(ss)$, $p_2(ss)$ and $p_3(ss)$:

$$p_1(ss) = p_2(ss) p_{1,2} + p_3(ss) p_{1,3} + p_1(ss) q_1$$
$$p_2(ss) = p_1(ss) p_{2,1} + p_3(ss) p_{2,3} + p_2(ss) q_2$$
$$p_3(ss) = p_1(ss) p_{3,1} + p_2(ss) p_{3,2} + p_3(ss) q_3.$$

It can be shown that any one of these equations can be derived from the other two (i.e., only two of the equations are independent). A third independent equation is provided by conservation:

$$p_1(ss) + p_2(ss) + p_3(ss) = 1.$$

The three independent equations can be solved simultaneously to yield the steady-state distribution.

Example 2.2. In the candidate lumped state depicted in Figure 2.2, find the steady-state probability distribution.

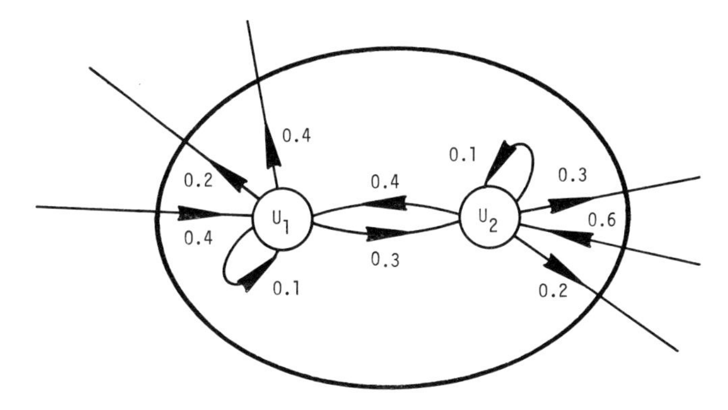

Fig. 2.2

Answer: First of all, the paths leading into and out of the candidate lumped state must be eliminated and the transition probabilities adjusted accordingly. Thus, each transition probability becomes the conditional probability that the corresponding path is taken, given that no path leading out of the lumped state is taken. The appropriate revised transition probabilities are given in Figure 2.3. Applying equation 2.1 to this revised set of probabilities, we have the following pair of equations

$$p_1(ss)=0.8\,p_2(ss)+0.25\,p_1(ss)$$
$$p_2(ss)=0.75\,p_1(ss)+0.2\,p_2(ss)$$

both of which reduce to the single equation

$$0.75\,p_1(ss)=0.8\,p_2(ss)$$

which can be solved simultaneously with the conservation relationship

$$p_1(ss)+p_2(ss)=1$$

to yield the distribution

$$p_1(ss)=0.516129$$
$$p_2(ss)=0.483871.$$

One question that the ergodic theorem and equation 2.1 fail to answer is how rapidly the steady-state distribution is approached. If, in a lumped set of primitive states, it is approached in times that are short relative to the mean or expected time that an object will remain in that set, then that set taken as a whole will approximate a single, *apparently Markovian state.* The process of approaching the steady-state distribution often is called *relaxation*, with the rate at which steady state is approached being called the *relaxation rate* and its reciprocal being called *relaxation time*. An extremely common assumption, often tacit, occasionally stated explicitly, in the modeling of physical systems comprising large populations of individual objects (e.g., populations of ions, molecules,

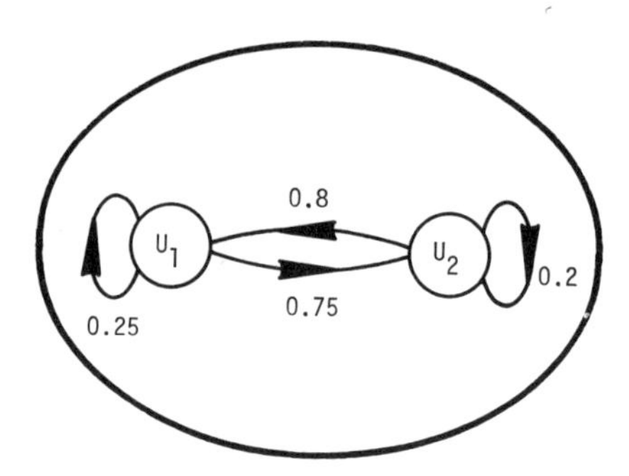

Fig. 2.3. States U_1 and U_2 of Fig. 2.2, taken to form a closed set (i.e., with extraneous paths removed and with transition probabilities adjusted accordingly)

and the like) is that the relaxation time in any given lumped state is extremely short compared to the times involved in the dynamic processes of interest.

If the $p_{i,j}$ in equation 2.1 are time dependent, then the situation is complicated slightly. The relaxation time will depend upon the $p_{i,j}$'s (see Figures 1.18 and 1.19 for examples of this dependence in a two-state system) and thus will in turn depend upon time. Therefore, in order to be *apparently Markovian*, a lumped state with time dependent $p_{i,j}$'s should exhibit relaxation times whose averages taken over times that are small compared to the times associated with the dynamics of interest are themselves small compared to those times. A connected set of such apparently Markovian lumped states with time-dependent $p_{i,j}$'s will comprise a nonhomogeneous Markov chain, as defined in section 1.14. The transition probabilities in this chain will vary with time as a result of the temporal variation in the steady-state distributions over the primitive states.

2.2. Shannon's Measure: Uncertainty in State Spaces and Lumped States

The previous section and sections 7 and 8 of chapter 1 have a common underlying theme, namely the randomization that occurs as an object progresses through its primitive state space. In the context of an entire primitive state space, as the time since the last observation of an object's primitive state increases, the uncertainty with respect to the object's present state also increases. Similarly, in the context of a lumped or aggregated collection of primitive states, as the time since an object was observed to have entered the collection increases, its present primitive state within the collection becomes increasingly uncertain. However, in the previous section it was shown that under certain rather general circumstances the uncertainty with respect to the object's present primitive state eventually will approach a unique constant level, corresponding to the unique steady-state probability distribution over all of the primitive states in question. At this point, it might be interesting to quantify uncertainty in general and this asymptotic uncertainty in particular. The means for doing so were provided by Shannon in his treatise on the transmission of information.

By requiring that it conform to certain, quite reasonable constraints, we can obtain a quantitative measure of uncertainty directly from Shannon's work. First of all, we shall require that the measure of uncertainty be based on the probabilities of occupancy of the various primitive states in the collection under consideration, when that collection is taken to be closed. Thus, the measure of uncertainty will be a function of the form

$$\mathit{uncertainty} = f(\Pr[\mathsf{U}_1], \Pr[\mathsf{U}_2], \Pr[\mathsf{U}_3], \ldots, \Pr[\mathsf{U}_k])$$

where there are k primitive states in the collection under consideration, and because the collection is taken to be closed

$$\sum_{i=1}^{k} \Pr[\mathsf{U}_i] = \mathbb{1}. \tag{2.2}$$

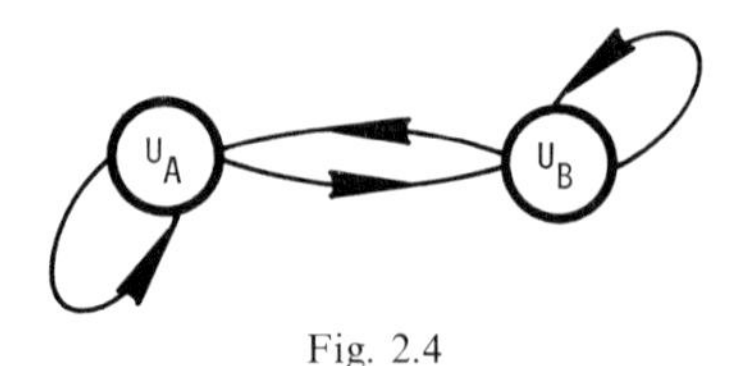

Fig. 2.4

Second, we shall require that the measure of uncertainty be maximum for a given collection of states when the probabilities of occupancy are equal. In other words, the maximum value of $f(\Pr[\mathsf{U}_1], \Pr[\mathsf{U}_2], \ldots, \Pr[\mathsf{U}_k])$ is that given by $f(\mathbb{1}/k, \mathbb{1}/k, \ldots, \mathbb{1}/k)$. Furthermore, as part of the second constraint, we shall require that between two collections of equally probable states, the collection with the greater number will have the greater value of uncertainty. In other words, $f[\mathbb{1}/(k+\mathbb{1}), \mathbb{1}/(k+\mathbb{1}), \ldots, \mathbb{1}/(k+\mathbb{1})]$ must be greater than $f(\mathbb{1}/k, \mathbb{1}/k, \ldots, \mathbb{1}/k)$ for all values of k. The third constraint involves the uncertainty of combined collections of primitive states. Consider the disjoint collections of primitive states depicted as lumped states U_A and U_B in Figure 2.4. Let $f(\mathsf{U}_A)$ and $f(\mathsf{U}_B)$ be the uncertainties of U_A and U_B respectively, when each is considered alone. In addition, let U_C be the collection of all the primitive states in U_A and all the primitive states in U_B, and let its uncertainty be $f(\mathsf{U}_C)$. We shall require that

$$f(\mathsf{U}_C) = f(\Pr[\mathsf{U}_A], \Pr[\mathsf{U}_B]) + \Pr[\mathsf{U}_A] f(\mathsf{U}_A) + \Pr[\mathsf{U}_B] f(\mathsf{U}_B), \tag{2.3}$$

where $\Pr[\mathsf{U}_A]$ and $\Pr[\mathsf{U}_B]$ are the probabilities of occupancy of collections U_A and U_B respectively when the two collections are considered together as a closed collection (U_C). In other words, if we know that an object is either in U_A or U_B (but we do not know which), then $\Pr[\mathsf{U}_A]$ is the probability that the object in fact is in U_A. What we are requiring in equation 2.3 is the following: if two lumped states are combined to form one, the measure of uncertainty over the combined state will be greater than that for either of its two component states alone (i.e., if we know that an object is in U_A we are less uncertain as to its present primitive state than we are if we merely know that it either is in U_A or U_B); and the measure of uncertainty over the combined state is the sum of three terms – the uncertainty as to whether U_A or U_B is occupied, the uncertainty as to which of the primitive states in U_A is occupied given that the object is in U_A, weighted by the probability that the object actually is in U_A, and the uncertainty as to which of the primitive states in U_B is occupied given that the object is in U_B, weighted by the probability that the object actually is in U_B. With a little reflection, the reader will find that equation 2.3 is an extremely reasonable constraint to impose upon our measure, and that in fact it should be extended to combinations of any number of disjoint collections of primitive states:

$$f(\mathsf{U}_C) = f(\Pr[\mathsf{U}_1], \Pr[\mathsf{U}_2], \ldots) + \Pr[\mathsf{U}_1] f(\mathsf{U}_1) + \Pr[\mathsf{U}_2] f(\mathsf{U}_2) + \cdots \tag{2.4}$$

where U_C is the combination of $\mathsf{U}_1, \mathsf{U}_2, \ldots$.

Example 2.3. In the special case of a collection of equally probable primitive states, show that equation 2.4 leads to an uncertainty that is *not* directly proportional to the number of states.

Answer: Consider the situation with m times n equally probable states. From equation 2.4 we have (taking m collections of n equally probable states each)

$$f(1/mn, 1/mn, \ldots, 1/mn) = f(1/m, 1/m, \ldots, 1/m) + f(1/n, 1/n, \ldots, 1/n). \tag{2.5}$$

Next, consider the sequence of values $mn = 2, 4, 8, 16$

$$\begin{aligned}
&f(1/2, 1/2) = f(1/2, 1/2) + f(1)\\
&f(1/4, 1/4, 1/4, 1/4) = f(1/2, 1/2) + f(1/2, 1/2) = 2f(1/2, 1/2)\\
&f(1/8, \ldots, 1/8) = f(1/4, 1/4, 1/4, 1/4) + f(1/2, 1/2) = 3f(1/2, 1/2)\\
&f(1/16, \ldots, 1/16) = 4f(1/2, 1/2)
\end{aligned}$$

from which we can see immediately that $f(1) = 0$ and that $f(1/2^n, \ldots, 1/2^n) = nf(1/2, 1/2)$, implying that f increases as the logarithm of the number of equally probable states.

Shannon demonstrated that only one function was able to meet the constraints described in the previous paragraph. That function is simply

$$f(\Pr[U_1], \Pr[U_2], \ldots, \Pr[U_k]) = \sum_{i=1}^{k} -\Pr[U_i] \log \Pr[U_i] \tag{2.6}$$

where the logarithm in the sum may be taken to any base. As Shannon himself pointed out, this formulation is identical to that deduced by Planck as a consequence of Boltzmann's principle, which states that the entropy of a particle is proportional to the logarithm of the probability of occupancy of each of that particle's states. Thus, our measure of uncertainty is essentially identical to the common measure of entropy, S, for a particle capable of occupying a certain number, N, of discrete states:

$$S = -k \sum_{i=1}^{N} p_i \log p_i$$

where k is the Boltzmann constant, and p_i is the probability that the particle occupies the ith state. The derivation of this relationship can be found in most textbooks on statistical mechanics (e.g., see Sommerfeld, 1956). In this way, *entropy* and *uncertainty* are very closely related concepts, a notion long ago recognized by statistical physicists. However, they are one and the same only under very special circumstances, namely, that the primitive Markovian states of the system are ergodic and equally probable in the limit (i.e., have equal steady-state probabilities of occupancy). According to generally accepted principles of thermodynamics, the entropy of a closed system increases monotonically with time, approaching a maximum value as the system approaches equilibrium. Uncertainty, as defined by equation 2.6, generally will increase as time progresses, but it need not take on its maximum value when the system is in equilibrium.

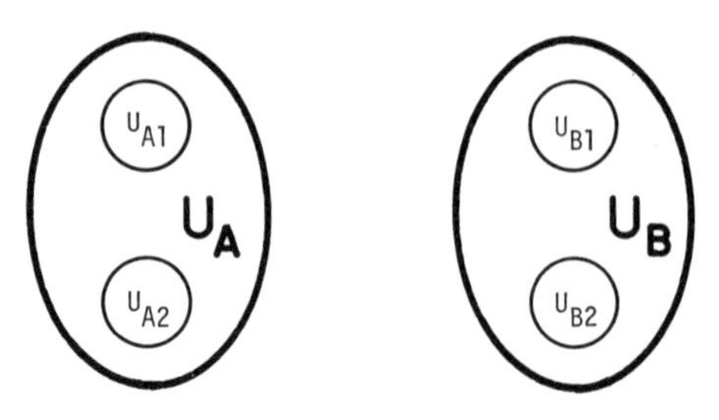

Fig. 2.5

Example 2.4. Consider the two lumped states, U_A and U_B, depicted in Figure 2.5. Each consists of two primitive states: U_{A1}, U_{A2} and U_{B1}, U_{B2}. When U_A and U_B are considered alone, the steady-state distributions over their states are

$$\Pr[\mathsf{U}_{A1}]_{ss}=1/3 \qquad \Pr[\mathsf{U}_{B1}]_{ss}=1/4$$
$$\Pr[\mathsf{U}_{A2}]_{ss}=2/3 \qquad \Pr[\mathsf{U}_{B2}]_{ss}=3/4.$$

When taken together, the steady-state distribution over the two lumped states is

$$\Pr[\mathsf{U}_A]_{ss}=1/4 \qquad \Pr[\mathsf{U}_B]_{ss}=3/4.$$

Find the steady-state or *asymptotic* uncertainties associated with each lumped state considered by itself and the asymptotic uncertainty associated with the two states taken together.

Answer: The asymptotic uncertainty of U_A taken by itself is simply

$$f(1/3,2/3)=-(1/3)\log[1/3]-(2/3)\log[2/3]$$

while that for state U_B by itself is

$$f(1/4,3/4)=-(1/4)\log[1/4]-(3/4)\log[3/4].$$

The asymptotic uncertainty for the combined states can be found by either of two routes. First, consider the combined state simply as the collection of the four primitive states U_{A1}, U_{A2}, U_{B1}, U_{B2}. If we are given that an object is somewhere in this collection, then we know that its steady-state probability of being either in U_{A1} or U_{A2} is 1/4, i.e., one quarter of the time it will occupy one or the other of those states, and during that one quarter of the time, its steady-state probability of occupying U_{A1} is 1/3 while its steady-state probability of occupying U_{A2} is 2/3. Evidently, therefore, when U_A and U_B are combined the steady-state probability that U_{A1} is occupied is $(1/4)\times(1/3)=1/12$, while the steady-state probability that U_{A2} is occupied is $(1/4)\times(2/3)=1/6$. Similarly, the steady-state probabilities for U_{B1} and U_{B2} are 3/16 and 9/16, respectively. Therefore, the asymptotic uncertainty for the collection of four states is

$$f(1/12,1/6,3/16,9/16)=-(1/12)\log[1/12]-(1/6)\log[1/6]-(3/16)\log[3/16]-(9/16)\log[9/16]. \tag{2.7}$$

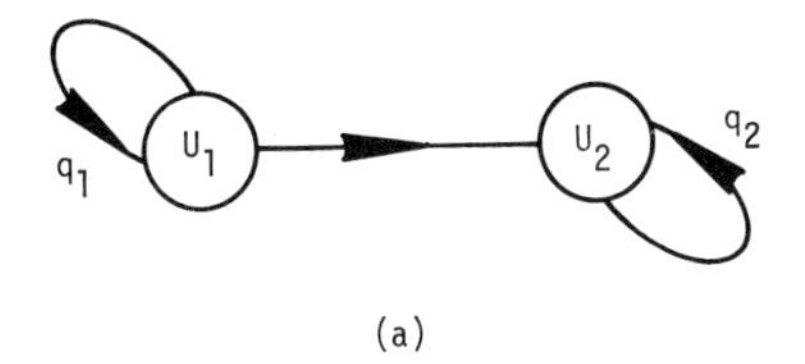

(a)

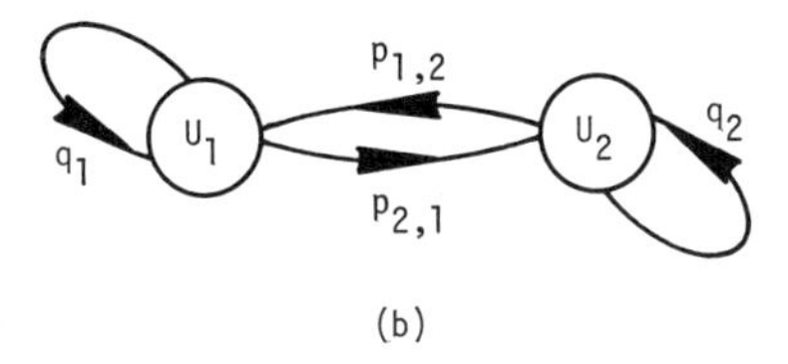

(b)

Fig. 2.6

The second method by which to obtain this uncertainty is the application of equation 2.4.

$$f(\mathsf{U}_C)=f(1/4, 3/4)+(1/4)f(1/3, 2/3)+(3/4)f(1/4, 3/4). \qquad (2.8)$$

Exercise. Demonstrate that the right-hand side of equation 2.8 indeed is identical to the right-hand side of equation 2.7 under Shannon's measure (as given in equation 2.6).

Before leaving the notion of uncertainty, let us return briefly to the notion that the uncertainty with respect to an object's present primitive state increases with increasing time since the state of that object last was observed. Now that we have an acceptable measure of uncertainty, we can follow its time course. This can be accomplished simply by application of equation 2.6 on an interval by interval basis to the dynamics of an object. However, qualitative trends can be detected even without an interval by interval analysis. For example, if initially we know that an object occupies state U_1 in Figure 2.6a, then the object's dynamics are given by

$$\Pr[\mathsf{U}_1(\tau)]=q_1^\tau$$
$$\Pr[\mathsf{U}_2(\tau)]=1-q_1^\tau$$

which, when combined with equation 2.6, lead to

$$f\{\Pr[\mathsf{U}_1(\tau)], \Pr[\mathsf{U}_2(\tau)]\}=-q_1^\tau \log[q_1^\tau]-(1-q_1^\tau)\log[1-q_1^\tau].$$

Initially, the value of uncertainty is zero, as it should be since we know that the object is in state U_1. Eventually, the value of uncertainty again will approach zero, as we become increasingly certain that the object has moved on to state U_2. In the meantime, uncertainty will increase and then gradually decrease again. In other words, uncertainty by no means is restricted to monotonic increase with time. Similarly, if an object initially is in state U_1 of Figure 2.6b, then initial uncertainty is zero. If p_{21} is greater than p_{12}, then from equation 2.1 we know that the limiting (equilibrium) value of $\Pr[U_2(\tau)]$ will be greater than that of $\Pr[U_1(\tau)]$. We also know that $\Pr[U_1(\tau)]$ and $\Pr[U_2(\tau)]$ both will pass through values close to $1/2$ as the dynamics proceed. At that time, the uncertainty with respect to the location of the object will be maximum, according to our original constraints on that measure; and it subsequently will decline again as the limiting values of the probabilities are approached.

2.3. Lumped Markovian States from Reducible Primitive Markovian State Spaces

If a primitive Markovian state space is reducible, then there will be at least one collection of states that has paths leading into it but no paths leading out of it (e.g., the collection making up lumped state U_B in Figure 2.7). Clearly, such a collection always can be combined to provide an absolutely Markovian lumped state. The transition probabilities associated with it are absolutely independent of history; once an object is in such a lumped state, the probability of transition out of it is zero, the probability of self-transition (e.g., from U_B to U_B) is one. This, of course, is a rather trivial case, involving a very special kind of lumped state (often called, for obvious reasons, an *absorbing state*). State U_A in Figure 2.7 is a different matter. In order to pass judgement on it, we must determine whether or not its primitive states are apparently ergodic and, if so, how its relaxation time compares to the dynamics of interest in the lumped system.

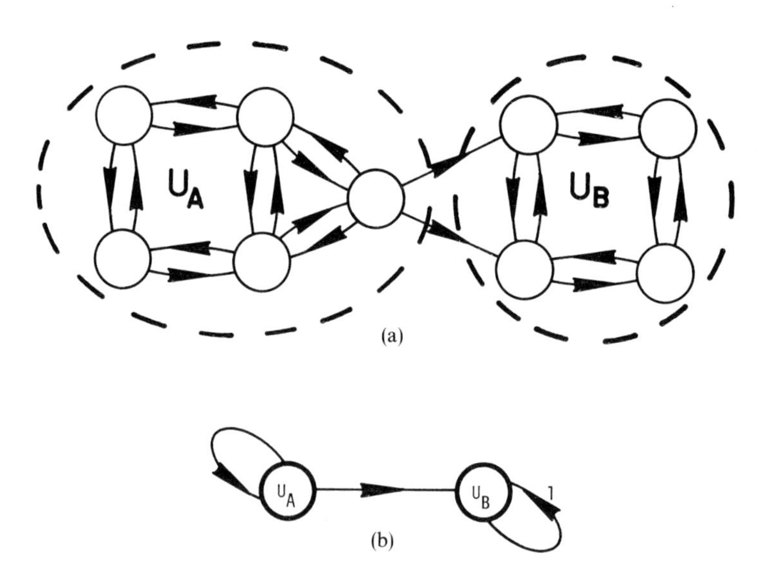

Fig. 2.7a and b. Candidate collections of primitive states for lumping (a), and the corresponding lumped states (b)

Of course, if we are willing to sacrifice resolution in state space, we also may be willing to sacrifice resolution in the parameter of state space, namely time. This can be done quite simply by lumping transition intervals; and it can lead both to analytical benefits and to serious artifacts, so it must be carried out judiciously. In the case of the system of Figure 2.7, if the state of an object is observed or calculated at sufficiently large lumped intervals of time, the probability of transition from U_A to U_B approaches one and the probability of self-transition in U_A approaches zero. In this manner, lumped state U_A approaches the Markovian ideal. However, if we use this method to convert U_A to a Markovian state, we essentially remove the dynamics from the system. On the other hand, we may not need to go to such an extreme. If we employ lumped intervals that simply are long compared to the relaxation time for U_A, then we will convert it to an apparently-Markovian state and still retain some of the slower dynamics, which may be of interest.

Example 2.6. Consider for an individual human being a state space in which there are only two resolvable states: (U_1) alive; (U_2) dead. In other words, all of the primitive states associated with life have been lumped into U_1 and all of the primitive states associated with death have been lumped into U_2. Can time be lumped in such a way as to make these states Markovian?

Answer: Being an *absorbing state*, U_2 automatically is Markovian. The nature of U_1, on the other hand, depends upon the transition interval that is selected for the lumped state space. Consider, for the sake of argument, a transition interval of one year. Clearly, in order for U_1 to be Markovian with this transition interval, the probability of death (transition to U_2) must not depend on the number of years that the human has been alive. From what we know about human biology it is clear that the probability of death does depend on the number of years that the human has been alive. Considering other transition intervals, we must come to the same conclusion for all of them that are less than approximately 130 years. On the other hand, if we select a transition interval greater than 130 years, the transition probabilities associated with U_1 effectively will be independent of the number of self-transitions; the probability of self-transition will be zero, the probability of transition to U_2 will be one.

If a reducible Markovian state space exhibits only unidirectional transitions from one primitive state to another and no self-transitions (as would be the case in any state space in which age were included as a state variable), then appropriate lumping of transition intervals can accompany aggregation of states to produce absolutely Markovian lumped states. There is a constraint on this type of lumping, which perhaps can best be appreciated with the aid of an example. Consider the primitive state space (or fraction thereof) depicted in Figure 2.8. Here we have twelve tiers of successive primitive states. If an object occupies one of the primitive states of a given tier during a certain transition interval, then during the subsequent interval it will, with probability one, occupy a primitive state in the succeeding tier. Therefore, if we lump all the states in a

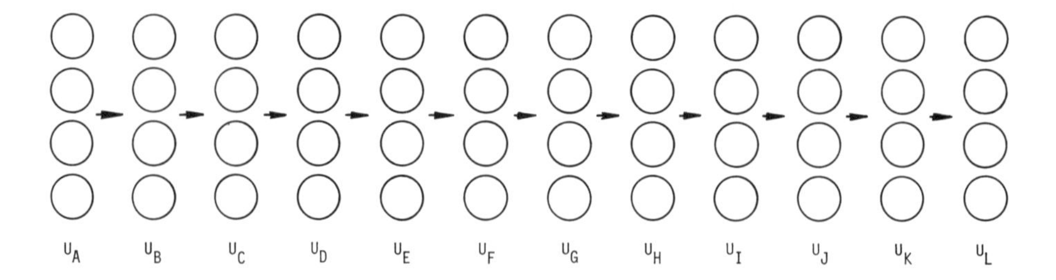

Fig. 2.8. Tiers of primitive states that can be lumped because of the unidirectionality of transitions

given tier and allow the transition interval to remain as it was (i.e., one transition interval of the primitive state space), then all of the lumped states will be absolutely Markovian. Suppose, however, that somewhere along the way we had lumped two tiers rather than one (e.g., formed a single lumped state by the combination of U_C and U_D in Fig. 2.8), yet left the transition interval as it was. In that case, an object that has just entered the two-tier state will, with probability one, remain in that state during the succeeding transition interval and then pass out of the state in the interval that follows. In other words, the object will undergo exactly one self-transition in the two-tiered state. Thus, the probability that an object passes on to the succeeding state is totally dependent upon how many self-transitions the object has made in the present (two-tiered) state. The two-tiered lumped state clearly is not Markovian under the primitive transition interval.

In order to make the two-tiered state absolutely Markovian we must do two things. First of all, *we clearly must divest it of any self-transitions;* otherwise the probability of transition out of the state will continue to be totally dependent upon the number of self-transitions already made. Note that because of its totally unidirectional transitions, the lumped state cannot be apparently ergodic or apparently Markovian; so self-transitions must be removed completely. Second, we must make the probabilities associated with the various transitions out of the state completely independent of the route by which the state was entered. To accomplish the first goal, we simply must form a lumped transition interval by combining two or more primitive transition intervals. In this way, we guarantee that an object occupying the two-tiered state during a given, lumped transition interval will have moved on and be occupying another state during the succeeding, lumped transition interval. Thus, by the simple means of lumping transition intervals, we can eliminate self-transitions in any collection of successive tiers (i.e., self-transitions in an n-tiered state will be eliminated by lumped transition intervals comprising n or more primitive transition intervals).

The second goal can be accomplished in several ways; but a large proportion of these will disrupt the essential structure of the state space, disconnecting one region from another. Suppose, for example, that all of the states in Figure 2.8 were aggregated into two-tiered lumped states (i.e., lumped states $U_A U_B$, $U_C U_D$, $U_E U_F$, $U_G U_H$, $U_I U_J$, and $U_K U_L$), as depicted in Figure 2.9. In order that the

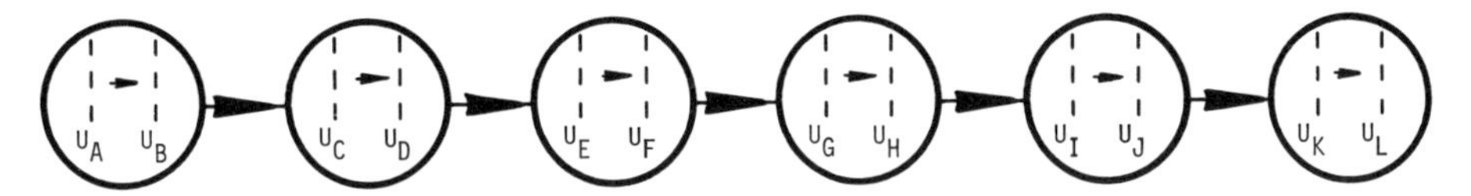

Fig. 2.9. Lumping of the states of Fig. 2.8 into groups of two tiers each; the corresponding transition interval must be twice the primitive transition interval in order for these lumped states to be Markovian

lumped states have no self-transitions, the lumped transition interval must comprise at least two primitive transition intervals. If the lumped transition interval were equal to two primitive transition intervals, then paths would exist only between successive lumped states; and each lumped state could be reached only from the one immediately to its left. Since there would be only one route by which each state could be reached, any dependence of exit-path probabilities on entry routes would be effectively eliminated; and all of the lumped states would be absolutely Markovian.

If the lumped transition interval were equal to three primitive transition intervals, on the other hand, most of the lumped states would not be Markovian. Consider, for example, state $U_E U_F$. An object entering that state from $U_C U_D$ in a single lumped transition interval (i.e., after three primitive transition intervals) would necessarily occupy the second tier (the first tier of $U_E U_F$ being unreachable in three primitive transitions from either tier of $U_C U_D$). After one more lumped transition interval (three more primitive transitions), that same object would have moved on to the first tier of $U_I U_J$. However, an object entering $U_E U_F$ from $U_A U_B$ would necessarily occupy the first tier and would necessarily move on to $U_G U_H$ in the subsequent lumped transition interval. Therefore, under this lumped transition interval, the path by which an object leaves a lumped state depends totally on the path by which it entered; and the conditions of the Markovian ideal thus are violated. If the lumped transition interval were equal to four primitive transition intervals, our state space would be decomposed into two independent sequences of lumped states. From state $U_A U_B$, an object would progress to state $U_E U_F$, then to state $U_I U_J$, and finally on to the absorbing state, $U_K U_L$, never occupying either of the other two states ($U_C U_D$ or $U_G U_H$). Similarly, an object in $U_C U_D$ would progress on to $U_G U_H$ and then to $U_K U_L$, missing the other states entirely. In spite of this decomposition of the state space, the states themselves would be Markovian under this lumped transition interval.

If the lumped transition interval were equal to five primitive transition intervals, then states $U_G U_H$ and $U_I U_J$ would not be Markovian (by the same logic applied to the situation with the lumped interval equal to three primitive intervals). With the lumped interval equal to six primitive intervals, all states would be Markovian, but the state space would be decomposed into three independent routes to the absorbing state ($U_K U_L$), one route being $U_A U_B$ to $U_G U_H$ to $U_K U_L$, another being $U_C U_D$ to $U_I U_J$ to $U_K U_L$, and the third being $U_E U_F$ to $U_K U_L$. Similarly, with the lumped interval equal to seven primitive intervals, $U_I U_J$ would be non-Markovian; and with the lumped interval equal

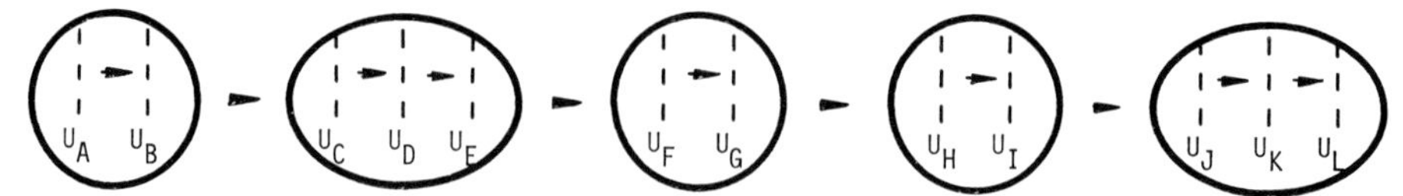

Fig. 2.10. Candidate lumping of the states of Fig. 2.8, with mixed numbers of tiers per lumped state. Such candidates should be eliminated out of hand

to eight or nine primitive intervals, all states would be Markovian, but the state space would be decomposed into four independent paths to the absorbing state and only one of those paths ($U_A U_B$ to $U_I U_J$ to $U_K U_L$) would have an intermediate step. Finally, with the lumped interval equal to ten or more primitive intervals, all states would be Markovian; but the state space would have become trivial with respect to dynamics, states $U_A U_B$, $U_C U_D$, $U_E U_F$, $U_G U_H$, and $U_I U_J$ all leading directly to absorbing state $U_K U_L$ in one lumped transition interval. Thus, with all of the nonabsorbing lumped states comprising two tiers, the only choice of lumped transition intervals that provided Markovian states and at the same time did not disconnect (or decompose) the original state space was two primitive intervals per lumped interval. In other words, generalizing by obvious induction, *the appropriate choice of lumped transition interval appears to be that whose number of primitive intervals is just equal to the number of tiers of primitive states in each nonabsorbing lumped state.*

Suppose that not all of the lumped states were made up of the same number of tiers; what would be the appropriate choice of lumped interval? Consider, for example, the situation depicted in Figure 2.10, where our system of tiers is lumped into two three-tier states ($U_C U_D U_E$ and $U_J U_K U_L$) and three two-tier states ($U_A U_B$, $U_F U_G$, and $U_H U_I$). If the lumped transition interval is less than three primitive intervals, the nonabsorbing three-tier state will be non-Markovian (absorbing states always are Markovian). If the lumped interval is equal to three, states $U_F U_G$ and $U_H U_I$ will be non-Markovian (note that the first state, $U_A U_B$, will be Markovian for all lumped transition intervals equal to or greater than its number of tiers, two): the exit probabilities for $U_F U_G$ will depend totally on whether the object came directly from $U_C U_D U_E$ or from $U_A U_B$ via $U_C U_D U_E$ (verification is left to the reader), while those for $U_H U_I$ depend on whether the object came directly from $U_C U_D U_E$ or directly from $U_F U_G$. If the lumped interval is equal to four intervals, state $U_C U_D U_E$ becomes non-Markovian (with exit probabilities that depend on whether the object had occupied $U_A U_B$ in the previous interval); and decomposition of the state space has begun (states $U_F U_G$ and $U_H U_I$ have become disconnected). As the number of primitive intervals in the lumped interval is increased further, further disconnection takes place until, once again, the state-space becomes completely trivial with respect to dynamics. Thus, by simple induction, it is clear that *the judicious choice of lumped states in systems such as that of Figure 2.10 is one with every state (with the possible exceptions of the first and last) having an equal number of tiers.*

If the lumped transition interval is equal to the number of primitive intervals required to move through each of the lumped states except possibly the first and the last (i.e., the number of primitive intervals in each lumped interval is equal to the number of tiers of primitive states in each lumped state), then the state projection matrix for a state space of the type depicted in Figure 2.8 will take on the following, extremely simple form:

$$
\boldsymbol{P}=\begin{bmatrix}
0 & 0 & 0 & 0 & 0 & 0 & 0 & 0 & . & . & . & 0\\
1 & 0 & 0 & 0 & 0 & 0 & 0 & 0 & . & . & . & 0\\
0 & 1 & 0 & 0 & 0 & 0 & 0 & 0 & . & . & . & 0\\
0 & 0 & 1 & 0 & 0 & 0 & 0 & 0 & . & . & . & 0\\
0 & 0 & 0 & 1 & 0 & 0 & 0 & 0 & . & . & . & 0\\
0 & 0 & 0 & 0 & 1 & 0 & 0 & 0 & . & . & . & 0\\
0 & 0 & 0 & 0 & 0 & 1 & 0 & 0 & . & . & . & 0\\
0 & 0 & 0 & 0 & 0 & 0 & 1 & 0 & . & . & . & 0\\
0 & 0 & 0 & 0 & 0 & 0 & 0 & 1 & . & . & . & 0\\
\vdots & \vdots & \vdots & \vdots & \vdots & \vdots & \vdots & \vdots & \vdots & \vdots & \vdots & \vdots\\
0 & 0 & 0 & 0 & 0 & 0 & 0 & 0 & 0 & 0 & 1 & 1
\end{bmatrix}. \tag{2.9}
$$

This projection matrix represents deterministic progression from state to state in ordered sequence, with no state being bypassed (see Example 1.8).

In the state space of Figure 2.8, at least one primitive state in the subsequent tier is reachable from each primitive state in any given tier (except the last one on the right). In other words, no absorbing primitive states occur in any of the tiers except the last one. If an absorbing primitive state did occur in a tier, it would have to be excluded from the lumped state incorporating that tier. Otherwise, the lumped state would have a self-transition (with probability less than one of being taken for the first time by any object, and a conditional probability one of being taken again given that it had been taken before), and the lumped state would be non-Markovian. Fortunately, there is no logical restriction on the lumping of absorbing states, so all of the absorbing states in a given tier may be lumped together with each other and, if so desired, with the absorbing states of any other tiers. Consider, for example, the state space depicted in Figure 2.11a. Here each tier of primitive states contains at least one absorbing state. In Figure 2.11b, we have lumped all of these together to form a single absorbing state, and we have lumped the remaining states into one-tier states. The corresponding state projection matrix is

$$
\boldsymbol{P}=\begin{bmatrix}
\mathbb{0} & \mathbb{0} & \mathbb{0} & \mathbb{0} & \mathbb{0} & \mathbb{0} & \mathbb{0} & \ldots & \mathbb{0}\\
p_{2,1} & \mathbb{0} & \mathbb{0} & \mathbb{0} & \mathbb{0} & \mathbb{0} & \mathbb{0} & \ldots & \mathbb{0}\\
\mathbb{0} & p_{3,2} & \mathbb{0} & \mathbb{0} & \mathbb{0} & \mathbb{0} & \mathbb{0} & \ldots & \mathbb{0}\\
\mathbb{0} & \mathbb{0} & p_{4,3} & \mathbb{0} & \mathbb{0} & \mathbb{0} & \mathbb{0} & \ldots & \mathbb{0}\\
\mathbb{0} & \mathbb{0} & \mathbb{0} & p_{5,4} & \mathbb{0} & \mathbb{0} & \mathbb{0} & \ldots & \mathbb{0}\\
\mathbb{0} & \mathbb{0} & \mathbb{0} & \mathbb{0} & p_{6,5} & \mathbb{0} & \mathbb{0} & \ldots & \mathbb{0}\\
\mathbb{0} & \mathbb{0} & \mathbb{0} & \mathbb{0} & \mathbb{0} & p_{7,6} & \mathbb{0} & \ldots & \mathbb{0}\\
\mathbb{0} & \mathbb{0} & \mathbb{0} & \mathbb{0} & \mathbb{0} & \mathbb{0} & p_{8,7} & \ldots & \mathbb{0}\\
. & . & . & . & . & . & . & . & .\\
p_{n,1} & p_{n,2} & p_{n,3} & p_{n,4} & p_{n,5} & p_{n,6} & p_{n,7} & \cdots & \mathbb{1}
\end{bmatrix} \tag{2.10}
$$

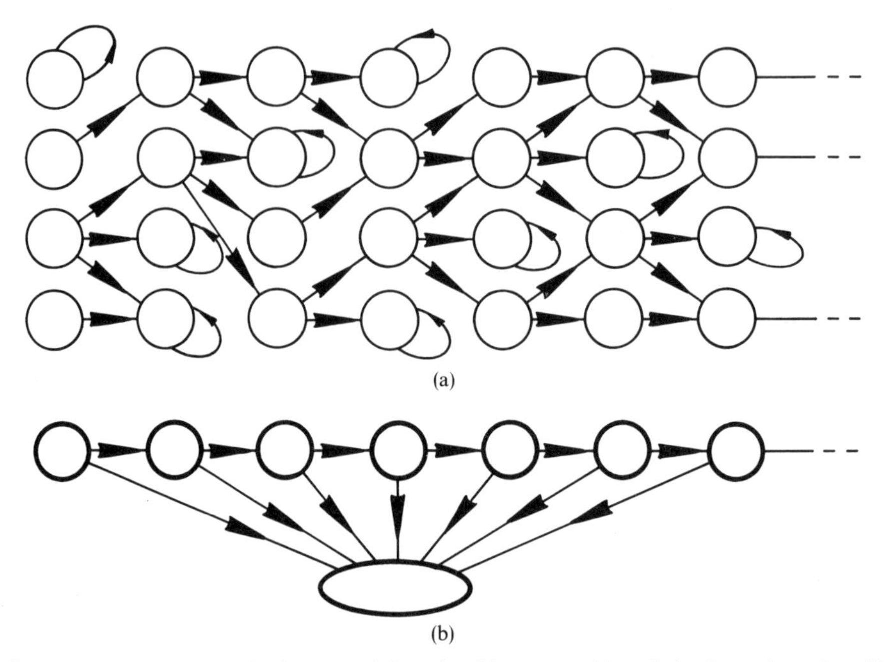

Fig. 2.11a and b. Lumpable tiers containing absorbing states (a) and the lumped version (b)

where the nth state (U_n) is the lumped absorbing state. Since an object that has reached the lumped absorbing state no longer participates perceptibly in the dynamics of the system one may choose to ignore the object altogether when it reaches that state, in which case the state itself, along with the bottom row in the projection matrix, may be omitted. However, when this is done, as it very often is in dynamic models, an appropriate adjustment must be made in the rules of conservation (e.g., see section 2.6).

Example 2.7. Consider for an individual human being a state space in which life and death are resolvable. Find lumped states that will be Markovian under transition intervals of one year, five years, and fifty years.

Answer: The candidate state *DEATH* already is Markovian, since it is an absorbing state. As we already have determined in Example 2.6, however, the candidate state *LIFE* is not Markovian under transition intervals of less than approximately 130 years. However, by subdividing *LIFE* into primitive states in which age is one of the state variables, we create a situation precisely analogous to that depicted in Figure 2.8. Regardless of what transitions occur between or among the other state variables, age progresses in a strictly deterministic fashion, leading to discrete states that exhibit unidirectional transition paths from one to the next and which can be arrayed in tiers, each corresponding to a given age resolved to one primitive transition interval. To accomplish the task set forth in Example 2.7, we simply must lump the tiers into groups of one year, five years, and fifty years, respectively. Thus, for any specified (lumped) transition interval, the lumped states of the individual could be designated as follows:

U_0 – born during the present transition interval (i.e., during τ)
U_1 – alive, born during the immediately previous transition interval (i.e., during $\tau - \mathbb{1}$)
U_2 – alive, born during $\tau - \mathbb{2}$
⋮
U_i – alive, born during $\tau - i$
U_n – dead

In this manner, if we find an individual in state U_i during interval τ, we are guaranteed of finding him or her either in state $\mathsf{U}_{i+\mathbb{1}}$ or state U_n during interval $\tau + \mathbb{1}$. Since there is only one path by which an individual can reach U_i, path dependence is not a factor, and the states all are absolutely Markovian.

Occasionally a modeler may wish to resolve two or more subsets of the nonabsorbing primitive states within a given tier. In some situations, this can be done in such a way as to preserve the Markovian nature of the lumped states, in other situations it cannot. Since judgement must be made on an *ad hoc* basis, we shall illustrate the problem by means of specific examples. Among the states of an individual human that we might wish to resolve are sex (*MALE*, *FEMALE*) several categories of height (e.g., *SHORT*, *MEDIUM*, *TALL*) several categories of mass (e.g., *LIGHT*, *HEAVY*); and, especially if we are interested in using the member state space to construct models of an entire human population, we very well might wish to resolve several categories of sexual maturity (e.g., *SEXUALLY IMMATURE*, *FERTILE*, *SEXUALLY SENILE*), several categories of reproductive state (e.g., *PREGNANT*, *NONPREGNANT*), and several categories of health (e.g., *INFECTED*, *INFECTIOUS*, *SUSCEPTIBLE TO INFECTION*). Each of these categories could be part of a description of one or more lumped states. Most of the categories must be accompanied by a state variable comparable to age in order for those lumped states to be Markovian, however.

For example, in a growing individual, the probability of transition from *LIGHT* to *HEAVY* or vice versa generally will depend upon how old the individual is. If *LIGHT* and *HEAVY* are considered in the absence of age, then the probability of the transition from *LIGHT* to *HEAVY* will depend upon the number of self-transitions that have taken place at *LIGHT*. The same thing applies to the transitions from *SHORT* to *MEDIUM* to *TALL*, from *IMMATURE* to *FERTILE* to *SENILE*, from *PREGNANT* to *NONPREGNANT*, and from *INFECTED* to *NONINFECTED*. In fact, among the candidate state variables listed in the previous paragraph, only *MALE* and *FEMALE* appear to define Markovian states without being accompanied by some measure of time; and some may even question those states. It is interesting to note that the probability of transition from *PREGNANT* to *NONPREGNANT* depends not so much upon age as it does upon the length of time since impregnation. Similarly, the transition from *INFECTED* to *NONINFECTED* depends not so much upon age as upon time since the infection was established. The "state variable" *AGE* simply is a running count of time initialized at the interval during which an individual was born. There is nothing preventing us from defining alternatives to *AGE* in the forms of running counts of time initialized at some other appropriate interval, such as the interval during which impregnation occurs or the interval during which infection occurs. With these alternatives to *AGE*, the states *PREGNANT* and *INFECTED* can be made Markovian.

Example 2.8. For an individual human being, construct a Markovian state space in which life, death, two resolvable categories of height (*SHORT* & *TALL*), and age resolved to ten-year intervals all are included as state variables.

Answer: We can begin by constructing a state space with approximately thirteen tiers, each representing a ten-year interval of age, assigning a transition interval of ten years, so that each tier will be connected to the next and no tier will exhibit self-transition. Each tier will have four states, defined in terms of the state variables as follows: 1. *AGEx*, *ALIVE*, *SHORT*, 2. *AGEx*, *ALIVE*, *TALL*, 3. *AGEx*, *DEAD*, *SHORT*, 4. *AGEx*, *DEAD*, *TALL*; which in Figure 2.12 we have denoted as U_{x1}, U_{x2}, U_{x3}, and U_{x4}, respectively. Next, we can ask whether or not these individual states all are Markovian. To answer that question, we will be forced to depend on generalizations from repeated observations rather than on first principles. Thus, we may be willing to accept as part of the definition of death that once an individual has entered a state labelled *DEAD*, he or she subsequently never will enter a state labelled *ALIVE*. Generally, the question of whether the states labelled *DEAD* are Markovian or not is moot as far as population dynamics are concerned; and all such states are lumped into a single absorbing state, U_n, which automatically is Markovian (by virtue of being absorbing). In this manner, the state space can be simplified to the structure shown in Figure 2.13. From repeated observation, we know that young human

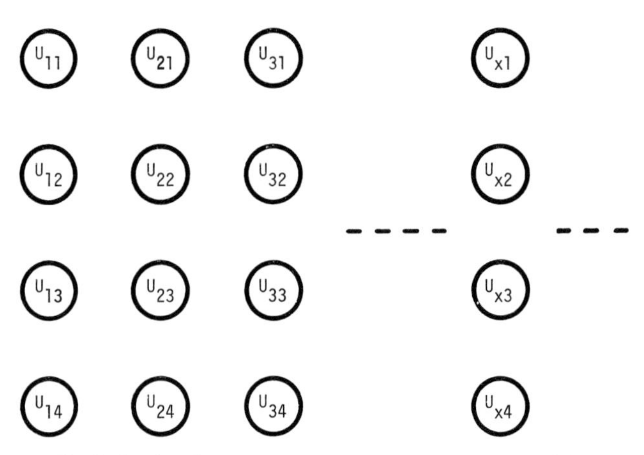

Fig. 2.12. Tiered state space corresponding to Example 2.8

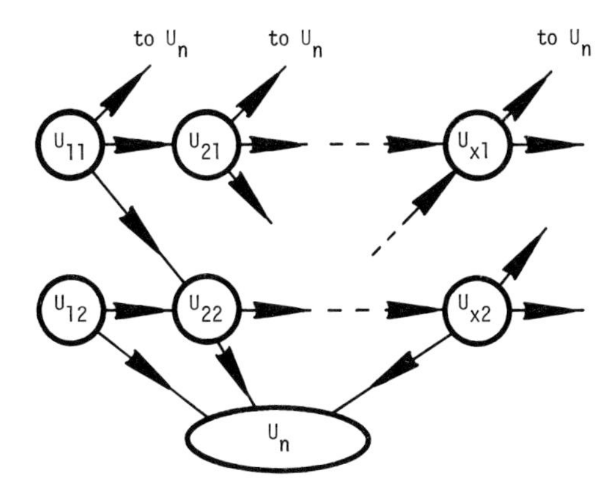

Fig. 2.13. Simplified version of the state space of Fig. 2.12. States representing death have been lumped into U_n

beings may become taller but are very unlikely to become shorter, while middle-aged and old human beings may become slightly shorter but are very unlikely to become taller. Therefore, for young human beings there is a finite probability of undergoing transition from a state labelled *SHORT* to one labelled *TALL*, but essentially zero probability of going from one labelled *TALL* to one labelled *SHORT* (of course we must establish a boundary between these two categories before we can assign specific probability values). Similarly, for middle-aged and older human beings, there is a finite probability of going from a state labelled *TALL* to one labelled *SHORT*, but essentially zero probability of going the other way. These empirical generalizations have been incorporated into the state space of Figure 2.13.

We have not yet settled the question as to whether or not the states are Markovian, however. To do so, we must decide whether or not the probability of transition from any state presently occupied to a subsequent state depends upon the history of states occupied by the individual in the past. For example, is the probability of transition from *TALL* to *SHORT* during the fourth decade of life dependent upon whether the individual progressed to *TALL* during the first, second, or third decade of life? Does the probability of death during the fifth decade depend upon the decade during which full height was reached? If, after viewing the empirical evidence, the modeler is satisfied that all such dependences are negligible, then he may decide to take the states in Figure 2.13 to be Markovian. If, on the other hand, he finds that such dependences are evident in the data, he might decide to forego resolution of *TALL* and *SHORT* in his state space, and instead combine the two states in each tier of Figure 2.13 to form a single lumped state representing the corresponding age-class (i.e., the corresponding decade of age, in this case). By doing so, he establishes lumped states that are absolutely Markovian.

Example 2.9. For an individual human being, construct a state space with transition paths, in which life, two categories of mass, (*HEAVY*) and (*LIGHT*),

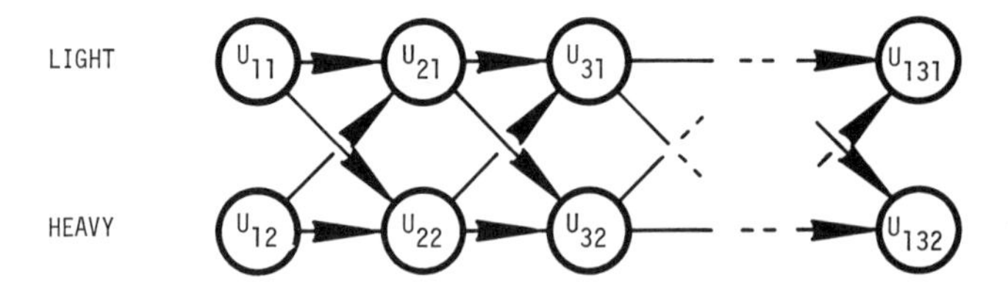

Fig. 2.14. Lumped states for Example 2.9. U_{x1} corresponds to age x (resolved to 10-year intervals) and light weight; U_{x2} corresponds to age x and heavy weight

and age are resolved, and the transition interval is ten years. Comment on the Markovian nature of the states.

Answer: If we consider an absolute life span of 130 years, then we have 13 categories of age, each with two categories of mass. We can lump (*DEATH*) into a single absorbing state with paths from every other state. Thus, we have a total of 27 states, 26 of which are associated with (*LIFE*). Unlike height, mass may have a significant probability of crossing the boundary between (*LIGHT*) and (*HEAVY*) in either direction in any age category. All of these generalizations are incorporated in Figure 2.14.

With humans, as with most organisms, mass will be strongly correlated with age, perhaps strongly correlated with season, and perhaps strongly dependent on extrinsic factors (via the food supply) that vary with time. None of these correlations or dependences in themselves prevents the states depicted in Figure 2.14 from being Markovian. On the other hand, it is quite possible that some individuals are intrinsically light, while others are intrinsically heavy. An intrinsically heavy individual presumably would spend most of its time in the states representating (*HEAVY*), but occasionally might slip to a state representating (*LIGHT*). In that case, it seems reasonable to assume that this individual would have a higher probability of undergoing a transition to (*HEAVY*) than would an intrinsically light individual who had spent most of its time in (*LIGHT*) states. Therefore, the transition probabilities very likely will be dependent on the path to the present state; so the states will not be Markovian.

Up to this point, we have considered lumped states formed by the aggregation of primitive states. This is not the procedure usually followed by modellers. Instead, candidate lumped states usually are formulated directly on the basis of repeated observation or experience with empirical data. In order to insure that such states are Markovian, the modeller attempts to incorporate in each one (either explicitly or implicitly) all of the variables that affect the probability of transition from that state to any other state. If transition probability depends on history (e.g., age of the individual), as it most often will, then that too must be incorporated in the state.

Example 2.10. For an individual, sexually mature human female, discuss the specification of Markovian states for a space that resolves life vs. death, and pregnant vs. nonpregnant, with a transition interval of one month.

Answer: Clearly, experience tells us that the probability of transition from pregnant to nonpregnant depends very strongly on the time since the beginning of pregnancy. Similarly, the probability of return to pregnancy depends on the length of time since the termination of the previous pregnancy (and, through cultural inputs, on the number of previous, successful pregnancies). Both of these probabilities depend also on age, as does the probability of transition from life to death. The latter also depends to some extent upon the history of pregnancies and the present state with respect to pregnancy. Thus, if the state space is to resolve life/death and pregnant/nonpregnant with a transition interval of one month, one reasonably might insist that in order to be Markovian the individual state descriptions must include age (in months), the number of previous pregnancies, and time (in months) since the last pregnancy or time (in months) since impregnation, depending on whether the state in question represents *PREGNANT* or *NONPREGNANT*. It should be obvious to the reader that a state space comprising such states will be exceedingly complex; and the rational modeller very well may decide to settle for somewhat less resolution.

2.4. Frequency Aliasing: The Artifact of Lumped Time

Basically, the lumping of time (i.e., the aggregation of primitive transition intervals) is accomplished by observing or deducing the state of an object during only a small proportion of the primitive transition intervals. That proportion can be distributed in a wide variety of ways over the entire set of primitive transition intervals; but usually the most convenient distribution is purely periodic, in which case the deduction or observation is made every *n*th primitive transition interval. Although this method of lumping time may be convenient, it also may produce artifacts resulting from the beating of the now periodic sampling of the system with any periodic phenomena that may exist in the system's dynamics. This type of artifact is called *frequency aliasing;* it tends to disguise the frequencies of components of the system's dynamics.

Consider the situation depicted in Figure 2.15. Here we have an ideal source of light, which produces instantaneous flashes (depicted as small stars in the figure) in an absolutely periodic manner (i.e., the interval between flashes is absolutely invariant). With continually alert, very rapidly responding light sensors (much more rapidly responding than the photoreceptors of our eyes, the response of which would be spread out over many tens of milliseconds), we detect each flash. Using a high-resolution clock, we assign to each flash a time given as the number of units of temporal resolution that have passed since the clock was started. For example, the clock could comprise another periodic sequence of events (denoted a, b, c, d ... in the figure); and the time of the flash could be determined by noting which pair of successive clock events bounded the occurrence of the flash. Thus, the first flash in the figure is bounded by clock events a and b and therefore is assigned the interval number or "time" 1. Note that the clock events could represent discrete physical events presumed to be periodic, or they could represent a presumedly constant, continuous physical phenomenon passing smoothly from one interval of resolution to the next (e.g., an object moving at a constant velocity passing from one unit of spatial

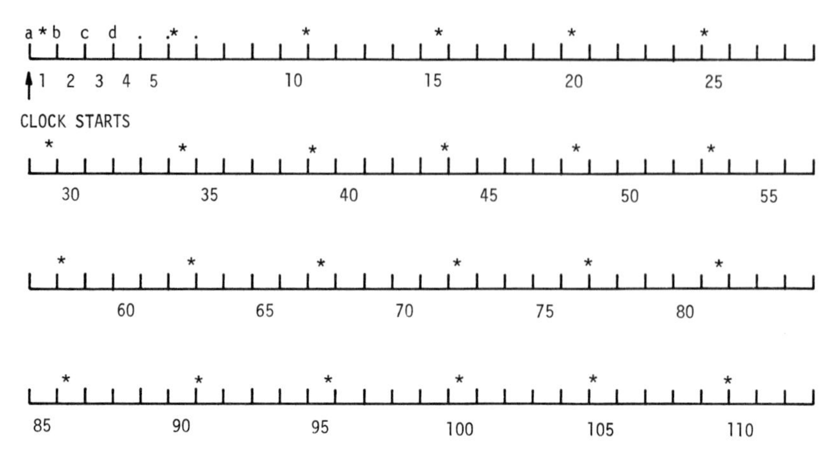

Fig. 2.15. Idealized, perfectly periodic light flashes (depicted as stars) occurring along a time line with periodic clock pulses (vertical lines) bounding intervals of temporal resolution. Each flash is assigned the value of discrete time corresponding to the interval in which it occurs

resolution to the next). In either case, if the observed flashes are periodic and the clock events are periodic, then unless the two periodic phenomena have precise harmonic relationships with one another (the likelihood of which is essentially zero unless the clock and the flashes are causally related), the phase of the flashes will drift with respect to the clock events. This drift will reduce the resolution with which one can determine the time interval between flashes on an interval by interval basis. In the case of the example of Figure 2.15, the observed times of the successive flashes are 1, 6, 10, 15, 20, 25, 29, 34, 39, 43, 48, 53, 58, 62, 67, 72, 76 ..., which correspond to the following sequence of intervals between flashes: 5, 4, 5, 5, 5, 4, 5, 5, 4, 5, 5, 5, 4, 5, 5, 4. The corresponding observed average interval is approximately 4.75. (Note that the method of construction of Figure 2.15, in which clock intervals were 0.25 inch and flash intervals were 3.0 cm, would lead to an average "observed" interval of 4.72440945, which could be deduced only after a very large number of flashes was observed.)

We cannot be certain on the basis of our observations that every flash interval is equal to the average interval, however; we only know that each successive interval lies somewhere between 4 and 5 units of temporal resolution. Thus, we see the hints of frequency disguise. According to our observations, the flashes may exhibit a simple periodicity (with interval 4.724 ... etc.), they may exhibit basically a simple periodicity with a slight, random wobble about their mean period, or their recurrence may be based on the complex interactions of two or more periodic phenomena (e.g., as in the case of the lunar cycle and the earth's rotational cycle combining to produce the complex tidal cycle, whose mean period is informative but does not tell the entire story). However, according to the well-known *Sampling Theorem*, which is directly applicable to this situation, we can only hope to discern periods that are at least twice as long as our unit of temporal resolution; our observations will be ambiguous with respect to the composition of periodic phenomena with shorter periods.

Acknowledging this limit to our resolution, we then can ignore altogether those periodicities that we cannot observe and consider only those periodicities that we can perceive unambiguously. Thus, if we are continuously alert for the flashes depicted in Figure 2.15, they will appear to us as being periodic, with interval 4.724 ..., regardless of the particular temporal resolution we are employing, *as long as that resolution meets the constraints of the sampling theorem.*

The situation changes, however, if we are not constantly alert for the flashes, but look for them only during every nth interval. In that case, a certain proportion of the flashes will be missed, and the composition of the flash periodicities will become ambiguous even if the sampling rate is high in comparison to the flash rate. For example, consider sampling for flashes on odd intervals only (i.e., $n=2$); we would observe the following flash times for the situation depicted in Figure 2.15: 1, 15, 25, 29, 39, 43, 53, 67, 81, ... etc. The corresponding observed intervals are 14, 10, 4, 10, 4, 10, 14, 14, ..., which disguise rather well the underlying, simple periodicity. Sampling with $n=3$, we might begin with clock interval 1, 2, or 3. Beginning with 1, we would observe the following flash intervals: 9, 15, 9, 9, 15, 9, 9, ...; beginning with 2 we would observe 9, 24, 9, 24, 9, 15, 9, ...; and beginning with 3 we would observe 9, 24, 9, 24, 9, 24, ... Sampling with $n=5$, we develop an acute beat phenomenon. For example, consider clock intervals that are integral multiples of 5 (e.g., 5, 10, 15, ...); of these, the flashes appear during 10, 15, 20, and 25, and then not again until 95, 100, and 105, indicating an underlying phenomenon with a period of approximately 80 units of temporal resolution. This is pure artifact and graphically illustrates the phenomenon of frequency aliasing.

To transfer the arguments surrounding Figure 2.15 to a state space situation, simply imagine that a flash during a given clock interval signals the occupancy of a certain periodic state during that interval. Thus, the lumping of time by means of periodic observation of such a state to see whether it is occupied or not will lead to precisely the artifacts described in the previous paragraph. Of course, there are alternative sampling schemes that could be used for the lumping of time. One might, for example, allow the separation between sampled clock intervals to be a random number with a predetermined mean. Such a scheme would eliminate the marked beating between sampling and sampled periodicities, but it would not preserve the Markovian nature of the aggregated tiers of primitive states (as discussed in the previous section), which was our motivation for lumping time to begin with. Alternatively, we could aggregate successive clock intervals into lumped intervals of equal lengths, and remain alert throughout each lumped interval for flashes or occurrences of states. For the flashes, as long as the constraints of the sampling theorem were met, this would be a perfectly workable scheme. For the occupancy of a given state by a given object, on the other hand, the scheme again will not preserve the Markovian nature of aggregated tiers and, in addition creates difficulties with respect to conservation. The primitive state space was based on empirical constraints that allowed us to observe the object in only one primitive state in any given primitive transition interval. In a succession of intervals, we very likely could and would observe the object in several states, all during the same lumped transition interval.

Thus, it would appear that in order to preserve the Markovian nature of aggregated tiers, we must use the periodic sampling scheme described in the first paragraph of this section; and if we do so, we must contend with the frequency-aliasing artifacts that this scheme might generate.

2.5. Idealizations: Thought Experiments and Hypothesis Testing

In section 2.2, we stated that a model can be valid within the specific analytical framework of this text if the empirical evidence supports our contention that the states we have selected for its state space are Markovian; and we subsequently concluded that it is quite possible for lumped states to meet this criterion of validity. There are often situations in which we wish to extend our considerations *beyond* a firm empirical base, however; and modeling in such situations is a perfectly valid enterprise. In fact, many of the modeling applications mentioned in the Introduction to this text virtually demand departure from the firm empirical base. Such departure very often leads to further reduction in the number of states in the state space and to further simplification of the models constructed on that space.

One of the most important group of situations demanding departure from the firm empirical base are those that fall in the general category of *thought experiments*. Such experiments often are generated by a question beginning with the phrase "what would happen if ...?"; and the object of the phrase usually is a hypothetical construct or an idealization. Here are some examples: "What would happen to the growth of the human population if every woman were limited to three successful pregnancies?" "What would happen to the growth of the human population if medicine advanced to the state that every person reached an absolute longevity limit of 115 years?", "What would happen to a certain exploited species if we captured every adult when it reached the age of two years?", "What would happen to the spread of a certain disease if we were able to identify and isolate every infected person within 48 hours?", "What would happen to a certain bothersome species of bird if we were able to destroy 90 percent of its eggs during every nesting season?".

Another important type of thought experiment often is generated by a question constructed as follows "how would ______ behave if ____________ were true?". The first blank is filled with an object, and the second with a hypothesis or idealization concerning that object. Here are some examples of this type of question: "How would a population behave if the probability of the death of its members were independent of their age?", "How would a certain population behave if the gestation period of its females were precisely one year?", "How would a certain population behave if each member became sexually mature and began participating in reproduction precisely two years after its zygote was formed?", "How would the population of a prey species respond to a predator that depended on random search for encounter?", "How would the prey-species population respond if predation were limited to the sexually immature?".

Certainly, questions of both types can be useful. Furthermore, both types of questions often can be answered through the use of models. The validity of the question and the validity of the model are *separate* issues. The validity of a particular question depends upon the question itself and upon the situation toward which it is addressed. However, it is quite clear that such questions are the quintessence of hypothesis formation, which always demands extrapolation, or generalization, from the firm empirical base. Thus, the validity of such questions in general hinges only on the validity of hypothesis-formation as part of the intellectual repertoire of an inquisitive being.

The validity of the model in this case depends only on its conformation to the hypothesis and its analytical consistency. Thus, we should feel free to generate state spaces and to construct analytically consistent models on them, all on the bases of hypothetical constructs, or idealizations. Each model itself becomes the deductive instrument by which we can close the classic cycle: OBSERVATION → GENERALIZATION AND HYPOTHESIS FORMATION → FORMALIZATION AND DEDUCTION → DECISION.

It is the valid model that allows us to make deductions in a logically consistent manner. Once these deductions are made, we can evaluate them (in the light of our expectations or in the light of empirical evidence) and then decide on appropriate action (e.g., rejection of the hypothesis or proposed management policy, further tests and observations, trial implementation of a policy, and the like).

By their very nature, hypothetical constructs usually are rather simple in form. In order to formalize such constructs into analytically consistent models, then, one needs to develop a repertoire of modeling methods for certain basic and rather simple idealizations. The remainder of this text will be devoted to some of the idealizations that apply to populations of organisms. We shall state these idealizations; then we shall attempt to develop lumped Markovian state spaces and construct models on them that are appropriate to those idealizations; and, finally, we shall discuss conceptual and analytical shortcuts, consistent with the logic of Markovian states, which will facilitate our deductive processes.

Consistent with our conclusions in Chapter 1, we generally will develop our lumped Markovian state spaces for the individual member of the population, then apply it directly to the population as a whole, usually invoking the Strong Law of Large Numbers in the process. Occasionally, when the situation demands it, we shall revert to a lumped population state space, which will allow us to deduce such things as the probability of extinction or the probability of reaching some other threshold with respect to population size or some other specific distribution over the member states.

2.6. Conservation: Defining Membership in a Given Population

Although the lumped state space upon which we construct our population model will be that of the individual member, its structure will be dictated by considerations related to the entire population. The most important of these considerations is the definition of *membership* in the population. With this definition, we shall be able to extend the basic principle of conservation of an individual object in its state space into the more complex principle of conservation of members of the population. Whereas, according to the discussion in section 1.10, conservation of the single object is reflected in its omnipresence in its state space, conservation applied to a population need not require that a particular, individual object always be a member of that population.

By the usual conventions, membership in a population of organisms generally requires that the candidate member be living (i.e., not be in a member state associated with death). One may wish to set other qualifications for membership, such as being of the same sex, or being in the same geographical location, or being in the same size range, or being sexually mature. Once one has established such qualifications, in order to construct a model consistent with them he or she must include in the state space clear demarcations between the states of the candidate member that imply membership in the population and the states that imply nonmembership.

Because of their apparently totally transitory nature, individual organisms belong to a rather special class of objects, which include such things as volcanoes, positrons, clouds, and many, many other things whose appearances as distinguishable entities occur suddenly and are difficult or impossible to predict in advance. One could incorporate this aspect into the state space of the object by including a state or set of states representing the quality of being (POTENTIALLY EXTANT BUT NOT YET DISTINGUISHABLE). To avoid the enormous task of bookkeeping the vast numbers of potentially distinguishable objects of this type, to say nothing of the philosophical problems, one almost always will exclude such states from membership in a population. Thus, in defining membership in a population of organisms, one usually excludes states representing the organism prior to its clear emergence as an individual entity. On the other hand, such states almost always are represented tacitly in the state space of the candidate member; and in certain cases, such as situations in which one wishes to ask questions or form hypotheses concerning the effects of shuffling of genes, the representation must be quite explicit.

The demarcations that one establishes between states of the candidate member that imply membership and those that imply nonmembership in the population become the statement of conservation for that population. That statement usually can be translated into a rather succinct verbal list of *all* the routes by which membership can be gained or lost; similarly, such verbal statements can be translated into demarcations in the state space of the candidate member. Perhaps the most common verbal statements of conservation are the following: (1) members are gained through birth *alone*, lost through death *alone;* or (2) members are gained *only* through birth or immigration, lost *only* through death or emigration. The translations of these statements into state-space demarcations is rather obvious: clearly, in either case (and in general) we must establish demarcations between states representing *LIFE* and those representing *DEATH* and between states representing *POTENTIAL FOR BEING BORN* and those representing *HAVING BEEN BORN*. In addition, we must establish hypothetical or observed transition paths between these states and assign hypothetical or observed probabilities to each path.

In other words, two of the most important processes that we must observe or form hypotheses about are the processes of *birth* (or, more generally, *reproduction*) and the processes of *death*. Most models of organism populations are formulated to generate deductions concerning the effects of those two processes on the population dynamics. In fact, those two processes generally form the keystones of population dynamics. It is to be expected, therefore, that much of the remainder of this text will deal with the modeling of idealized birth and death situations.

2.7. Conservation and Constitutive Relationships for a Single State

When we construct our population models on the lumped state space of the individual member, then each state, U_i, is occupied by a certain number, n_i, of the members of the population as a whole. In other words, each state is occupied by a subset of the total population. Just as one can establish a principle of conservation for the population as a whole, one also can establish a conservation principle for this subset. To do so, one simply establishes all of the paths by which individuals can enter or leave the state in question and excludes the possibility of individuals appearing in the state without having passed over one of the established entry paths or disappearing without having passed over one of the exit paths. In other words, conservation is simply a matter of absolutely thorough bookkeeping.

Conventionally, conservation is treated in terms of *contents* and *flows*. For example, we might consider a reservoir of water. We could say that the net change in the *contents* of the reservoir from one time to the next is precisely equal to the total *flow* into the reservoir less the total *flow* out of the reservoir between those two times. It would be very convenient to apply similar logic to the contents of states in our models, and we shall do just that. However, because we are dealing with a state space and time both of which are made discrete by precision limits, the usual notion of continuous flow must be tempered somewhat. In the primitive state space of the individual, we should be able to observe the occupancies of the individual states during each primitive transition interval (of course the primitive transition interval most likely will have to be lengthened, i.e., the temporal resolution decreased, as the number of individuals we are following increases). Then, by observation of sequential occupancies of the various states, we can infer from conservation a net flow into or out of each state from one transition interval to the next. In other words, we *observe the contents* of the primitive states and *infer the flows* from state to state. For example, if we observe $n_i(\tau)$ individuals occupying state i during primitive transition interval τ, and $n_i(\tau+\mathbb{1})$ individuals occupying that same state during interval $\tau+\mathbb{1}$, then by conservation we shall say that there was a net *flow* of J_i individuals into state i, where

$$J_i \equiv n_i(\tau+\mathbb{1}) - n_i(\tau). \tag{2.11}$$

By analogy with the reservoir, we might say that this particular value of J_i applies to the interval of time between τ and $\tau+\mathbb{1}$. However, since we are dealing with units of temporal resolution (of which τ in this case is a running count), we simply cannot distinguish an interval *between* τ and $\tau+\mathbb{1}$. Therefore, we can denote J_i as belonging either τ, to $\tau+\mathbb{1}$, or to both, and simply remember and be consistent in whichever convention we choose. In this text, we shall use the convention

$$J_i(\tau) \equiv n_i(\tau+\mathbb{1}) - n_i(\tau). \tag{2.12}$$

Notice that by defining flow in this manner, we have made it commensurate with state contents, or occupancy. In other words, flow here is given in terms of numbers of individuals, with *per transition interval* implied. This is in contrast with the conventional definition of flow (as a continuous time process), where it is given explicitly in terms of contents per unit time (e.g., gallons per second flowing from a reservoir). It will be convenient to retain this nonconventional notion of flow, even for situations in which primitive transition intervals have been aggregated to provide lumped time.

At this point, it is convenient to label the flows from individual state to individual state. For that purpose, we shall make the following definition:

> $J_{j,i}(\tau)$ is defined to be the total number of individuals that are in state U_i at τ *and* will be in state U_j at $\tau+\mathbb{1}$, where τ is a running count of either the number of *primitive* transition intervals or the number of *lumped* transition intervals.

Clearly, the net change in the number of individuals occupying a given state, taking place from interval τ to interval $\tau+\mathbb{1}$, is simply the sum of the flows into that state during τ, less the sum of the flows out of it during τ. Thus, we can express our conservation relationship in terms of a simple equality:

CONSERVATION FOR STATE U_i

$$n_i(\tau+\mathbb{1}) - n_i(\tau) = \sum_h J_{i,h}(\tau) - \sum_j J_{j,i}(\tau) \tag{2.13}$$

where the sums are taken respectively over all paths into and all paths out of U_i. As long as the time convention is remembered and followed, this conservation relationship is valid both for Markovian and for non-Markovian member states.

If the member states are Markovian, then we can complement this conservation equation with rather simple relationships based on the transition probabilities. If we observe the transitions repeatedly under the same extrinsic and intrinsic conditions, then we *may* find a mean, or expected value of each J in the conservation equation. Because it can depend on intrinsic conditions (i.e., the total population or the distribution of members over the member states), the variable J should include $\mathbf{n}$ in its argument: $J_{j,i}(\mathbf{n},\tau)$ and $\mathrm{Ex}[J_{j,i}(\mathbf{n},\tau)]$. Since we presume knowledge of all the elements of $\mathbf{n}$ (including n_h, n_i and n_j), if we know the expected value of $J_{j,i}(\mathbf{n},\tau)$, then we automatically know the expected proportion $\mathrm{Ex}[p_{j,i}(\mathbf{n},\tau)]$ of individuals in state U_i during interval τ that will have moved on to state U_j and be there during the following interval:

$$\mathrm{Ex}[J_{j,i}(\mathbf{n},\tau)] = \mathrm{Ex}[p_{j,i}(\mathbf{n},\tau)] \times n_i(\tau). \tag{2.14}$$

Since the expected value of a sum is equal to the sum of the expected values (see section 1.18), we can rewrite our conservation relationship as follows:

$$\mathrm{Ex}[n_i(\tau+1)] = \mathrm{Ex}[n_i(\tau)] + \sum_h \mathrm{Ex}[J_{i,h}(\mathbf{n},\tau)] - \sum_j \mathrm{Ex}[J_{j,i}(\mathbf{n},\tau)]. \tag{2.15}$$

Now, if we combine equations 2.14 and 2.15, we obtain a relationship that is similar to those summarized at the end of Chapter 1. Our new relationship falls short, however, because it contains both n_i and $\mathrm{Ex}[n_i]$. In order to be a useful analytical algorithm, it must contain only one or the other. Therefore, our next goal would be to replace n_i everywhere with its expected value $\mathrm{Ex}[n_i]$, for all values of the index i.

Unfortunately, according to the arguments in section 1.19, we can do this only if one or both of two conditions prevails: (1) the values of n_i (for all i) are very large, or (2) the transition probabilities (the $p_{j,i}$'s) are independent of $\mathbf{n}$ (i.e., independent of the size and distribution of the population). If neither of these conditions is met, then in order to evaluate $\mathrm{Ex}[p_{j,i}(\mathbf{n},\tau)]$, we *must* have the actual set of values of the elements of the vector $\mathbf{n}$ (i.e., n_i for all i). The expected values of the elements will not suffice. Furthermore, unless two random variables are totally independent of one another, the product of their expected values generally will not be equal to the expected value of their products. Now, the expected value of $J_{j,i}$ is equal to the expected value of the product of $p_{j,i}$ and n_i. As long as n_i is known, the expected value of the product is given by the right-hand side of equation 2.14; but we cannot replace n_i in that product by $\mathrm{Ex}[n_i]$, because in general

$$\mathrm{Ex}[p_{j,i} \times n_i] \neq \mathrm{Ex}[p_{j,i}] \times \mathrm{Ex}[n_i].$$

Therefore, we cannot replace n_i with its expected value in either the argument of $p_{j,i}$ or the product $p_{j,i}n_i$ unless one of the two conditions prevails.

If the transition probabilities are independent of the total number of members and of the distribution of those members over the member states, then we can rewrite equations 2.14 and 2.15 in completely compatible forms:

CONSERVATION EQUATION

$$\mathrm{Ex}[n_i(\tau+1)] = \mathrm{Ex}[n_i(\tau)] + \sum_h \mathrm{Ex}[J_{i,h}(\tau)] - \sum_j \mathrm{Ex}[J_{j,i}(\tau)] \tag{2.16}$$

CONSTITUTIVE EQUATION

$$\mathrm{Ex}[J_{j,i}(\tau)] = \mathrm{Ex}[p_{j,i}(\tau)]\,\mathrm{Ex}[n_i(\tau)]. \tag{2.17}$$

Equation 2.17 is called a *constitutive* equation because it provides *basic* information, distinct from conservation, concerning the relationships between the variables.

If the occupancy, n_i, of every state is sufficiently large, then we can invoke the Strong Law of Large Numbers and rewrite equations 2.16 and 2.17 in different forms:

CONSERVATION EQUATION

$$n_i(\tau+1) = n_i(\tau) + \sum_h J_{i,h}(\mathbf{n},\tau) - \sum_j J_{j,i}(\mathbf{n},\tau) \tag{2.18}$$

CONSTITUTIVE APPROXIMATION

$$J_{j,i}(\mathbf{n},\tau) \simeq p_{j,i}(\mathbf{n},\tau)\, n_i(\tau). \tag{2.19}$$

According to the Strong Law of Large Numbers, the constitutive approximation improves as the size of the population increases.

If we combine equations 2.16 and 2.17 and we combine equations 2.18 and 2.19, we obtain the following forms:

STATE EQUATION

$$\begin{aligned} \mathrm{Ex}[n_i(\tau+1)] = \mathrm{Ex}[n_i(\tau)] &+ \sum_h \mathrm{Ex}[p_{i,h}(\tau)]\,\mathrm{Ex}[n_h(\tau)] \\ &- \sum_j \mathrm{Ex}[p_{j,i}(\tau)]\,\mathrm{Ex}[n_i(\tau)] \end{aligned} \tag{2.20}$$

for the situation in which the p's and J's are independent of $\mathbf{n}$, and

STATE EQUATION

$$n_i(\tau+1) \simeq n_i(\tau) + \sum_h p_{i,h}(\mathbf{n},\tau)\, n_h(\tau) - \sum_j p_{j,i}(\mathbf{n},\tau)\, n_i(\tau) \tag{2.21}$$

for the situation in which the population is very large. Noting that

$$q_i(\mathbf{n},\tau) = 1 - \sum_j p_{j,i}(\mathbf{n},\tau)$$

we can see that these forms are equivalent to those of equations 1.45 and 1.50 in the summary of Chapter 1, which were developed without the assistance of the variable J. In subsequent sections, however, it will become quite clear that the variable J is not superfluous. In fact, it provides an extremely valuable conceptual tool for the construction of population models.

2.8. Reproduction, Death and Life as Flow Processes

According to the discussion in section 2.3, when age or some other initialized-time variable is introduced into the state description, the states associated with life can be aggregated by tiers to form sequences of Markovian lumped states. Each lumped state would represent a certain span of age or initialized time (e.g., months since birth, months since impregnation, etc.); and the transition interval would be precisely equal to that span. Using age alone, we can depict life graphically as a simple, linear concatenation of lumped states, as illustrated in Figure 2.16. Here we have two terminal states: U_0 on the left-hand end, representing the first interval in the life of an individual, and U_v on the right-hand end, representing the last interval, beyond which (age) no individuals have been observed to survive. Thus, the states span the observed *absolute longevity* of the individual. The interval of life represented by each lumped state often is called an *age-class*. The labelling of age classes is variable. For example, we might employ the Western convention, whereby an individual enters the 0th age class at the time of birth, or the Oriental convention, whereby

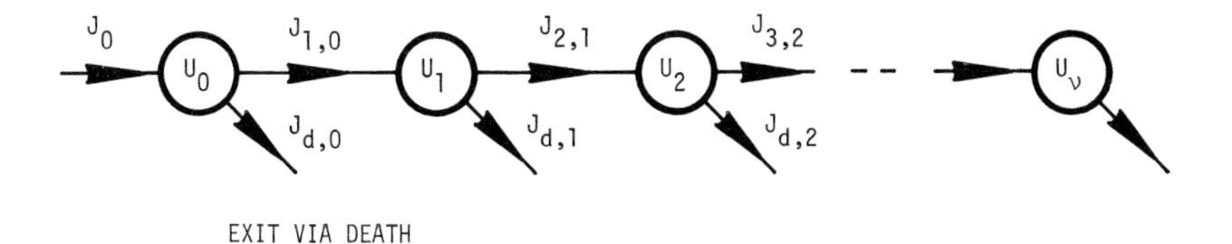

Fig. 2.16. Life, depicted as a linear concatenation of age classes, each represented as a lumped Markovian state

an individual enters the 1st age class at the time of birth, or the race-horse convention, whereby an individual enters the 1st age class at a specific date (e.g., January 1) following its birth, with that date being the same for all individuals. As far as the logic of our deductions is concerned, the labelling convention is irrelevant as long as all of the age classes represent equal time spans and the transition interval is equal to that time span. In that case, all individuals in age class i (e.g., the ith year of life) during interval τ (e.g., the year τ) either will have died or will have graduated to age class $i+1$ and be there during interval $\tau+1$.

Thus, we can picture the individual as *flowing* through the age classes in sequence, beginning in class U_0 and progressing one class per transition interval until it dies. Of course, not every individual is expected to survive to the limits of absolute longevity; so some must pass out of the sequence before they reach class U_ν. To depict this in Figure 2.16 we have drawn arrows leading down and out of each state and representing the death of individuals in that age class. The arrow leading into U_0 from the left represents entry into life (birth, hatching, sprouting, etc.). Again, we can picture the individual as *flowing* into life at birth and *flowing* out of it at death. Thus, every arrow in Figure 2.16 represents a possible path along which the individual members can flow, progressing into life, through it from age class to age class, and then out of it again.

Once again, the flow of individuals can be represented by the variable J. Consistent with equation 2.12, $J_0(\tau)$ will be the number of individuals that flowed into the first state (age class) of life, U_0, and survived to be observed in that state during interval $\tau+1$:

$$n_0(\tau+1)=J_0(\tau) \tag{2.22}$$

$J_{1,0}(\tau)$ is the number of individuals that graduated from U_0 to U_1 and survived to be observed or observable in U_1 during $\tau+1$ (having been in U_0 during τ); and $J_{d,0}(\tau)$ is the number of individuals in U_0 during τ that did not survive to be observed or observable in U_1 during $\tau+1$:

$$n_i(\tau+1)=J_{1,0}(\tau). \tag{2.23}$$

In general, since there is only one path into any age class other than U_0 in Figure 2.16, the following conservation relationships hold:

CONSERVATION EQUATIONS

$$n_i(\tau) = J_{i,\,i-1}(\mathbf{n}, \tau - 1) \qquad (i > 0) \tag{2.24}$$

$$J_{i+1,\,i}(\mathbf{n}, \tau) + J_{d,\,i}(\mathbf{n}, \tau) = J_{i,\,i-1}(\mathbf{n}, \tau - 1) \qquad (i > 0) \tag{2.25}$$

$$J_{1,\,0}(\mathbf{n}, \tau) + J_{d,\,0}(\mathbf{n}, \tau) = J_0(\mathbf{n}, \tau - 1). \tag{2.26}$$

We still can use the constitutive equations from the previous section:

CONSTITUTIVE EQUATIONS

$$\mathrm{Ex}[J_{i,i-1}(\tau)] = \mathrm{Ex}[p_{i,i-1}(\tau)]\,\mathrm{Ex}[n_{i-1}(\tau)] \tag{2.27}$$

when J and p are independent of $\mathbf{n}$; and

$$J_{i,i-1}(\mathbf{n}, \tau) \simeq p_{i,i-1}(\mathbf{n}, \tau)\, n_{i-1}(\tau) \tag{2.28}$$

when the size of the population is very large. These can be combined with the conservation relationships to provide the state equations:

STATE EQUATIONS

$$\mathrm{Ex}[J_{i,\,i-1}(\tau)] = \mathrm{Ex}[p_{i,\,i-1}(\tau)]\,\mathrm{Ex}[J_{i-1,\,i-2}(\tau - 1)] \tag{2.29}$$

when J and p are independent of $\mathbf{n}$, and

$$J_{i,\,i-1}(\mathbf{n}, \tau) \simeq p_{i,\,i-1}(\mathbf{n}, \tau)\, J_{i-1,\,i-2}(\mathbf{n}, \tau - 1) \tag{2.30}$$

when the size of the population is very large; the elements of the vector $\mathbf{n}$ in the latter case are given by equation 2.28 in terms of the J's.

Thus, we can write the complete state equations in terms of the flows. Using equation 2.27 or 2.28, we also can translate these directly into equations in terms of occupancies (or contents). Each of these choices will prove to be a valuable tool in our repertoire of population modeling methods, as will be the hybrid equations, involving both occupancies and flows.

Although the size and structure of a population may be affected by many different processes, depending on one's definition of membership, the one process that almost universally is included is reproduction. In Figure 2.16, this process is represented by the flow variable, $J_0(\tau)$, which is the number of newly produced individuals that first appear in the initial age class during $\tau + 1$. Clearly, this flow of new individuals will depend very strongly on the size and composition of the existing population. If reproduction is sexual, it will depend on the distribution of individuals over a state variable that often (but not always) is independent of age, namely sex. Keeping in mind that reproduction may depend on these several factors, we nonetheless can represent it simply by the variable J_0, and adapt the argument of that variable to fit the situation being modeled.

Since J_0 is the major input to our populations, it is appropriate to relate the other variables of our models to it. This can be done quite simply by application of equations 2.23 through 2.30. Before proceeding, however, it will be useful to simplify our notation. Since there is only one path into each state other than U_0, there is no need to carry a double subscript on each $p_{i,i-1}$ and $J_{i,i-1}$. Instead, we shall use a single subscript, so that $p_{i,i-1}$ will become p_{i-1} (the probability of graduation or survival *out* of state $i-1$) and $J_{i,i-1}$ will become J_i (the flow into state i). With this notation, we have the following relationships when J and p are independent of $\mathbf{n}$:

$$\mathrm{Ex}[J_i(\tau)] = \mathrm{Ex}[p_{i-1}(\tau)]\,\mathrm{Ex}[p_{i-2}(\tau-1)] \ldots \mathrm{Ex}[p_0(\tau-i+1)]\,J_0(\tau-i), \tag{2.31}$$

$$\mathrm{Ex}[n_i(\tau)] = \mathrm{Ex}[p_{i-1}(\tau-1)]\,\mathrm{Ex}[p_{i-2}(\tau-2)] \ldots \mathrm{Ex}[p_0(\tau-1)]\,J_0(\tau-i-1) \tag{2.32}$$

and the following when the population is very large:

$$J_i(\mathbf{n},\tau) \simeq p_{i-1}(\mathbf{n},\tau)\,p_{i-2}(\mathbf{n},\tau-1)\,p_{i-3}(\mathbf{n},\tau-2) \ldots p_0(\mathbf{n},\tau-i+1)\,J_0(\tau-i), \tag{2.33}$$

$$n_i(\tau) \simeq p_{i-1}(\mathbf{n},\tau-1)\,p_{i-2}(\mathbf{n},\tau-2)\,p_{i-3}(\mathbf{n},\tau-3) \ldots p_0(\mathbf{n},\tau-i)\,J_0(\tau-i-1) \tag{2.34}$$

where, to summarize, $J_i(\tau)$ is the flow of individuals from age class $i-1$ to age class i (to be present in age class i during $\tau+1$); $J_0(\tau)$ is the flow of new offspring into age class 0 (to be present in that class during $\tau+1$) from all of the reproductive age classes; and $n_i(\tau)$ is the number of individuals in age class i during interval τ.

If we consider individually the contributions to $J_0(\tau)$ from each reproductive age class, we can write

$$J_0(\tau) = \sum_i J_{0,i}(\tau) \tag{2.35}$$

where $J_{0,i}$ is the contribution from the ith age class, and summation is carried out over all reproductive age classes. All that remains is to establish constitutive relationships between the $J_{0,i}(\tau)$ and the $n_i(\tau)$. Let us define an age-class reproductive rate, $b_i(\mathbf{n},\tau)$ such that

$$\mathrm{Ex}[J_{0,i}(\mathbf{n},\tau)] = \mathrm{Ex}[b_i(\mathbf{n},\tau)\,n_i(\tau)] \tag{2.36}$$

or, when J and b are independent of $\mathbf{n}$,

$$\mathrm{Ex}[J_{0,i}(\tau)] = \mathrm{Ex}[b_i(\tau)]\,\mathrm{Ex}[n_i(\tau)] \tag{2.37}$$

and, when the size of the population is very large,

$$J_{0,i}(\mathbf{n},\tau) \simeq b_i(\mathbf{n},\tau)\,n_i(\tau). \tag{2.38}$$

The state projection matrix now takes on a classic form, first introduced by Lewis (1942) and Leslie (1945):

$$\mathbf{p} = \begin{bmatrix} b_0 & b_1 & b_2 & b_3 & \ldots & b_v \\ p_0 & 0 & 0 & 0 & 0 & 0 \\ 0 & p_1 & & 0 & \ldots & 0 \\ 0 & 0 & p_2 & 0 & \ldots & 0 \\ \vdots & \vdots & \vdots & & & \vdots \\ \cdot & \cdot & \cdot & \cdot & \cdot & p_v \end{bmatrix}. \tag{2.39}$$

Here, the first row represents reproduction and the subdiagonal elements represent survival. On death, individuals are projected out of the population. This matrix, especially the form in which the *b*'s and *p*'s are constant, has been discussed in great detail in the literature, with an excellent recent review of its properties being given by Parlett (1970).

2.9. Further Lumping: Combining Age Classes for Simplified Situations and Hypotheses

Although it does not include a separation of sexes, the model of the previous section does represent the processes of reproduction, life and death in considerable detail, in fact much more detail than our data usually justify or our hypotheses require. There is one way to reduce the amount of detail, and that is to lump together selected *sequences* of age classes (i.e., selected states), as illustrated in Figure 2.17. Each age class has associated with it three variables: its occupancy, the flow of survivors out of it into its neighbor on the right, and the flow of survivors into it from its neighbor on the left. When several age classes are lumped together, their variables also are lumped, so that there are only three variables associated with the entire sequence: the flow into the sequence, the flow of survivors out of the sequence, and the total occupancy of the sequence (i.e., the sum of the occupancies of the age classes included in the lumped sequence).

There are many situations in which lumping of age-class sequences is a thoroughly rational procedure in population modeling. Consider, for example, a situation in which one is concerned only with reproduction and death, and the organism involved does not begin to participate in reproduction until it has graduated several age classes beyond U_0. As long as one can keep track of

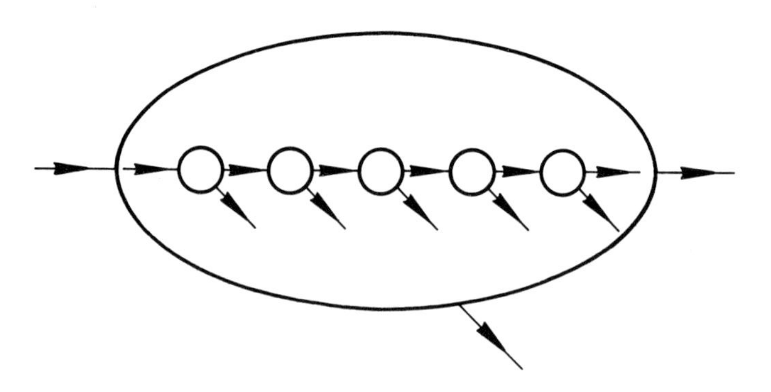

Fig. 2.17. An entire sequence of age classes may be lumped to form a single modeling element, with on input flow path representing entry into the sequence, one output flow path representing graduation from the sequence, and one output flow path representing death while within the sequence

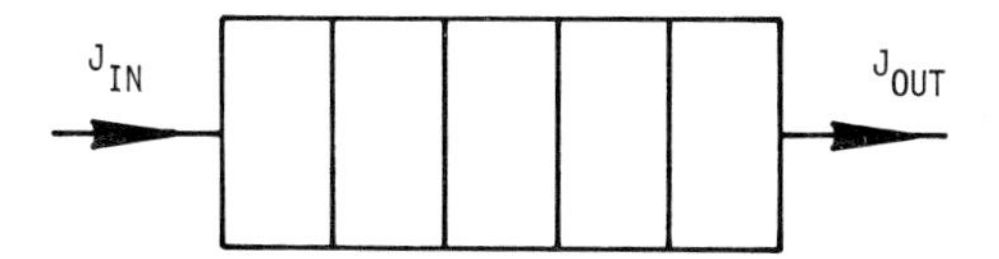

Fig. 2.18. Another way to represent a lumped sequence of age classes graphically

mortality in those age classes, he or she does not need to worry about other details, such as the occupancies of the individual states or the detailed flows from state to state. All one really needs to keep track of is the probability that the organism will survive to graduate from the last of the nonreproducing age classes and the transition interval of that graduation. Therefore, the only variables that one needs to consider are the flow into the sequence of nonreproducing states and the flow of survivors out of that sequence. For simplicity, the entire sequence can be represented pictorially as a single block, with a path for input flow and a path for output flow, as illustrated in Figure 2.18.

Under the assumption that the population is large, the input/output relationship for the block (the relationship between the input flow and the output flow) can be written simply as follows:

$$J_{\mathrm{out}}(\mathbf{n}, \tau) = \gamma(\mathbf{n}, \tau)\, J_{\mathrm{in}}(\mathbf{n}, \tau - T) \tag{2.40}$$

where T is the number of transition intervals represented by the states that were lumped together and are depicted by the block; and $\gamma(\mathbf{n}, \tau)$ is the probability that an individual organism entering the sequence of states during interval $\tau - T$ has survived to emerge from the sequence during interval τ. If the first state in the lumped sequence is U_0 and the last state is U_i, then

$$T = i + \mathbb{1}, \tag{2.41}$$

$$\gamma(\mathbf{n}, \tau) = p_i(\mathbf{n}, \tau)\, p_{i-\mathbb{1}}(\mathbf{n}, \tau - \mathbb{1})\, p_{i-\mathbb{2}}(\mathbf{n}, \tau - \mathbb{2}) \ldots p_0(\mathbf{n}, \tau - T + \mathbb{1}) \tag{2.42}$$

where $p(\mathbf{n}, \tau - i)$ is to be interpreted as the value of p at $\tau - i$ given the values of the vector $\mathbf{n}$ at $\tau - i$. The bookkeeping of mortality is taken care of completely by the factor γ, which usually is called the *survivorship factor*.

Now, suppose that once the members of the population reach the age of reproduction they begin to reproduce and continue to do so in essentially the same manner over a sequence of several age classes. Suppose further that the probability of death for the individual organism does not vary significantly over those same age classes (or, at least, if it does we are unaware of it and thus unable to specify the manner of variation) and that $b(\mathbf{n}, \tau)$ also does not vary over the age classes (or, if it does, we are unaware of it). In other words, as far as mortality and reproduction are concerned, all of the age classes in the sequence are equivalent. In that case, we do not need to keep track of the number of individuals occupying each state in the sequence, all we really need to follow is the total number, $N(\tau)$, of individuals occupying the entire sequence. To do so, we must follow the flow into the sequence, the flow of survivors out of it, and the flow out of it as a result of death.

The algorithm for keeping track of the occupants of the sequence is a very simple *difference equation:*

$$\Delta N(\tau+1)=J_{\text{in}}(\tau)-J_{\text{out}}(\tau)-J_d(\tau), \tag{2.43}$$

$$J_{\text{out}}(\tau)=\gamma(\tau)\,J_{\text{in}}(\tau-T) \tag{2.44}$$

where $\Delta N(\tau+\mathbb{1})$ is the net number of members added to the sequence from interval τ to interval $\tau+\mathbb{1}$. $J_{\text{in}}(\tau)$ is the flow of new occupants into the sequence; $J_d(\tau)$ is the flow of members out of the sequence due to death; and $J_{\text{out}}(\tau)$ is the output flow of survivors. Since we have stipulated that the probability of survival be the same for all age classes in the lumped sequence, we can express it as a single variable $p(\mathbf{n},\tau)$, which is the probability that any member in the sequence survives from interval τ into interval $\tau+\mathbb{1}$. If the first state in the sequence is U_i and the last state is U_j, then

$$T=j-i+\mathbb{1}, \tag{2.45}$$

$$\gamma(\mathbf{n},\tau)=p(\mathbf{n},\tau)\,p(\mathbf{n},\tau-\mathbb{1})\,p(\mathbf{n},\tau-\mathbb{2})\ldots p(\mathbf{n},\tau-T+\mathbb{1}), \tag{2.46}$$

$$J_d(\mathbf{n},\tau)=[\mathbb{1}-p(\mathbf{n},\tau)]\,N(\tau) \tag{2.47}$$

where $[1-p(\mathbf{n},\tau)]$ is the probability that any member in the sequence during interval τ does not survive into interval $\tau+\mathbb{1}$.

Equations 2.43 through 2.47 can be summarized in a difference equation that relates N to J_{in}:

$$\Delta N(\mathbf{n},\tau+\mathbb{1})=J_{\text{in}}(\tau)-\gamma(\mathbf{n},\tau)\,J_{\text{in}}(\tau-T)-[\mathbb{1}-p(\mathbf{n},\tau)]\,N(\mathbf{n},\tau). \tag{2.48}$$

With this equation, if we know $J_{\text{in}}(\tau)$ for every interval, $p(\mathbf{n},\tau)$ for every interval, and $N(\mathbb{0})$ for any initial interval, we can compute the expected value of $N(\mathbf{n},\tau)$ for all subsequent intervals.

We now have described two types of lumped sequences. The simplest of the two merely imposes a delay of several transition intervals on the flow of surviving individuals. The other, which is slightly more complicated and more restricted in its application, accumulates the surviving individuals in a certain sequence of age classes. The lumped sequence that merely imposes a delay involves only two variables, J_{in} and J_{out} and can be represented rather well in diagrams by a single block with two paths, one for each variable. The lumped sequence that accumulates, on the other hand, involves four variables, J_{in}, J_{out}, J_d, N, and thus is rather more complicated to represent in diagrams.

It is important to note that lumping in each of these cases has not produced a single Markovian state. Both types of lumped sequences in fact represent sequences of Markovian states, with individual objects represented implicitly (but not explicitly) as progressing through them from state to state.

2.10. The Use of Network Diagrams to Construct Models

Already we have seen that network diagrams can be very useful for visualization of Markov chains. As we also shall see, network diagrams can be extremely useful for the construction, modification, and analysis of population models in general. However, the basic Markov-chain type of network, in which each state is represented by a single element, often is cumbersome and not especially heuristic in population situations, and therefore often is inappropriate. In the previous section, we introduced a slightly more compact graphical symbol for lumped Markovian states (age classes); but even this can be improved upon.

To do so, we shall borrow one of the more distinctive conventions of systems theory, namely the use of diagrams constructed with very simple graphical symbols, or *elements*. The diagrams themselves may range from extremely simple to extremely complex, and each represents an equation describing the dynamics of a hypothetical or modeled system. Each element of a diagram conventionally represents a simple *operation* (such as addition or multiplication) with a *single result* (e.g., a single sum or a single product) formed from one or more variables being operated upon. Thus, a system diagram may be viewed as a *graphically parsed equation*. The real beauty of the scheme, however, lies not in its application to the parsing of established equations, but in the heuristic mechanism that it provides for the development of the complicated descriptions of systems for which equations have not yet been established.

The elements of system diagrams, or *networks*, can be divided into two classes. The elements of the first class represent instantaneous operations with no dependence on or "memory of" the history of the variable or variables being operated on and often are referred to as *static* elements. Examples of generalized static elements are illustrated in Figure 2.19. Conventionally, the result of the

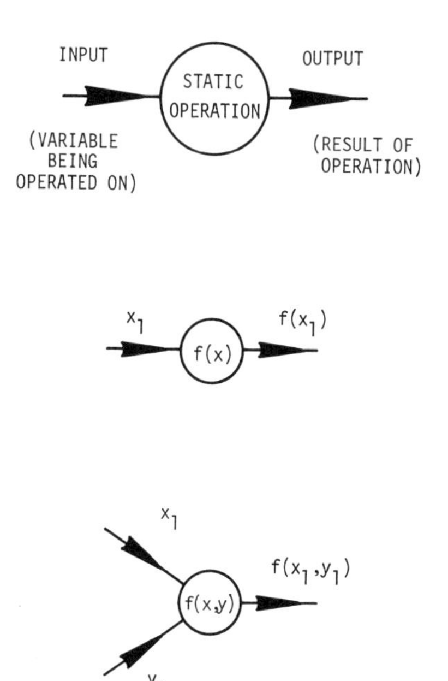

Fig. 2.19. Static elements of network models

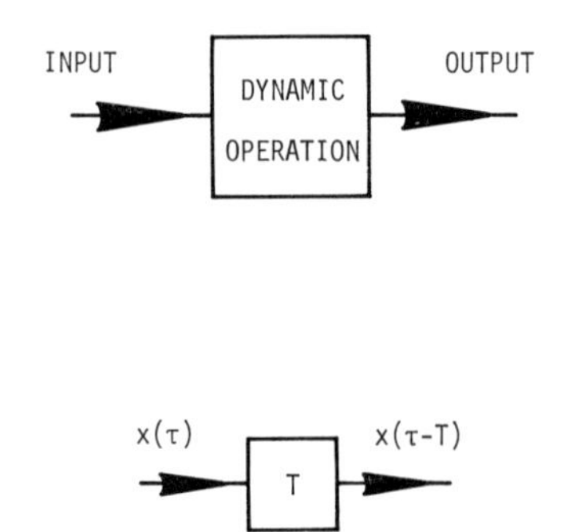

Fig. 2.20. Dynamic element of network models

operation represented by the element is called the *output* of the element and is denoted graphically by a single path directed out of the element; and the variable or variables being operated upon are called *inputs* and are denoted by one or more paths directed into the element. The elements of the second class represent operations that involve the history of the variable or variables being operated upon. Such elements often are said to have "memory" and are called *dynamic* elements. In this text, we shall direct our attention entirely toward discrete-time systems; and for that purpose we shall require only one type of dynamic element, namely the *time delay*, which is depicted in Figure 2.20. The parameter T in this element is the number of transition intervals (lumped or primitive) by which the input is delayed. One can picture the *time delay* especially conveniently in terms of flows. For example, one can picture it as accumulating individual members as they flow into it and later expelling them in the same order that they entered. It is quite analogous to a conveyor belt. Individual organisms might be pictured a being loaded on the belt at one end, only to emerge T transition intervals later at the other end. During any interval, the belt would have a certain accumulation of individuals distributed over its length. Thus, the *time delay* stores or "remembers" a segment of its input function's history T units of time long. It is this aspect of its nature that makes it a *dynamic element*. Because it represents a delay of T intervals, the *time delay* corresponds to a sequence of T separate Markovian states. Thus, we might consider distance along the conveyor belt as being divided into T segments each representing a resolvable state with the belt itself carrying its passengers from state to state in sequence.

The simplest of the static elements probably are those representing basic algebraic operations, which are depicted in Figure 2.21. Actually, graphical symbols such as those in Figure 2.21 seldom are seen in conventional systems diagrams. Conventionally, the symbol for addition, for example, is one of those illustrated in Figure 2.22. Throughout this text, we shall employ the symbol of Figure 2.22a; and we shall allow as many input terminals to converge on it as we need and can draw conveniently, as demonstrated in Figure 2.22d. The conventional symbols for multiplication are illustrated in Figure 2.23. When direct multiplication of two system variables is to be represented in the diagrams

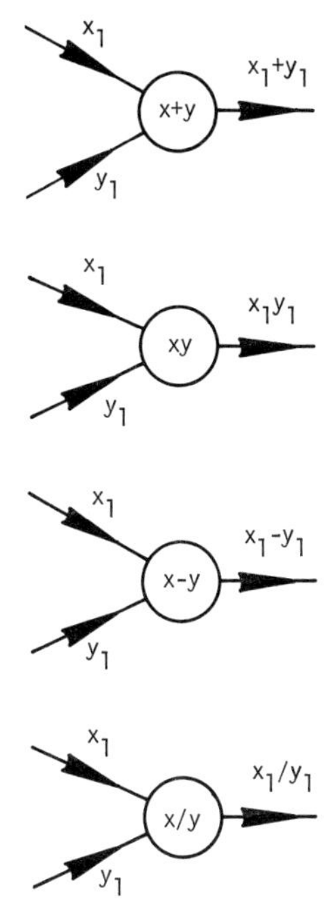

Fig. 2.21. Static elements representing the four basic operations of algebra

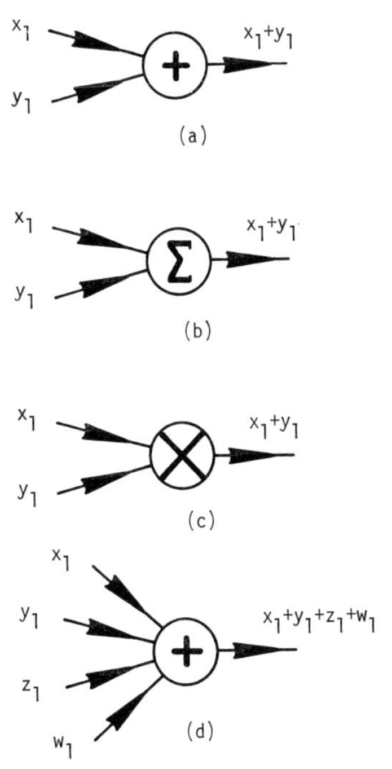

Fig. 2.22a–d. Various network symbols used to represent the process of addition. In this text we shall use the symbol of (a) and (d)

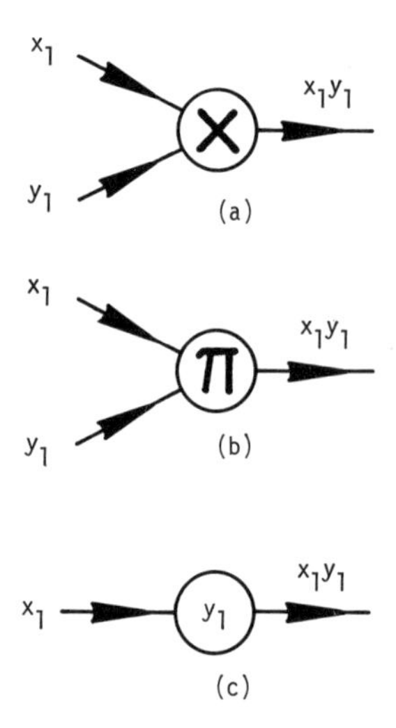

Fig. 2.23a–c. Various network symbols used to represent the process of multiplication. In this text we shall use symbol (a) to represent the product of two variables and symbol (c) to represent the product of a variable, x_1, and a parameter, y_1

of this text, we shall employ the symbol of Figure 2.23a. When multiplication of a system variable by a parameter is to be represented, we shall employ the symbol of Figure 2.23c. For all other instantaneous operations, we shall employ either combinations of the three symbols selected so far, or explicit statement of the operation within the symbol (as we have done in the symbols of Figure 2.21). The three selected symbols are presented again in Figure 2.24, along with their conventional systems-theory labels: the *adder*, the *scalor*, and the *multiplier*. In the diagrams of this text, the *adder* will be used almost exclusively to depict graphically the convergence or merging of two flows into a single path. The *scalor* will represent the multiplication of a flow variable (J) or an accumulation (N) of individuals by a certain parameter, or scale factor (such as a and b in Figure 2.24). The multiplier will be used to represent the formation of the product of two accumulations, two flows, or two functions thereof.

The output of each element in a system usually is a result needed as an input to another element, or it is a result whose value the investigator wishes to follow (i.e., a variable representing the dynamics of the system being modeled). When it is to be an input for another element, a graphical connection is made between the appropriate output and input paths. Thus several or many elements can be connected to represent complex operations or sequences of operations. Simple examples of this process are illustrated in Figure 2.25. In Figure 2.25a, a *scalor* and an *adder* are combined to represent subtraction; and in Figure 2.25b a reciprocating element is combined with a *multiplier* to represent division.

Often we shall wish to present one output to several inputs at once, in which case we shall be required to distribute it over several paths. This can be accomplished graphically by the simple expedient of constructing the required number of divergent branches from a single point on the appropriate path, as

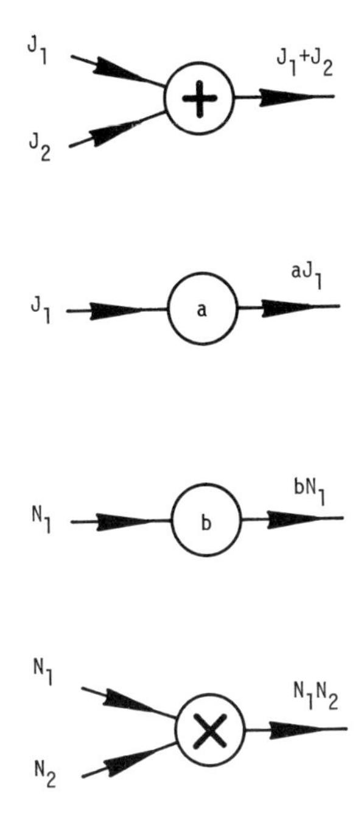

Fig. 2.24. Static-element symbols used throughout this text: top to bottom: the *adder*, two *scalors*, the *multiplier*

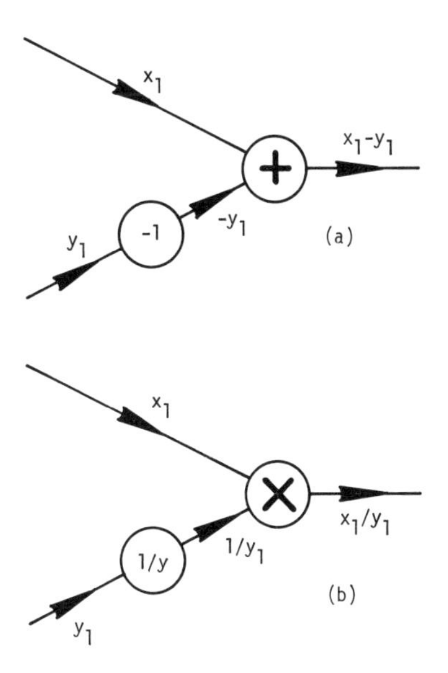

Fig. 2.25. Simple connections of static-element symbols, formed to represent sequences of operations

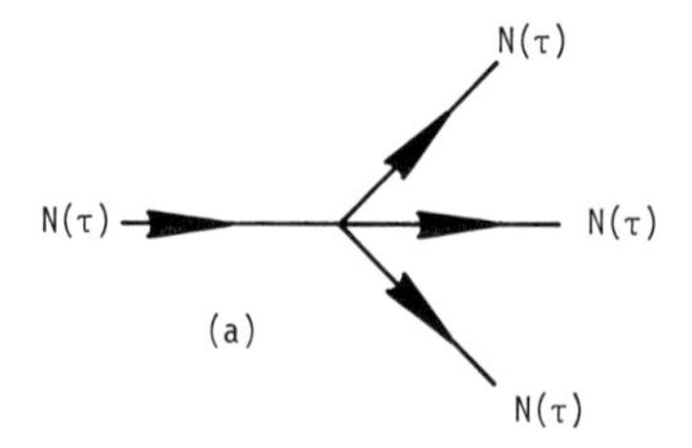

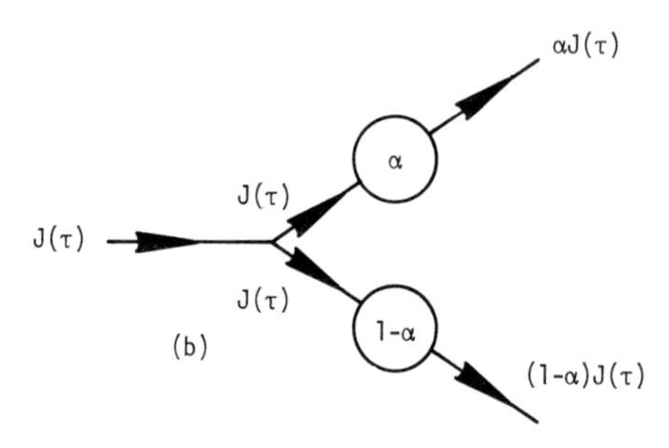

Fig. 2.26a and b. Two types of branch points. Duplicating (a), and conservative (b)

illustrated in Figure 2.26. Occasionally, the various branches emanating from such a point will connect directly to *scalors* the sum of whose scale factors is precisely equal to one, as illustrated in Figure 2.26b. In that case, the branch point ultimately represents a partitioning of the variable associated with it. Since this partitioning is conservative in the sense that the sum of the accumulations or flows emerging from the *scalors* is precisely equal to the accumulation at or flow into the branch point, the branch point itself will be denoted as *conservative*. Not all branch points in a diagram will be conservative, however, and those that are not we shall denote simply as *duplicating* branch points in reference to their function of representing duplication of a particular variable. In Figure 2.27, a duplicating branch point is used in a simple diagram representing the function $xz(x+y)$.

At this point, we have almost all of the elements and structures we shall require for the construction of networks representing populations in discrete time. However, we have yet to represent in terms of these elements and structures the two lumped sequences of states described in the previous section. The simpler of the two lumped sequences can be represented by the following equation:

$$J_{\text{out}}(\mathbf{n}, \tau) = \gamma(\mathbf{n}, \tau)\, J_{\text{in}}(\mathbf{n}, \tau - T). \tag{2.49}$$

We can parse the right-hand side of this equation into a concatenation of two of the operations represented by the elements of Figure 2.24: *delay* by T intervals, and *scaling* by the parameter $\gamma(\mathbf{n}, \tau)$. The order of the operations is crucially important. If the *delay* operation is carried out first, followed by the *scaling* operation, then the correct result is obtained. On the other hand, if the order of

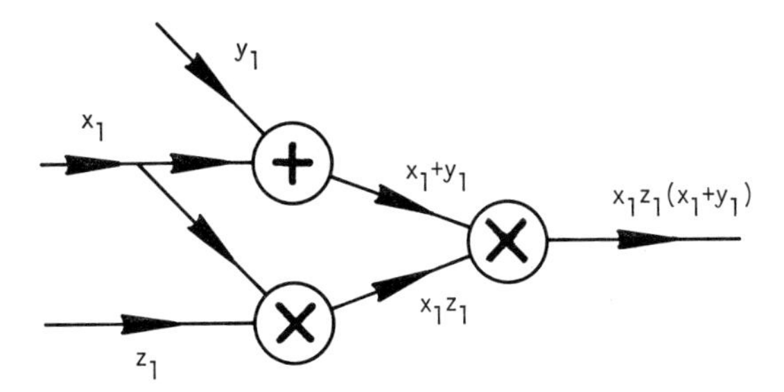

Fig. 2.27. A duplicating branch, two *adders* and a *multiplier* combined in a network to represent the generation of the function $xz(x+y)$

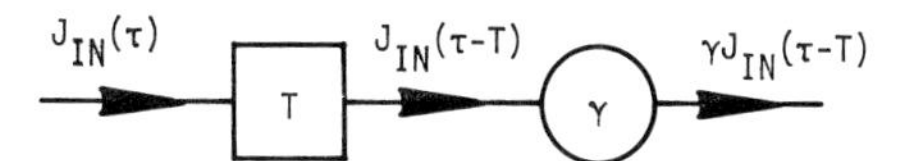

Fig. 2.28. The standard configuration of a *time delay-scalor* combination. The *scalor*, representing survival through the lumped age classes in T, follows the *time delay* representing those age classes

the operation is reversed, then we obtain the result

$$J_{\text{out}}(\mathbf{n}, \tau)=\gamma(\mathbf{n}, \tau-T)\, J_{\text{in}}(\mathbf{n}, \tau-T) \tag{2.50}$$

which is incorrect unless the parameter γ happens to be a fixed constant, independent both of $\mathbf{n}$ and of τ. Of course the parameter could have been defined in such a way as to make equation 2.50 correct (i.e., differently from the definition of equation 2.42); but for situations in which γ depends upon $\mathbf{n}$ or τ, or both, the execution of the scaling operation indicated in equation 2.50 would require anticipation of future values of γ, which in turn might require anticipation of the vector $\mathbf{n}$. Therefore, the more reasonable approach is to employ equation 2.49 and the definition of equation 2.42, in which case the *delay* operation is carried out first, followed by the *scaling* operation, as depicted in Figure 2.28. This is the conventional graphical representation of the simpler of the two lumped sequences of section 2.9; it comprises a *time delay* equal to the number of lumped transition intervals, followed by a *scalor* (survivorship scalor) equal to the proportion of individuals expected to survive through the lumped intervals.

The more complicated of the two lumped sequences of section 2.9 can be represented by the following equation:

$$N_{\text{out}}(\mathbf{n}, \tau)=p(\mathbf{n}, \tau-\mathbb{1})\, N_{\text{out}}(\mathbf{n}, \tau-\mathbb{1})+J_{\text{in}}(\mathbf{n}, \tau-\mathbb{1})-\gamma(\mathbf{n}, \tau-\mathbb{1})\, J_{\text{in}}(\mathbf{n}, \tau-T-\mathbb{1}) \tag{2.51}$$

which was obtained by simple modification of equation 2.48. The right-hand side of this equation can be parsed rather simply with the symbols of Figure 2.24 to yield the network of Figure 2.29a. Through the use of duplicating branch points, this diagram can be simplified to the extent that it has a single input path and a single output path, as illustrated in Figure 2.29b. J' in each network can

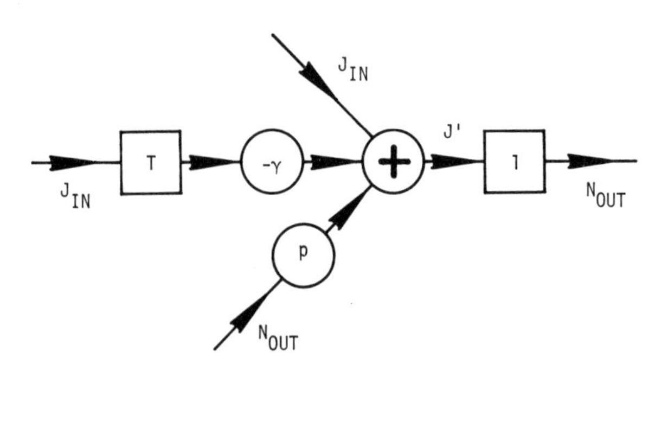

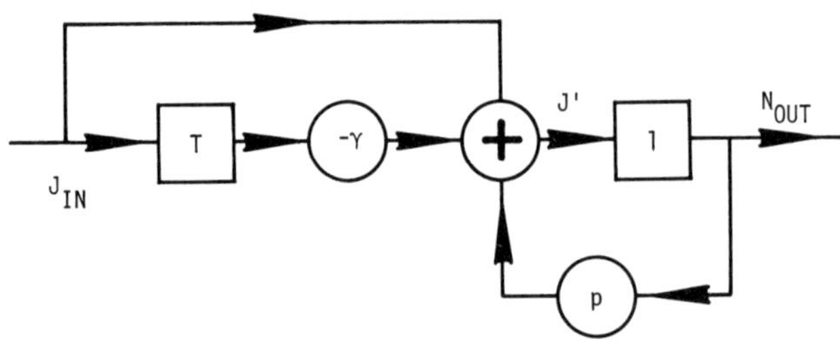

Fig. 2.29a and b. A combination of symbols that represents accumulation of individuals in a set of lumped age classes (i.e., provides a running total of the individuals in those classes). (a) The basic structure; (b) simplified version employing branch points and loops

be seen to represent the following sum:

$$J' = J_{in}(\mathbf{n}, \tau) + p(\mathbf{n}, \tau)\, N_{out}(\mathbf{n}, \tau) - \gamma(\mathbf{n}, \tau)\, J_{in}(\mathbf{n}, \tau - T). \tag{2.52}$$

The one-interval *time delay* (*unit delay*) simply serves to replace τ by $\tau - 1$ everywhere that the former appears in this sum. In Figure 2.29b, we have a rather simple representation of the process of accumulation described at the end of section 2.90. Individuals are represented as flowing into a loop, where the survivors continue to cycle for T transition intervals and then are removed. The output variable (N_{out}) of the loop provides a running count of the number of individuals accumulated in it. This simple representation of accumulation is made possible by our original definitions of N and J, whereby the two types of variables are commensurate. One might picture the diagram of Figure 2.29b as representing a reservoir for living organisms, with new individuals flowing into it through the path on the far left and with a meter on the far right providing an interval-by-interval count of the reservoir's contents. The one thing that is lacking is a path whereby surviving individuals can flow out of the reservoir. This is accomplished in the network of Figure 2.30. It is left as an exercise for the reader to verify the consistency between this network and that of Figure 2.29b.

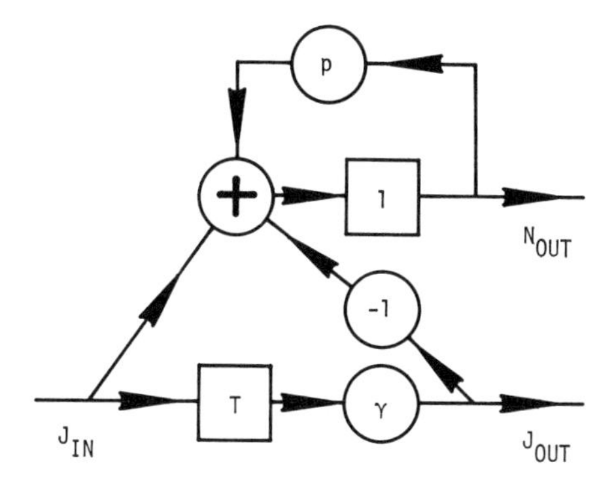

Fig. 2.30. The *accumulator* of Fig. 2.29b with a path provided for output flow

2.11. Basic Principles of Network Construction

In certain types of network models, often called *circuit* models, each path may represent several important network variables. The typical number is two, one being something akin to a force, the other being akin to a flow (e.g., voltage and current in an electrical circuit diagram, pressure and volume velocity in a hydraulic network diagram). In such networks, the elements generally are not oriented with respect to cause and effect or their direction of action. One cannot say whether the force causes the flow or the flow causes the force; one merely knows that the two variables are postulated to interact according to the rules associated with the network elements. Furthermore, the action postulated to be taking place at each element is markedly affected by the action of the elements on *both* sides of it; the elements do not have a path that one can call "input" and another that one can call "output." Because of this lack of orientation, the analysis of such networks often is extremely complicated and time consuming.

The elements introduced in the previous section all are oriented. Each exhibits a definite input path or paths and a definite output path. The output is

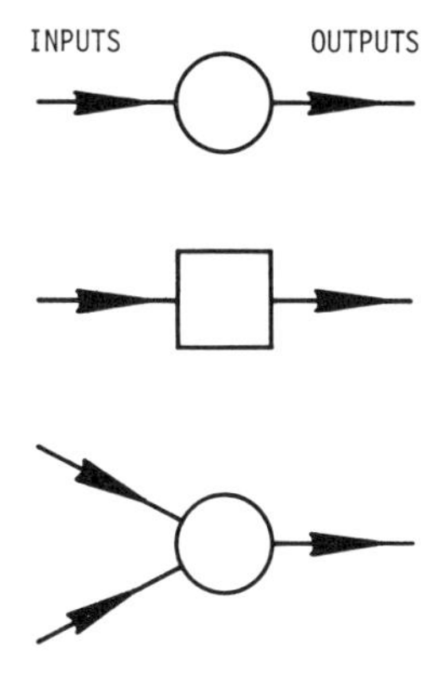

Fig. 2.31. "Oriented" elements. The direction of causality is indicated by the direction of the arrows

affected only by the input; it is totally unaffected by the action of elements connected to the output path unless they also are connected back to the input path. The direction of causality is completely specified: the input is cause, the output is effect. Consistent with this simple structure, each line in the network is directed (i.e., has an arrow indicating the direction of causality) and represents one and only one network variable. That variable is the input of the element *toward* which the line is directed, the output of the element *from* which it is directed. Analysis of such networks is considerably less tedious and time consuming than analysis of networks whose elements are not oriented.

Our plan at this point is to construct models of populations based on the states of the individual member. The dynamic element we shall employ (the *time delay*) is based on the lumping of member states whose specifications include age or some other measure of initialized time (e.g., time initialized at the moment (interval) of impregnation, or the moment of parturition, or the moment of infection). To a great extent, then, we shall be basing our models upon the *life cycles* of the modeled organisms, occasionally, perhaps, incorporating some extraordinary events, such as infection. The skeletons of such models can be constructed simply by considering the bookkeeping of individuals as they pass through their life cycles, which is an excellent starting point for construction of the entire model. The bookkeeping skeleton must adhere strictly to the established conservation principles (the source and destiny of every individual must be accounted for at all times).

In our network models, we shall be dealing principally with two types of variables: accumulations of individuals (N) and flows of individuals (J). Generally, the most convenient and heuristic variable for the bookkeeping skeleton will by the J-type. The elements and structures of the bookkeeping skeleton usually will be the *time delay*, the *scalor*, the *adder*, and the *conservative branch point*. The latter will represent the conservative apportioning of flow among two or more paths. Since *scalors* are not in themselves conservative elements (not all individuals represented in the flow into a *scalor* are represented in the flow out of it) every *scalor* in the bookkeeping skeleton will be associated either with an explicit branch point or with an implicit branch point at which one of the branches (e.g., a branch representing death) is not represented graphically because it carries individuals permanently out of the population. Since all explicit branch points in the bookkeeping skeleton will be of the conservative type, each branch will have a *scalor* associated with it; and the sum of the scale factors associated with a given branch point will be precisely equal to one at all times. Since we generally will not represent accumulations in the bookkeeping skeleton, only positive flows will occur. Therefore, the scale factors associated with the *scalors* all will be limited to the range from zero to one.

The bookkeeping skeleton by itself is merely that, a skeleton of the model; its flesh is provided by all of the constitutive relationships of the model. For the class of models that has been studied most thoroughly, namely those in which all parameters are fixed constants, independent of $\mathbf{n}$ and of τ, filling in much of the flesh will involve assignment of constant scale factors to each of the *scalors* in the skeleton and assignment of durations to the *time delays*. An important part of the completed model, however, will be that which represents the constitutive relationships between the variables of the skeleton and the *input flow of new offspring* to the earliest state of the life cycle. In some cases, these relationships will be represented simply by duplicating branch points and *scalors* with appropriate, constant scale factors. In other cases, duplicating branch points and *accumulators* may be required.

For convenience of discussion, the network structures that are added to the bookkeeping skeleton to represent the production of new offspring will be called collectively the *natality portion* of the network. If parameters in the bookkeeping skeleton or the natality portion of the network are to be dependent upon τ, but not $\mathbf{n}$, then they can be inserted into the model through *scalors*, requiring no additional network structures. On the other hand, when parameters are to be dependent upon $\mathbf{n}$, auxiliary networks may be used to represent the generation of the appropriate functions. These auxiliary networks will be labelled *function-generating* portions of the network model, and quite often they will include *accumulators*.

Since the structures of the three parts of a network model (the bookkeeping skeleton, the natality portion, and the function-generating portion) are not independent, when one constructs the bookkeeping skeleton, he or she should have the rest of the model well in mind (although modification of the bookkeeping skeleton to make it compatible with the other two portions usually is quite easy to accomplish). A very good procedure, one we shall follow in this text, is to construct the three portions simultaneously, thus insuring their compatibility. At this point, however, in order to emphasize their distinction, we shall consider two examples in which they are developed in sequence, beginning with the bookkeeping skeleton. In constructing the network models, we shall follow a useful dictum: THE SIMPLEST WAY TO CONSTRUCT A NETWORK MODEL OF A POPULATION OFTEN IS TO BEGIN WITH THE FLOW OF NEWLY FORMED MEMBERS INTO THE POPULATION AND FOLLOW IT THROUGH THE LIFE CYCLE; IF REPRODUCTION IS SEXUAL, THE FLOW TO FOLLOW GENERALLY IS THAT OF THE FEMALES.

Example 2.11. Simple Binary Fission. Consider a population of idealized, identical bacteria that reproduce solely by the process of binary fission. The culture medium on which these bacteria live is sufficiently well regulated that the time between successive fissions always is the same. In other words, the products of fission (the daughter cells) require a fixed period for growth and development before they reach maturity, at which time they divide immediately. Develop a bookkeeping skeleton for the population.

Answer: Before we begin, we must establish an explicit statement of conservation, which of course must be based on the definition of membership in the population. Usually this is quite straight forward. In the case of binary fission, however, we have a minor problem. Does the parent cell cease to exist at the time it undergoes fission, being replaced at that time by the two daughter cells; or does the parent cell continue to exist as one of the daughter cells, with only the other daughter being considered an offspring. Although our decision in this

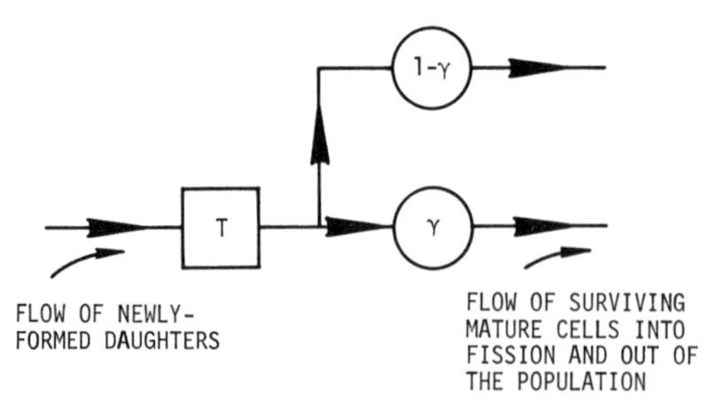

Fig. 2.32. Bookkeeping network for binary fission. T represents the time required for maturation of a daughter cell, γ is the proportion of daughters expected to survive to maturity

matter cannot affect our final deductions concerning the population dynamics, it will affect the basic structure of the model, as we shall see.

Let us begin by decreeing that the parent cell ceases to exist at the time of fission. We can represent it in network form, as illustrated in Figure 2.32. Here we have begun on the left with the flow of newly formed daughter cells into the population. After a fixed time delay, a certain proportion, γ, of these daughters survive to undergo fission and thereby leave the population; the remaining proportion, $\mathbb{1}-\gamma$, are (have been) removed by death. If we are interested only in surviving bacteria, we can eliminate the path representing death.

Now, let us take the other approach to our conservation statement, decreeing that the parent cell survives fission, producing a single daughter cell. Beginning with the flow of newly formed daughters, we have the same sequence of *time delays* followed by conservative branch point (or implied conservative branch point); but those parent cells that emerge alive from the *delay* simply must enter the same sequence all over again. We can represent this in either of two ways. We can merge their flow with that of the newly formed daughters and run them through the same *time delay* over and over again until they die; or, if we want to keep track of them separately, we can repeat the sequence of delay and branch point over and over again, ad infinitum. These two alternatives are depicted in Figure 2.33.

Example 2.12. Sexual Reproduction with Periodic Ovulation. Consider a population of idealized cetaceans with the following life cycle, observed with a temporal resolution of one month: A newly weaned female requires T_1 months to reach sexual maturity and her first ovulation. If she survives to maturity, she ovulates periodically, once every month, and at each ovulation faces a certain probability of conception. If she does not conceive, she faces a certain probability of surviving the month to the next ovulation. If she does conceive, she will enter a gestation period of T_2 months, with a certain probability of survival. Surviving gestation, she enters a T_3-month period of lactation, during which survival probability is reduced owing to her increased exposure to whalers. On termination of lactation, she ovulates for the first time since becoming pregnant, reentering the periodic ovulation cycle. Construct a bookkeeping skeleton for a network model of this population under the assumption that the survivorships and impregnation probability are independent of age.

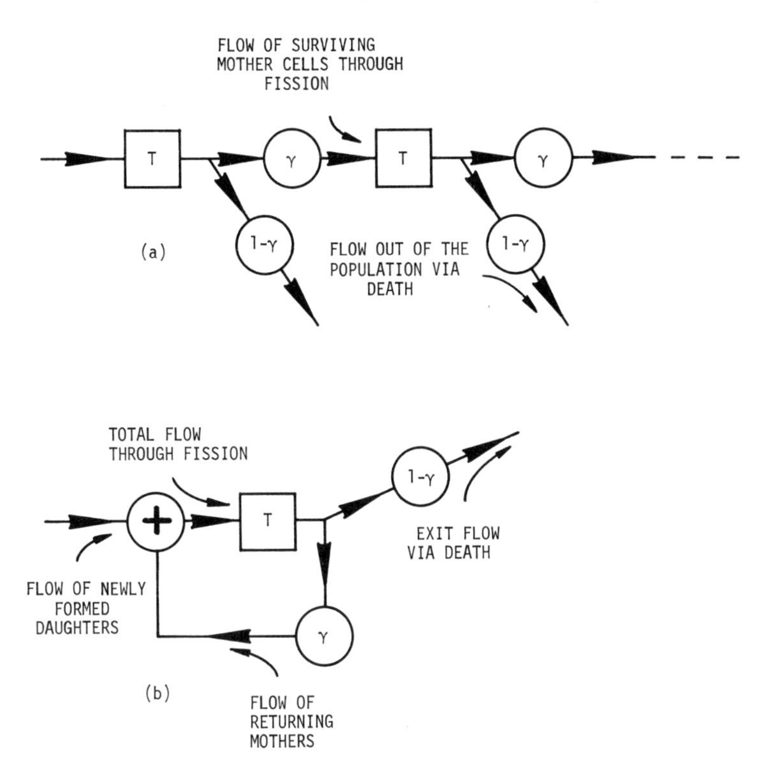

Fig. 2.33. Alternative configurations for bookkeeping binary fission

Answer: In this example, the conservation statement already is embodied reasonably completely in the life-cycle statement. We merely need to add the usual definition of membership in the population: in this case we shall include only living females.

Although there are several phases and associated survival probabilities in this life cycle, construction of the bookkeeping network is quite straightforward if one begins at the beginning (with the new-born females) and progresses step by step from there, as illustrated in Figures 2.34 and 2.35. The flow of newborn females enters an T_3-month *time delay*, followed by an implied conservative branch point, with γ_0 of those that entered emerging alive as weaned females. The flow of newly weaned females enters a T_1-month *time delay*. Emerging from this *time delay* and its associated survivorship scalor is the flow of newly ovulating females, of which β become pregnant and enter gestation and $\mathbb{1}-\beta$ do not. From this point on in our idealized life cycle, the flow of females faces a perpetual array of alternative paths. We could make a token attempt to represent each alternative path separately; but the resulting network diagram will be extremely complicated (see Figure 2.34). On the other hand, we can replace the perpetual array with simple loops, just as we did in the case of the bacteria of Example 2.11. In fact, we can do this very heuristically by adhering to the following rule:

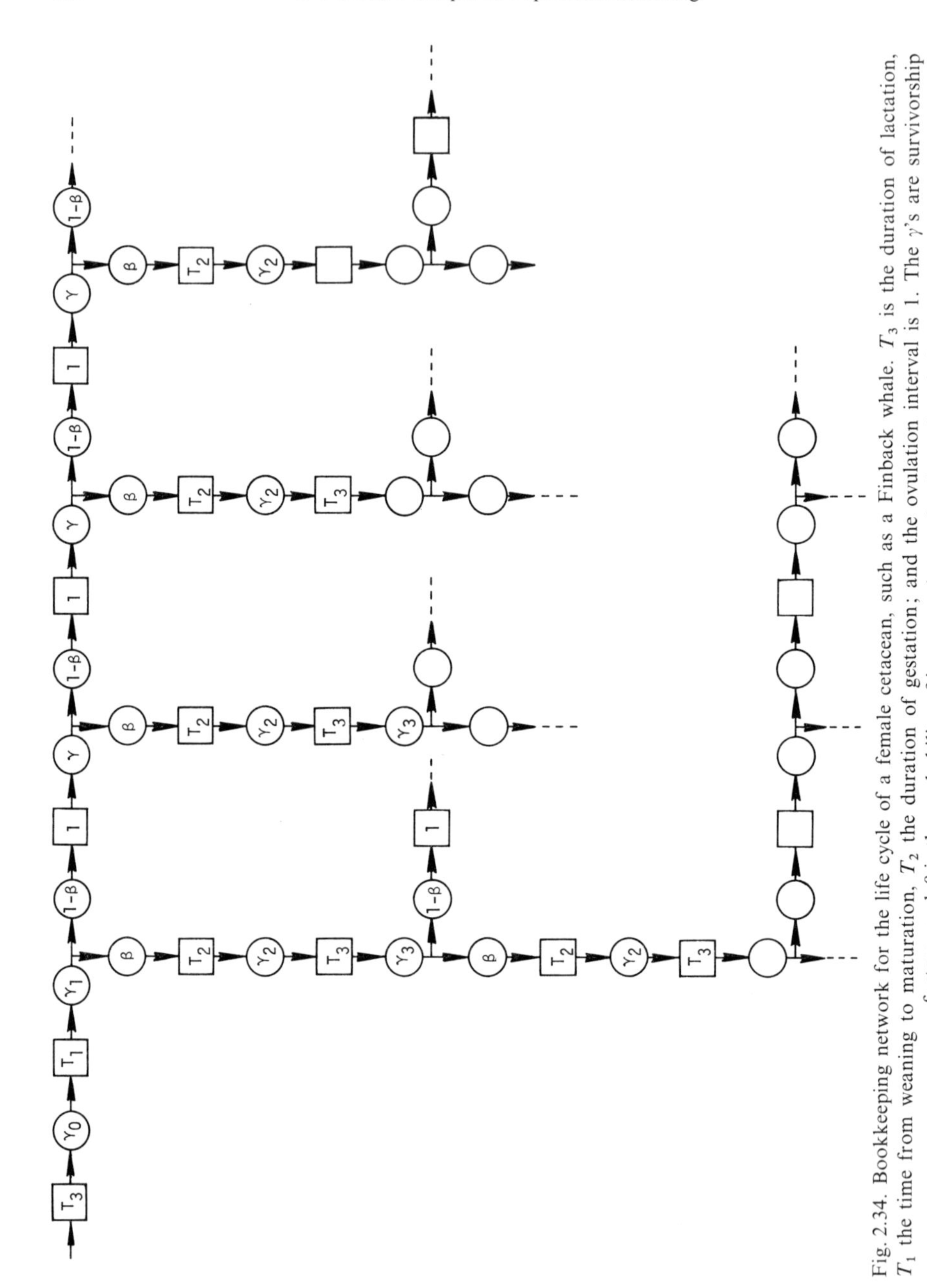

Fig. 2.34. Bookkeeping network for the life cycle of a female cetacean, such as a Finback whale. T_3 is the duration of lactation, T_1 the time from weaning to maturation, T_2 the duration of gestation; and the ovulation interval is 1. The γ's are survivorship factors, and β is the probability of impregnation at the time of ovulation

NEVER DRAW A NEW *TIME DELAY* IF A PREVIOUSLY DRAWN *TIME DELAY* CAN PERFORM THE SAME FUNCTION.

We must draw one *time delay* for the interval between successive ovulations (one month), another to represent the gestation period (T_2 months), and a third to represent the lactation period (T_3 months). If we follow our dictum, however, we shall attempt to get by without repeating these delays in the network. It is quite

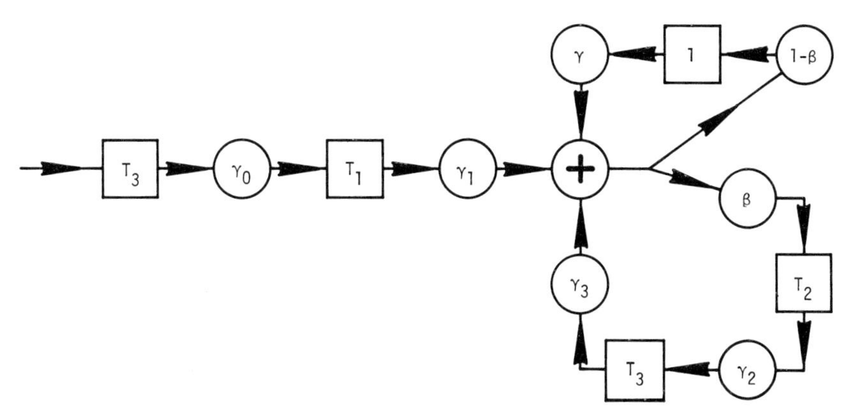

Fig. 2.35. Simplified version of the network of Fig. 2.34, employing feedback loops to merge flows of individuals in equivalent states

clear that (owing to our assumption of age-independent survivorship and impregnation probability) we can succeed.

Let us begin by considering the $\mathbb{1} - \beta$ of the newly ovulating females that do not conceive. They immediately enter a $\mathbb{1}$-month *time delay* with its associated survivorship scalor; and those that survive to emerge ovulate again. Now, according to our idealized life-cycle statement, these second-time ovulators face exactly the same future prospects as the first-time ovulators (the same probabilities of survival, the same probabilities of conception). Therefore, we are perfectly free to merge their flow with that of first-time ovulators, which we have done in Figure 2.35 with the *adder* just to the left of the conservative branch point representing conception. In this manner, we have generated a perpetual loop to replace part of the perpetual array of alternative paths. As long as neither impregnation nor death occurs, our representation of the female will continue to pass round and round the loop.

If she becomes pregnant, she will pass into the T_2-month gestation period and, if she survives, on into the T_3-month lactation period, both represented by *time delays* with associated survivorship *scalors*. Termination of lactation is signaled by ovulation, and from that point on, the female once again faces the same prospects as a first-time ovulator. Therefore, we can return the flow path directly to the *adder*, converging the flow of females completing lactation with the rest of the flow of ovulating females.

With the bookkeeping skeletons in hand, we can proceed to the natality portions of the network models. In the case of binary fission, the appropriate constitutive relationship is implied in the name of the process itself: for each bacterium surviving to proceed into fission, two bacteria will emerge from fission. In the bookkeeping skeleton of Figure 2.32, both of these are taken to be daughter cells; therefore the natality portion of the network simply will comprise a *scalor* with scale factor $\mathbb{2}$ connected from the path representing the flow of mature cells into fission back to the path representing the flow of newly

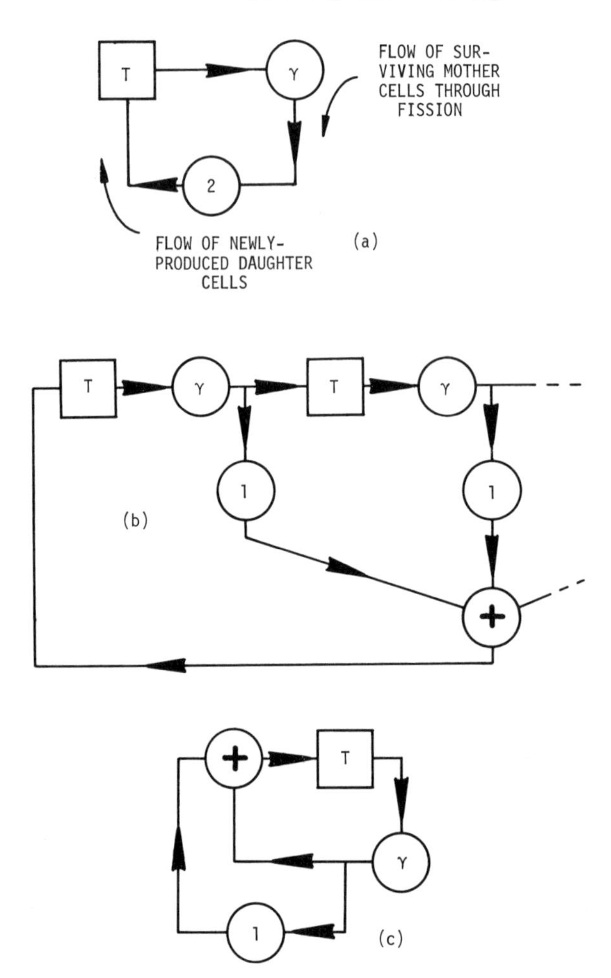

Fig. 2.36. Completed networks (bookkeeping portions plus natality portions) representing binary fission

formed daughter cells, as illustrated in Figure 2.36a. In the bookkeeping skeletons of Figure 2.33, one of the cells emerging from fission is taken to be the surviving mother cells, the other to be the newly formed daughter. In that case, the appropriate scalors in the natality portions will have scale factors of 𝟙 and will be connected from the paths representing the flow of surviving mother cells through fission to the paths representing the flow of newly formed daughter cells, as illustrated in Figures 2.36b and c. Where several *scalors* converge on a single path, an *adder* is used.

In the case of the idealized cetacean population, natality can be represented by a *scalor* connected from each path representing the flow of females emerging from gestation to the path representing the flow of newborn female offspring, as illustrated in Figure 2.37. Once again, an *adder* is used to represent the convergence of flows. The scale factor, b, is the expected number of female

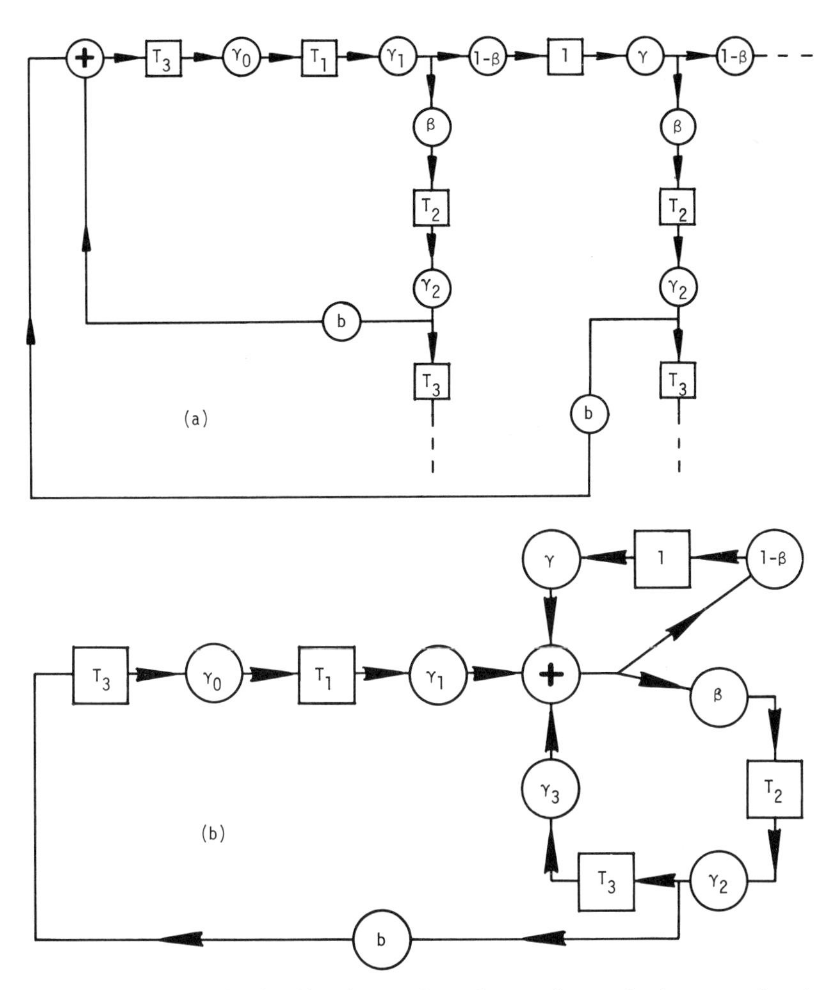

Fig. 2.37. Completed networks (bookkeeping portions plus natality portions) representing the cetacean life cycle

offspring per parturition. It is represented as being independent of age in the network models of Figure 2.37.

Finally, if the various time delays (T's) and various scale factors (γ's, β's and b's) in the network are stipulated to be constants, independent of $\mathbf{n}$ and of τ, then the network models will be completed when those constant values are inserted; or they may be considered complete as they are, with the various parameters taken to be of constant, but unknown magnitude. In either case, no further network structures (i.e., function-generating portions) are required. Similarly, if the parameters are stipulated to be functions of τ but not of $\mathbf{n}$ (e.g., exhibiting seasonal variations), the appropriate functions can be represented within the context of the existing elements, without the addition of further structures. On the other hand, additional structures may be useful for parsing the time functions; and, in general, whenever dependence of a parameter on $\mathbf{n}$ or

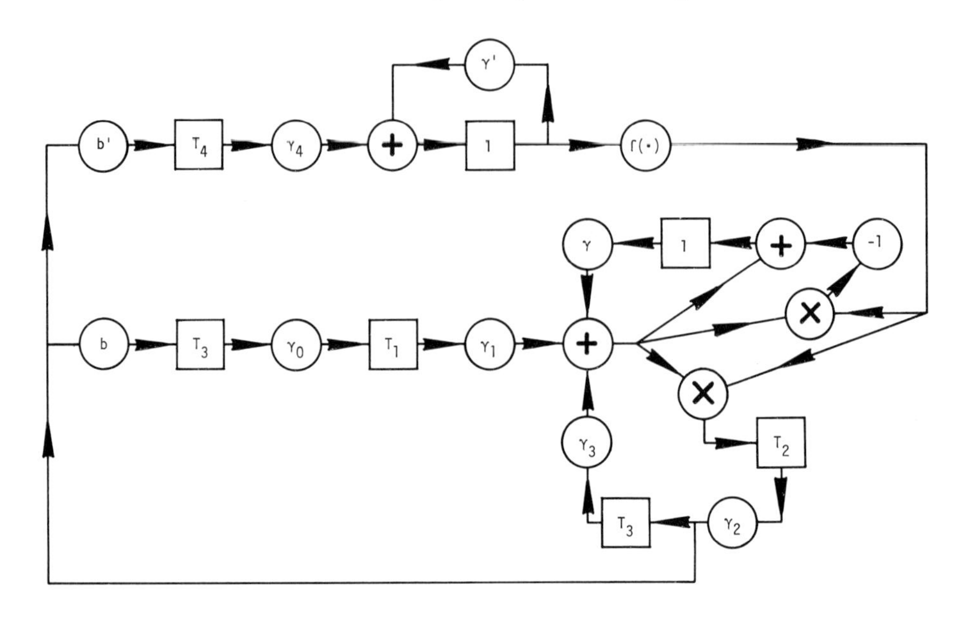

Fig. 2.38. Variation on the network model for a cetacean life cycle. Males are represented as flow through the upper segment, females as flow through the lower segment. The probability of impregnation at the time of ovulation is $f(N_m)$, an arbitrary function of the total number of mature males. The multipliers form part of a conservative branching process

some aspect of **n** is to be represented, additional structures will be used to generate the appropriate dependence. Consider, for example, the likely possibility that the probability of impregnation (β) depends on the total number of available adult males in our population of idealized cetaceans. In fact, let β be equal to $f(N_m)$, where N_m is the number of adult males and where $f(N_m)$ varies from zero to one as N_m increases from zero to very large numbers. This functional dependence is represented in Figure 2.38, where a function-generating portion has been attached to the network of Figure 2.37b. The function generating portion consists of a path representing the flow of newborn male offspring, passing into a *time delay* (T_4) for maturation and a corresponding survivorship *scalor* (γ_4), and then into a loop representing the accumulation of surviving adult males, with the month to month survivorship given by γ'. An element labelled f represents the transformation of N_m to $f(N_m)$. Since $f(N_m)$ now is a network variable, rather than simply a parameter, its application to the scale factors in the bookkeeping skeleton can be represented most conveniently through *multipliers*. Therefore, the *scalor* β in Figure 2.37b has been replaced by a *multiplier*, and the *scalor* $\mathbb{1}-\beta$ has been replaced by a *multiplier* in conjunction with a *scalor* and *adder*. The principal assumption made in the construction of the function-generating portion was that adult male mortality is independent of age. It is left as a simple exercise for the reader to verify that this new configuration correctly represents the postulated scaling of the flow into ovulation.

Clearly, *multipliers* can serve useful roles in network representations, replacing *scalors* when the scale factors of the latter depend on an output of a function-generating portion of the network and allowing us to depict explicitly the *closing* of the underlying dynamic equations into a self-contained system. Multipliers also can be used to depict open-ended aspects of system dynamics, replacing scalors whose scale factors depend on extrinsic and intrinsic phenomena and are unknown or unpredictable and are intended to be supplied later, if possible.

Example 2.13. Reef Fish. Consider a population of idealized tropical marine fish with the following life cycle, resolved to one day: the fertilized eggs are quite well protected, so that the probability of hatching is essentially constant and the same for all eggs at 0.98, with hatching taking place in one day. The early larval form exists and grows for five days on the energy provided in an oil globule from the egg. Survivorship during this period depends upon the conditions in the pelagic realm, including weather and the presence of predators. At the end of the five-day early-larval period, the survivors progress into the normal larval form and begin to feed on zooplankton. This continues for fifty days, during which survivorship depends on the amount of zooplankton available for food and on the intensity of predation. At the end of the fifty-day period, the late-larval fish enter shallow rocky or reef areas and tide pools where they undergo the transformation to the juvenile stage. The transformation requires four days, during which time survivorship depends on the intensity of wave action, the density of larvae moving into a given shallow-water area, and the intensity of predation. The young fish emerge from this four-day transformation as alga-eating juveniles. The juvenile stage continues for 240 days, at the end of which the young adult fish begin to participate in reproduction. Survivorship during those eight months depends primarily on the intensity of wave action, the intensity of predation, and the density of juveniles in a given shallow-water area. The adult fish spawn cyclically, in synchrony with the lunar cycle. The probability that a given female spawns during a given cycle depends upon the conditions of the water (surf, temperature, etc.) as well as on social factors involving the male and female fish in the adult population. The expected number of fertilized eggs produced by each adult female is 500,000, regardless of size and age. The survivorship of the adults is essentially independent of size and age, but depends on wave action and on the intensity of predation (which varies markedly from time to time as various schools of large predators sweep over the shallow-water home of the fish).

Construct the bookkeeping skeleton of a network model for this population, and fill in as many constitutive details as you can.

Answer: The bookkeeping skeleton is shown in Figure 2.39. We have begun at the flow of new eggs into a one-day hatching *delay;* the survivors flow on into a five-day nonfeeding period, then into a 50-day normal-larval period, on into a four-day transformation period, into a 240-day juvenile period, and, finally, into a

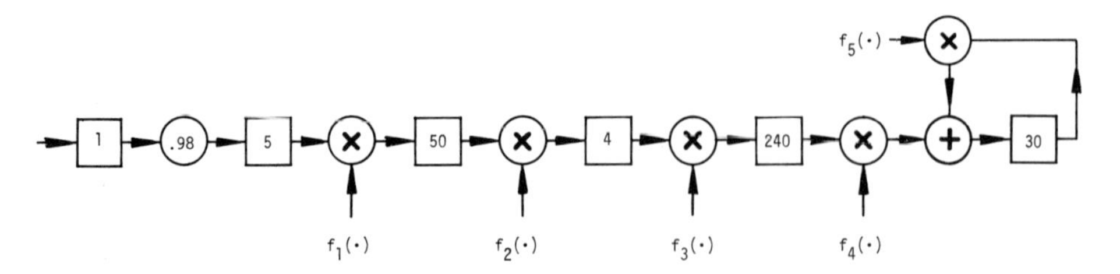

Fig. 2.39. Bookkeeping skeleton for the reef fish model

thirty-day perpetual lunar cycle. Only the first survivorship is known. The rest are unknown and for convenience are represented by *multipliers* rathers than *scalors.*

Clearly, before specific conclusions can be deduced concerning the dynamics of a population such as that depicted in Figure 2.39, the open ends of the multipliers and the open path to the first *time delay* all must be connected to something. In other words, each factor in the system must be specified and the source of flow for every path must be specified. Once this is done, the model becomes self-contained, or *closed*, and detailed analysis or simulation can proceed. The problem of closure in models is an important one, and has received considerable attention in the literature. According to the discussion in section 2.7, one important method of achieving closure is by hypothesis or stipulation.

2.12. Some Alternative Representations of Common Network Configurations

Before proceeding further with the construction of network models based on specific life cycles, we shall spend a brief moment developing a small repertoire of alternative representations that will allow us to simplify our network structures as we develop them. A few of these alternatives already have been introduced in the previous section; here we shall develop them more explicitly. To begin with, we shall consider small network segments each of which has one or more inputs and a single output. If two such segments exhibit the same relationship between the input variables and the output variable, then we shall consider them to be equivalent alternatives. Virtually all of these equivalences are based on the algebraic properties of real numbers, the most basic of which are sufficiently self-evident that the network alternatives stemming from them can be presented without comment or explanation. For example, real numbers obey the associative and commutative laws for multiplication. Therefore, the four structures shown in Figure 2.40a all exhibit the same relationship between input flow and output flow and thus are equivalent alternatives. Figure 2.40 depicts several, similar sets of structures; the equivalence of the members of each set should be obvious to the reader.

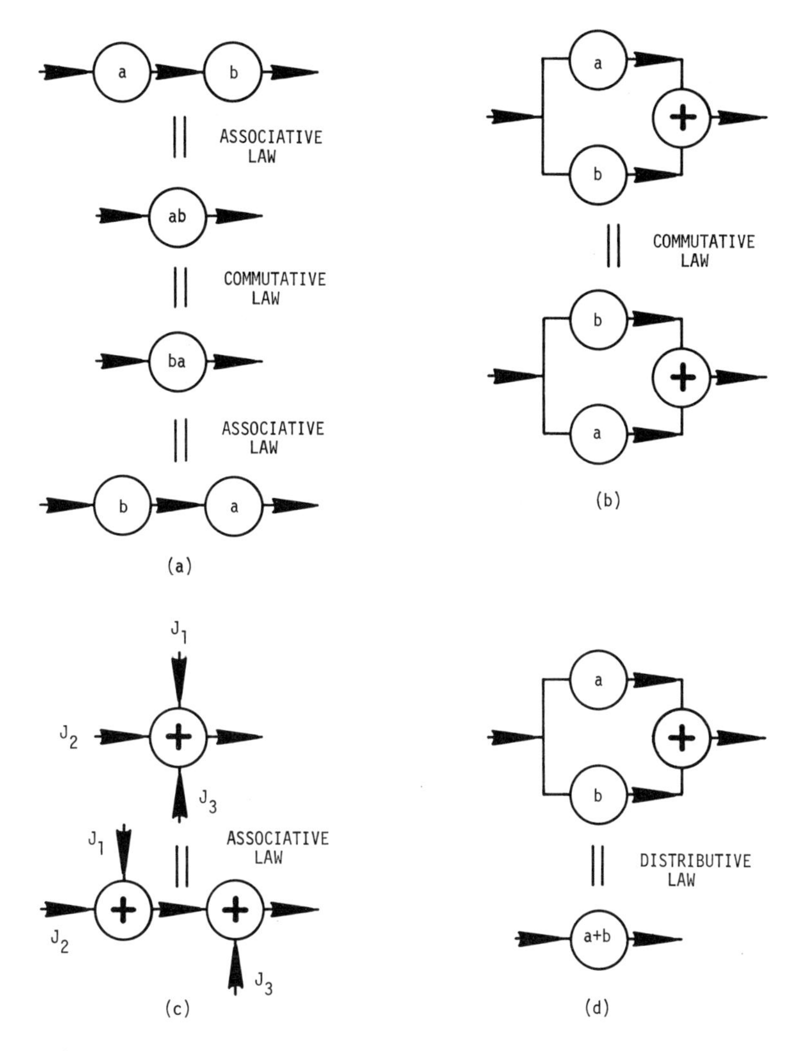

Fig. 2.40a–h. Equivalences among network elements and element combinations

Another set of equivalences, employed already in section 2.11, allows abbreviated representation of repetitive structures involving *time delays*. The most commonly used of these are summarized in Figure 2.41. The first (Figure 2.41a) simply involves the same type of lumping described in section 2.9. The lumped scale factor, γ', is found by application of the principles of equation 2.42:

$$\gamma' = \gamma_1(\tau - T' + T_1)\gamma_2(\tau - T' + T_1 + T_2)\gamma_3(\tau - T' + T_1 + T_2 + T_3)\ldots\gamma_{v-1}(\tau - T_v)\gamma_v(\tau). \tag{2.53}$$

In the special case of the survivorships (γ_i's) in the original network segment being independent of $\mathbf{n}$ and τ, the lumped scale factor becomes the simple

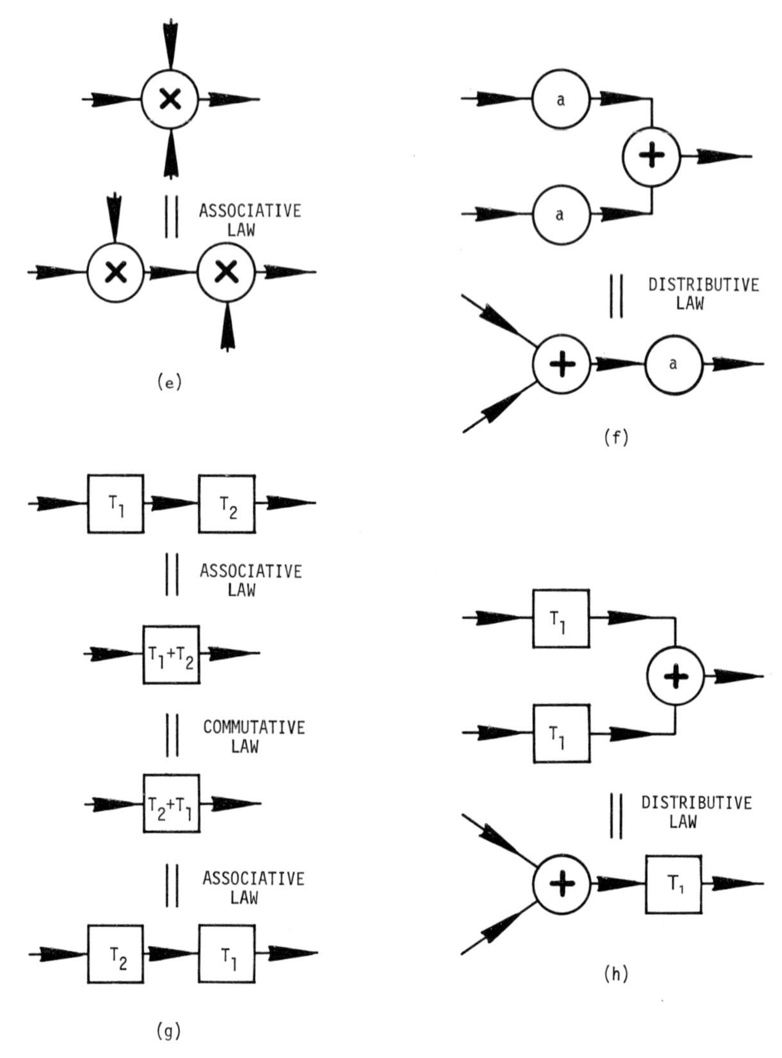

Fig. 2.40e–h

product of the original scale factors. When identical *time delays*, T, occur repetitively in a segment, the lumped *time delay* becomes

$$\sum_{i=1}^{v} T_i = v\,T. \tag{2.54}$$

For perpetual repetition of identical *time delays* coupled with identical *scalors*, as depicted in Figure 2.41b, the abbreviated network becomes a perpetual loop. This abbreviation is valid whether or not γ is constant with respect to τ and $\mathbf{n}$. Instead of passing through an indefinite sequence of identical time *delays* and *scalors*, the flow passes an indefinite number of times through the same *time*

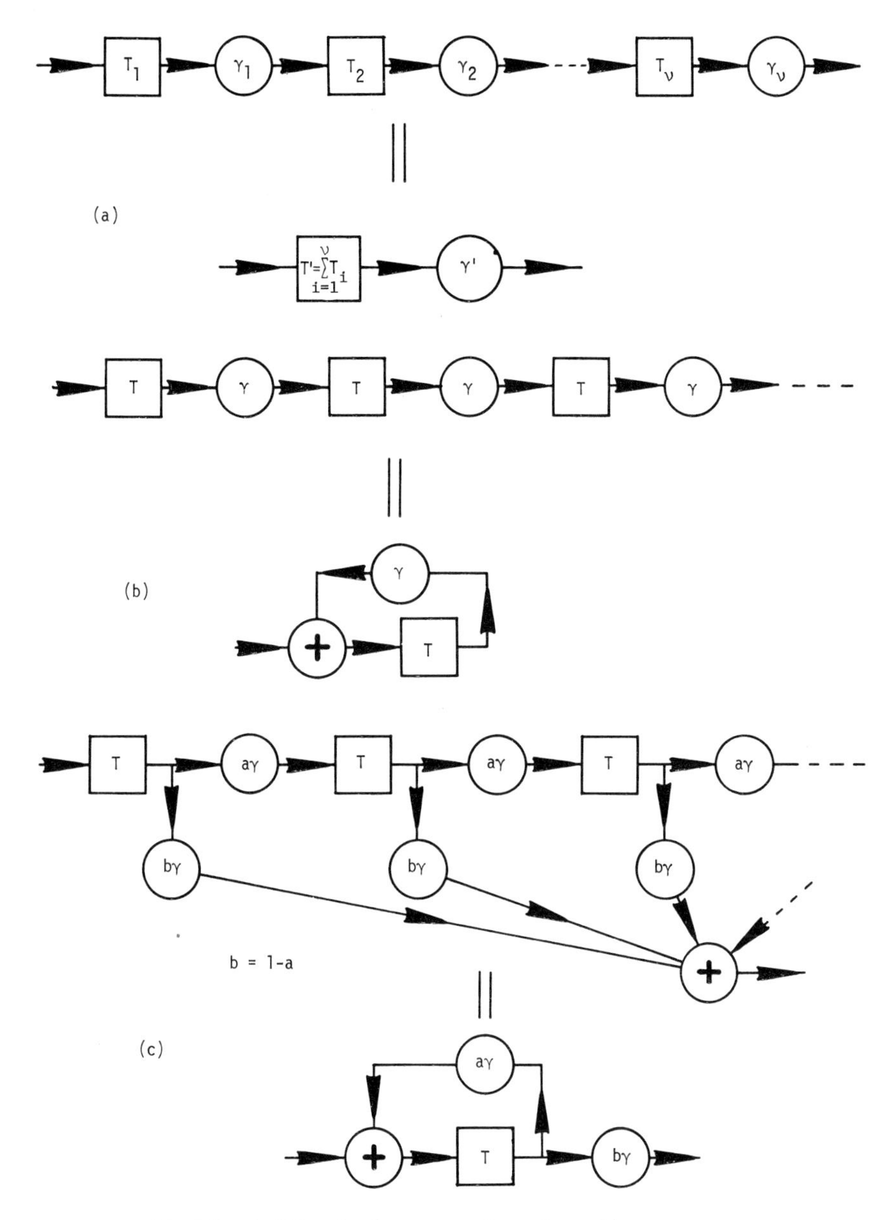

Fig. 2.41 a–d. Abbreviated representations of repetitive structures

delay and *scalor*. If, after each delay, some proportion of the flow passes out of the perpetual sequence, then we have the equivalence shown in Figure 2.41c, which also is valid whether or not γ is dependent on $\mathbf{n}$ or τ. In the abbreviated form, all of the flows are accumulated and are passing through a single *time delay;* so we form the output flow simply by replacing the sum of the output proportions by the same proportion of the sum (distributive law). Finally, if the

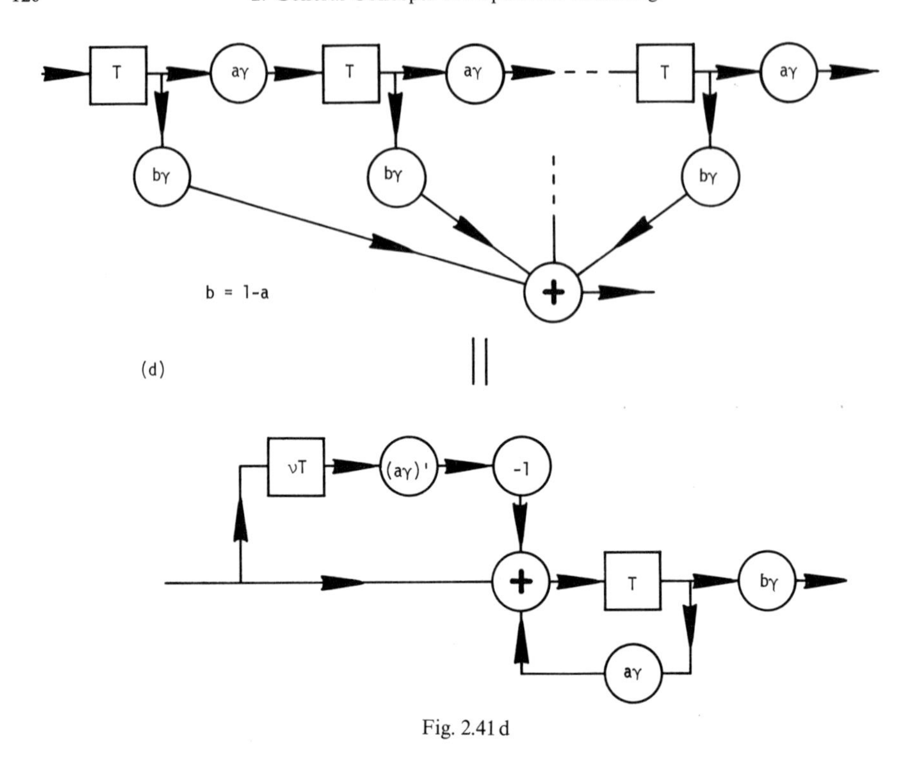

Fig. 2.41 d

sequences depicted in Figures 2.41b and c comprise finite numbers of repetitions, we can insert a *time delay* in the corresponding abbreviation to keep track of the time that each component of flow enters the loop and withdraw what is left of it after the appropriate time has passed, as illustrated in Figure 2.41d. In this way, we can convert a perpetual loop into a finite loop. Once again $(a\gamma)'$ is given by equation 2.42 and becomes $(a\gamma)^{\nu}$ when a and γ are constants.

2.13. Some References for Chapter 2

The numbers in the following lists refer to entries in the Bibliography at the end of the text. The lists themselves are designed to guide the interested reader into some of the literature relevant to particular topic areas treated in chapter 2.

On Markovian States, Ergodicity, Uncertainty, Lumping

10, 14, 20, 26, 62, 63, 72, 89, 101, 148, 217, 238, 257, 258, 269, 270.

On Models and Their Roles in Science

29, 30, 60, 62, 63, 73, 79, 90, 107, 114, 124, 155, 156, 174, 182, 186, 189, 235, 239, 241, 244, 249, 250, 255, 278, 295.

On Network Diagrams and Their Applications

32, 33, 51, 52, 56, 110, 141, 147, 158, 159, 179, 180, 206, 218, 257.

General References on Population Biology and Demography

1, 2, 22, 25, 49, 53, 54, 61, 64, 66, 78, 117, 140, 142, 161, 170, 173, 199, 200, 205, 220, 221, 230, 240, 260, 265, 266, 294, 296, 298, 299.

General References on Population Modeling

9, 12, 13, 27, 31, 35, 36, 38, 50, 55, 59, 69, 71, 74, 76, 83, 88, 128, 133, 134, 166, 170, 175, 176, 185, 221, 222, 228, 229, 245, 262, 263, 266, 273, 281, 287, 294, 296.

3. A Network Approach to Population Modeling

List of Titled Examples

3.1. Introduction to Network Modeling of Populations

Now we have the basic equipment with which to construct network models of a very wide variety of populations of organisms on the basis of their life cycles. Even when these life cycles are relatively simple, it very often is quite difficult without networks to keep track of the conservation and constitutive relationships implied in them and to develop appropriate state equations. As we soon shall see, not all life cycles are amenable to network modeling; but for those that are, the network models provide a simple, heuristic mechanism by which constitutive and conservation relationships can be represented. Furthermore, the rather well-developed analytical techniques for networks provide straightforward algorithms by which the networks themselves can be converted directly into state equations. When the state equations themselves promise to be too complex for useful, general solution, the network provides a ready-made flow chart for the construction of a digital-computer program for simulation of the population. If an analog computer with time delays or a digitally simulated analog computer is available, the network is more than a flow chart, it is essentially the program itself. Every key parameter and variable is conspicuously apparent and accessible in the network model. Therefore, modifications usually are extremely easy to make; and this in turn greatly facilitates the construction of the model itself.

To summarize, we can state that the major advantages to the use of network models where they are applicable are (1) their ease of construction, (2) their ease of modification, and (3) the ease with which they can be analyzed or simulated.

We have alluded to the heuristic nature of network models. This perhaps deserves some further discussion. *Heuristic* means "helping to discover or learn," and that is precisely what the network model can do. When its application is feasible, the network model almost inevitably is sufficiently concise and can be laid out in a sufficiently straightforward manner that the modeler is drawn through it automatically, with almost no chance of becoming lost or confused by details. As the network is being constructed, each element and each line automatically suggests elements and lines to follow. Once it is complete, the network itself suggests changes in the underlying hypotheses, suggests new data that should be gathered about the life cycle, suggests new relationships that were not at all apparent in the confusion that existed before its construction. As an example of this aspect of the network model, the reader is invited to reconsider Example 2.12 of section 2.11. Compare the lengthy and rather complicated verbal statement of the life cycle with its very concise representation in Figure 2.35.

In Chapter 3 we shall employ the network elements and techniques introduced in Chapter 2 to construct network models of specific population situations. We shall attempt to draw a sharp boundary between those situations that are amenable to networks and those that are not. We also shall introduce some new network elements that will be useful in some of the more complicated situations.

In constructing our models, we shall continue to follow the very useful dictum of Chapter 2: THE SIMPLEST WAY TO CONSTRUCT A NETWORK MODEL OFTEN IS TO BEGIN WITH THE FLOW OF NEWLY-FORMED MEMBERS INTO THE POPULATION AND FOLLOW IT THROUGH THE LIFE CYCLE; IF REPRODUCTION IS SEXUAL, THE FLOW TO FOLLOW GENERALLY IS THAT OF THE FEMALES.

3.2. Network Models for Some Basic, Idealized Life Cycles

It should be clear from section 2.11 that construction of the bookkeeping skeleton of a network model without knowledge of the constitutive relationships is a rather futile exercise, since that portion may well be modified considerably as those relationships are incorporated. This is true not only of reproductive processes and mortality processes, but also of any other processes that affect the population as we have defined its membership. Of particular importance are the timed processes related to reproduction. Such processes generally will include maturation or growth of newly produced offspring into reproducing adults. Beyond that, reproduction will vary so much from species to species that generalizations are of little use.

3.2.1. Simple Life Cycles

In this section, we shall consider some examples of these variations in highly idealized situations; and we shall develop network models for each of them. From the outset, we shall develop each model step by step, including as we go both conservation and constitutive relationships. Thus, we shall construct the bookkeeping, natality, and function-generating portions of our models together. At the moment, our task will be made easier by the fact that the constitutive relationships in our highly idealized life-cycle statements are very simple. We shall begin with several variations on the theme of Figure 2.36a.

Example 3.1. Deterministic Binary Fission with Poisson Death Process. Consider a population of identical protozoans that reproduce solely by the process of binary fission and require no conjugation. The culture medium is sufficiently well regulated that the time between successive fissions is a constant, T_F transition intervals. The probability of surviving from one transition interval to the next is the same for all the protozoans for all time (owing to constant pressure from a nonselective predator). That probability is γ. Construct a network model that embodies the dynamics of this situation.

Answer: We can begin by considering the flow of newly formed daughter cells into the population. After a delay of T_F transition intervals, a certain proportion of these daughters will emerge as surviving, mature protozoans, ready to undergo fission. The proportion that survive each of the T_F intervals is γ; so the proportion surviving all T_F intervals is γ raised to the power T_F. Thus, we can construct directly the network model of Figure 3.1. Maturation is represented by the *time delay* (T_F), survival by the *scalor* (γ^{T_F}), and fission by the *scalor* (2). The products of fission are themselves newly formed daughter cells and the flow representing their production can be connected back into the *time delay*.

The scale factor of the survivorship scalor in this model takes on its particular value (γ^{T_F}) because it is the probability of the occurrence of T_F sequential events, each with the probability γ of occurring. Although one should be familiar with such probabilities, one need not be in order to arrive at the same value for the scale factor. In fact, all one would have to do is construct the network model with maximum temporal resolution (one *time delay* and one survivorship *scalor* for each transition interval from fission to fission) and then proceed to lump the *time delays* and *scalors* according to the rules in Figure 2.41. When the lumping was completed, one would have a single *time delay*, T_F,

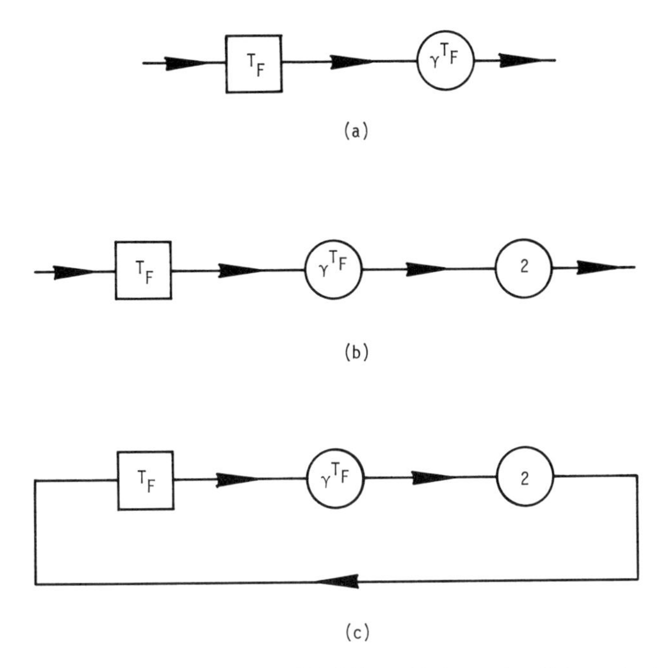

Fig. 3.1. Development of a network model of binary fission with a Poisson death process

and a single *scalor*, γ^{T_F}. Thus, by the very process of lumping, one would have carried out one of the logical processes of probability theory, elementary as it might be. Some of the logical processes of probability theory are far more complicated, yet, as we shall see, these too often can be carried out through similar, simple manipulation or reduction of the network model.

Suppose, now, that our deductions require an interval-by-interval estimate of the total number of protozoa in our population. How do we generate this from our model? We could begin by keeping a running total of the number of protozoa represented in the *time delay*. However, since mortality is not accounted for until the flow leaves the time delay, such a count would be an overestimate. What we actually need is an *accumulator* (see Figure 2.30) connected in the configuration of Figure 3.2, which would sum, interval by interval, surviving immature protozoans and dividing protozoans (assuming that each division produces two new daughter cells and annihilates the mother cell).

Note that the network structure added in Figure 3.2 over and beyond that in Figure 3.1c is not essential to the dynamics of the system. Those dynamics are embodied entirely in the network of Figure 3.1c. The added network merely represents our computation of a particular function, namely $N(\tau)$, that is not a state variable in the model itself. Thus, the added network is a nonessential function generator. It would become essential if, for example, γ were some function of $N(\tau)$. As it is, however, it merely dangles from the essential network and provides a disconnected output function.

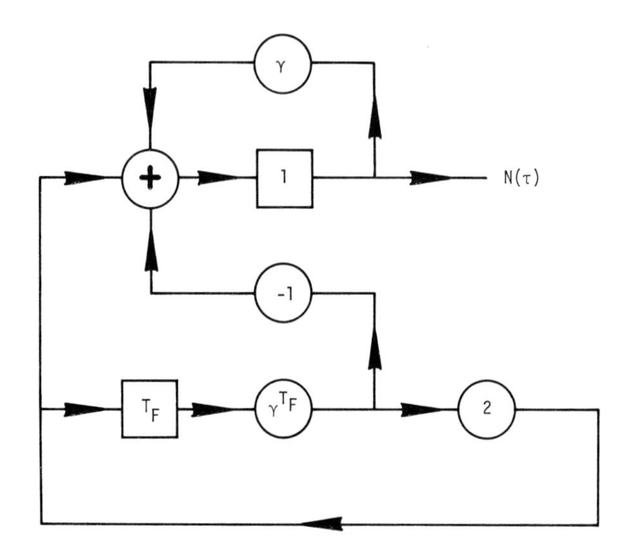

Fig. 3.2. An *accumulator* attached to the network of Fig. 3.1 to keep a running total of the population

Example 3.2. Deterministic Binary Fission with Age-Dependent Death. Consider a population of identical protozoans that reproduce by binary fission, requiring no conjugation and with time between fissions equal to T_F transition intervals. The probability of surviving depends markedly on the age, χ, of the individual, given in transition intervals since formation by fission. Thus, a protozoan that has survived χ transition intervals since it was formed has a probability $\gamma(\chi)$ of surviving one more. Construct a network model that embodies the dynamics of this situation.

Answer: The basic dynamics of this situation can be embodied in a network that is no more complicated than that in Figure 3.1c. We simply begin by considering the flow of newly formed daughter cells. After a delay of T_F units, a certain proportion of those daughters will emerge as surviving adults, ready to undergo fission immediately. As before the proportion, γ', that survive is equal to the product of the $\gamma(\chi)$'s taken over all of the T_F age classes between formation and maturation, but in this case, $\gamma(\chi)$ is not a constant. The resulting network is shown in Figure 3.3.

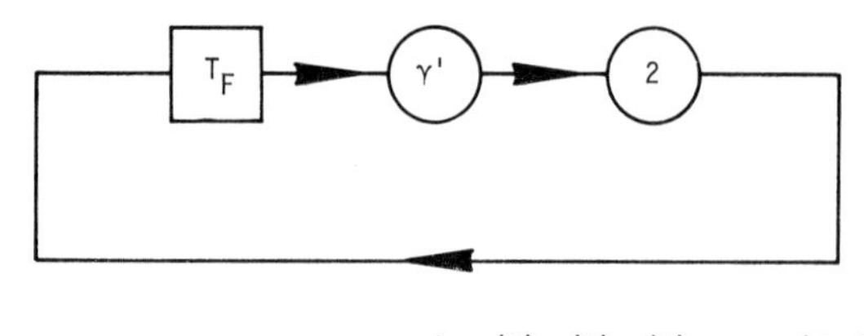

$$\gamma' = \gamma(0)\cdot\gamma(1)\cdot\gamma(2)\cdot\ldots\cdot\gamma(T_F-1)$$

Fig. 3.3. Network model for binary fission with age-dependent death

As long as we do not need $N(\tau)$, either for our own information or for constitutive relationships involving $N(\tau)$, (i.e., parameters that depend on $N(\tau)$), the simple network model of Figure 3.3 will be sufficient to describe the population dynamics. On the other hand, if we need an estimate of $N(\tau)$, matters will become more complicated, with the extent of the complication depending on the accuracy required in our estimate. For example, if we are willing to settle for a crude estimate, then we simply might keep a running total of the contents, N_{T_F}, of the *time delay* in our basic network before the survivorship *scalor* has been applied (i.e., before subtraction of the proportion expected to die). Regardless of the function $\gamma(\chi)$, we know that the actual value of $N(\tau)$ will be limited to the following range:

$$\gamma' N_{T_F}(\tau) \leqq N(\tau) \leqq N_{T_F}(\tau) \tag{3.1}$$

where the lower limit applies if all of the individuals that die do so during age 0 (i.e., when they first emerge as new-formed daughter cells); and the upper limit applies if all individuals that die do so as they mature, just prior to or during fission. Clearly the truth will lie somewhere between these extremes. However, if γ' is not much different from one, then we nonetheless have a fairly good estimate of $N(\tau)$. To represent the computation of $N_{T_F}(\tau)$, we simply can add a *scalor*, and *adder* and an *accumulator* to the basic network model, as illustrated in Figure 3.4a.

In Example 3.1, we kept the contents of the *accumulator* completely up to date and accurate within the limits of temporal resolution merely by placing a survivorship *scalor* in the loop (see Figure 3.2). We cannot use that method to improve our accuracy in the present example, because we simply do not have a single mortality factor applicable to the entire population (i.e., applicable to the entire contents of the *accumulator*). Instead, we have $\gamma(\chi)$, a set of T_F different survivorships. Therefore, if we want an estimate of $N(\tau)$ that is as accurate as possible in the face of the existing temporal resolution, we must divide the *time delay* into T_F subunits, each with its associated survivorship, as illustrated in Figure 3.4b. The interval-by-interval contents of each *time delay* is given by its input flow. Therefore, one merely needs to sum the input flows to obtain the most accurate interval-by-interval estimate of N.

Estimates of $N(\tau)$ with accuracies intermediate between those given by Figures 3.4a and b can be obtained by intermediate divisions of T_F, such as that shown in Figure 3.4c. Of course the subunits resulting from our divisions of T_F do not need to represent equal time spans. In fact, if T_F is a prime integer, the subunit *time delays* cannot all be equal if they are greater than one interval. An *accumulator* can keep a running total of the contents of each *time delay*, and the sum of the contents of all *accumulators* becomes our estimate of $N(\tau)$.

Example 3.3. Poisson Binary Fission with Poisson Death. Consider a population of identical protozoans that reproduce by binary fission. Each new daughter cell requires precisely T_M transition intervals to reach maturity. During the next transition interval, she faces the constant probability β of dividing. If she does not divide, but progresses into the subsequent transition interval, she again

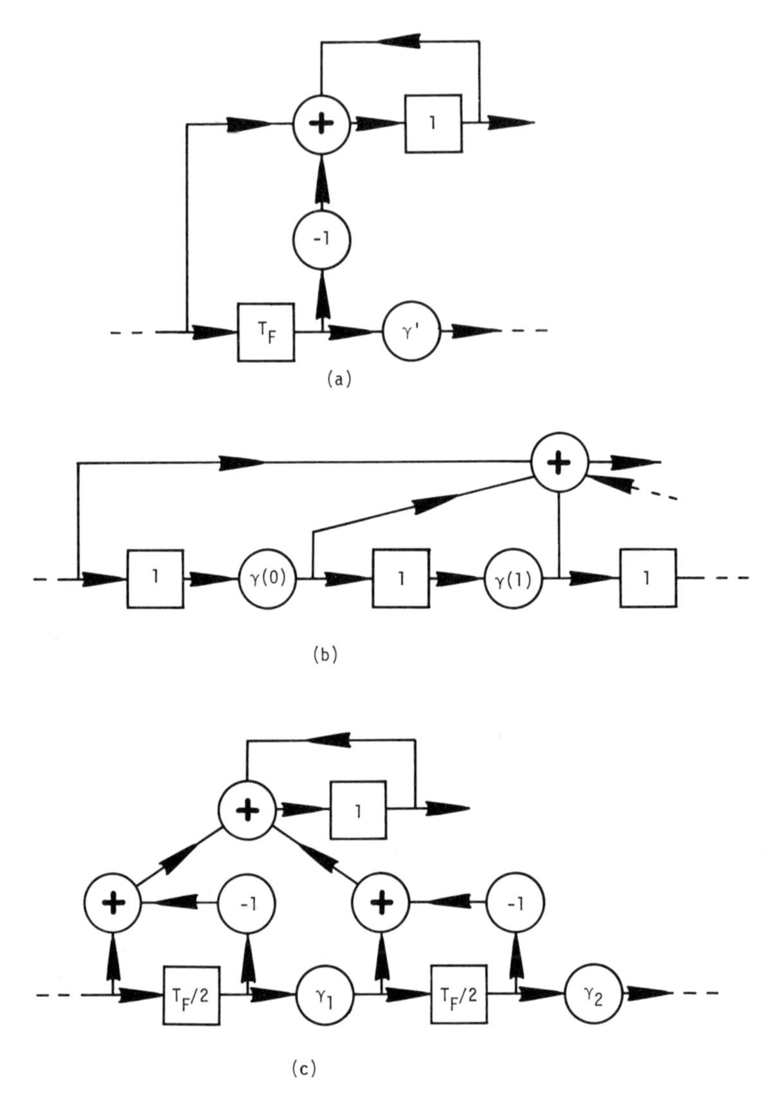

Fig. 3.4a–c. Three schemes for keeping running totals of the population in the model of Fig. 3.3. Configuration (a) provides the least accurate total, (b) provides the most accurate, and (c) provides intermediate accuracy

faces the probability β of dividing. This process continues until she finally divides (during each interval that she enters undivided, she faces the same probability of dividing). (Note: stochastic processes such as this, where the probability of occurrence is constant from one interval to the next, are known as *Poisson processes*). Furthermore, the probability that a newly formed daughter survives to maturity is γ_1, thereafter, for each interval during which she is alive, the probability is constant, γ_2, that she will survive (divided or undivided) to the next interval. In other words, for the mature protozoa, death too is a Poisson

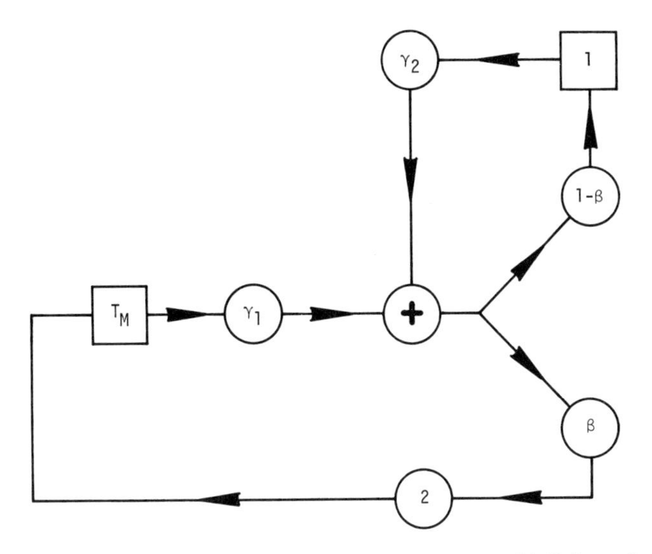

Fig. 3.5. A network model for a Poisson fission process with Poisson death

process. Construct a network model that embodies the dynamics of this situation.

Answer: Now we have a new wrinkle to consider. Nonetheless, we simply should begin by following the flow of newly produced daughters. This will pass into a *time delay*, T_m, from which we take the proportion γ_1 as emerging survivors. Of the proportion that survive, a second proportion, β, divide and a proportion $\mathbb{1}-\beta$ do not. Stipulating that division results in the production of two daughter cells and the annihilation of the mother cell, we can insert a *scalor* with scale factor $\mathbb{2}$ in the line representing division and connect its output flow (the flow of newly formed daughters) back to the input of the *time delay*, T_m. Those that have survived and gone undivided enter the next transition interval to repeat the process. However, because both death and division are Poisson processes, every interval is just like the one that preceded it as far as the surviving adult is concerned; so we simply merge the flow of survivors with the influx of newly maturing adults. In other words, all surviving adults are identical, regardless of age. The resulting network is shown in Figure 3.5.

This model provides directly an interval-by-interval estimate of the total number of mature protozoa that is as accurate as possible for the specified temporal resolution (i.e., the transition interval). On the other hand, if we wish to obtain an accurate estimate of the number of maturing protozoans, we have a bit of a problem, inherent in our life-cycle specifications. We know only γ_1 for the entire period of maturation; we do not know how mortality is divided over that period. Therefore, the best we can do is keep a running total of the contents of *time delay* T_m and establish the range within which our desired number must fall:

$$\gamma_1 N_{T_m}(\tau) \leqq N(\tau) \leqq N_{T_m}(\tau). \tag{3.2}$$

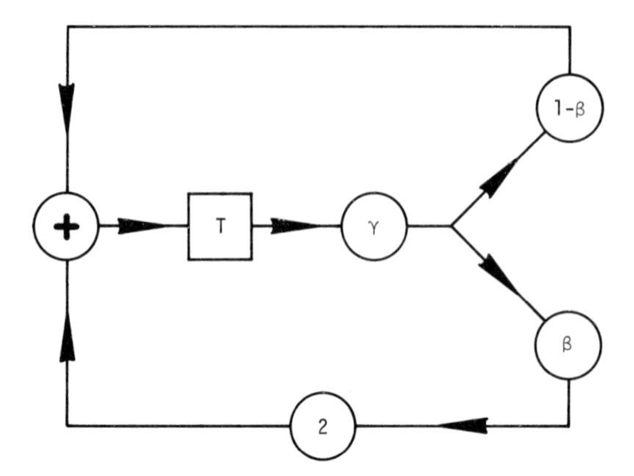

Fig. 3.6. A network model for synchronous, periodic binary fission

In this case, our estimate is not as accurate as it could be for the specified transition interval, but is limited by the reduced temporal resolution of our determination (or specification) of mortality for the maturing protozoa. Notice that this deficiency does not affect the network embodying the dynamics of the population unless $N(\tau)$ is required in the constitutive relationships.

Example 3.4. Synchronous Periodic Binary Fission. Consider a population of identical protozoans that reproduce by binary fission, induced by a periodic extrinsic stimulus (e.g., the diurnal cycle) whose period is fixed and equal to T transition intervals. At a particular phase of each cycle (i.e., the νth transition interval of each cycle), a proportion β of the protozoans divide once; the rest do not. No divisions occur at any other phase of the cycle, so once the division phase has passed, all individuals must wait until the same phase of the subsequent cycle to have another opportunity to divide. The probability of surviving from the division phase of one cycle through the division phase of the next (i.e., through one entire cycle) is γ for all individuals. Construct a network model embodying the dynamics of the population.

Answer: Once the division phase is past, the newly formed daughter cells join the cells that did not divide and wait for the next cycle. This wait is represented by the *time delay* T, which represents the interval beginning at the end of one division phase (the end of the νth transition interval in one cycle) and extending through the next (through the νth transition interval of the next cycle). Of the flow emerging from the *time delay*, a proportion γ represents survivors; and of these a proportion β just have divided, $\mathbb{1}-\beta$ have not and recycle for another try. For each division, two new daughters enter the cycling flow and one mother cell leaves. The resulting network is shown in Figure 3.6.

As in the previous example, our ability to estimate $N(\tau)$ interval by interval will be limited by the fact that we have not specified mortality on an interval-by-interval basis. Once again, the best we can do is keep a running count of the contents of the *time delay* and thus establish the range of $N(\tau)$.

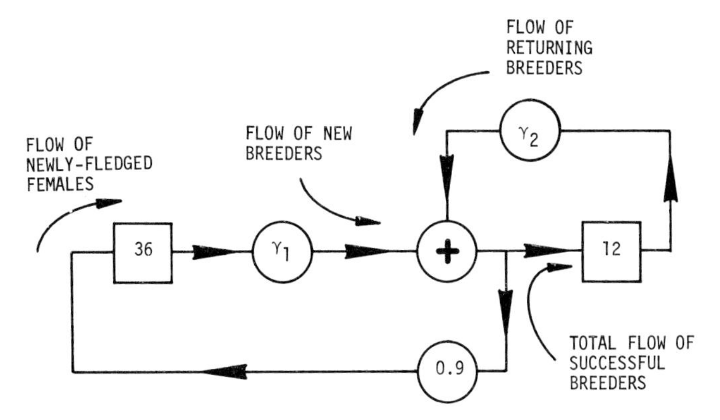

Fig. 3.7. A network model for an idealized Herring Gull life cycle

Example 3.5. Simple Periodic Sexual Reproduction with Age-Independent Death. Construct a network embodying the dynamics of a population of idealized herring gulls with the following life cycle, resolved to an interval of one month: Nesting occurs in the fourth month of every year. Each newly fledged female has the same probability (γ_1) of surviving to age 36 months, at which time she is sexually mature and has just produced her first fledged brood. She faces the probability γ_2 of surviving the next twelve months to produce another fledged brood, and the same probability of surviving each twelve-month period thereafter, producing one fledged brood at the end of each. Out of each brood, the expected number of female fledglings is 0.9.

Answer: According to the life-cycle statement, every adult female participates in breeding, implying that there is no scarcity of nesting partners. Therefore, the dynamics of the population are embodied entirely in the life cycle of the female. Once again, we can begin construction of the model by considering the flow of newly formed individuals, in this case newly fledged females. Our representation of this flow enters a 36-month *time delay;* and of those emerging we take a proportion γ_1 to represent survivors that have just bred for the first time and have produced a fledged brood. Following the first brood production, the flow continues into a twelve-month *time delay* representing the interval between productions of fledged broods. Of the emerging flow, we take a proportion γ_2, to represent survivors who have bred again. Since age makes no difference with respect to survivorship or brood production, we simply can merge the flow representing newly mature females with the flow representing returning breeders, forming a sum that represents the total flow of females through the nesting process. The flow of newly fledged females is equal to 0.9 times the flow through nesting. Thus, we can complete (close) the network with a simple loop (Figure 3.7).

With these specifications, the resolution of our estimate of the total female population is limited to the range

$$\gamma_1 N_{36}(\tau)+\gamma_2 N_{12}(\tau) \leqq N(\tau) \leqq N_{36}(\tau)+N_{12}(\tau) \tag{3.3}$$

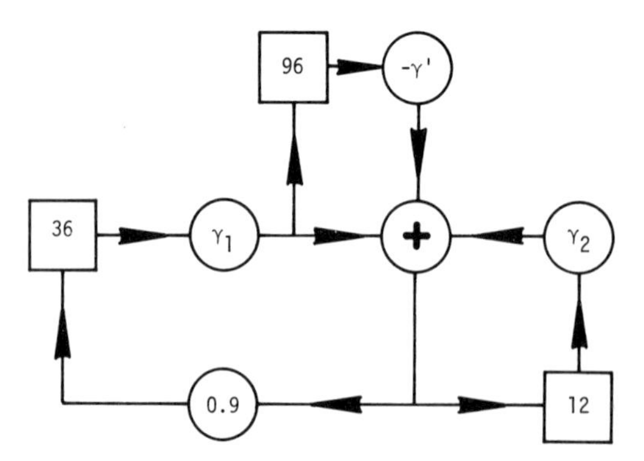

Fig. 3.8. The model of Fig. 3.7 for idealized Herring Gulls, modified to include 11-year (132-month) absolute longevity

where $N_{36}(\tau)$ is the content of the 36-month *time delay* during the τth month and $N_{12}(\tau)$ is the content of the 12-month *time delay* during the same month. Once again, this lack of resolution has no effect on the dynamics deducible from the model unless constitutive relations within it depend upon $N(\tau)$, which in this particular case has been stipulated not to be true.

Example 3.6. Periodic Reproduction with Fixed Longevity. Construct a network model for the idealized gulls of Example 3.5 with age-independent survivorship, γ_2, up to age 11, the absolute longevity of the gulls, beyond which there are no survivors.

Answer: Figure 3.8 shows the appropriate modification of the network of Figure 3.7. The absolute longevity is represented by a simple, double-bookkeeping scheme. The flow of emerging 3-year old females passes into a 96-month *time delay;* and the proportion, γ' (see equation 2.53), that survive the corresponding 8 years (i.e., to the limit of longevity) is subtracted from the flow of nesting adults. Comparing Figure 3.8 with Figure 2.29 in section 2.10, one can see that the portion of the network immediately surrounding the *adder* has exactly the same configuration as an *accumulator.* However, it is not operating on a transition-interval by transition-interval (i.e., month to month) basis and therefore might be considered an *accumulator* of reduced resolution. In fact, its resolution is reduced by a factor of twelve from that of a conventional accumulator.

Example 3.7. Periodic Reproduction with Random Time to First Mating. Consider the idealized gulls of Example 3.5 with the following modification: A newly fledged female faces probability γ_1 of surviving to age 36 months, at which time she is sexually mature and capable of participating in nesting for the first time. The probability that she will do so is β_1. Once she has participated in nesting, she will continue to do so every spring for the rest of her life. If she did not nest during her first adult season, the probability is β_2 that she will do so in her second, if she survives; failing this, the surviving female is virtually certain to

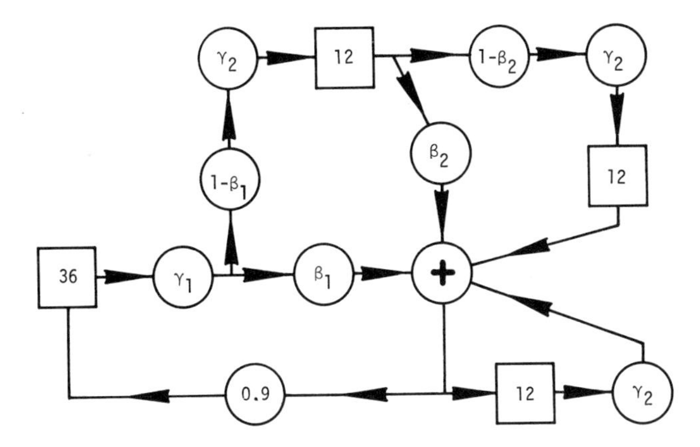

Fig. 3.9. The idealized Herring Gull model modified to include random time to first mating in the female

nest in her third adult season. Construct a network model embodying the dynamics of this idealized gull population.

Answer: The appropriate modifications of the network model of Figure 3.7 are shown in Figure 3.9. Here, the flow of females through nesting comprises four components: the flow of three-year old nesters, the flow of first-time, four-year old nesters, the flow of first-time, five-year old nesters, and the flow of previous nesters that have survived from last season and are returning to nest again.

Example 3.8. Periodic Reproduction with Mortality that is Piecewise Independent of Age. Consider the idealized gull population of Example 3.6 with the following modification: During the first four years of adulthood, the female faces probability γ_2 of survival; during the second four years, she faces a reduced probability, γ_3. Construct a corresponding network model.

Answer: Up to this point, we have considered for the most part idealized gulls in which the probability of death (i.e., the mortality rate given in deaths per gull per year) was independent of the age of the gull. The one exception was Example 3.6, where the mortality rate was independent of age for eight years of adulthood, then abruptly rose to $\mathbb{1}$, so that no adults survived to their ninth year of adulthood (twelfth year of total age). Now we are complicating this slightly by considering a situation in which mortality rate is age-independent over each of two segments, or "pieces," of the span of adulthood, but takes on different values for each piece. Realization of this specification in a network model is quite simple and is illustrated in Figure 3.10.

Here we carry out double bookkeeping in each of the two segments of adulthood (employing two reduced-resolution *accumulators* in place of the one that appears in Figure 3.8). The flow out of the first *accumulator* passes into the second.

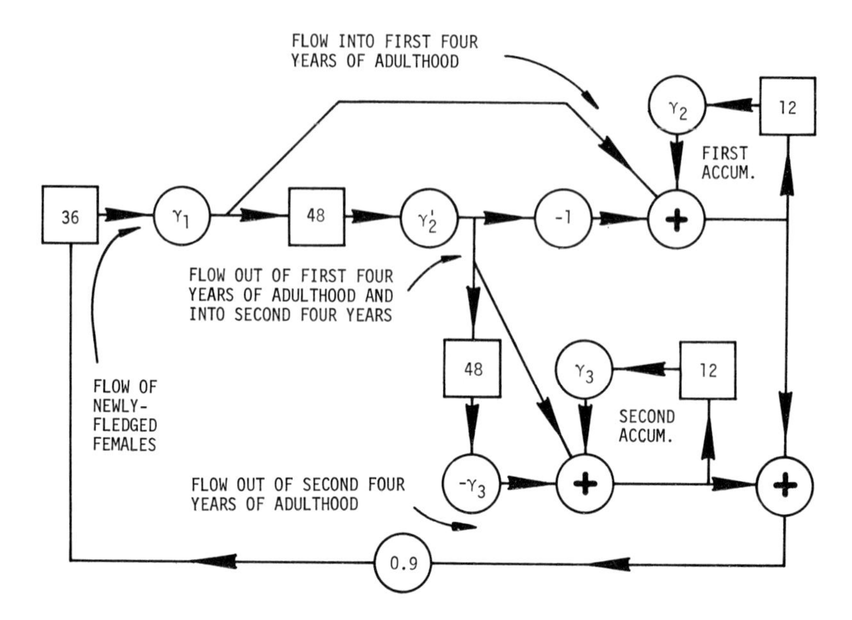

Fig. 3.10. The idealized Herring Gull model modified to include simple age-dependent mortality (age being resolved into two four-year classes)

Example 3.9. Periodic Reproduction with Natality that is Piecewise Independent of Age. Consider the idealized gull population of Example 3.6 with the following modification: During the first four years of adulthood, the female produces broods with 0.8 female offspring expected to survive to fledge; during the second (final) four years of adulthood, she produces broods with 1.0 female offspring expected to fledge. Construct an appropriate network model.

Answer: The same double-bookkeeping scheme can be employed that was used in Example 3.8. The network realization is illustrated in Figure 3.11.

Example 3.10. Periodic Reproduction with Age-Dependent Mortality and Natality. Consider the idealized gull population of Example 3.6 with the following modifications: The probability that an adult survives from the end of her ith nesting season to the end of her $i+1$ nesting season is p_i; and the number of fledged female offspring that she is expected to produce in her ith nesting season is b_i. Construct a network model embodying the dynamics of this population.

Answer: A network realization of this situation is shown in Figure 3.12. Except for its first *time delay*, this network simply is a graphical equivalent of the matrix of equation 2.39 in section 2.8 (the "Leslie Matrix"). The factors p_i become subdiagonal elements of that matrix; the factors b_i become elements of the first row. It is left as an exercise for the reader to determine the first three elements of the first row and the first three elements of the subdiagonal.

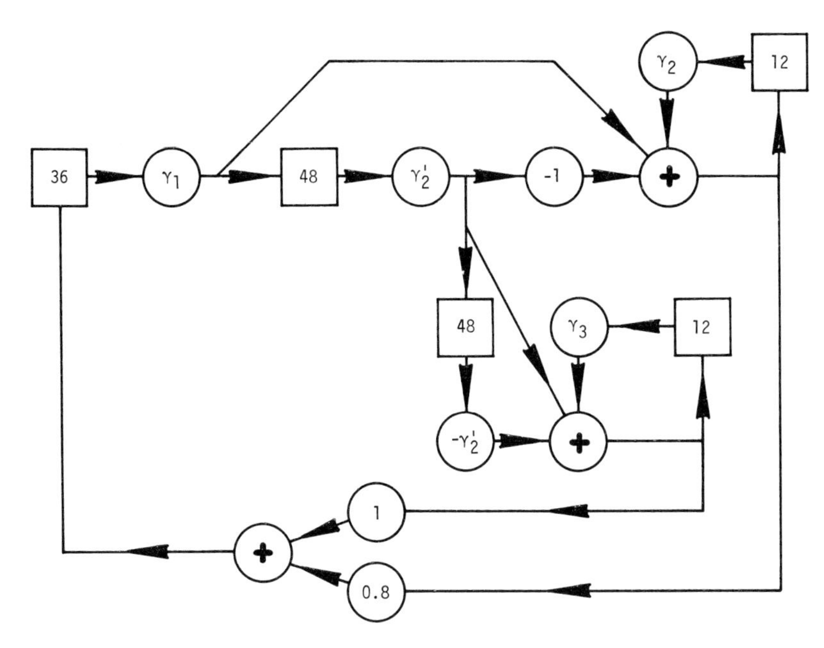

Fig. 3.11. The idealized Herring Gull model with simple age-dependent natality

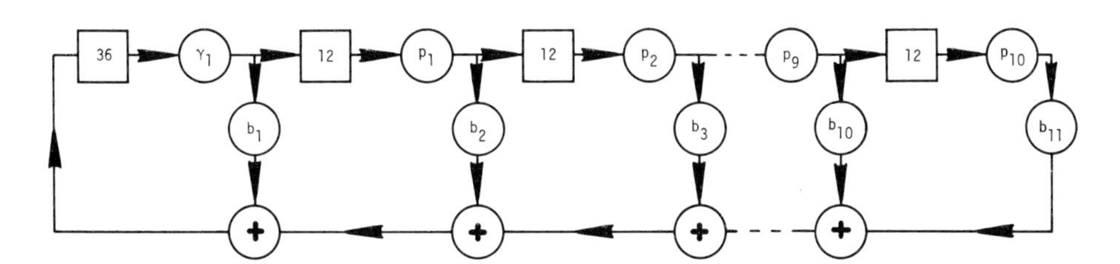

Fig. 3.12. The idealized Herring Gull model with age-dependent mortality and natality, age resolved to one-year intervals

Example 3.11. Probabilistic Sexual Reproduction with Periodic Ovulation and Poisson Death. Consider a population of idealized mammals with the following life cycle, resolved to one day: Each newborn female offspring requires 300 days to reach sexual maturity, which is signaled by her first ovulation. Subsequently, she ovulates once every twenty days; and at each ovulation she faces the same, fixed probability (β) of successful conception. Gestation requires 38 days and the expected number of newborn female offspring per adult female that has conceived and survived gestation is 3.7. Mandatory lactation continues for 25 days after parturition, after which time the offspring can be weaned and expected to survive. During mandatory lactation, survival of the offspring requires survival of the mother. The twenty-day ovulation cycle begins again, with the first ovulation occurring fifteen days after parturition. Throughout her life, after

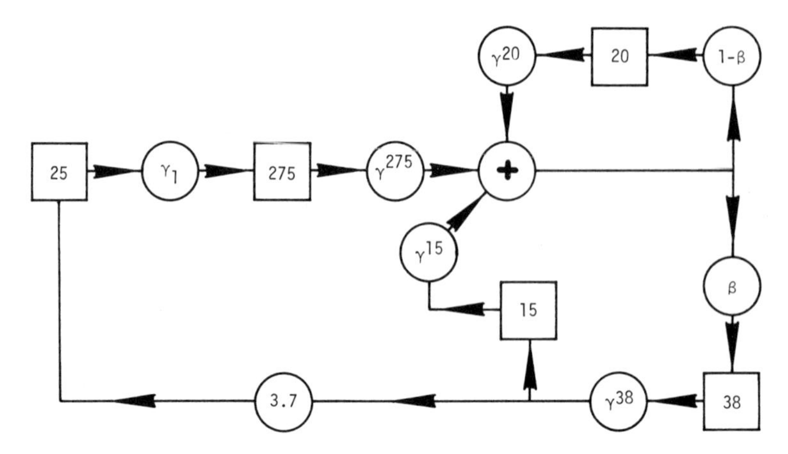

Fig. 3.13. Network model of an idealized mammal population with periodic ovulation and Poisson death

weaning, the female faces the same, fixed, day by day probability (γ) of survival. Her probability of surviving from birth to weaning is γ_1. Construct a network model embodying the dynamics of the population.

Answer: Once again, we can begin by considering the flow of newborn female offspring (Figure 3.13). Immediately after birth, each new individual requires 25 days of mandatory lactation, during which time she faces double jeopardy—her own innate probability of dying and the probability that her mother dies. The factor γ_1 must include both of these possibilities. After mandatory lactation, she completes the rest of the 300-day maturation period (i.e., the 275 days remaining) and her probability of surviving is γ^{275}. If she survives, she ovulates for the first time and faces the probability β of conceiving and passing into the 38-day gestation period, with the probability γ^{38} of surviving to parturition. After parturition, she enters a 15-day waiting period (with probability γ^{15} of surviving) before reentering the ovulation cycle. If, at the time of ovulation, she does not conceive, then she remains in the ovulation cycle, entering a twenty-day waiting period (probability γ^{20} of surviving) before ovulating again. For each female completing gestation an average of 3.7 newborn females begin mandatory suckling.

Example 3.12. Probabilistic Sexual Reproduction with Induced Ovulation, Mortality Dependent on Nonoverlapping Time Classes. Consider a population of idealized mammals with the following life cycle, resolved to one day: a newly weaned female requires 300 days to reach sexual maturity, after which she is receptive and capable of ovulation. Ovulation does not occur spontaneously, but is induced by copulation. If copulation does not lead to conception, then there is a 25-day period of regression, during which ovulation cannot recur and the female is not receptive. Gestation requires 35 days, and is followed by a 20-day period of mandatory lactation, during which time the female is neither receptive

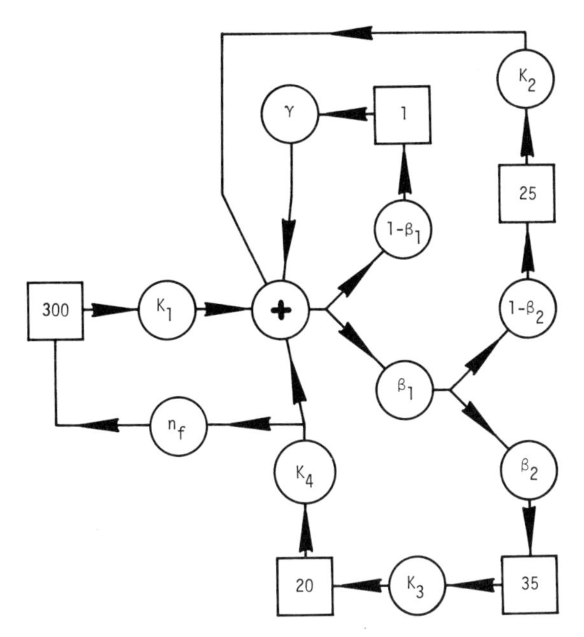

Fig. 3.14. Network model of an idealized mammal population with induced ovulation, mortality dependent upon non-overlapping time classes

nor capable of ovulation, and during which time the survival of the suckling youngsters is completely contingent upon survival of the lactating mother. The probability that a receptive female copulates in a given day is β_1. The corresponding probability of conception is β_2. The expected number of female offspring in a newly weaned litter is n_f. The day by day survival of the female depends upon her time class, being $\gamma_1(\chi)$ for the χth day of maturation, $\gamma_2(\chi)$ for the χth day of regression, $\gamma_3(\chi)$ for the χth day of gestation, $\gamma_4(\chi)$ for the χth day of mandatory lactation, and simply γ for all days during which she is receptive and capable of ovulation. Construct a network model embodying the dynamics of the population.

Answer: Since the output of the reproductive process is given in term of newly weaned females, we can begin with that (see Figure 3.14). As long as none of the constitutive relationships depend upon the total population or any segment of it, we can construct a model embodying the complete dynamics of the population by lumping the survivorships and intervals corresponding to each period in the life cycle. Thus, we have a single, 300-day *time delay* representing maturation, followed by a single survivorship *scalor*, K_1. The flow of newly mature females converges with the flows of other receptive females all of whom face a probability β_1 of copulating in a given day, with those that do not returning to the flow of receptive females the following day (through the 1-day *time delay* and its survivorship *scalor*). The flow of females that have copulated passes a conservative branch point, with a proportion β_2 moving on into gestation and a proportion $1-\beta_2$ returning through regression (represented by a 25-day *time delay* and its associated, lumped survivorship *scalor*, K_2) to the flow of receptive females. The flow into gestation passes through a 35-day *time delay* with its lumped survivorship scalor, K_3, and then on through a twenty-day *time delay* with its survivorship scalor, together representing mandatory lactation. On leaving the latter, the flow returns to join that of receptive females. The flow of newly weaned female offspring is equal to n_f times the flow of mothers completing mandatory lactation.

If one or more parameters of the model depends upon the accumulations of females in one or more of the life-cycle stages, the interval-by-interval specification of mortality will allow us to make accurate interval-by-interval estimates of that accumulation. To do so, we must follow the pattern of Figure 3.4b, dividing the corresponding *time delay* and *scalor* into a series of one-interval *time delays*, each with its appropriate *scalor*, then adding the inputs to the *time delays* to form the total accumulation in all of the *time delays*.

3.2.2. Models with Overlapping Time Classes

Over and over again in this section we have employed loops in our network models to represent the recycling flows of individuals through various processes in their life cycles. When such loops are employed, the ages of the individuals represented as passing around them are thoroughly obscured (e.g., in the network model of Figure 3.14, the age of an individual is represented for the first 300 days of its life, as it passes through the 300-day *time delay;* from that point on, its age is not represented at all). As long as it is cycling within a loop or set of loops, the state of the individual is represented as being independent of its precise age (the precise number of transition intervals since birth, hatching, or the like), but dependent instead on the number of transition intervals that have passed since some other event in the life cycle.

When we keep track of an individual's age, we essentially are assigning it to an age class (there being one such class for each transition interval in the life of an organism). On the other hand, individuals represented as passing around loops have been assigned to time classes of a different sort (i.e., time counted from events other than birth or hatching). Thus, in the model of Figure 3.13 there are three different types of time classes to which an individual might be assigned, defined by the following: 1) Time since birth (for the first 300 days after birth), 2) time since last ovulation (for mature individuals, not pregnant), and 3) time since impregnation (for 53 days after impregnation). There are 300 classes (one for each transition interval) of the first type, 20 of the second type, and 53 of the third type, making a total of 373 time classes (lumped states) that an individual could occupy. The model of Figure 3.14 has four types of time classes defined as follows: 1) by the time since birth; 2) by the time since impregnation; 3) by the time since unsuccessful copulation, and 4) by being receptive and capable of ovulation. There are 300 classes of the first type, 55 classes of the second type, 25 classes of the third type, and one class of the fourth type.

In most of the models presented in this section, individuals have been represented as belonging to only one time class at a time. The exceptions were the models of Figures 3.8, 3.10, and 3.11, where double bookkeeping was employed to keep track of age and time since nesting for idealized herring gulls. Individual gulls were represented as cycling around a loop for a certain number of years and then being removed from that loop. It was possible to represent removal of the gulls from the loop in an easy manner simply because, at the time of removal, a surviving gull was certain to be in a particular time class. If that had not been the case, if the gull had nonzero probabilities of being in any of several time classes at the time of removal, representation of removal would have been more difficult. Thus, the applications of *accumulators* and reduced-resolution *accumulators*, such as those in Figures 3.8, 3.10, and 3.11, are limited to the lumping of a sequence of equivalent age classes or a sequence of equivalent age sequences (e.g., equivalent passes through an annual breeding cycle). Except for situations such as this, we shall not be able to use the double-bookkeeping scheme to keep track of an individual's progress through two or more sets of time classes at once. In other words, it generally will be the case that the time classes in our network models will not be overlapping (i.e., an individual generally may not occupy more than one time class at a time). When the specifications of an idealized population include overlapping time classes, these generally will have to be expanded into single sets of combined time classes; and this expansion often will result not only in an enormous number of combined classes, but also in an extremely complicated, multibranched network model.

Example 3.13. Mortality Dependent on Overlapping Time Classes. Consider the idealized mammals of Example 3.11 with the following modification: Mortality among the females depends upon the age of the individual as well as whether or not she is pregnant; the absolute longevity is approximately 2000 days.

Answer: Because they obscure age, the 20-day ovulation loop and the 53-day gestation/post-partum loop in the network of Figure 3.13 cannot be used now. To construct the network without these loops, we simply can begin again with the flow of newborn female offspring and follow it into the life cycle. The first ovulation occurs after 300 days, at which point, the female either becomes pregnant and enters the 53-day gestation/post-partum delay or fails to become pregnant and enters the 20-day ovulation delay. Thereafter, at each ovulation, she faces the same choice, leading to the multiply branched tree-like network of Figure 3.15. *Scalors* appropriate to age and time class can be placed in series with each *time delay* (they are omitted from the figure for simplicity). Although the loops of Figure 3.13 cannot be used, it is possible that somewhere in this tree-like model two branches will represent females flowing into ovulation at precisely the same ages, in which case those two branches could be merged at an *adder* and the network could be simplified. What we would require is that

$$53x_1 + 20y_1 = 53x_2 + 20y_2 \tag{3.4}$$

where x_1 and x_2 are integers equal to the number of times path 1 and path 2 respectively passed through 53-day *time delays;* and y_1 and y_2 are integers equal

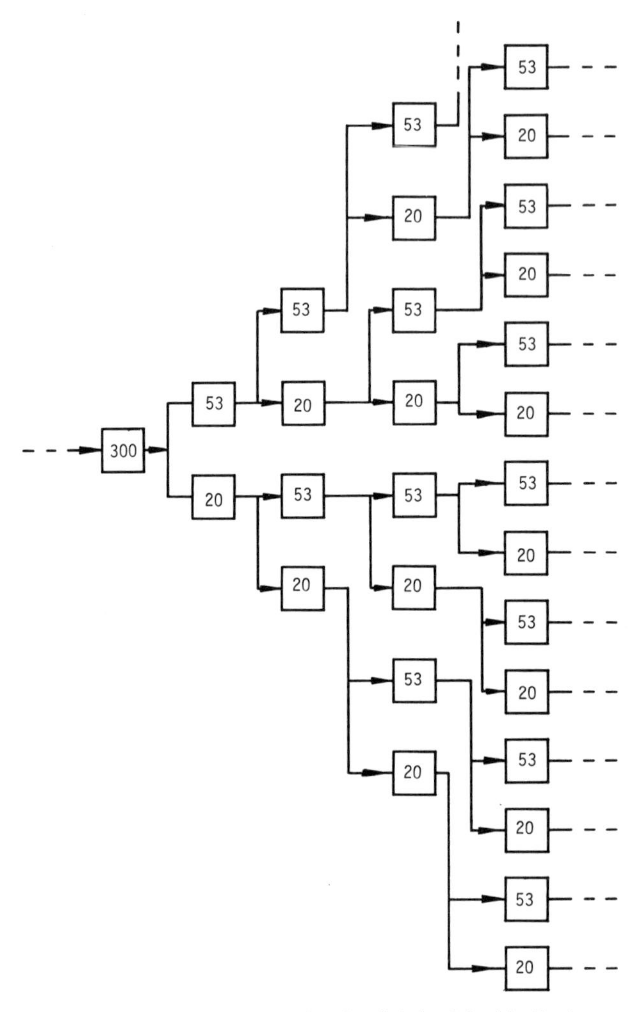

Fig. 3.15. Network model (with *scalors* left out for simplicity) of the idealized mammal population of Fig. 3.13, modified to include mortality dependent upon overlapping time classes

to the number of times those same paths respectively passed through 20-day *time delays*. Rearranging terms, we have

$$53(x_1-x_2)=20(y_2-y_1). \tag{3.5}$$

Factorizing both sides to remove common prime factors from 53 and 20, one immediately can see that there are no such factors (since 53 is prime and therefore 20 and 53 are relatively prime), so that (x_1-x_2) must be equal to $20n$ and (y_1-y_2) must be equal to $53n$, where n is an integer. It is left as a simple exercise for the reader to demonstrate that zero and one are the only values of n that are consistent with an absolute longevity of 2000 days. In other words, path 1 will be of the same length (in terms of total time delay) as path 2 if both paths

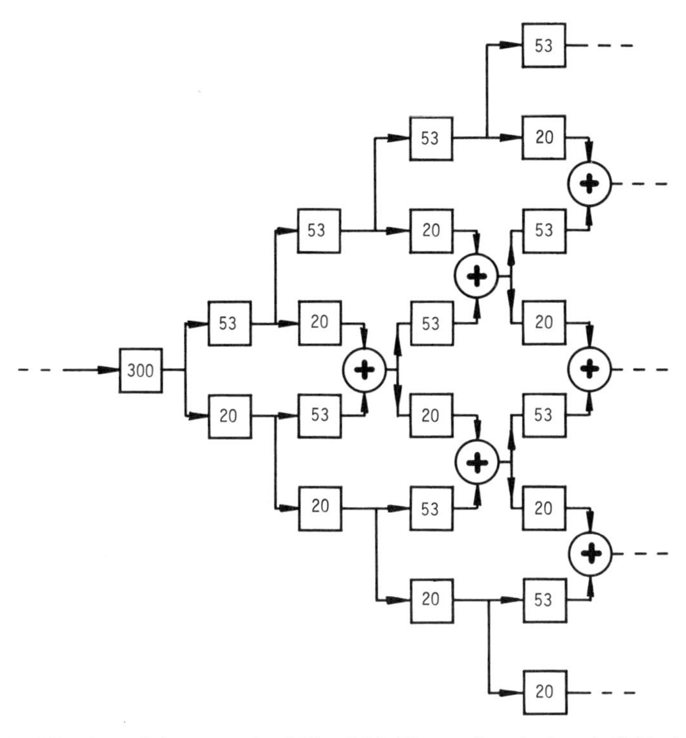

Fig. 3.16. Simplification of the network of Fig. 3.15. Flows of equivalent individuals have been merged in *adders*

have the same number of 20-day *delays* and the same number of 53-day *delays*, or if path 1 has twenty more 53-day *delays* than path 2 *and* path 2 has fifty-three more 20-day delays than path 1, or vice versa. Thus we are led to the simplified network of Figure 3.16. At the end of the tree corresponding to an age of 2000 days, this network will have 73 branches, 53 of which were formed at age multiples of 20 and 20 of which were formed at age multiples of 53. At every age of the type $53x+20y$, two paths will converge and two new paths will be formed, leading to no increase in the total number of branches. If branches of the type $53x+20y$ were not merged in the model, then the total number of branches at the end of the tree corresponding to 2000 days would be extremely large (as if 73 were not large enough already!). To calculate the number of branches that merging eliminates, consider the total number of different sequential combinations of 53-day and 20-day *time delays* that would lead to ovulation at age $53x+20y$. That number simply is the number of ways that x 53-day *delays* can be distributed over $x+y$ possible positions. For example, there are three sequential combinations leading to ovulation on day 93: 20, 20, 53; 20, 53, 20; and 53, 20, 20. Thus, without merging, the number of paths leading into ovulation at age $53x+20y$ is $\binom{x}{x+y}$ and the number of paths leading out is just twice that number. Summing over all such ages, we would find

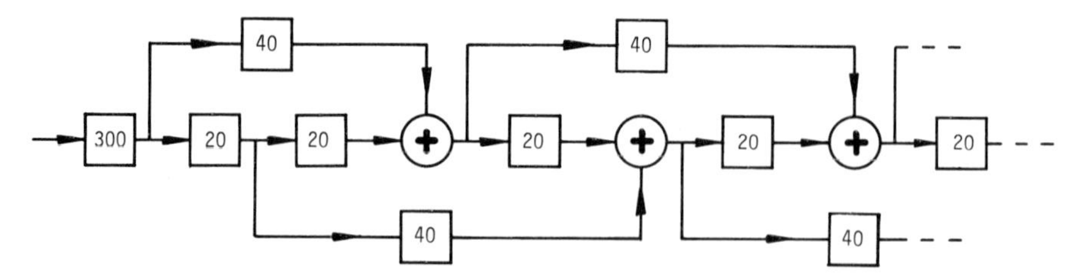

Fig. 3.17. Simplification of the network of Fig. 3.15 that is made possible by the new ovulation and gestation intervals with their large GCD

the total number of additional paths. To see how large this number will be, consider ovulation at age 1985 days, which is reached by the various combinations of thirty-three 20-day delays and twenty-five 53-day delays. The number of such combinations is approximately 2×10^{16}.

Example 3.14. Consider the population of Example 3.13 with the following modifications: The ovulation interval is 20 days; the gestation/post-partum interval is 40 days.

Answer: The network now can be made quite simple by the process of merging paths, as illustrated in Figure 3.17. The simplicity is possible because the numbers of time classes (i.e., 20 days and 40 days) in the two divergent sets of classes (ovulation vs. gestation) share large common factors. It is rather an easy matter, in fact, to show that the eventual number of parallel paths in a network model in which all possible mergings have been made is given by

$$\text{Eventual Number of parallel paths} = \frac{T_1 + T_2}{(\text{GCD}[T_1, T_2])} \tag{3.6}$$

where T_1 and T_2 are the number of time classes in set 1 and set 2; and $\text{GCD}[T_1, T_2]$ is the greatest common divisor of T_1 and T_2. In the previous example, the greatest common divisor was one; in the present example it is twenty. Thus, in the present example, the number of paths (achievable by maximum merging of paths) is $(20+40)/20=3$.

Example 3.15. Consider a population of idealized mammals in which ovulation is monthly and occurs immediately after parturition, allowing the female to become pregnant at that time, and in which the mortality is independent of age but depends instead upon whether the female is pregnant, lactating, or both. The gestation period and the lactation period both are three months, with the latter beginning immediately at the end of the former. Maturation requires four months, at which time the first ovulation occurs. Construct a network model embodying the dynamics of this system.

Answer: Once again, the specifications for our population include overlapping time classes. Let us depict these as follows: the label *Pi* indicates an individual in the *i*th month of pregnancy; the label *Li* indicates an individual in

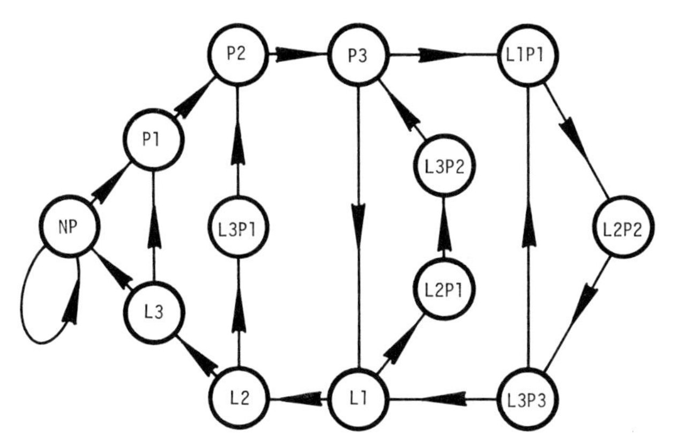

Fig. 3.18. Markovian state space for a population of idealized mammals with mortality being dependent on lactation and pregnancy. *Li* depicts the ith month of lactation, *Pi* the ith month of pregnancy, and *NP* the state of nonpregnant. States corresponding to death are omitted

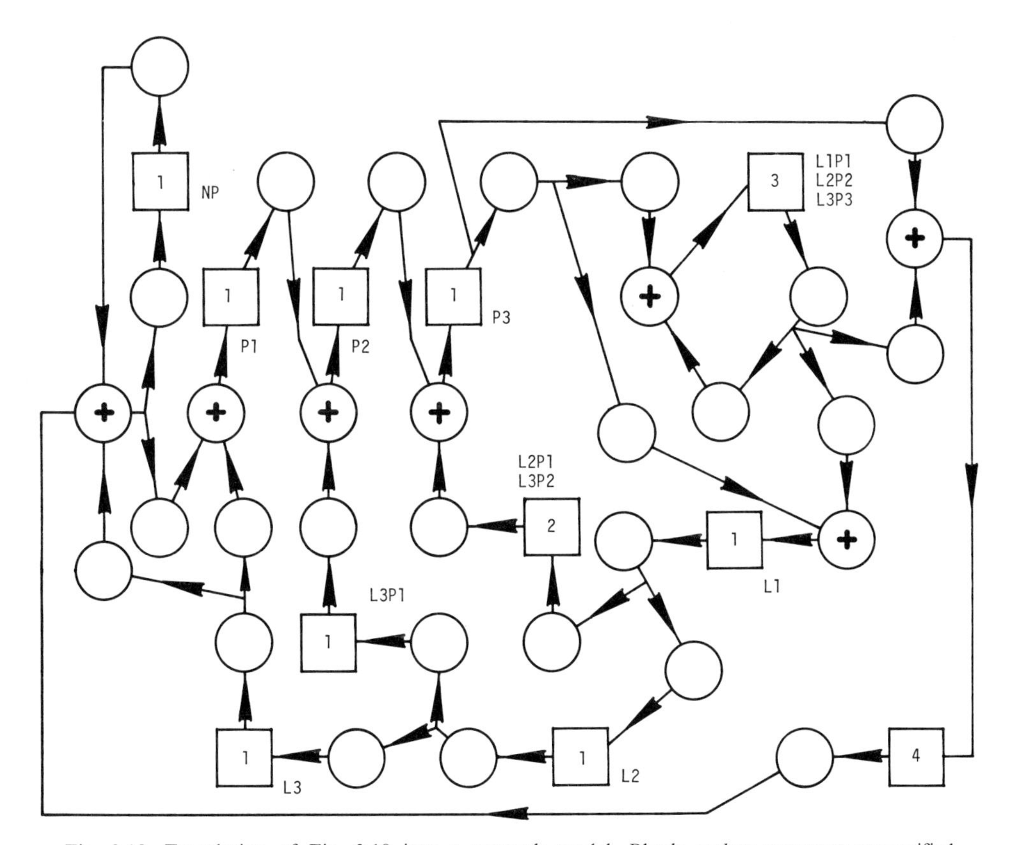

Fig. 3.19. Translation of Fig. 3.18 into a network model. Blank *scalors* represent unspecified survivorships and unspecified probabilities of impregnation. The Markovian states have been translated into *time delays* (labelled appropriately) and conditional probabilities of transition are represented by the *scalors*

the ith month of lactation; and the label NP indicates an individual that is neither pregnant nor lactating. Considering the various combinations of labels that can be applied to an individual, we can define a set of lumped Markovian states for any surviving adult female:

$$P1,\ P2,\ P3,\ L1,\ L2,\ L3,\ L1P1,\ L2P2,\ L3P3,\ L2P1,\ L3P2,\ L3P1,\ NP.$$

The connections between these states are rather limited and are depicted in the network model of Figure 3.18. The transition interval for this network is one month. In the network of Figure 3.19, the Markovian states have been lumped where possible and replaced by *time delays*, as indicated; and a natality portion has been added, with a four-month *time delay* representing maturation.

This final example illustrates a process of network construction that may be of use when overlapping time classes occur: exhaustive listing of the lumped Markovian states of the individual according to the listed specifications, followed by construction of a network of those Markovian states, then lumping into a network of the standard form, with *time delays*, *adders*, and *scalors*. Of course, the specifications may lead to Markovian states that are far to numerous to represent conveniently in this manner, in which case direct application of network modeling would be impractical. As an illustration of such a situation, consider the possibility of induced ovulation and a transition interval of one day in Example 3.15.

3.3. *Scalor* Parameters and *Multiplier* Functions

The examples in the previous section should be ample to illustrate the basic principles of combining *scalors* or *multipliers* together with *time delays*, *adders*, and branch points to construct network models for single populations. Little space has been devoted so far, however, to the specification of the parameters of the models, namely, the scale factors of the *scalors* and the durations (T's) of the *time delays*, or to the specification of the functions that might be applied to *multipliers* in the models. In this section we shall focus our attention on *scalors* and *multipliers*, and in the subsequent section we shall focus on the parameter of the *time delay*.

As they are employed in the bookkeeping skeletons of network models, *scalors* and *multipliers* serve simply to apportion flows among the branches emanating from conservative branch points or implied conservative branch points (in the case of survivorship). Thus, the parameter of the *scalor* or the function applied to a *multiplier* in this part of the network will represent the probability at the moment (i.e., in the present transition interval) that an individual passing a branch point will take a particular path. Of course, such a probability might depend upon one or more of many factors, which in turn may fall into one or more of the following categories: a) factors that depend upon time in a periodic manner (e.g., the season of the year, the phase of the lunar cycle); b) factors that depend upon time in an unpredictable or aperiodic manner (e.g., the weather); c) factors that depend directly upon some aspect of the population being modeled (e.g., the number of individuals represented as entering the branch point); d) factors that depend upon the immediate physical environment of the population (e.g., the availability of certain types of shelter or nesting sites); and e) factors that depend upon populations of other organisms in the immediate environment (e.g., the availability of food, the presence of predators).

Furthermore, the probability associated with a particular branch might depend also upon one or more unrepresented or lumped states of the individuals entering the branch point and thus vary from individual to individual. In that case, the apparent probability will be the mean taken over all individuals entering the branch point (see section 1.15) and will vary with time if the composition of those individuals varies. Finally, the probability associated with a branch might be more or less independent of all such factors and thus appear to be essentially constant.

3.3.1. Single-Species Models

Example 3.16. Periodic Variation of Parameters. Consider a population of ideal schistosomes with the following life cycle: Paired adult worms in mammalian hosts produce eggs at a constant rate. After migration through the circulatory system and tissues, a certain, fixed proportion of the eggs is excreted; a certain, fixed proportion of the excreted eggs passes into the local water system and hatches to become miracidia, a larval form of the worm. A certain, fixed proportion of the miracidia manages to find and invade snail hosts and then to progress into the sporocyst stage. During the months of November, December, January, February and March, however, the rate of maturation is very slow compared with the death rate of the snails, so no sporocysts survive. During the remaining seven months a fixed proportion of the sporocysts survives and each surviving sporocyst requires one month to mature in the snail host and then undergoes division into a fixed number of free-swimming cercariae. A certain proportion of the cercariae manages to find and invade wading or swimming mammalian hosts, at which point they progress into the schistosomule stage. After two more months, a fixed proportion of the schistosomules has survived and now becomes adults, ready to find mates and settle down to egg production. Pairing takes place at a staging center, where half of any incoming group of new adults is expected to find a mate and half is expected to be of the same sex as those already present at the center and therefore must wait for a subsequent group of new adults in order to find a mate. The probability that an adult survives from one month to the next is constant with respect to time and does not vary from individual to individual. Widows and widowers do not remate. Construct a network model embodying the dynamics of this population.

Answer: The network model of Figure 3.20 was constructed in the usual manner, beginning with a flow path for newly produced eggs, and was based on a temporal resolution of one month. According to the specifications of the population, all of the parameters but one are fixed constants. The one exception is γ_1, the probability that a sporocyst survives to undergo division. Taking the

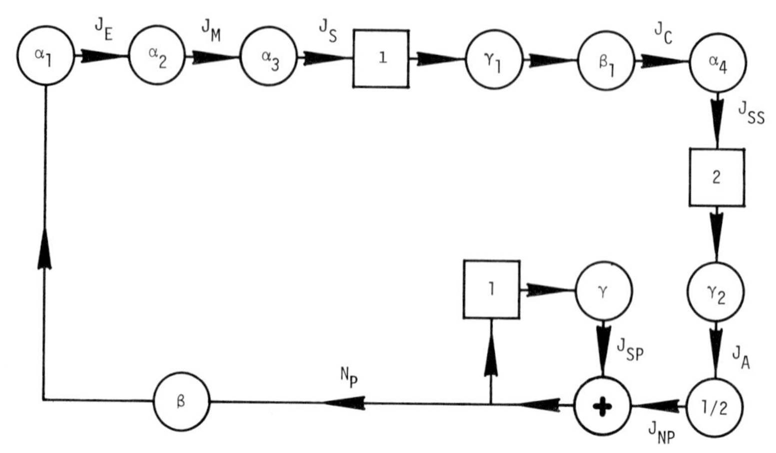

Fig. 3.20. A network model of the life cycle of idealized schistosomes (blood flukes). J_E is the flow of eggs from mated adult pairs; J_M is the flow of newly hatched miracidial larvae into the freshwater system; J_S is the flow of sporocyst larvae into the intermediate (snail) host; J_C is the flow of cercarial larvae out of the snail into the freshwater system; J_{SS} is the flow of schistosomule larvae into the primary (human) host; J_A is the flow of worms into the adult phase; J_{NP} is the flow of newly-mated pairs into the reproductive pool; J_{SP} is the flow of surviving pairs from the pool during the previous month; and N_P is a running total of the population of mated pairs in the pool

first transition interval ($\tau=0$) to correspond to April, we can describe the specified time variation of γ_1 quite compactly:

$$\begin{aligned} \gamma_1 &= K \quad \text{when } 12n \leqq \tau \leqq 6+12n \\ \gamma_1 &= 0 \quad \text{when } 7+12n \leqq \tau \leqq 11+12n \end{aligned} \tag{3.7}$$

where K is the probability that a sporocyst survives to undergo division in the seven months from April through October; n is any integer.

If one were willing to sacrifice temporal resolution to the extent that the transition interval became one year, then γ_1 simply would take on its average value over its annual cycle:

$$\gamma_1 = (7/12)K. \tag{3.8}$$

This is a commonly used gambit for eliminating periodic variations in parameters (i.e., simply failing to resolve them). Clearly, it generally requires selection of a unit of temporal resolution that is an integral number of cycles of the periodic parameter. If several periodic parameters are present, each exhibiting a different period, then the minimum transition interval required to eliminate all parameter periodicities clearly would be the least common multiple of the various periods. If those periods exhibited few common prime factors, then the required transition interval could be quite large. For the case at hand, with the new transition interval of one year, a network model is shown in Figure 3.21. The parameter $(\alpha'\gamma_1')$ of the lumped *scalor* is equal simply to the product of the scale factors across the top of the network of Figure 3.20, with γ_1 specified by

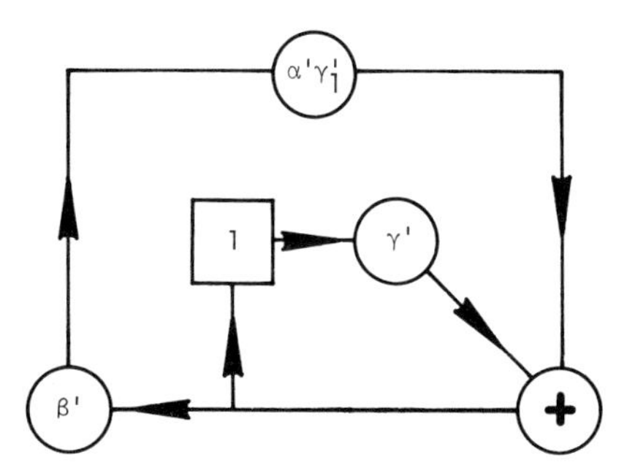

Fig. 3.21. Simplified version of the model of Fig. 3.20, with temporal resolution (transition interval) equal to one year, thus eliminating the need to keep track of annual fluctuations of parameters

equation 3.8; and the parameter γ' simply is equal to γ^{12} (representing twelve successive passes through the scalor γ). On the other hand, the parameter β' is a bit more complicated. What we must do in order to determine that parameter is estimate the number of eggs that some number of pairs alive at the beginning of a year would be expected to produce by the end of that same year (numbers of this type often are called *reproductive potentials*). For the first month, we would expect β eggs per original pair, as before. However, we would expect only γ of the original pairs to survive to produce eggs during the second month, so we would expect $\gamma\beta$ eggs per original pair for that month. Similarly, for the nth month, we would expect $\gamma^n\beta$ eggs per original pair. Thus, over the entire year, the reproductive potential of each original pair is

$$\beta' = \sum_{n=0}^{11} \gamma^n \beta = \frac{1-\gamma^{12}}{1-\gamma}\beta. \tag{3.9}$$

Example 3.17. Mass Action. Consider the population of idealized schistosomes with the following modification: instead of being the same for all worm pairs, the expected rate of egg production varies from pair to pair; but the distribution of expected rates over all pairs and the rates themselves both remain constant from interval to interval. Modify the bottom of the network of Figure 3.20 accordingly.

Answer: As illustrated in Figure 3.22a, we simply can represent egg production by a set of n parallel paths, one for each value of expected rate, β_i. The pairs are apportioned over the paths by the *scalors* p, with p_i being the proportion of pairs with expected rate β_i. Invoking the equivalences of Figures 2.40a and b, we immediately can simplify the structure of Figure 3.22a to that of Figure 3.22b, where the value of the scale factor ($\bar{\beta}$) in the single equivalent *scalor* is simply

$$\bar{\beta} = \sum_{i=1}^{n} p_i \beta_i \tag{3.10}$$

which, because the p_i (representing the distribution of rates) and the β_i are constant, is itself a constant, equal to the mean rate of egg production, taken over the entire population of pairs.

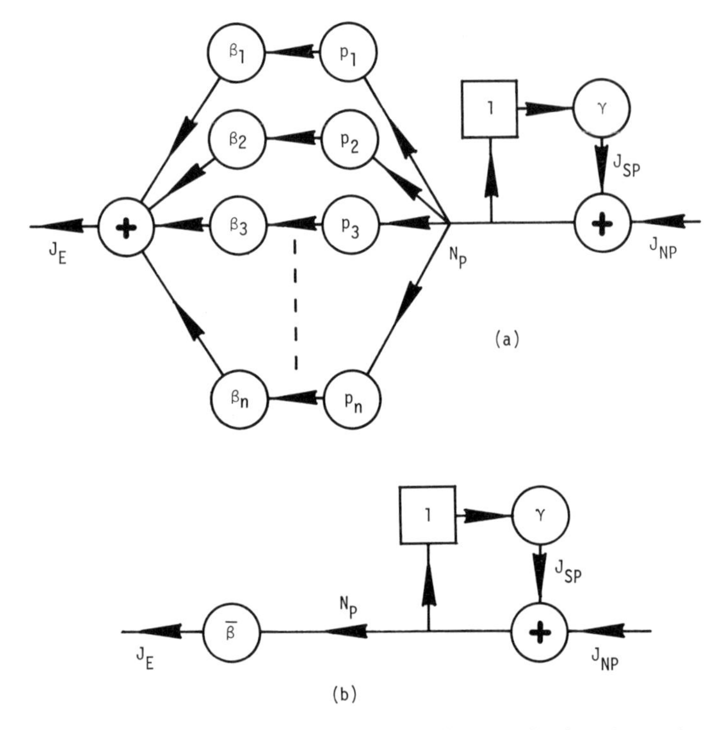

Fig. 3.22a and b. Illustration of a mass-action principle for reproduction. A certain proportion, p_i, of the mated pairs, N_P, exhibit egg-production rate β_i, as depicted in (a). The many parallel paths in (a) can be combined into the single path in (b), with the single scale factor $\bar{\beta}$, the average egg-production rate taken over all pairs

This situation is quite analogous to mass action in chemistry, whose underlying concomitant is an invariant distribution of states of a given reactant (i.e., a distribution that recovers extremely rapidly as a result of relaxation and therefore is not altered by the removal of individuals in a given range of states). In the present case, however, the constraint is not really so rigid; only the mean of the distribution must be invariant in order that the rate of egg production be related to the number of pairs by a constant coefficient:

$$J_e = \bar{\beta} N_p. \tag{3.11}$$

Example 3.18. Selective Mortality. Consider the population of idealized schistosomes with the following modification: instead of being the same for all mated adult worms, the probability of survival varies from worm to worm: but the probability for a given worm remains constant from interval to interval. Modify the bottom of the network of Figure 3.20 accordingly.

Answer: By analogy with the network for Example 3.17, we might use the structure of Figure 3.23a, with p_i being the proportion of pairs with survivorship γ_i. The survivorship of a pair is simply the product of the survivorships of each of its two members (i.e., the probability that *both* survive a transition interval).

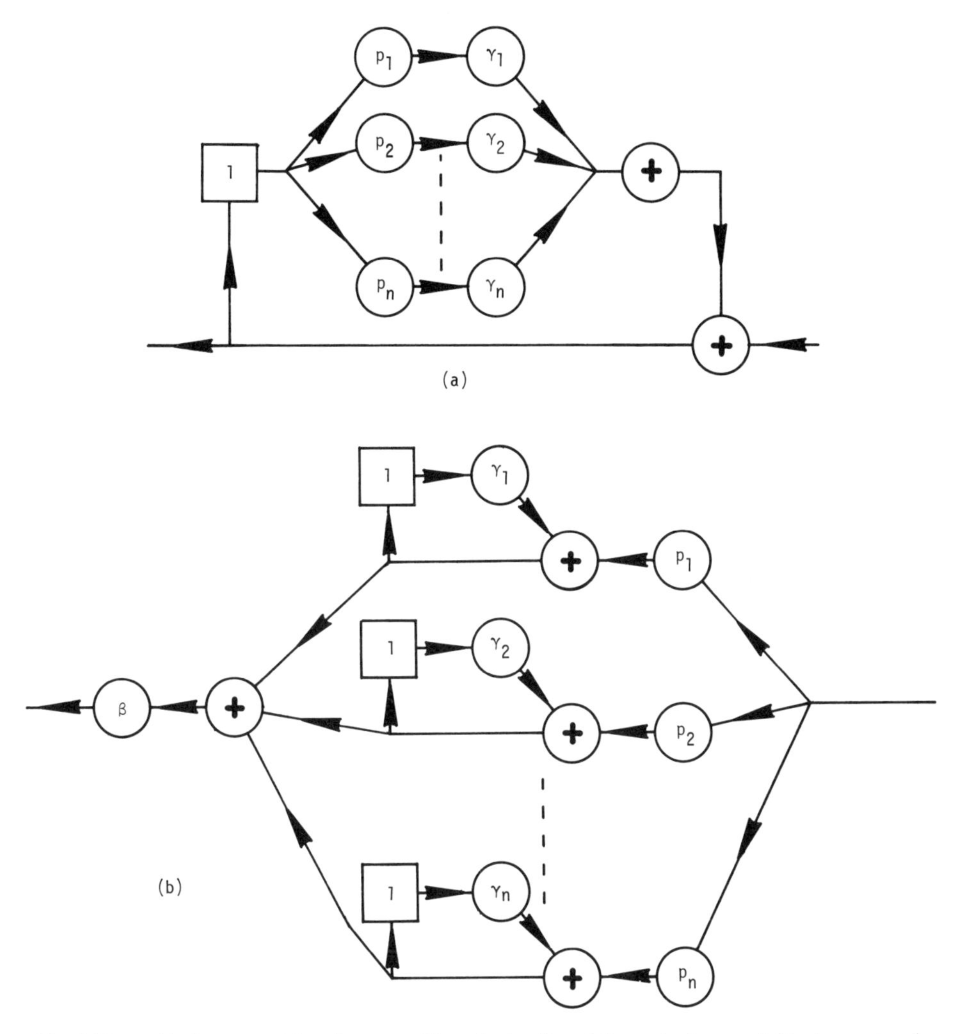

Fig. 3.23a and b. Incorrect (a) and correct (b) methods of modeling selective mortality in a network structure

However, consider what would happen to a large number (*cohort*) of new pairs entering at the beginning of a year. By the end of the year, larger proportions of those with higher survival probability will, in fact, have survived; and the values of the p_i will have been altered over the cohort. In other words, mortality will have acted selectively over the population of mated adults. Actually, the variations of the p_i are systematic and can be included rather easily as part of the dynamics represented by the model. To accomplish this, however, we must keep track of the number of pairs associated with each survivorship, γ_i, as we have done in the network of Figure 3.23b. Here, the flow of newly paired adult worms is apportioned by the p_i's into a separate accumulator for each survival probability. In this case, the p_i's will be constant if the distribution of survival probabilities does not vary from interval to interval for the incoming cohorts of newly paired worms. Of course, if the propensity toward longevity is heritable, the selective nature of mortality will be reflected in variations in the p_i's of Figure 3.23b. In other words, as preferential accumulation of longer-lived adults progresses, more and more of the offspring will tend toward longer lives.

Example 3.19. Heritable Selection. a) Consider the population of idealized schistosomes with the following modification: Instead of being the same for all female worms, the expected rate of egg production varies from female to female, but remains constant from interval to interval for a given female; the offspring of a given female will have the same egg-production rate as their mother. Modify the network of Figure 3.20 accordingly.

Answer: Because, in this idealization, we have ignored totally the genetic mixing that should occur in mating, the population can be considered to consist of an independent subpopulation for each value of egg-production rate. Thus, an appropriate modification would comprise n independent networks of the form of that in Figure 3.20, where n is the number of different egg-production rates.

b) Consider the population of idealized schistosomes with the following modification: Egg-production rate by the female worm is controlled by a single locus with two alleles, the three resulting genotypes each corresponding to a different, fixed rate; mates and gametes are not selected on the basis of these genotypes (i.e., the population is *panmictic* with respect to this locus): the sex ratio is 1:1 and mortality and the genotype distributions (gene frequencies) are independent of sex. Construct an appropriately modified version of the network of Figure 3.20.

Answer: Suddenly, in spite of the rather simple description, we have a much more complicated situation. There are nine types of adult pairs producing three types of offspring, which, in turn, combine to form newcomers to each of the various pair types. The stipulated variation in egg-production rate and marked overlap of generations require that we account separately for the populations of each of these types. Let us begin by considering the flow of newly produced eggs of both sexes of a particular homozygous genotype (A_1A_1). These individuals flow through the miracidium, sporocyst, cercaria and schistosomule stages as represented by the top portion of the network of Figure 3.20. Of those emerging from the schistosomule stage as adults, half mate immediately and the other half join a pool of bachelor worms waiting for future inputs of new, unmated adults. This pool of bachelor worms comprises the accumulations of all three genotypes of interest (A_1A_1, A_2A_2 and A_1A_2), which must be followed separately. This is done in Figure 3.24 by means of a separate *accumulator* for each of the three genotypes. Of the half of the new adults mating immediately, half are male and half are female, so the flow divides again. The flow of females is apportioned among three pair types: A_1A_1/A_1A_1, A_1A_1/A_2A_2, and A_1A_1/A_2A_1 (where the notation is given as female-genotype/male-genotype); and the scale factors are the accumulated gene frequencies among the bachelors (including the most recent influx of new bachelors). The flow of males also is apportioned among

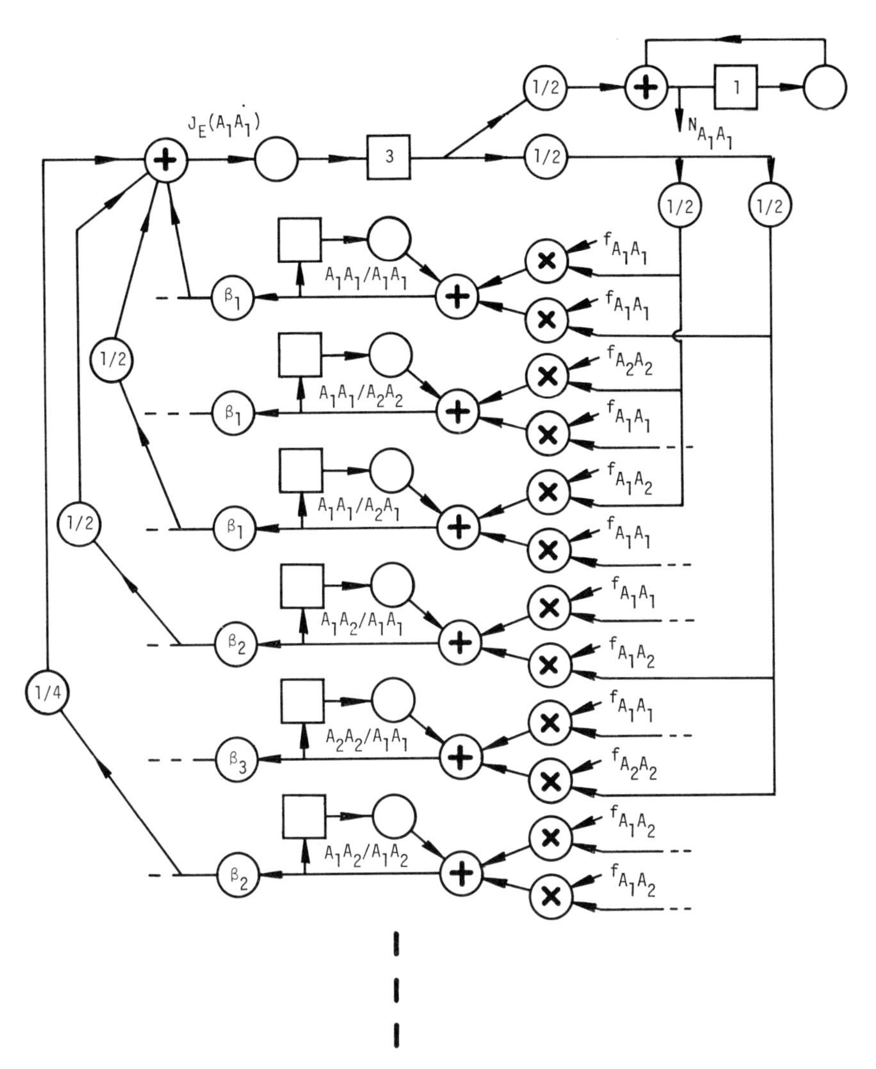

Fig. 3.24. Part of a network model incorporating heritable selection at one locus with two alleles (A_1 and A_2). Panmixia is assumed, leading to the scale factors on the far left; but egg-production rate (β) is assumed to depend upon the genotype of the female with respect to the locus under consideration. $N_{A_1A_1}$ is the number of unmated adults of genotype A_1A_1; $f_{A_iA_j}$ is the frequency of the A_iA_j genotype in the pool of unmated adults; and $J_E(A_1A_1)$ is the flow of fertilized eggs (zygotes) of genotype A_1A_1

three pair types. The egg production rate of a pair type is determined by the female, and is denoted in the network as β_1 for type $A_1A_1/--$, β_2 for type $A_1A_2/--$, and β_3 for type $A_2A_2/--$. All of the eggs produced by pair types A_1A_1/A_1A_1 are of genotype A_1A_1; half of those produced by pair types A_1A_1/A_1A_2 and A_1A_2/A_1A_1 and one-quarter of those produced by types A_1A_2/A_1A_2 are of genotype A_1A_1. Thus, four pair-type pools contribute to the flow of type A_1A_1 eggs, as depicted in the network of Figure 3.24. Completion

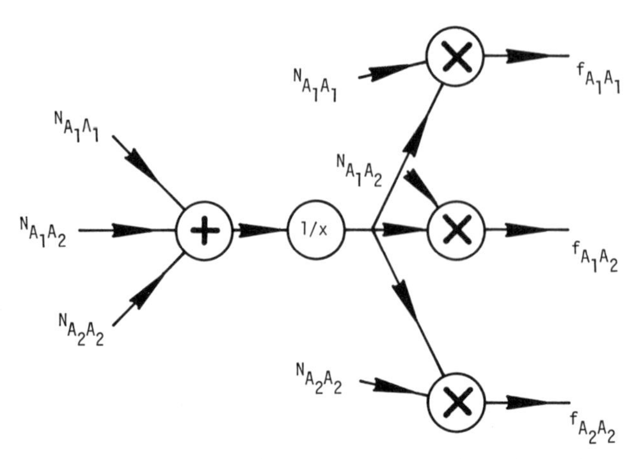

Fig. 3.25. Network to generate genotype frequencies from total numbers of unmated adults of each genotype

of the bookkeeping and natality portions of the network is left as an exercise for the reader. Clearly, one can simplify the network slightly by combining the flows of mating males and females into a single path. The genotype frequencies to be applied to the multipliers can be generated by a simple function-generating portion, as illustrated in Figure 3.25:

$$f_{A_iA_j} = N_{A_iA_j}/(N_{A_1A_1} + N_{A_1A_2} + N_{A_2A_2}). \tag{3.12}$$

Example 3.20. Random Variation of Parameters. a) Consider the population of idealized schistosomes with the following modification: During the seven months from April through October, the probability that a schistosome egg passes into the local water system is a function of the recent history of rainfall and therefore is an unpredictable parameter. Modify the network of Figure 3.20 appropriately.

Answer: In the network of Figure 3.20, the *scalor* α_2 represents the probability in question. We simply may define α_2 to be a function of time, assigning for each interval a value chosen randomly from a set of values with appropriate, specified distribution. The deductions for those particular values then could be made. Since the deductions are *ad hoc*, applying specifically to the assigned values, the process would have to be repeated several times, with new assigned values each time, in order to lead to general conclusions about the dynamics. This procedure often is called a *Monte Carlo* analysis. In this particular network, a simple alternative may be open to us. Since all of the other parameters in the model are fixed constants, we may, under certain circumstances, replace the random parameter α_2 with its expected value during each interval, in which case our analysis will lead us directly to the expected values of the population dynamics. The required circumstances are independence of the value of α from one interval to the next (i.e., lack of correlation of the interval to interval rainfall).

b) Consider the population of idealized schistosomes of Example 3.20a with the following modification: In addition to unpredictable passage of eggs into the local water system, the probability that an individual miracidium finds a snail host is a function of the local water temperature, and therefore is an unpredictable parameter. Modify the network of Figure 3.20 appropriately.

Answer: Now we must deal both with α_2 and with α_3 in Figure 3.20. If we decide to use the Monte Carlo method, we must insure that any specified correlation between rainfall and air temperature (as reflected in correlation between α_2 and α_3) is included as we assign values to the two random parameters. If we wish to take advantage of the fact that the remaining parameters of the model are fixed constants, and deduce directly the expected dynamics, then we can replace the two *scalors*, α_2 and α_3, by a single *scalor* whose parameter is the interval by interval expected value of the product $\alpha_2\alpha_3$:

$$\mathrm{Ex}[\alpha_2\alpha_3] = \mathrm{Ex}[\alpha_2]\,\mathrm{Ex}[\alpha_3] + \mathrm{Cov}(\alpha_2, \alpha_3) \tag{3.13}$$

where Cov(–,–) is the covariance of the two random parameters. In other words, if the two parameters are correlated, we may not simply replace each of them by its expected values.

Example 3.21. Macdonald's Random Pairing Model (Monogamous Mating). Consider the population of idealized schistosomes of Example 3.16, with the following modification: Newly maturing adults pair only with members of their own cohort (i.e., with other newly maturing adults); pairing possibilities in every cohort are exhausted (i.e., in a given host, only worms of one sex remain unmated); the worm sex ratio is $\mathbb{1}$; and every host is equally likely to be infected. Construct an appropriately modified network model.

Answer: We now come face to face with perhaps the quintessential aspect of many intraspecific and interspecific interactions, namely the problem of search and encounter. In this case, we have postulated an adult organism that is strictly territorial, being absolutely confined to a given host, and strictly monogamous, taking at most one mate for life. What we seek is the probability that a given member of a cohort takes a mate; this probability will replace the factor $\mathbb{1}/\mathbb{2}$ in the *scalor* leading to J_{NP} in the network of Figure 3.20. Although it is not represented explicitly as such in the model of Figure 3.20, the flow J_A of new adult worms actually is divided among the mammalian hosts. According to our new specifications, in fact, it is apportioned equally among those hosts. As a first step in our calculations, we can consider a fixed number N_a of new adults in a given host and determine the proportion of those adults that we expect to find paired. Because the sex ratio is one, we can consider each adult to be a trial (very much analogous to the toss of a coin), with a probability $\mathbb{1}/\mathbb{2}$ of being male and a probability $\mathbb{1}/\mathbb{2}$ of being female. Assuming that the sexes of the individuals making up N_a are independently determined (an assumption that should be reasonably valid as long as an individual mammalian host has not been infected by a clone of cercariae from a single sporocyst), we have a set of N_a *Bernoulli trials* (N_a independent trials each with two possible outcomes and the same outcome probabilities). The probability that the outcome of every trial is a female is simply $(1/2)^{N_a}$, as is the probability of all males. Therefore, the probability that out of N_a new adults we find no pairs at all simply is

$$\Pr[\text{no pairs}] = \Pr[\text{all males or all females}] = 2(1/2)^{N_a}. \tag{3.14}$$

The probability of finding exactly one pair simply is equal to the probability of finding exactly one male or one female, with all of the rest of the worms being of the opposite sex. The probability that a particular adult is male and all the rest are female again is simply $(1/2)^{N_a}$ (in fact, that is the probability of any situation in which the outcome of each of these N_a trials is specified individually). However, there are N_a choices of the particular adult that will be male; therefore, the probability of exactly one male is $N_a(1/2)^{N_a}$, and

$$\Pr[\text{exactly one pair}] = \Pr[\text{exactly one male or one female}] = 2\,N_a(1/2)^{N_a}. \tag{3.15}$$

Generalizing to exactly n males or n females, we have

$$\Pr[\text{exactly } n \text{ pairs}] = \Pr[\text{exactly } n \text{ males or } n \text{ females}] = 2\binom{N_a}{n}(1/2)^{N_a} \tag{3.16}$$

where $\binom{N_a}{n}$ is the number of ways to have exactly n successes out of N_a trials (see section 1.14) and is given by the following expression:

$$\binom{N_a}{n} = N_a!/(N_a - n)!\,n!. \tag{3.17}$$

If N_a is an even number, then we shall have considered all possible cases when we have taken every value of n from 0 to $N_a/2$. However, the case in which n equals $N_a/2$ is special in that it applies to $N_a/2$ males and $N_a/2$ females at the same time, so that

$$\Pr\begin{bmatrix}\text{exactly } N_a/2\\ \text{pairs}\end{bmatrix} = \Pr\begin{bmatrix}\text{exactly } N_a/2 \text{ males}\\ \textit{and } N_a/2 \text{ females}\end{bmatrix} = \binom{N_a}{N_a/2}(1/2)^{N_a}. \tag{3.18}$$

Therefore, for N_a even, the mean or expected number of pairs simply is

$$\mathrm{Ex}\begin{bmatrix}\text{Number of pairs}\\ \text{in } N_a \text{ (even) adults}\end{bmatrix} = \left[(N_a/2)\binom{N_a}{N_a/2} + 2\sum_{n=0}^{(N_a/2)-1} n\binom{N_a}{n}\right](1/2)^{N_a}. \tag{3.19}$$

If N_a is odd, then we shall have considered all possible cases when we have taken every value of n from 0 to $(N_a - 1)/2$; and the situation with equal numbers of males and females does not arise.

$$\mathrm{Ex}\begin{bmatrix}\text{Number of pairs}\\ \text{in } N_a \text{ (odd) adults}\end{bmatrix} = \left[2\sum_{n=0}^{(N_a-1)/2} n\binom{N_a}{n}\right](1/2)^{N_a}. \tag{3.20}$$

Since two worms are involved in each pair and there are N_a total worms, we now can find the proportion of worms that are mated (i.e., the probability that a

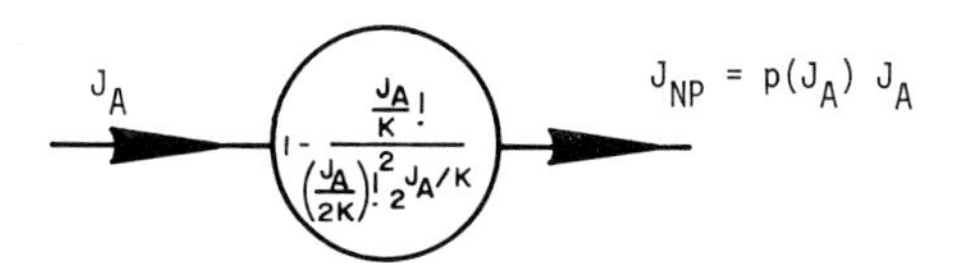

Fig. 3.26. Nonlinear scalor representing the relationship between the flow of newly-mature adults (J_A) and the flow of newly-mated pairs (J_N) according to the model of Macdonald

particular worm out of the N_a is mated):

$$p = \Pr\begin{bmatrix}\text{particular worm out} \\ \text{of } N_a \text{ adults is mated}\end{bmatrix} = \frac{2}{N_a}\,\mathrm{Ex}\begin{bmatrix}\text{number of pairs} \\ \text{in } N_a \text{ adults}\end{bmatrix}. \tag{3.21}$$

This, of course, is the probability that we seek. Macdonald (1965) combined equations 3.19 and 3.20 with equation 3.21 and then simplified the result to obtain the following expression for p when N_a is even:

$$p = 1 - N_a!/\{[(N_a/2)!]^2\, 2^{N_a}\}. \tag{3.22}$$

The value of p for odd N_a turns out to be precisely the same as that for $N_a - 1$ (i.e., the next lower even value of N_a). Now, if there are K mammalian hosts among whom the J_A newly mature adult worms are divided, then we can set N_a equal to J_A/K and use the resulting right-hand side of equation 3.22 as the scale factor we have sought. The result is a rather extreme example of a scale factor dependent on the input to the *scalor*, and is illustrated in Figure 3.26. Values of the scale factor, calculated by Macdonald, are given in Table 3.1. We can draw two rather obvious conclusions about our idealized population from these results. First of all, the probability that a given member of a cohort finds a mate increases monotonically as the size of the cohort increases. With a little reflection, the reader should be able to convince himself that this same conclusion should be applicable generally to organisms with monogamous mating. Second, the modified scalor now carries the factor K, which is the number of host organisms present. Thus, implicitly, we have represented the interaction of two populations, that of the parasite and that of the host. This representation will be made more explicit in a subsequent example.

So far, we have made the rather optimistic assumption that the flow, J_A, of new adults is apportioned into precisely equal shares among the K hosts that are equally likely to be infected. Actually, even within the idealized specifications of this example, such an assumption is justified only if the number J_A/K is very large, so that the law of large numbers applies (see section 1.20) and

$$J_A/K \cong \mathrm{Ex}[J_A/K]. \tag{3.23}$$

In that case, we may substitute the expected value of J_A/K for N_a in equation 3.22 and thereby obtain a very good approximation to the expected value of p. On the other hand, if J_A/K is not large, then substitution of its expected value into equation 3.22 may not provide an especially good approximation to the expected value of p. More generally, if x is a random variable, and $f(x)$ a function of that variable, the relationship

$$\mathrm{Ex}[f(x)] \cong f(\mathrm{Ex}[x]) \tag{3.24}$$

is a good approximation if the law of large numbers applies to x and if $f(x)$ does not exhibit disproportionately large jumps (section 1.20), but may be a bad approximation otherwise. This was the core of the difficulty discussed in section 1.19.

Since every host is equally likely to be infected by a particular worm, the probability that a particular worm infects a particular host is $1/K$, the reciprocal of the number of hosts. Therefore, the J_A worms in a cohort represent J_A Bernoulli trials, with probability of success (infection by a particular worm) equal to $1/K$ and probability of failure (noninfection by a particular worm) equal to $1-1/K$. The probability that exactly n of the J_A worms infect a given host simply is

$$\Pr\begin{bmatrix} n \text{ members of cohort} \\ \text{of size } J_A \text{ infect} \\ \text{particular host} \end{bmatrix} = \binom{J_A}{n} (1/K)^n (1-1/K)^{J_A-n} \tag{3.25}$$

which follows from the discussion related to equation 3.16 but is derived explicitly in Appendix B. Now, the generally correct form of equation 3.24 is

$$\mathrm{Ex}[f(x)] = \sum_j f(j) \Pr[x=j] \tag{3.26}$$

where the summation is carried out over all of the values (j) that the random variable x may take; and $\Pr[x=j]$ has its usual meaning, i.e., the probability that x takes on the particular value of j. From equation 3.22, we have the function whose expected value we wish to determine: for even values of x

$$f(x) = f(x+1) = 1 - \frac{x!}{\left(\frac{x}{2}!\right)^2 2^x} \tag{3.27}$$

and from equation 3.25 we have $\Pr[x=n]$, where x (the random variable of concern) here is the number of worms from a given cohort infecting a particular individual. Substituting into equation 3.26, we would obtain a rather complicated expression for the expected value of our scale factor. Once again, however, Macdonald has carried out the numerical calculations, the results of which are presented in Table 3.2. In making these calculations, he assumed that J_A and K both were fairly large, in which case equation 3.25 can be approximated quite well by

$$\Pr[x=n] = \frac{m^n}{n!} e^{-m} \tag{3.28}$$

where m is the mean number of new adult worms per host (i.e., J_A/K).

Clearly, for small values of J_A/K, there is a vast difference between the expected value of p as given in Table 3.2 and the value derived under the assumption of

Table 3.1. Probability (p) of pairing by a particular adult among J_A/K adults entering a particular host as a cohort

J_A/K	p	J_A/K	p
0 or 1	0	18 or 19	0.816
2 or 3	0.5	20 or 21	0.825
4 or 5	0.625	30 or 31	0.855
6 or 7	0.687	40 or 41	0.875
8 or 9	0.726	60 or 61	0.897
10 or 11	0.754	80 or 81	0.911
12 or 13	0.774	100 or 101	0.921
14 or 15	0.790	400 or 401	0.960
16 or 17	0.804	1000 or 1001	0.975

Table 3.2. Expected proportion ($\bar{p}$) of adults finding mates when J_A new adults infect K hosts, each equally susceptible to infection

J_A/K	$\bar{p}$	J_A/K	$\bar{p}$
0.1	0.048	1.6	0.428
0.2	0.091	1.8	0.453
0.3	0.130	2.0	0.476
0.4	0.166	3.0	0.560
0.5	0.199	4.0	0.614
0.6	0.229	5.0	0.652
0.7	0.256	6.0	0.681
0.8	0.281	7.0	0.704
0.9	0.305	8.0	0.722
1.0	0.326	9.0	0.738
1.2	0.365	10.0	0.751
1.4	0.398		

absolutely equal partitioning of the worms (given in Table 3.1). For example, if each host had precisely one new adult worm, no new pairs would be formed; whereas when the same number of new adult worms (one per host) is distributed by Bernoulli trials over the hosts, nearly one-third of them are expected to find mates. On the other hand, by the time the number of new adults has reached ten per host, the probability in Table 3.1 and the expected probability in Table 3.2 are very close to one another.

3.3.2. Two-Species Models

Example 3.22. The Nicholson-Bailey (Uniform Gantlet) Model of Search and Encounter. Consider once again the population of idealized schistosomes, this time with the following modifications: A miracidium always penetrates the first snail that it encounters; all miracidia are equally good searchers and all snails, whether infected or not, are equally likely targets for any particular searching miracidium; the number of snails varies from interval to interval, and the lifetime of a miracidium is short compared to a transition interval. Find an appropriate modification of the network of Figure 3.20.

Answer: Here, we have a classic, idealized search and encounter situation with equally effective searchers and equally likely targets. In this particular case, the number of targets (snails) may vary with time, but it is not affected by the number of successful encounters (i.e., a penetrated snail remains a perfectly good target). One can picture this situation as follows: The targets form a gantlet through which each searcher must pass; only a fraction of the searchers pass through the gantlet without running into a target and thus being absorbed. Let $N_s(\tau)$ be the number of snails in interval τ, and let q_{sm} be the probability that a particular miracidium can search for its entire lifetime and fail to encounter a particular snail. Volume or area models often are employed to estimate such probabilities. Thus, one might consider the snail as occupying an essentially infinitesimal fraction of the total water-system volume, and each miracidium (presumably capable of sensing targets remotely by means of chemoreceptors) as sweeping out a small, but finite effective fraction f of the volume in its search. q_{sm} in that case would be the probability $(\mathbb{1}-f)$ that a particular snail lay outside the potential search volume of a particular miracidium. Such models become a bit more complicated when the search volumes and the target volumes both are finite or if searcher or target distribution is not uniform. In the present case, however, and in many similar cases, we can bypass the volume or area models and simply assume that a probability, q_{sm}, exists, which is the same for all snails and all miracidia and, for a particular miracidium, does not depend upon the number of snails already missed (i.e., the snails can be considered independent trials for a given miracidium). Under that assumption, we immediately can deduce the probability that a particular miracidium misses all $N_s(\tau)$ snails during its lifetime:

$$\Pr\begin{bmatrix}\text{a particular} \\ \text{miracidium fails} \\ \text{to find any snail}\end{bmatrix} = q_{sm}^{N_s(\tau)}. \tag{3.29}$$

The only alternative to this outcome is finding and penetrating a snail. The probability of that alternative is simply

$$p_m(\tau) = \Pr\begin{bmatrix}\text{a particular} \\ \text{miracidium penetrates} \\ \text{a snail}\end{bmatrix} = \mathbb{1} - (q_{sm})^{N_s(\tau)}. \tag{3.30}$$

Noting that

$$q_{sm}^{N_s(\tau)} = \exp[(\log_e q_{sm})\, N_s(\tau)] \tag{3.31}$$

and that because q_{sm} always is less than one, $\log_e q_{sm}$ always is negative and can be expressed

$$\log_e q_{sm} = -|\log_e q_{sm}| = -\alpha \tag{3.32}$$

we can restate equation 3.30 in another common and convenient form

$$p_m(\tau) = \mathbb{1} - \exp[-\alpha N_s(\tau)]. \tag{3.33}$$

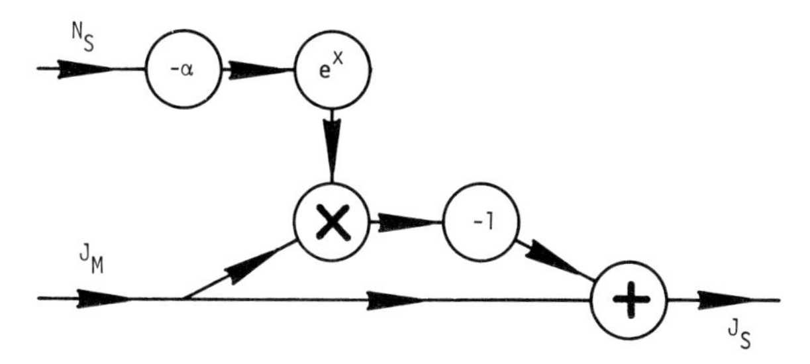

Fig. 3.27. Network representation of the Nicholson-Bailey model of search and encounter. J_M is the flow of newly-hatched miracidia; N_S is the number of available snail hosts; and J_S is the flow of larvae into the sporocyst phase

Equation 3.30 and its alternative, equation 3.33, often are referred to in population modeling as the Nicholson-Bailey model. They arise in many other areas, however, and might be referred to more generally as a uniform gantlet model.

Our present task will be completed when we replace the scale factor α_3 in Figure 3.20 with the factor $\mathbb{1}-\exp[-\alpha N_s(\tau)]$. This can be done by generating $N_s(\tau)$ with a network model for the snail population, then coupling it through a function-generating network structure into the network of Figure 3.20, as depicted in Figure 3.27. In this manner, we have explicit representation of the dynamics of one population influencing directly the dynamics of a second. In principle, the postulated gantlet model of encounters can be tested for validity; and, if it proves to be reasonably valid, the coefficient α can be estimated directly, without reversion to volume models.

Example 3.23. Simple Saturation; Another Gantlet Model. Consider the idealized schistosome population of the previous example, but with the following additional modifications: The number of snails available in any given interval always is relatively small, so that the vast majority of miracidia fail to encounter any snails at all; once a snail has been penetrated by a single miracidium, it becomes impenetrable by other miracidia; the number of miracidia emerging in a given interval is very very large compared with the number of snails available during that interval. Construct an appropriate modification of the network structure of Figure 3.27.

Answer: There are several ways to apply the gantlet model here. First of all, we might consider the snails as the targets and the miracidia as searchers running the gantlet of snails; in which case the number of targets is diminished by one for each miracidium that runs into a snail. When the number of targets diminishes in this manner, the formulation of the probability of encounter becomes complicated. Therefore, we should attempt to state the situation (or any other, similar situation) in a manner in which the number of targets is not diminished through the process of encounter (i.e., in a manner in which the targets are not used up). Since the number of miracidia is taken to be very large compared to the number of snails, successful penetrations will have relatively little effect on the number of miracidia. If we assume that that number of miracidia is essentially constant, then we can arrive at a simple formulation simply by reversing the roles in the gantlet model. First of all, it is left as a simple exercise for the reader to show that the probability that a particular snail fails to run into a particular miracidium is very nearly equal to q_{sm} under the conditions of this example. Let the miracidia be the targets, through which the snails must run, and let the probability that a given target is missed be independent of the targets already missed (i.e., the miracidia can be considered independent trials for a given snail). Using the notation of Example 3.22, we can write

$$\Pr\begin{bmatrix}\text{a particular snail}\\ \text{fails to run into}\\ \text{any miracidia in}\\ \text{interval } \tau\end{bmatrix} = [q_{sm}]^{(J_m(\tau))}. \tag{3.34}$$

From which

$$p_s(\tau) = \Pr\begin{bmatrix}\text{a particular snail}\\ \text{is penetrated by}\\ \text{a miracidium in}\\ \text{interval } \tau\end{bmatrix} = 1 - [q_{sm}^{J_m(\tau)}] \tag{3.35}$$

or

$$p_s(\tau) = 1 - \exp[-\alpha J_m(\tau)] \tag{3.36}$$

which again, in principle, can be tested for validity by experiment (with α being estimated through the same experiments). From equation 3.36, we can estimate the expected number of newly penetrated snails in interval τ,

$$\mathrm{Ex}\begin{bmatrix}\text{number of}\\ \text{newly penetrated}\\ \text{snails}\end{bmatrix} = N_s(\tau)p_s(\tau) = N_s(\tau)(1 - \exp[-\alpha J_m(\tau)]) \tag{3.37}$$

and from this, the expected probability that a given miracidium will penetrate a snail:

$$p_m(\tau) = \Pr\begin{bmatrix}\text{a particular}\\ \text{miracidium}\\ \text{penetrates a}\\ \text{snail}\end{bmatrix} = N_s(\tau)(1 - \exp[-\alpha J_m(\tau)])/J_m(\tau) \tag{3.38}$$

which simply is equal to the expected number of snails penetrated per miracidium. An appropriate modification of Figure 3.27 is shown in Figure 3.28.

Example 3.24. Double Saturation. Consider the previous example, but with a relatively small number of miracidia emerging in any given interval. Find the appropriate scale factor to replace α_3 in Figure 3.20.

Answer: In this case, we can consider the gantlet model on an encounter-by-encounter basis. After each encounter, the number of targets and the number of searchers both will be diminished by one. We can proceed as follows: From

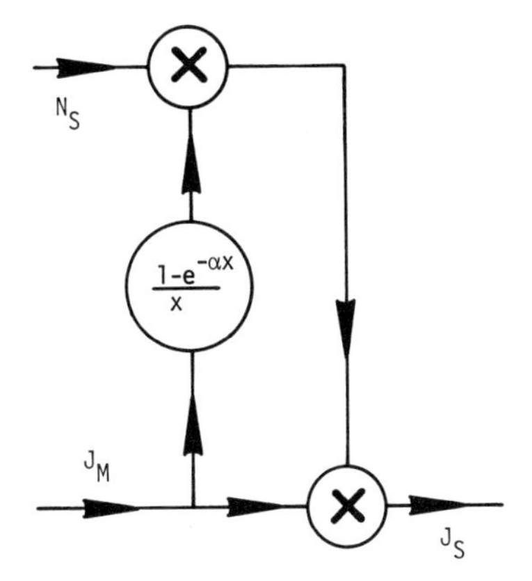

Fig. 3.28. Modification of Fig. 3.27 to incorporate simple saturation of the snail host

equation 3.34,

$$\Pr\begin{bmatrix}\text{a particular snail}\\ \text{fails to run into}\\ \text{any miracidia}\\ \text{at } \tau\end{bmatrix} = [q_{sm}]^{[J_m(\tau)]},$$

which leads to

$$\Pr\begin{bmatrix}\text{none of the}\\ \text{snails run into}\\ \text{any miracidia}\\ \text{at } \tau\end{bmatrix} = [q_{sm}]^{[J_m(\tau)\,N_s(\tau)]}, \tag{3.39}$$

and

$$\Pr\begin{bmatrix}\text{at least one}\\ \text{snail runs into}\\ \text{a miracidium}\\ \text{at } \tau\end{bmatrix} = \mathbb{1} - [q_{sm}]^{[J_m(\tau)\,N_s(\tau)]}. \tag{3.40}$$

After one such encounter, the number of targets and the number of searchers both are reduced, leading to

$$\Pr\begin{bmatrix}\text{at least one more}\\ \text{encounter at } \tau\text{, given}\\ \text{that one already}\\ \text{has occurred}\\ \text{at } \tau\end{bmatrix} = \mathbb{1} - [q_{sm}]^{[J_m(\tau)-\mathbb{1}]\,[N_s(\tau)-\mathbb{1}]} \tag{3.41}$$

which easily is generalized to

$$\Pr\begin{bmatrix}\text{at least one more}\\ \text{encounter at } \tau\text{, given}\\ \text{that } k \text{ already}\\ \text{have occurred at } \tau\end{bmatrix} = \mathbb{1} - [q_{sm}]^{[J_m(\tau)-k]\,[N_s(\tau)-k]}. \tag{3.42}$$

The probability that exactly k encounters take place simply is

$$\Pr\begin{bmatrix}\text{exactly } k\\ \text{encounters}\\ \text{at } \tau\end{bmatrix}=\Pr\begin{bmatrix}k \text{ encounters}\\ \text{have}\\ \text{occurred at } \tau\end{bmatrix}\Pr\begin{bmatrix}\text{no more}\\ \text{encounters}\\ \text{occur at } \tau\end{bmatrix}=p(k) \tag{3.43}$$

$$p(k)=[1-q_{sm}^{J_m N_s}][1-q_{sm}^{(J_m-1)(N_s-1)}]\ldots[1-q_{sm}^{(J_m-k+1)(N_s-k+1)}]\,q_{sm}^{(J_m-k)(N_s-k)} \tag{3.44a}$$

which can be abbreviated as follows

$$p(k)=q_{sm}^{(J_m-k)(N_s-k)}\prod_{j=0}^{k-1}(1-q_{sm}^{(J_m-j)(N_s-j)}). \tag{3.44b}$$

The expected number (N_{enc}) of encounters at interval τ simply is

$$\mathrm{Ex}[N_{enc}]=\sum_{k=1}^{\mathrm{Min}(N_s,\,J_m)} k\,p(k) \tag{3.45}$$

where the upper limit of summation, $\mathrm{Min}(N_s, J_m)$ is the smaller of the number of snails and the number of miricidia. For convenience we have left out explicit representations of the time variable in equations 3.44 and 3.45; nonetheless, J_m and N_s remain variable with respect to time. The factor α_3 in Figure 3.20 should be replaced by the expected number of encounters per emerging miracidium:

$$\alpha_3=\mathrm{Ex}[N_{enc}]/J_m. \tag{3.46}$$

Clearly, the resulting expression (when equation 3.44 is combined with equation 3.45 to form $\mathrm{Ex}[N_{enc}]$ which then is inserted into equation 3.46) is quite complicated.

Example 3.25. Single Saturation at Higher Levels. Consider the idealized schistosome population of Example 3.23, but with the following additional modifications: rather than becoming impenetrable after being penetrated by a single miracidium, each snail may be penetrated by k miracidia (where k times the number of snails still is very small compared to the number of miracidia) and then it becomes impenetrable. Find an appropriate scale factor to replace α_3 in Figure 3.20.

Answer: Once again, we can consider the snails as the runners of the gantlet and the miracidia as an essentially undiminishable supply of targets. Let $p(j)$ be the probability that a snail encounters j miracidia. The expected number of miracidia penetrating a particular snail is

$$\mathrm{Ex}[N'_{enc}]=\sum_{j=1}^{k} j\,p(j). \tag{3.47}$$

The expected number of penetrations of all snails is

$$\mathrm{Ex}[N_{enc}]=N_s\,\mathrm{Ex}[N'_{enc}] \tag{3.48}$$

so the expected proportion of miracidia penetrating snails (i.e., the expected total number of penetrations per miracidium) is

$$\alpha_3 = \frac{N_s}{J_m} \sum_{j=1}^{k} j p(j). \tag{3.49}$$

All that remains is to determine $p(j)$. Since the number of miracidia is very large compared to the number that may penetrate snails, the probability that a particular miracidium fails to penetrate any snails is very close to one. Therefore the probability (p_{sm}) that a particular miracidium penetrates a particular snail must be very nearly equal to $\mathbb{1} - q_{sm}$, the complement of the probability that it could search for its lifetime and fail to encounter that snail:

$$p_{sm} \approx \mathbb{1} - q_{sm}. \tag{3.50}$$

Thus, for each snail, we have J_m nearly Bernoulli trials (one for each miracidium) with probability p_{sm} of one outcome (penetration) and probability q_{sm} of the alternative outcome (no penetration). The corresponding probability of j penetrations, $p(j)$, simply is

$$p(j) = \binom{J_m}{j} p_{sm}^{j} q_{sm}^{J_m - j} \tag{3.51}$$

where $p_{sm}^{j} q_{sm}^{J_m - j}$ is the probability of a particular combination of exactly j penetrations and $J_m - j$ (all of the rest of the trials) failures; and $\binom{J_m}{j}$ is the number of such combinations.

Example 3.26. Simple Saturation with Delayed Recovery (Discrete-Time Holling Model). Consider the idealized schistosome population of Example 3.23 with the following modifications: Once a snail has been penetrated by a single miracidium, it becomes impenetrable to other miracidia; however, after T transition intervals it has shed all of its cercariae and has recovered its penetrability. Construct an appropriate network to replace the scalor α_3 in the network of Figure 3.20.

Answer: We can begin by constructing a portion of a network model for the snail population, as shown in Figure 3.29. Here we have the available (penetrable) snails, N_{sa}, cycling round and round an accumulator. On each pass (corresponding to each transition interval) a fraction ($\exp[-\alpha J_m]$) manage to run the gantlet of miracidia without being penetrated, and the rest ($\mathbb{1} - \exp[-\alpha J_m]$) are penetrated by single miracidia. Those that are penetrated are represented as passing into *time delay* T, from which the proportion γ_2 survive to emerge and rejoin the pool of available snails. The interval-by-interval survivorship of the available snails is represented by the *scalor* γ_1. The open path on the left represents possible inputs to the snail population as results of natality and migration. The scale factor α_3 can be replaced by the function on the right-hand side of equation 3.38, with $N_s(\tau)$ replaced by $N_{sa}(\tau)$. A corresponding modification of the network of Figure 3.28 is depicted in the lower half of Figure

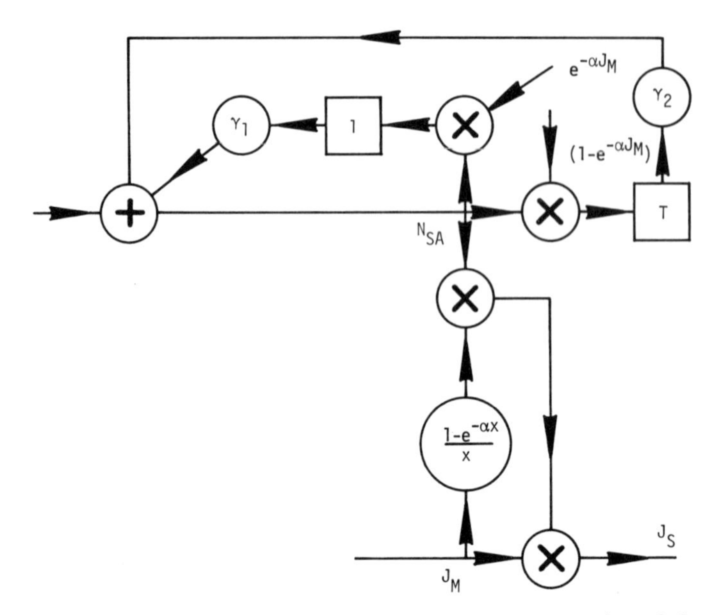

Fig. 3.29. Further modification of Fig. 3.27 to incorporate simple saturation of the snail host, with fixed duration, T, of snail infection

3.29. Holling was one of the first to incorporate delayed recovery from saturation (satiation of a predator, in Holling's case) in population models; and for that reason, the model of Figure 3.29 might very well be labelled a "Holling-type" model.

If J_m is relatively constant over the short term (i.e., over several transition intervals), and if the turnover rate of snails is relatively low (i.e., low mortality/low natality), then the number of available snails can be expressed as a simple fraction of the total number of snails (N_s). In that case, the penetrated snails would be distributed approximately uniformly over the *time delay*, T, so that

$$N_s = N_{sa} + T(\mathbb{1} - \exp[-\alpha J_m])\, N_{sa} \tag{3.52}$$

where the first term on the right represents the contents of the accumulator (the presently available snails), and the second term on the right represents the contents of *time delay* T [i.e., $(\mathbb{1} - \exp[-\alpha J_m])\, N_{sa}$ for every transition interval in T]. Rearranging terms, we find

$$\frac{N_{sa}}{N_s} = \mathbb{1}/[\mathbb{1} + T(\mathbb{1} - \exp[-\alpha J_m])]. \tag{3.53}$$

According to this function, the proportion of available snails ranges from one (when J_m is very small) to $\mathbb{1}/(\mathbb{1} + T)$ when J_m is very large. It is left as an exercise for the reader to become convinced that these limits are appropriate. What the function on the right-hand side of equation 3.53 does is describe one of many possible *transitions* from one limiting situation to another. Applying this description to the lower portion of the network of Figure 3.29, we obtain a rather interesting description of J_s:

$$J_s = \{N_s(\mathbb{1} - \exp[-\alpha J_m])\}/\{\mathbb{1} + T(\mathbb{1} - \exp[-\alpha J_m])\} \tag{3.54}$$

which ranges from $N_s/(\mathbb{1}+T)$, when J_m is so large that the snail hosts are thoroughly saturated, to $N_s J_m$ when J_m is so small that only a tiny fraction of the snails are infected (penetrated). At one limit, the production of sporocysts (J_s) is directly proportional to the snail population (i.e., a "first-order reaction"); at the other limit, the production of sporocysts is directly proportional to the snail population and to the miracidium population (i.e., a "second-order reaction"). The right-hand side of equation 3.54 describes the transition that should occur between the two types of "reaction" under the stipulated conditions.

Example 3.27. The Uniform Gantlet with Very Small Transition Intervals: Lotka-Volterra Models. Consider the modified schistosome population of Example 3.22 with transition intervals that are sufficiently short that only a tiny fraction of the miracidia present are able to encounter snails and only a tiny fraction of the snails are encountered in any particular interval, and the miracidia survive for several or many intervals. Find an appropriate substitute for the scale factor α_3 in Figure 3.20.

Answer: Let q be the probability that a particular miracidium, having searched for one transition interval, has failed to find a particular snail. By the statement of the specifications of the idealized population, q is the same for all miracidia and all snails. The probability that a particular miracidium misses all snails is q^{N_s}; the probability that all miracidia miss a particular snail is q^{N_m}. The probability that a particular miracidium encounters a snail is $\mathbb{1} - q^{N_s}$; the probability that a particular snail encounters a miracidium is $\mathbb{1} - q^{N_m}$. Because both probabilities are very small (as only a tiny fraction of either population is involved in encounters in one transition interval),

$$\begin{aligned} \mathbb{1} - q^{N_s} &\approx (\log_e q)\, N_s \\ \mathbb{1} - q^{N_m} &\approx (\log_e q)\, N_m. \end{aligned} \tag{3.55}$$

It follows that the rate (J_s) of sporocyst production is

$$J_s \approx N_m (\log_e q)\, N_s = \alpha N_s N_m \tag{3.56}$$

and the rate of snail infection is $N_s(\ln q)\, N_m$, which is the same. An appropriate replacement for α_3 is shown in Figure 3.30.

The second-order relationship of equation 3.56 is the essence of the well-known Lotka-Volterra models and very well might be labelled the *Lotka-Volterra* relationship. Thus, the structure in Figure 3.30 is a network equivalent of the basic Lotka-Volterra relationship. Furthermore, if the transition interval were made sufficiently small, every one of the five previous examples would lead to the basic Lotka-Volterra relationship, or various combinations thereof. For example, the double-saturation effect of Example 3.24 could be represented by the network structure of Figure 3.31; and the higher-level saturation effects of

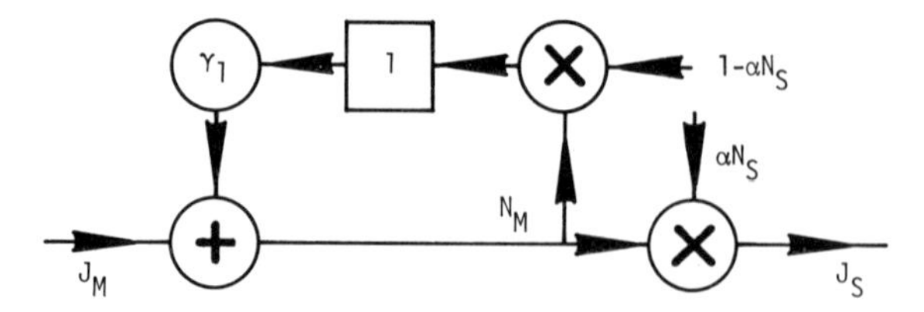

Fig. 3.30. Network representation of a discrete-time version of the Lotka-Volterra model for one species (the larval schistosome)

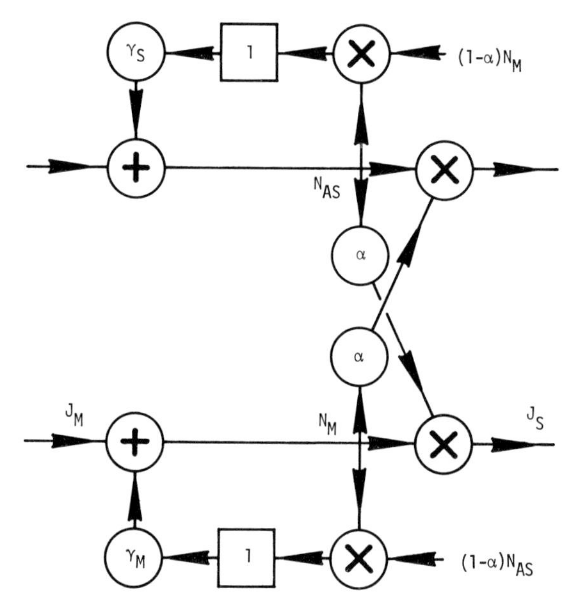

Fig. 3.31. Discrete-time version of the Lotka-Volterra model for two-species interaction. N_{AS} is the number of snails available for penetration; N_M is the number of miracidial larvae; J_M is the flow of newly-hatched miracidia; and J_S is the flow of larvae into the sporocyst phase

Example 3.25 could be represented by the structure in Figure 3.32. In essence, these networks represent algorithms for estimating the right-hand sides of equations 3.46 and 3.49 respectively.

It should be clear from these examples that the Lotka-Volterra relationship is fundamental to all gantlet-type models of encounter in which all searchers are equally effective and all targets equally likely to be struck. The relationship emerges, however, only when the transition interval is sufficiently small.

Example 3.28. Nonuniform Search and Encounter. Consider the population of idealized schistosomes of Example 3.16, with the following modifications: A miracidium always penetrates the first snail that it encounters; all snails are equally likely targets for a particular miracidium, but not all miracidia are equally effective searchers (some being innately handicapped, others being

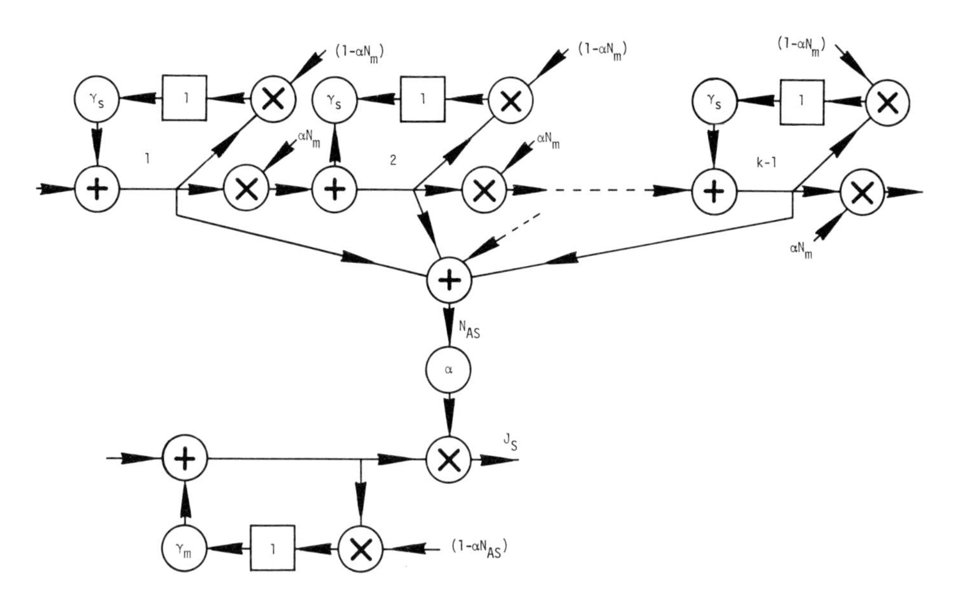

Fig. 3.32. Modified version of Fig. 3.31, incorporating saturation of the snail host at a level of k miracidial infections

handicapped by virtue of having emerged in a location not favored by snails); the number of snails varies from interval to interval, and the lifetime of a miracidium is short compared to a transition interval. Find an appropriate substitute for α_3 in the network of Figure 3.20.

Answer: We can begin by considering a newly emerging miracidium to have a finite number (k) of possible states, each representing a different snail-search effectiveness, with a probability p_i that the miracidium occupies the ith such state. Let $q_{sm}(i)$ represent the probability that a miracidium in a state i can search for its entire lifetime and fail to find a particular snail. Because all snails are equally likely targets, we can take $q_{sm}(i)$ to be the same for all snails. Finally, let $N_s(\tau)$ be the total number of snails in interval τ. Clearly, for a miricidium in state i, the Nicholson-Bailey model would apply in the following form:

$$p_{mi}(\tau)=1-\exp[-\alpha_i N_s(\tau)] \tag{3.57}$$

where

$$\alpha_i=|\log_e q_{sm}(i)| \tag{3.58}$$

and p_{mi} is the probability that a particular miracidium in state i manages to find and penetrate a snail (see equation 3.33). The probability that a newly emerging miracidium finds itself in state i and proceeds to penetrate a snail simply is the

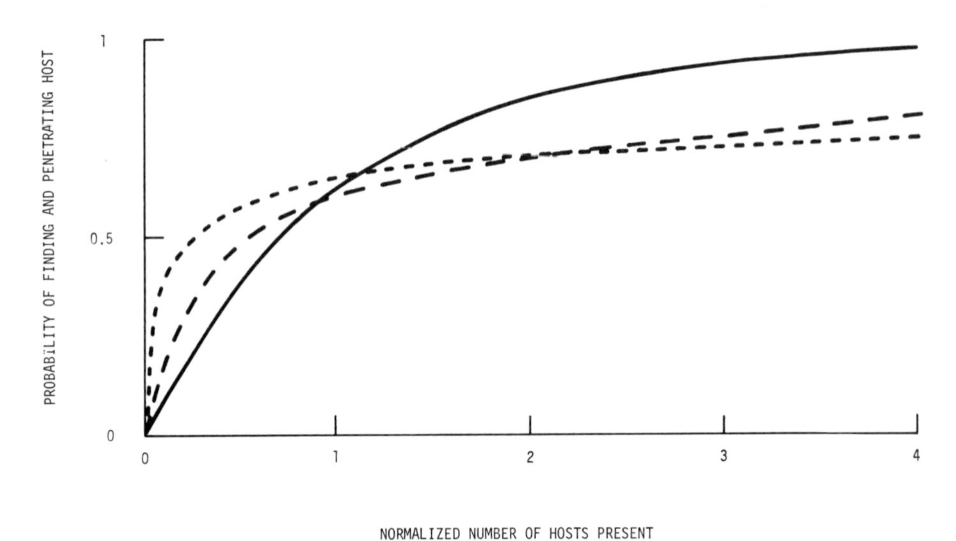

Fig. 3.33. Success rate as a function of target population in three examples of nonuniform search and encounter. Solid line: $p_1=1$, $\alpha_1=1$; long-dashed line: $p_1=1/2$, $p_2=1/2$, $\alpha_1=1$, $\alpha_2=2$; short-dashed line: $p_1=1/2$, $p_2=1/2$, $\alpha_1=4$, $\alpha_2=1/4$. The function can take any shape that starts at the origin, approaches 1 asymptotically, and has a monotonically decreasing positive slope, depending upon the distributions of the p_i and α_i in equation 3.59

product $p_i p_{mi}$. To find the total probability, $p_m(\tau)$, that the newly emerging miracidium finds and penetrates a snail, we simply must sum that product over all of the k states:

$$\alpha_3 = p_m(\tau) = \sum_{i=1}^{k} p_i\{1-\exp[-\alpha_i N_s(\tau)]\}. \tag{3.59}$$

This function is capable of taking on a wide range of appearances, depending on the values of α_i and the corresponding probabilities, p_i. A few of its variations are displayed in Figure 3.33.

If, in this example, the snails were not all considered to be equally likely targets, then we could proceed by letting $q_{sm}(i,j)$ be the probability that a particular miracidium in state i can search for its entire lifetime and fail to find a particular snail in state j. The value of $q_{sm}(i,j)$ would be the same for all snails in state j and all miracidia in state i. If the expected value of the geometric mean of $q_{sm}(i,j)$ taken over all snails were independent of the total number of snails, N_s, then equation 3.59 will be valid as written, but with $q_{sm}(i,j)$ in the expression for α_i (equation 3.58). It is left as a simple exercise for the reader to demonstrate that this is so. If the expected geometric mean of $q_{sm}(i,j)$ does vary with N_s, then one simply can consider α_i in equation 3.59 to be a corresponding function of N_s:

$$\alpha_i(N_s) = |\log_e[\mathrm{Ex}[(\prod_{\text{all snails}} q_{sm}(i,j))^{1/N_s}]]| \tag{3.60}$$

where $\prod$ indicates a product of terms (see equation 3.44), with one term for every snail.

If the transition interval is made sufficiently short in a nonuniform search and encounter situation such as this, where either the searcher is removed after striking one or more targets or the target is removed after being struck one or more times, then the process will become selective, preferentially removing those searchers with highest probability of striking a target or those targets with highest probability of being struck. Thus, if the transition interval in this example were made short compared to the lifetime of a miracidium, as was done in Example 3.27, the process of snail finding would be selective on the pool of searching miracidia; and the composition of that pool would not remain constant. For that reason, the simple Lotka-Volterra relationship of equation 3.56 would not hold. In fact, it generally will not hold for nonuniform search and encounter schemes.

Example 3.29. Cooperative Search and Encounter. Consider the population of idealized schistosomes of Example 3.22, but with the following modifications: Because of the increased effectiveness of their combined chemotaxic stimuli, snails that are clustered are more likely targets for a searching miracidium than are isolated snails. Find an appropriate substitute for α_3.

Answer: In this case, the specifications are rather general. What we have here is another type of situation to which the Lotka-Volterra relationship usually will not apply. The degree of clustering among the snails (i.e., the probability that a given snail is part of a cluster) may increase, remain the same, or even decrease with increasing snail population. If it either increases or decreases, q_{sm} will be a function of N_s, in turn making α (equation 3.32) a function of N_s. With this in mind, we can substitute the factor $\mathbb{1}-\exp[-\alpha N_s]$ for α_3. Taking very small transition intervals, we could approximate this factor by αN_s; which, by virtue of the dependence of α on N_s would not provide a simple Lotka-Volterra relationship.

One of the important attributes a modeler can have is the ability to detect analogies. A particular formulation or complicated relationship may be rather difficult to derive the first time, but at the same time it may apply equally well to a broad range of hypothetical situations (i.e., it may represent some common essence among those situations). In that sense, the situations are *analogous* to one another. If the analogy is recognized, much labor may be saved in a subsequent formulation. Furthermore, if the consequences of a particular hypothetical situation have been deduced, perhaps through considerable effort, those same deductions (or simple modifications thereof) should be applicable to the analog situations. Thus, once again, considerable time and effort can be saved.

An example that comes immediately to mind is the generalized, linear force-flow relationship of nonequilibrium thermodynamics. The commonly postulated models of diffusion of matter, flow of heat, flow of fluids, first-order chemical reactions, and the like are precisely analogous to the usual model of an electrical circuit made up of resistors and capacitors. For a number of reasons, the properties of such electrical circuit models are exceptionally well known, having been examined in great detail and at considerable expense for many years. By analogy, then, the same basic properties apply to all of those other models. Thus, for example, one would be forced to conclude that chemical oscillators (such as those that might underly biological clocks) cannot be constructed from sequences of first-order chemical reactions, as those reactions normally are modeled, no matter how they are coupled or fed back upon one another.

The examples of this section are rather specific, and may appear much too specialized to the reader. It was hoped and intended, however, that they would be taken in the spirit of analogy, with our idealized schistosomes merely providing a convenient vehicle. Thus, for example, the various postulated interactions between the miracidia and the snails could be taken to be disguised versions of a wide variety of postulated intraspecific or interspecific interactions. The miracidia of Example 3.22 might be female garter snakes roaming about searching for males of their species (the snails). On the other hand, the miracidia of that example might be small insects running the daily gantlet of virtually insatiable insect-eating birds. The miracidia of Example 3.23 might be small gazelles, running a daily gantlet of idealized cheetahs, each of which is satiated by one gazelle per day. The miracidia of Example 3.24 might be male birds searching for mates, or they might be bird pairs competing for a limited number of nesting sites, or mammals competing for a limited number of shelters. The snails in Example 3.25 might be predators that become satiated after devouring k prey (the miracidia). In Example 3.26, the snails might be striated cones requiring several days for digestion of one fish before they are ready or able to catch another. And, of course, the cooperative phenomena discussed so briefly in Example 3.29 might represent chorus formation by male frogs, cooperative pheromone emission by corn-ear-worm moths, hunting coalition formation by wolves, defensive coalitions by muskoxen, and so on and on. Clearly, the examples in this section are far from being exhaustive. In fact each modeler must develop a catalog of relationships to suit his or her own needs; and that catalog should be treated as a repertoire to be called upon in future performances.

3.4. *Time Delay* Durations

Up to this point, we have incorporated into our network models such life-cycle parameters as time to maturity, ovulation interval, gestation period, duration of regression, duration of nonreproductive lactation, and the like. In this section, we shall extend the network approach to include these same types of parameters when their values are not necessarily fixed, but may vary from interval to interval or from one member of a population to another. We might picture a *time delay* in discrete-time models as being analogous to a sequence of compartments, such as those depicted as squares in Figure 3.34, with each compartment representing a time class and containing the corresponding cohort. Thus, the cohorts in the various compartments would form a queue; and at each transition interval, the entire queue would shift one compartment to the right. In

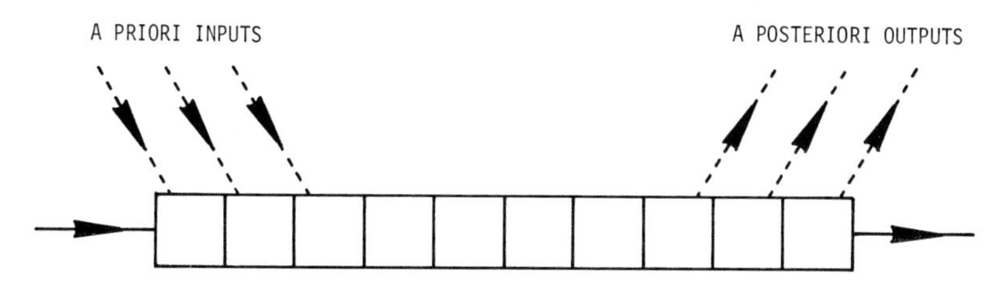

Fig. 3.34. A *time delay* depicted as a sequence of compartments, each representing a time class one transition interval in duration

a *time delay* whose duration is fixed, the output always would be the last cohort on the right, and input cohorts always would enter at the first compartment on the far left. For *time delays* whose durations are variable, we can consider two alternatives: the *time delay* whose duration is determined *a posteriori* (i.e., is not determined for a cohort until the interval of emergence from the *time delay*) and the *time delay* whose duration is determined *a priori* (i.e., is determined for a cohort at the interval it enters the *time delay*). For *a posteriori time delays*, we could introduce all incoming cohorts into the first compartment on the far left and select from all of the other compartments, on an interval by interval basis, those cohorts that are ready to emerge. For *a priori time delays*, on the other hand, we would know the appropriate duration in advance for each cohort and could adjust the point of entry correspondingly, taking the output for each interval from the last compartment on the far right. Thus, in the case of an *a priori time delay*, a given cohort could be augmented by new inputs as it progressed from compartment to compartment. In the case of a *time delay* that varies from individual to individual, a proportion of the occupants of each compartment might emerge at each interval, with all inputs being applied at the far left; or the input cohort might be distributed over several compartments, with the output always being from the far right.

Clearly, such a representation of a *time delay* is a bit awkward, but then so are variable *time delays* themselves. An alternative representation, in terms of the network elements already introduced, is displayed in Figure 3.35. Here, the scalors can represent an apportioning either of the inputs (for durations determined *a priori*) or of the outputs (for durations determined *a posteriori*). A question that arises immediately with this network structure is whether or not the bookkeeping of survivorship should be separated from or included in the apportioning of flows through the various *time delays*. As we already have mentioned in section 2.10, survivorship *scalors* generally should be placed after their corresponding *time delays;* therefore, we generally would not want to incorporate survivorship into the *scalors* on the left side of the structure of Figure 3.35. On the other hand, we might very well represent survivorship in the *scalors* on the right-hand side of that structure. Alternatively, we might choose to keep track of survivorship outside of the structure (i.e., with a separate *scalor*), just as we keep track of it outside of a simple *time delay* element, in which case all individuals represented as flowing into the structure eventually would be represented as flowing out of it. To distinguish between these two approaches,

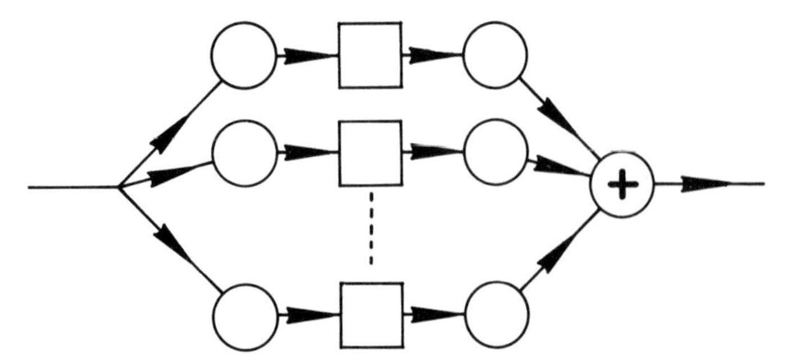

Fig. 3.35. One possible way to represent a variable *time delay* (i.e., with one path for each possible value of delay time, with scalors on the left for *a priori* distribution and scalors on the right for *a posteriori* distribution)

Fig. 3.36. Network symbol for a conservative *time delay* whose duration may vary from individual to individual, may depend upon the state (n) of the population, and may otherwise be dependent upon time

we might label the latter a *strictly conservative variable time delay* (or a *conservative delay*, for short) and the former a *nonconservative variable time delay* (or simply a *nonconservative delay*). It is the *conservative delay* that is most closely analogous to our original *time delay* element; in fact, it can be represented very simply by the graphical symbol of Figure 3.36, where the duration, $T(i, \boldsymbol{n}, \tau)$, might depend on the individual, i, the state of the population, $\boldsymbol{n}$, and the time, τ. When the structure in Figure 3.35 is employed as a *conservative delay*, the apportioning represented by the *scalors* is carried out over the individuals that have survived to emerge from the *delay*. Thus, if the same number of individuals enterred the delay interval after interval, the scale factor associated with duration T_j would be the proportion of individuals emerging from the entire structure (the entire *variable delay*) now that had entered it T_j transition intervals ago; and the sum of the scale factors over all of the durations represented in the structure must be equal to one.

Example 3.30. Periodically Varying *Time Delay*. Consider once again the idealized schistosomes of Example 3.16, but with the following modifications: The time required for the sporocysts to mature in the snail hosts varies from month to month through the year, being one month for those whose miracidia entered the snails in the months from May through November, being five months for those that entered in December or January, four months for those that entered in February, three months for those that entered in March, and two months for those that entered in April. Construct an appropriate *conservative delay* to replace the one-month *time delay* between α_3 and γ_1 in Figure 3.20.

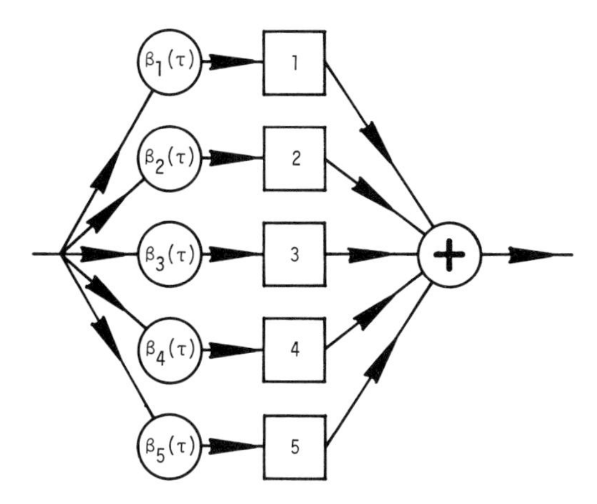

Fig. 3.37. Variable *time delay* with *a priori* specification of duration

Answer: Since the durations are specified on an *a priori* basis, we can use the structure of Figure 3.35 with the *scalors* on the left, as illustrated in Figure 3.37. The time-dependent scale factors will take on the following values:

	Jan	Feb	Mar	Apr	May	Jun	Jul	Aug	Sep	Oct	Nov	Dec
β_1 ...	0	0	0	0	1	1	1	1	1	1	1	0
β_2 ...	0	0	0	1	0	0	0	0	0	0	0	0
β_3 ...	0	0	1	0	0	0	0	0	0	0	0	0
β_4 ...	0	1	0	0	0	0	0	0	0	0	0	0
β_5 ...	1	0	0	0	0	0	0	0	0	0	0	1

In this case (and in other similar situations), the probability of survival of a given sporocyst undoubtedly depends upon how long it must reside in the snail and thus face the same high risks of mortality that the snails face. Therefore, rather than lumping survivorship bookkeeping into a single scalor to the right of the structure in Figure 3.37, it might be more convenient to construct a nonconservative delay, with an appropriate survivorship *scalor* associated with each duration, as depicted in Figure 3.38. In this new structure, the apportioning of flows over the various durations is accomplished by the scale factors β on the left, while the survivorship bookkeeping is accomplished by the scale factors γ' on the right. The latter would replace γ_1 in Figure 3.20. It is interesting to note that with the durations specified in this example, the cohorts entering snails in January, February, March, April and May all will emerge in June. Such merging of cohorts is to be expected in any *time delay* whose duration is varying.

Example 3.31. *Time Delay* Dependent on Environmental Factors. Suppose it were found that the rate of maturation of the sporocyst increased linearly with water temperature above a certain threshold, with no progress toward maturity being made at all when the temperature was below threshold. In other words, when the temperature is above threshold, the amount of progress made toward

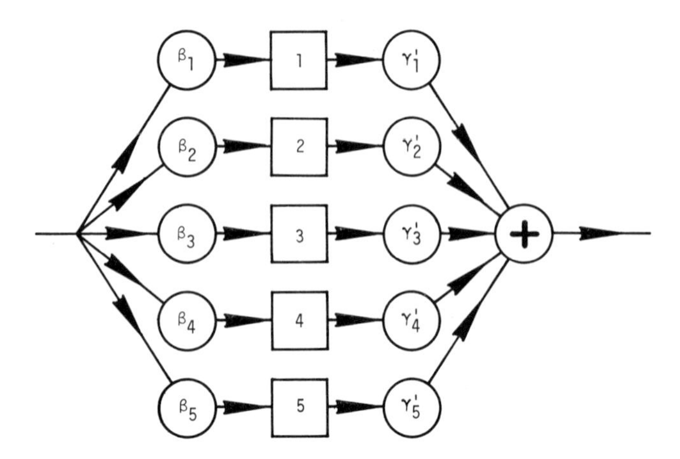

Fig. 3.38. Variable *time delay* with *a priori* specification of duration (the distribution over which is given by the β's) and with survivorship *scalors* (scale factors being the γ's)

maturity in a particular interval is directly proportional to the difference $t - t_0$ between the temperature, t, during that interval and the threshold temperature, t_0. How could one incorporate such a finding into the network model of Figure 3.20?

Answer: There are two fundamentally different ways to deal with this problem: First, we might decide to postulate, *a priori*, a temperature history, thereby closing the model. Likely choices would be the water-temperature history observed in a particular year or the mean water-temperature history observed over several years. By selecting a specific history, we would transform the problem into one essentially identical to that of the previous example. Our second choice would be to leave the system open, applying water-temperature data to the model on an *a posteriori*, interval by interval basis. There are many ways to carry out the second approach. Perhaps the easiest is to revert to a basic Markov chain, as depicted in Figure 3.39a. On penetration of a snail, a miracidium would enter state U_1, corresponding to the state of least maturity. Then, depending upon the resolution of water temperature, it would progress a certain number of steps to the right per degree above threshold at each subsequent transition interval. In this way, each larval worm would carry its own temperature history along with it as it progressed through the chain. From the last state, U_n, in the chain, it would progress on to the cercaria state (i.e., state U_n would represent the last state of immaturity). Although this scheme is rather simple conceptually, it does not lead to a particularly simple network structure, since every state in the chain must be connected by a path to every other state that is beyond it to the right, and the parameter associated with each path will vary from interval to interval, being dependent upon temperature. In terms of the network elements we have been using, the ith Markov state of Figure 3.39a can be represented as an accumulator, as depicted in Figure 3.39b, with the associated scalor b_j leading to the accumulator j steps to the right and

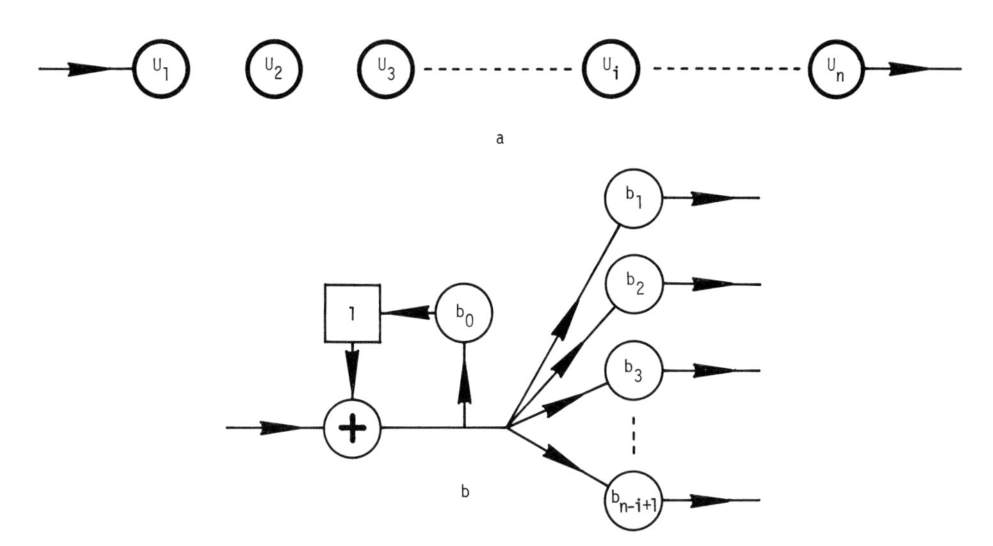

Fig. 3.39a and b. Markov-chain representation (a) of the system described in Example 3.31, along with a network model (b) for any one of the individual Markovian states. Conceptually, one can imagine the sporocyst progressing from state to state at a rate directly proportional to the difference between the ambient temperature and the threshold temperature, whenever the ambient temperature exceeds threshold, and making no progress at all when the ambient temperature is below threshold. Progress along such a chain may be construed as accumulated heat or as accumulated time (with the individual time increments having been weighted by a temperature factor). The latter notion has led to the term "physiological time"

being dependent on temperature as follows:

$$b_j = \mathbb{0} \quad \text{when } t - t_0 \neq j \text{ (units of temp. resolution)}$$

$$b_j = \mathbb{1} \quad \text{when } t - t_0 = j$$

if suvivorship bookkeeping is carried out separately from the time delay process, or

$$b_j = \gamma \quad \text{when } t - t_0 = j$$

if suvivorship bookkeeping is incorporated in the delay process.

Example 3.32. Stochastic *Time Delay*. In Example 3.16, it was postulated that all schistosomules require two months to mature in the mammalian host. Suppose that, in fact, this maturation time (resolved to an interval of one week) varied from schistosomule to schistosomule as follows: No schistosomules mature in less than five weeks, and all that survive to reach adulthood have matured by ten weeks; a certain proportion, p_i, of the schistosomules surviving to adulthood reach maturity in i weeks, where i is equal to or greater than five, less than or equal to ten. Find an appropriate structure to replace the $\mathbb{2}$-month *time delay* in the network of Figure 3.20.

Answer: What we have here is a process (maturation) whose duration *appears* to be a random variable. In other words, when a cercaria first enters a

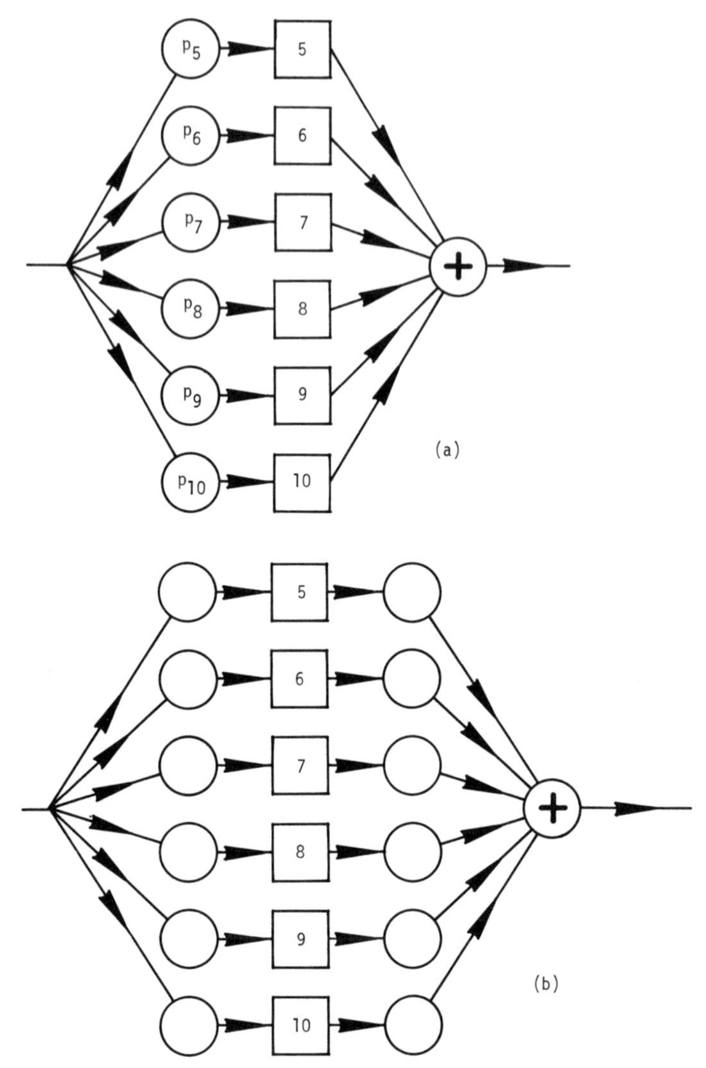

Fig. 3.40a and b. Conservative stochastic *time delay* (a), in which the sum of the p_i's over all paths is equal to 1; and a nonconservative stochastic time delay (b), in which survivorship scalors are incorporated on the right-hand side

mammalian host, we cannot say (according to the observations we are attempting to incorporate into our model) how long it will require to pass through the schistosomule stage and reach adulthood. We can say, however, that according to the given proportions, its probability of reaching adulthood in i weeks is p_i. Perhaps if we were able to observe the physiological, anatomical, or behavioral determinants underlying maturation, the duration of the process would be quite predictable. Lacking those, however, we perfectly well may treat the process as being probabilistic and represent it as a *stochastic time delay* (i.e., a *time delay* with a randomly distributed duration).

Once again we can employ the basic structure of Figure 3.35 as either a *conservative delay* (Figure 3.40a) or a *nonconservative delay* (Figure 3.40b), depending upon how we wish to incorporate survivorship bookkeeping.

3.5. Conversion to a Stochastic Model

There certainly are situations in which deductions concerning mean or expected behavior of a population's dynamics are inadequate, even misleading. Among the most prominent of these are situations in which the dynamics may proceed along one of two or more quite distinct courses, and the probabilities associated with the various alternatives are not negligible (i.e., one of the alternatives is not overwhelmingly more likely than the others). The expected dynamics in such cases will comprise middle roads (weighted sums of the various possible routes), which very likely never actually are taken by the system. Consider, for example, the dynamics depicted graphically in Figure 3.41. Here we have only two paths *(state trajectories)* available to the system, one leading sharply downward and one leading sharply upward. If both trajectories were equally probable, then the mean, or expected, trajectory would be horizontal, lying halfway between the two. Not only is this expected trajectory not one that the system ever would take, it also would be quite misleading with respect to the system's actual capabilities if it were all that were available by way of deductions. On the other hand, if one of the trajectories were overwhelmingly more likely than the other, then the expected dynamics would follow that trajectory rather well, and thus represent the system's behavior rather well.

What will concern us in this section, however, are situations in which there is not a single, overwhelmingly likely trajectory to dominate the expected behavior. In such situations, we very likely will wish to deduce all of the reasonably probable trajectories, each along with its probability of being taken. We can distinguish two distinct cases. In one case, a given trajectory becomes overwhelmingly likely once an initial commitment is made to it, but the initial commitments themselves are probabilistic. Thus, in Figure 3.41a, if we began at the extreme left-hand end of either of the two trajectories, we would remain on that same trajectory; our only problem is that we are uncertain as to where we actually started. This is by far the easier of the two cases to handle. The necessary deductions can be carried out directly with models based on the law of large numbers. In the second case, the trajectories themselves diverge, as in Figure 3.41b. In such cases knowledge of an initial commitment is not sufficient to determine a single, overwhelmingly likely trajectory. Thus, in Figure 3.41b, if we start at the far left-hand end of the trajectory, we are uncertain as to which of the two final courses it will take. In order to deduce the trajectories and their associated probabilities in such cases, we shall be forced to revert to stochastic models, giving up to a great extent our dependence upon the law of large numbers and its benefits.

Stochastic models themselves can be divided into two classes: those in which functions of two or more correlated variables are required and those in which such functions are not required. The latter often can be derived by direct and simple conversion of the corresponding network model based on large numbers.

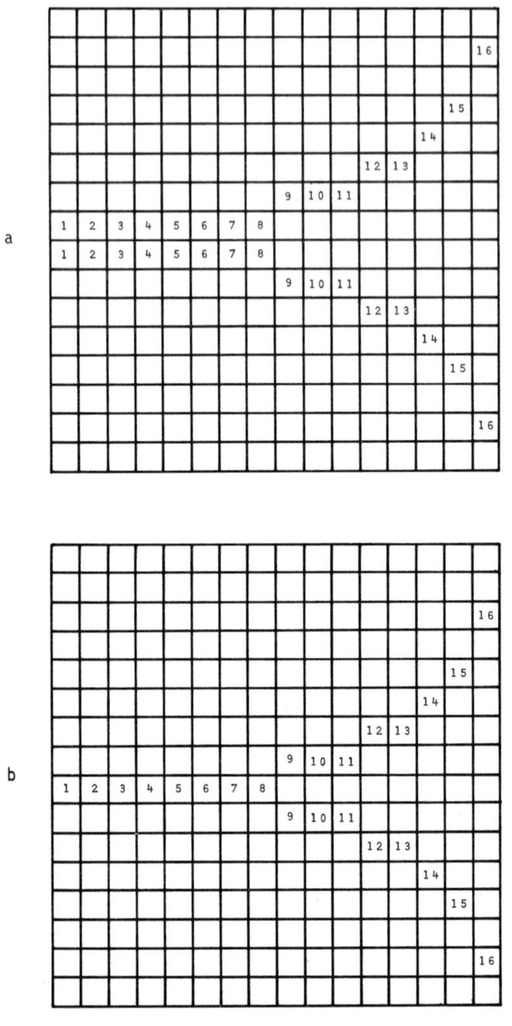

Fig. 3.41a and b. Two situations for which stochastic models are appropriate. Each square in the grid represents a state; and progress of the system through its state space is depicted by a sequence of numbers, one for each transition interval, placed in the corresponding square. Thus, in (a) there are two trajectories that begin on the far left (interval 1) and progress to the right. At transition interval 9 the trajectories begin to diverge; and by the time of interval 16 they are well separated

The former, on the other hand, generally will require considerable modification of the large-numbers network model. In our large-numbers network models, we have considered two types of variables, flows (J's) and accumulations (N's); and we took each variable to be its mean or expected value at each interval. In the stochastic versions, we shall consider the same types of variables; but instead of taking each to be its mean value, we shall consider the range of all possible values or the range of all values whose likelihoods are not negligible. In the models based on large numbers, the dynamics of a particular variable comprised an interval by interval history or prediction of its expected values; in the stochastic version, the dynamics will comprise an interval-by-interval history or prediction of the *probabilities* associated with each value that the variable might assume. Thus, in the stochastic model, instead of deducing a single value for each variable for each interval, we shall deduce an entire array of probabilities. In converting our networks based on large numbers to stochastic network models, therefore, we must convert the operations represented by the elements from those applicable to single-valued, or scalar variables to equivalent operations applicable to arrays of probabilities.

3.5.1. Models without Interactions of Correlated Variables

If functions of two or more correlated variables are not required, then the probability array with which we must deal is the simple probability vector. As in the examples of the previous sections, a network model based on large numbers can be constructed. Then each variable, J or N, in that model can be replaced by the vector representing the distribution of probabilities over the values that the variable might assume. Thus, a variable $J(\tau)$ would be replaced by

$$\mathbf{Pr}[J(\tau)] = \begin{bmatrix} \Pr[J(\tau)=0] \\ \Pr[J(\tau)=1] \\ \Pr[J(\tau)=2] \\ \vdots \\ \Pr[J(\tau)=n] \end{bmatrix} \tag{3.61}$$

where n either is the largest value that $J(\tau)$ may assume or is the largest value that has what the modeler defines to be a nonnegligible probability. The converted operations of the network elements are completely analogous to their operations in the large-numbers networks. A *time delay* of duration T simply imposes a delay of T intervals upon a vector, so that if the input is $\mathbf{Pr}[J(\tau)]$, the output is $\mathbf{Pr}[J(\tau-T)]$. The converted *scalor* transforms one probability vector (e.g., $\mathbf{Pr}[J_1(\tau)]$) into another (e.g., $\mathbf{Pr}[J_2(\tau)]$) on an interval-by-interval basis. As we already have seen in section 1.18, such a transformation is represented very conveniently as the inner product of the vector and an appropriate matrix. In this case, the matrix simply is an array of conditional probabilities of the form $\Pr[J_2(\tau)=k;\ J_1(\tau)=j]$, which can be read *the probability that the value of* J_2 *is equal to* k *in interval* τ, *given that the value of* J_1 *is equal to* j *in that same interval.* We simply can label this conditional-probability matrix $\mathbf{Pr}[J_2(\tau);\ J_1(\tau)]$ and display it as follows:

$$\mathbf{Pr}[J_2(\tau); J_1(\tau)] = \begin{bmatrix} \Pr[\mathbb{0};\mathbb{0}] & \Pr[\mathbb{0};\mathbb{1}] & \Pr[\mathbb{0};\mathbb{2}] & \ldots & \Pr[\mathbb{0};n_j] \\ \Pr[\mathbb{1};\mathbb{0}] & \Pr[\mathbb{1};\mathbb{1}] & \Pr[\mathbb{1};\mathbb{2}] & \ldots & \Pr[\mathbb{1};n_j] \\ \Pr[\mathbb{2};\mathbb{0}] & \Pr[\mathbb{2};\mathbb{1}] & \Pr[\mathbb{2};\mathbb{2}] & \ldots & \Pr[\mathbb{2};n_j] \\ \cdot & \cdot & \cdot & \cdot & \cdot \\ \Pr[n_k;\mathbb{0}] & \Pr[n_k;\mathbb{1}] & \Pr[n_k;\mathbb{2}] & \ldots & \Pr[n_k;n_j] \end{bmatrix} \tag{3.62}$$

where $\Pr[k;j]$ is an abbreviation of $\Pr[J_2(\tau)=k;\ J_1(\tau)=j]$. Thus, every scale factor in the large-numbers network model is replaced by a conditional-probability matrix of this form in the stochastic network model. Using the usual vector/matrix notation, the converted operation of a *scalor* can be abbreviated as follows:

$$\mathbf{Pr}[J_2(\tau)] = \mathbf{Pr}[J_2(\tau); J_1(\tau)] \cdot \mathbf{Pr}[J_1(\tau)]. \tag{3.63}$$

Note that according to our present constraints, the conditional-probability matrix must not be a function of (i.e., its elements must not depend upon) any variable (other than J_1 itself) that is correlated with J_1.

Each column in the matrix $\mathbf{Pr}[J_2(\tau);\ J_1(\tau)]$ is a vector array of the probabilities for the various values of J_2 in an interval τ, for a specified value of J_1 in that same interval (e.g., the first column gives the probabilities for the various values of J_2 in τ under the condition that J_1 is equal to zero in interval τ). If we knew the actual value of the cause-variable, $J_1(\tau)$, then by selecting the appropriate column of the matrix, we would have the corresponding probabilities for the various possible values of the effect-variable, $J_2(\tau)$. However, we do not claim to know the actual values of any variable. Instead, we have estimates of its probability of assuming each of its possible values. Therefore, what we actually do is sum the columns of $\mathbf{Pr}[J_2(\tau);\ J_1(\tau)]$, each weighted by the estimated probability of the corresponding value of $J_1(\tau)$. In this way, we form an array of *expected* or *mean probabilities* for the various values of the variable J_2, taken over all of the possible values of the variable J_1. In other words, we form a vector of expected probabilities, which we might very well call an *expected probability distribution vector.*

Finally, we come to the *adder.* When two variables converge to form a third, we want to obtain the distribution of probabilities over the values that their sum can assume. Thus, the converted *adder* must transform two or more input vectors, $\mathbf{Pr}[J_1(\tau)]$, $\mathbf{Pr}[J_2(\tau)], \ldots,$ into a single output vector, $\mathbf{Pr}[J_1(\tau)+J_2(\tau)+\cdots]$. As long as the variables being summed are not correlated (i.e., as long as we meet our present constraint), the distribution over the sum is generated by a simple process called *convolution.* The process itself can be appreciated most easily when considered for a single element of the resulting distribution vector, when two input vectors are convolved:

$$\Pr[J_1+J_2=k] = \sum_{j=0}^{k} \Pr[J_1=k-j]\,\Pr[J_2=j] \tag{3.64}$$

or alternatively,

$$\Pr[J_1+J_2=k] = \sum_{j=0}^{k} \Pr[J_2=k-j]\,\Pr[J_1=j]. \tag{3.65}$$

Thus, the probability that J_1+J_2 is equal to k is the sum of the probabilities of all combinations J_1, J_2 that add up to k. This would be the kth element of the resulting vector. The process depicted in equations 3.64 and 3.65 therefore would be repeated for every possible or significant value of J_1+J_2. The operation of repeated application of the process (i.e., the operation of convolution) often is represented by an asterisk (*). Thus, we can abbreviate the operation as follows:

$$\mathbf{Pr}[J_1(\tau)+J_2(\tau)] = \mathbf{Pr}[J_1(\tau)] * \mathbf{Pr}[J_2(\tau)]. \tag{3.66}$$

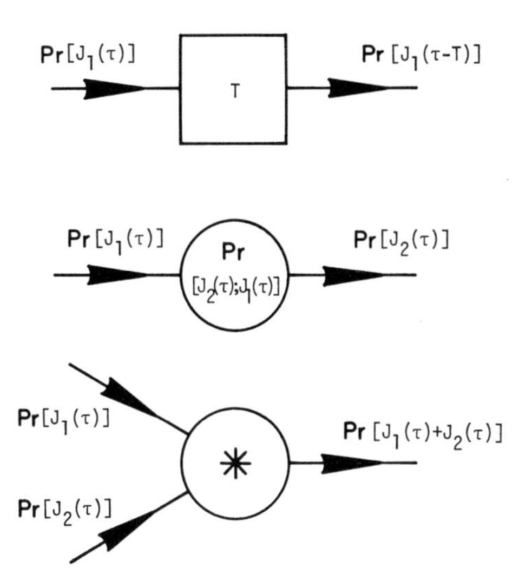

Fig. 3.42. Elements that can be used to construct stochastic network models when functions of two or more correlated variables are not required. The *time delay* remains as it was, with the interpretation indicated in the upper symbol. The *scalor* is replaced by the center symbol, which represents inner-product formation between the input array and a conditional-probability array. The *adder* is replaced by the lower symbol, representing convolution of the input arrays

Since the right-hand sides of equations 3.64 and 3.65 are equivalent, it is clear the operation of convolution commutes, so that

$$\mathbf{Pr}[J_1(\tau)+J_2(\tau)]=\mathbf{Pr}[J_2(\tau)]*\mathbf{Pr}[J_1(\tau)]. \tag{3.67}$$

Furthermore, the operation conforms to an associative law, so that

$$\mathbf{Pr}[J_1+J_2+J_3]=\mathbf{Pr}[J_3]*\mathbf{Pr}[J_1+J_2]=\mathbf{Pr}[J_3]*\mathbf{Pr}[J_2]*\mathbf{Pr}[J_1] \tag{3.68}$$

and it therefore is a very easy matter to extend it to sums of several variables.

In summary, if functions of two or more correlated variables are not required in the model, then the stochastic version can be developed by direct conversion of the large-numbers version by replacement of all variables with their expected probability distribution vectors, replacement of scale factors by conditional-probability matrices, and replacement of the operation of addition with the operation of convolution (which, like addition, obeys associative and commutative laws). These replacements are illustrated in Figure 3.42.

3.5.2. Models with Interactions of Correlated Variables

When two random variables are correlated, computation of the distribution of probabilities over their sum must be modified accordingly. Generally, there are three cases that will arise in network modeling. First of all, two variables (J_1 and J_2) may be correlated and the conditional probability matrix ($\mathbf{Pr}[J_1;J_2]$) may be known, in which case the distribution of probabilities over their sum is given by

$$\Pr[J_1+J_2=k]=\sum_{j=0}^{k}\Pr[J_1=k-j;J_2=j]\Pr[J_2=j] \qquad (3.69)$$

or, alternatively,

$$\Pr[J_1+J_2=k]=\sum_{j=0}^{k}\Pr[J_2=k-j;J_1=j]\Pr[J_1=j]. \qquad (3.70)$$

The equivalence of these two expressions lies at the heart of a commonly applied rule from probability theory known as Bayes's Rule.

BAYES'S RULE

$$\Pr[x=k;y=j]=\Pr[y=j;x=k]\Pr[x=k]/\sum\Pr[y=j;x=i]\Pr[x=i] \qquad (3.71)$$

where the summation in the denominator is taken over all possible values, i, of the variable x. Thus, if the conditional probability matrix $\mathbf{Pr}[J_1;J_2]$ is known, Bayes's Rule can be used to generate from it the matrix $\mathbf{Pr}[J_2;J_1]$. Often, the conditional-probability matrix directly relating two correlated variables is not available, but the two may be causally determined by a common set of one or more other variables, in which case the computation of the probability distribution over the sum can be carried out in two steps:

$$(1)\ \mathbf{Pr}[J_1+J_2;\boldsymbol{X}]=\mathbf{Pr}[J_1;\boldsymbol{X}]*\mathbf{Pr}[J_2;\boldsymbol{X}] \qquad (3.72)$$

$$(2)\ \mathbf{Pr}[J_1+J_2]=\mathbf{Pr}[J_1+J_2;\boldsymbol{X}]\cdot\mathbf{Pr}[\boldsymbol{X}] \qquad (3.73)$$

where $\boldsymbol{X}$ represents the set of one or more other variables. Thus, if $J_1(\tau)$ and $J_2(\tau)$ both are causally determined by the single variable, $J_3(\tau)$, $\boldsymbol{X}$ would be replaced by $J_3(\tau)$:

$$\mathbf{Pr}[J_1+J_2]=\mathbf{Pr}[J_1+J_2;J_3]\cdot\mathbf{Pr}[J_3]. \qquad (3.74)$$

Similarly, if the two variables are causally determined respectively by two disjoint or overlapping sets of correlated variables, $\boldsymbol{X}$ and $\boldsymbol{Y}$, then

$$(1)\ \mathbf{Pr}[J_1+J_2;\boldsymbol{X},\boldsymbol{Y}]=\mathbf{Pr}[J_1;\boldsymbol{X}]*\mathbf{Pr}[J_2;\boldsymbol{Y}] \qquad (3.75)$$

$$(2)\ \mathbf{Pr}[J_1+J_2;\boldsymbol{X}]=\mathbf{Pr}[J_1+J_2;\boldsymbol{X},\boldsymbol{Y}]\cdot\mathbf{Pr}[\boldsymbol{Y};\boldsymbol{X}] \qquad (3.76)$$

$$(3)\ \mathbf{Pr}[J_1+J_2]=\mathbf{Pr}[J_1+J_2;\boldsymbol{X}]\cdot\mathbf{Pr}[\boldsymbol{X}]. \qquad (3.77)$$

Thus, if $J_1(\tau)$ were determined causally by $J_3(\tau-T_1)$ and $J_2(\tau)$ were determined causally by $J_3(\tau-T_2)$, then equations 3.75 through 3.77 could be applied together as follows:

$$\mathbf{Pr}[J_1(\tau)+J_2(\tau)]=\{\{\mathbf{Pr}[J_1(\tau);J_3(\tau-T_1)]*\mathbf{Pr}[J_2(\tau);J_3(\tau-T_2)]\}\cdot\mathbf{Pr}[J_3(\tau-T_1);J_3(\tau-T_2)]\}\cdot\mathbf{Pr}[J_3(\tau-T_2)]. \qquad (3.78)$$

In each of the three cases, a crucial part of the process is *deconditioning* a conditional-probability matrix to convert it to an expected total probability vector. This is accomplished by calculating the mean values of the appropriate

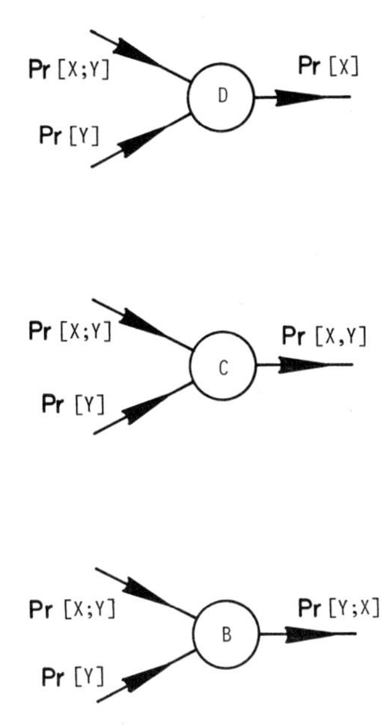

Fig. 3.43. Elements that can be used to construct stochastic network models when functions of two or more correlated variables are required. Upper symbol represents the operation of deconditioning an array; center symbol represents the operation of combining two arrays to form a joint-probability array; lower symbol represents the operation of Bayes's rule to reverse the conditioning of an array

conditional probabilities taken over the conditioning variable. In the first case, e.g., equations 3.69 and 3.70, deconditioning is carried out simultaneously with convolution, and the two processes are inextricable. In the second and third cases, however, convolution and deconditioning are carried out in separate steps. In equations 3.73 and 3.74, deconditioning is a one-step process; in equations 3.76, 3.77 and 3.78 it is carried out in two steps, the first being partial deconditioning. The basic operation involved in deconditioning simply is the formation of the inner product of two arrays (see Appendix A). In every case, one of the arrays is of higher dimension than the other and the operation serves to reduce that dimensionality. In the stochastic replacement for a *scalor*, we are representing the formation of the inner product of a vector (the input to the *scalor*), and a matrix (the modified scale factor). Now it will be convenient to have a stochastic network element analogous to a *multiplier*, where both factors are treated as inputs. We might label such an element a *deconditioner*, and represent it as a D enclosed in a circle, as illustrated in Figure 3.43. It also will be convenient to have an element that represents the operation corresponding to Bayes's Rule (equation 3.71). Finally, in order to execute Bayes's Rule, we occasionally shall require an element that represents the combination of $\mathbf{Pr}[X;Y]$ with $\mathbf{Pr}[Y]$ to form $\mathbf{Pr}[X,Y]$, the probability of particular values of both X and Y. Such elements also are introduced and defined in Figure 3.43 and in Appendix A.

The conditional-probability matrices required for the computations depicted in equations 3.69 through 3.78 may be available directly in the converted network model; or they may not be, in which case their generation often can be represented by additional network structures. It is left as a simple exercise for the reader to become convinced that similar conditional-probability matrices will be required whenever two correlated variables are brought together in a network model by any operation or function (e.g., multiplication of two variables, scaling one variable by a function of another, etc.).

3.5.3. Examples of Stochastic Network Models

Example 3.33. Stochastic Input to Model Based on Large Numbers. Consider the idealized schistosome population of Example 3.21, with the following additions: By a combined effort of molluscicide treatment and chemotherapy, all larval worms in the snails and all schistosomules in the mammals have been killed; the adult pairs all were equally susceptible to the chemotherapy, and purely by chance a few remain; even if only one pair remains, however, the number of offspring per month surviving to adulthood will be sufficiently large to allow the law of large numbers to apply. How would one go about representing these additional specifications in the network model.

Answer: First of all, we must assign a probability to each possible starting condition (i.e., to the possibility of starting with no pairs, with one pair, with two pairs, etc.). To do so, we simply can consider each of the N_{po} pairs that existed prior to chemotherapy as a Bernoulli trial, with a certain probability, p, of surviving and the complementary probability, $\mathbb{1}-p$, of having one or both partners killed. In that case, the probability of k surviving pairs simply is given by the binomial distribution (see Appendix B):

$$\Pr[k \text{ surviving pairs}] = \mathrm{b}(k; N_{po}, p). \tag{3.79}$$

Next, we set the inital contents of all *time delays* in our network model equal to zero, then set the initial number of pairs in the accumulator equal to k and carry out the analysis (see chapter 4). This is repeated for each value of k that is of interest; and to each of the correspondingly deduced dynamics is assigned the appropriate probability, from equation 3.79.

Example 3.34. A Stochastic Model without Interaction of Correlated Variables. Consider a population of idealized anadromous fish with the following life cycle: Females require three years to progress from the egg stage to sexual maturity and spawning. Each female egg faces the same probability, γ, of surviving to spawn; and each female that does so, produces n_f female eggs. Having spawned, the female dies. Construct a large-numbers network model and from it a stochastic network model of the population.

Answer: The large-numbers network model is generated by the usual procedure (see Figure 3.44a). Beginning with the flow J_E of female eggs we have a three-year *time delay* followed by a survivorship *scalor* (γ), the output of which is

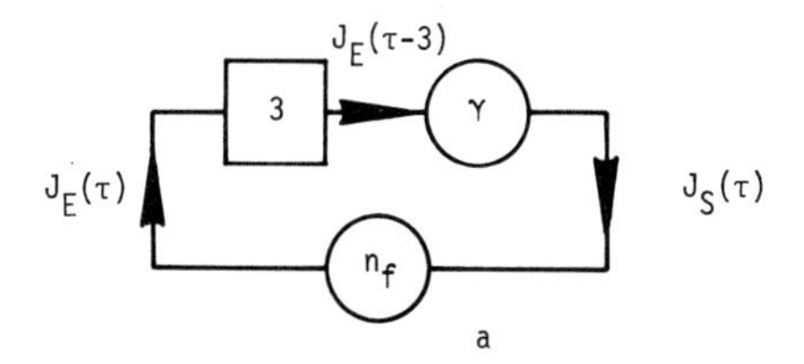

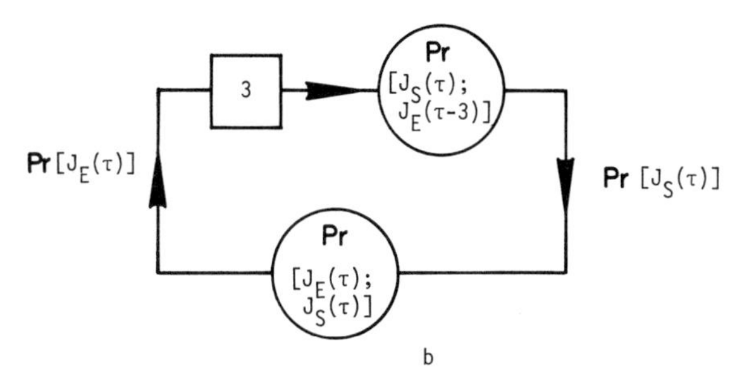

Fig. 3.44a and b. Network models of a population of idealized anadromous fish: (a) the large-numbers version, (b) the stochastic version (constructed with the elements of Fig. 3.42)

the flow J_S of females into spawning. A *scalor* (n_f) relates this flow to the flow of newly produced eggs. Because this network exhibits no interaction between correlated flows or accumulations, we can convert it directly to the stochastic version of Figure 3.44b, simply by employing the replacements displayed in Figure 3.42. What remains is to determine the elements of the conditional-probability matrices, $\mathbf{Pr}[J_S(\tau); J_E(\tau-3)]$ and $\mathbf{Pr}[J_E(\tau); J_S(\tau)]$.

According to the specifications of the life cycle, a spawning female produces n_f eggs. Therefore, if there are k spawning females, there will be kn_f eggs:

$$\begin{aligned} \Pr[J_E = kn_f;\ J_S = k] &= \mathbb{1} \\ \Pr[J_E \neq kn_f;\ J_S = k] &= \mathbb{0}. \end{aligned} \tag{3.80}$$

If there were J_E female eggs three years ago, each with probability γ of surviving, then we have J_E Bernoulli trials, each with probability γ of success. The corresponding probability of k surviving adults is given by the binomial distribution:

$$\Pr[k \text{ surviving adults};\ J_E \text{ eggs}] = \mathrm{b}(k; J_E, \gamma), \tag{3.81}$$

$$\mathbf{Pr}[J_S(\tau); J_E(\tau-\mathbb{3})] = \begin{bmatrix} b(\mathbb{0};\mathbb{0},\gamma) & b(\mathbb{0};\mathbb{1},\gamma) & b(\mathbb{0};\mathbb{2},\gamma) & \dots \\ b(\mathbb{1},\mathbb{0},\gamma) & b(\mathbb{1},\mathbb{1},\gamma) & b(\mathbb{1},\mathbb{2},\gamma) & \dots \\ b(\mathbb{2},\mathbb{0},\gamma) & b(\mathbb{2},\mathbb{1},\gamma) & b(\mathbb{2},\mathbb{2},\gamma) & \dots \\ b(\mathbb{3},\mathbb{0},\gamma) & b(\mathbb{3},\mathbb{1},\gamma) & b(\mathbb{3},\mathbb{2},\gamma) & \dots \\ b(\mathbb{4},\mathbb{0},\gamma) & b(\mathbb{4},\mathbb{1},\gamma) & b(\mathbb{4},\mathbb{2},\gamma) & \dots \end{bmatrix}. \tag{3.82}$$

Clearly, we have two quite different matrices here. That of equation 3.80 represents a process taken to be essentially deterministic (n_f eggs per spawning female), whereas that of equation 3.82 represents a process taken to be stochastic (survival).

Because of the stochastic nature of the process it represents, the matrix of equation 3.82 exhibits an unusual property; the sum of the elements in any column is equal to one. Matrices with this property (regardless of how or why they were derived) often are called *stochastic matrices*.

Example 3.35. A Stochastic Model with Interaction of Correlated Variables. Consider the idealized situation of Example 3.33, but with the following modifications: Each adult pair produces an enormous number (β) of eggs each month, each with equal probability γ' of passing through the larval stages and reaching adulthood; however, that probability is very low, so that only a very few (perhaps considerably less than one) surviving adults are expected to be produced per month per pair. Construct an appropriately modified version of the network of Figure 3.20.

Answer: To begin with, our revised specifications provide a single probability of passing through all of the various larval stages. Therefore, we can lump all of the *scalors* and *time delays* between J_E and J_A in the network of Figure 3.20, as illustrated in Figure 3.45. Next, because we are operating under the specifications of Example 3.21 (see Example 3.33), we must replace the scale factor $\mathbb{1}/\mathbb{2}$ (following J_A) with a function of J_A derived from equations 3.25 and 3.27 and giving the probability that a particular worm finds a mate. The rest of the network of Figure 3.45 remains as it was in Figure 3.20.

In converting this large-numbers network model to a stochastic network model, we must face the fact that N_p is formed by the addition of two variables, J_{SP} and J_{NP}, that are correlated, both being causally determined by N_p at earlier intervals. Therefore, we cannot simply replace the process of addition by the process of convolution. We certainly can make the other substitutions of Figure 3.42, however, and in doing so, we form two separate arms, as illustrated in Figure 3.46. One arm represents the computation of $\mathbf{Pr}[J_{NP}(\tau)]$ from $\mathbf{Pr}[N_p(\tau-\mathbb{3})]$, and the other represents the computation of $\mathbf{Pr}[J_{SP}(\tau)]$ from $\mathbf{Pr}[N_p(\tau$

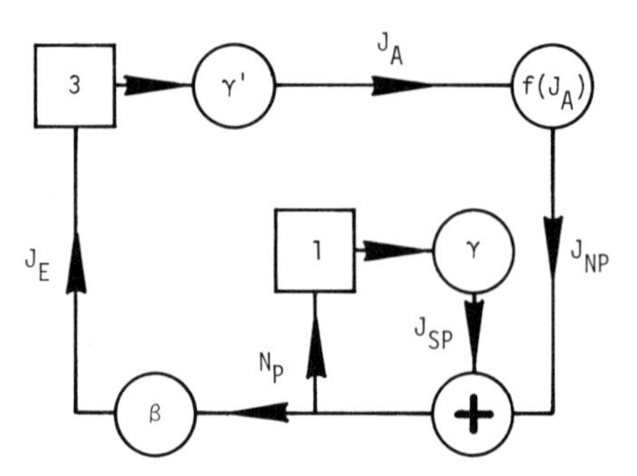

Fig. 3.45. Large-numbers network model of an idealized schistosome population (J_E=flow of eggs; J_A=flow of newly-mature adults; J_{NP}=flow of newly-mated pairs; J_{SP}=flow of surviving pairs; and N_P=total number of pairs

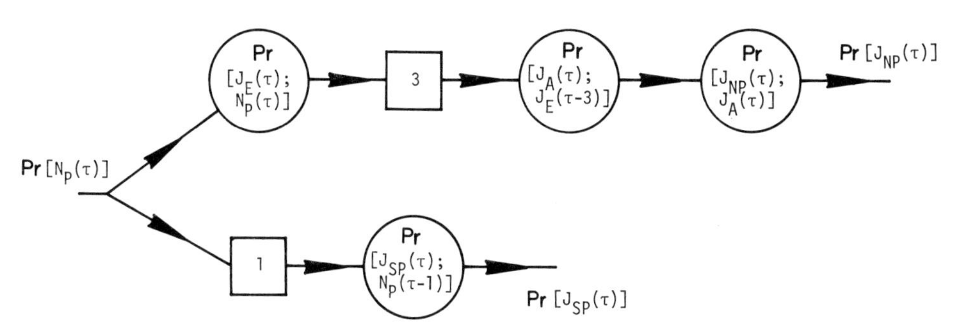

Fig. 3.46. First step in converting the model of Fig. 3.45 to a stochastic network model. So far, only the elements of Fig. 3.42 are required

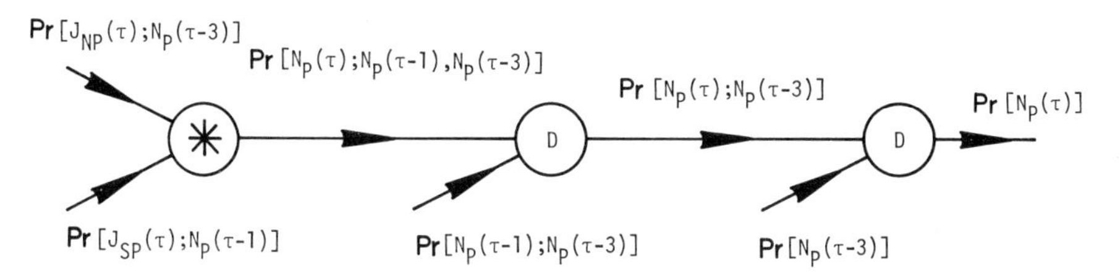

Fig. 3.47. Second step in the conversion to a stochastic network model. Now we must begin to use the elements of Fig. 3.43

$-\mathbb{1})]$. Lumping the *scalor* replacements in the upper arm, we have a single conditional-probability matrix $\mathbf{Pr}[J_{NP}(\tau);\ N_p(\tau-\mathbb{3})]$. The single conditional-probability matrix in the lower arm is $\mathbf{Pr}[J_{SP}(\tau);\ N_p(\tau-\mathbb{1})]$. Using the convolution and two-step deconditioning operations indicated in equations 3.77 and 3.78, we can convert these two conditional-probability matrices to the probability distribution vector $\mathbf{Pr}[N_p(\tau)]$. This process is represented in network form in Figure 3.47. The conditional-probability matrices on the far left are available from the population specifications (or the two arms of the network of Figure 3.46). We need one more conditional-probability matrix, $\mathbf{Pr}[N_p(\tau-\mathbb{1});\ N_p(\tau-\mathbb{3})]$, and a probability vector, $\mathbf{Pr}[N_p(\tau-\mathbb{3})]$, in order to achieve closure of the model. The latter is obtained by passing the output, $\mathbf{Pr}[N_p(\tau)]$, through a *time delay* of duration $\mathbb{3}$. The former, on the other hand, is not so directly available from the model as it stands.

To find a route by which we can obtain the matrix $\mathbf{Pr}[N_p(\tau-\mathbb{1});\ N_p(\tau-\mathbb{3})]$ and thereby obtain closure, we simply should begin by considering the probability arrays already available and the transformations that are available to convert those arrays to new forms. Then, step by step, we can proceed backward through a candidate route until either we have exhausted all avenues, at which point we must reconsider our model, or until we find an array recurring, at which point closure may be achieved. In the present case, we are given

$$\Pr[N_p(\tau)=W;\ N_p(\tau-\mathbb{1})=X,\ N_p(\tau-\mathbb{3})=Z]$$

which we can abbreviate as $\Pr[W;X,Z]$. We somehow must generate an estimate of

$$\Pr[N_p(\tau-\mathbb{1})=X;N_p(\tau-\mathbb{3})=Z]$$

which we can denote $\Pr[X;Z]$. The potentially useful transformations available to us are combination (C), deconditioning (D), Bayes's Rule (B), and the time delay. If we stipulate that W specifically is the postulated value of $N_p(\tau)$, X specifically is the postulated value of $N_p(\tau-\mathbb{1})$, Y specifically is the postulated value of $N_p(\tau-\mathbb{2})$, and Z specifically is the postulated value of $N_p(\tau-\mathbb{3})$, then delay by one interval will convert W to X, X to Y, and Y to Z. Therefore, through the time delay, we can generate $\Pr[X;Z]$ directly from $\Pr[W;Y]$. In other words, after one transition interval, $\mathbf{Pr}[N_p(\tau);N_p(\tau-\mathbb{2})]$ becomes $\mathbf{Pr}[N_p(\tau-\mathbb{1});N_p(\tau-\mathbb{3})]$. Through Bayes's Rule, we can interchange the conditioning and conditioned variables. Thus, by application of Bayes's Rule, we can convert $\Pr[X;Z]$ to $\Pr[Z;X]$; but to do so we also must have an estimate of $\Pr[Z]$. In our present model, $N_p(\tau-\mathbb{1})$ and $N_p(\tau-\mathbb{3})$ together determine $\mathbf{Pr}[N_p(\tau)]$; knowledge of $N_p(\tau-\mathbb{2})$ adds *nothing* to our knowledge of $\mathbf{Pr}[N_p(\tau)]$. Therefore

$$\Pr[W;X,Y,Z]=\Pr[W;X,Z]. \tag{3.83}$$

We seek

$$\Pr[X;Z]$$

which can be obtained by time delay from

$$\Pr[W;Y]$$

which can be obtained by deconditioning of $\Pr[W;X,Y,Z]$ with

$$\Pr[X,Z;Y]$$

which can be obtained by time delay from

$$\Pr[W,Y;X]$$

which can be obtained by deconditioning

$$\Pr[W,Y;X,Y,Z]$$

(obtained by expanding the dimensionality of $\Pr[W;X,Y,Z]$, see Appendix A) with

$$\Pr[Y,Z;X]$$

which can be obtained by time delay from

$$\Pr[X,Y;W]$$

which can be obtained by Bayes's Rule from

$$(1)\quad \Pr[W;X,Y] \quad \text{along with} \quad \Pr[X,Y]$$

the first of which can be obtained by deconditioning $\Pr[W; X, Y, Z]$ with

$$\Pr[Z; X, Y]$$

which can be obtained by time delay from

$$\Pr[Y; W, X]$$

which can be obtained by Bayes's Rule from

$$(2) \quad \Pr[W, X; Y] \quad \text{along with } \Pr[Y]$$

the first of which can be obtained by deconditioning

$$\Pr[W, X; X, Y, Z]$$

(obtained by expanding the dimensionality of $\Pr[W; X, Y, Z]$)
with

$$\Pr[X, Z; Y]$$

which already has occurred. At this point, it might very well appear that we require prior knowledge of $\Pr[X, Z; Y]$ in order to compute $\Pr[X, Z; Y]$, in which case we merely have developed a chain of circular logic. Actually, the two closing ends of our circle are buffered by time delays; we apply the present value of $\Pr[X, Z; Y]$ in order to compute $\Pr[X, Z; Y]$ three intervals later. Similarly, the time-delay buffers allow us to apply the presently computed value of $\Pr[Y]$ to point (2) in the chain, since it will be used in the calculation of *subsequent* values of $\Pr[Y]$. All that remains now is to close the chain at point (1). To do so, we require

$$\Pr[X, Y]$$

which can be obtained by time delay from

$$\Pr[W, X]$$

which can be obtained by deconditioning

$$\Pr[W, X; Y] \quad \text{with } \Pr[Y]$$

both of which already are available.

Employing the elements of Figure 3.43 and the *time delay*, we can construct a network to depict the entire logical chain from $\Pr[X, Z; Y]$ to $\Pr[W]$ and $\Pr[Y]$. To do so, we simply follow our route backward, beginning at $\Pr[X, Z; Y]$. The result is illustrated in Figure 3.48. The only paths remaining open in this network are those carrying $\Pr[W; X, Z]$ and its immediate de-

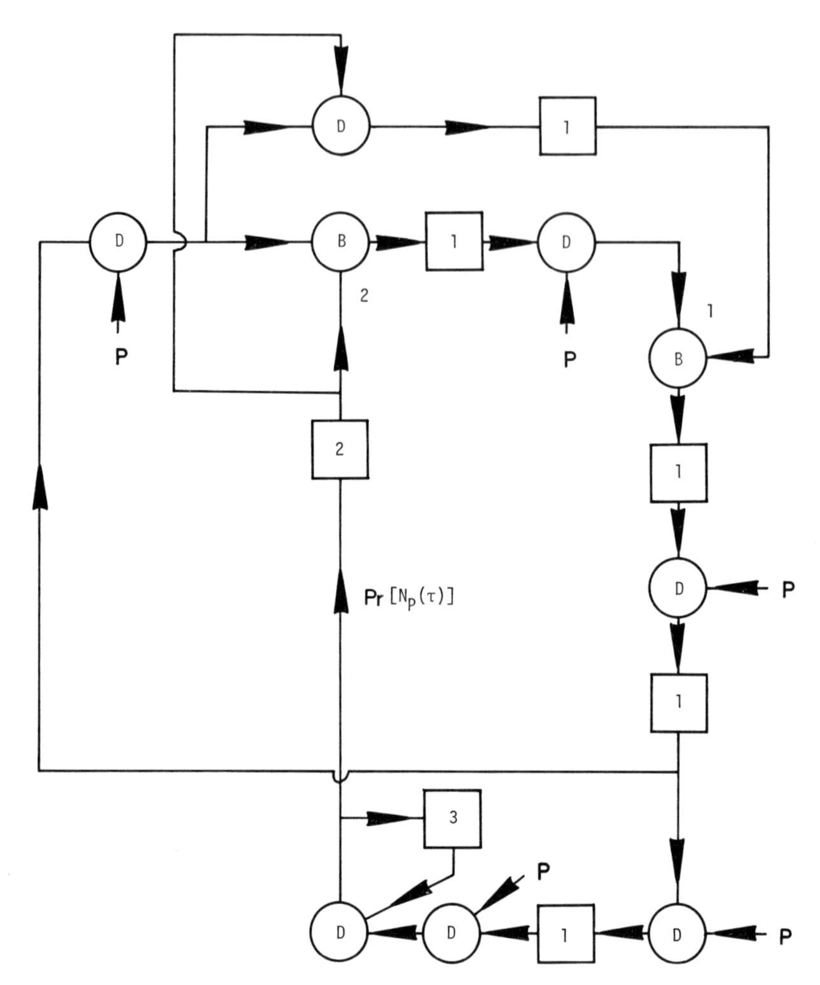

Fig. 3.48. Final step in the conversion to a stochastic network model. The segment of Fig. 3.47 is incorporated at the bottom, immediately to the left of the one-interval *time delay*. P represents the array $\Pr[W;X,Z]$ and its immediate derivatives. The dynamics of the system are represented by $\Pr[N_P(\tau)]$, which is available at the central arm of the network

rivatives to deconditioning operators. $\Pr[W;X,Z]$ is calculated by the convolution of $\mathbf{Pr}[J_{NP}(\tau);\ N_p(\tau-\mathbb{3})]$ and $\mathbf{Pr}[J_{SP}(\tau);\ N_p(\tau-\mathbb{1})]$. Thus, with the addition of the convolution operator of Figure 3.41, the network of Figure 3.48 becomes a closed, stochastic version of the network of Figure 3.45. Notice that most of the structure in this stochastic network is related to the manipulations of probability arrays required because of the correlation between J_{NP} and J_{SP}. In essence, this structure keeps track of the history of the system dynamics that is required to deduce the present dynamics. If the correlation between J_{NP} and J_{SP} were ignored, so that a structure such as that of Figure 3.48 could be avoided, then the stochastic model would lead to fallacious deductions. For example, if one assumed that initially there were no larval worms and that there was a certain probability (p_E) that all pairs had been exterminated, then the probability that the population was extinct to begin with and therefore would remain so would be p_E. If one computed, interval by interval, the probability of extinction, it never should become less than p_E. When the correlation between J_{NP} and J_{SP} is properly accounted for, this expectation is fulfilled. On the other hand, when the correlation is ignored, the probability of extinction gradually declines, to values well below p_E. Such an outcome, of course, is a logical absurdity.

3.6. Some References for Chapter 3

The numbers in the following lists refer to entries in the Bibliography at the end of the text. The lists themselves are designed to guide the interested reader into some of the literature relevant to particular topic areas treated in Chapter 3.

On Life Cycles, Life-History Data

1, 8, 17, 23, 37, 47, 49, 58, 80, 103, 116, 127, 143, 145, 146, 164, 177, 178, 195, 196, 204, 207, 209, 214, 227, 253, 260, 268, 283.

On *n*-Dependence

25, 31, 39, 40, 41, 44, 49, 67, 68, 84, 97, 140, 142, 145, 154, 173, 188, 190, 191, 203, 205, 219, 222, 225, 226, 228, 229, 233, 246, 266, 271, 275, 284, 294, 297, 299.

On Search and Encounter, Population Interactions

5, 9, 11, 12, 13, 16, 25, 44, 67, 77, 83, 84, 85, 91, 95, 99, 100, 104, 105, 106, 107, 111, 115, 131, 136, 140, 152, 167, 170, 172, 183, 185, 186, 190, 193, 199, 201, 202, 205, 211, 212, 223, 229, 231, 234, 240, 247, 248, 251, 266, 286, 290, 294, 296.

On Time Delays in Population Models

36, 55, 84, 112, 127, 129, 135, 147, 149, 154, 158, 159, 182, 183, 186, 187, 223, 279, 289.

On Periodic and Aperiodic Environmental Factors

6, 7, 15, 21, 25, 42, 44, 78, 108, 126, 138, 142, 147, 165, 169, 170, 173, 182, 188, 205, 210, 288, 294.

On Epidemiology, Parasite-Host Interactions

3, 9, 10, 11, 12, 13, 14, 16, 20, 42, 84, 85, 86, 91, 93, 94, 99, 100, 115, 118, 120, 131, 147, 172, 174, 192, 197, 207, 248, 268, 276, 294.

On Heritable Selection

20, 24, 28, 38, 45, 49, 54, 66, 80, 96, 115, 116, 117, 137, 139, 153, 156, 160, 161, 162, 186, 188, 205, 209, 213, 228, 230, 231, 232, 256, 296.

On Stochastic Population Models

9, 10, 13, 14, 20, 27, 46, 72, 85, 98, 115, 118, 125, 130, 182, 198, 217, 229, 237, 238, 243.

Additional, General References on Populations and Population Modeling

2, 22, 53, 59, 61, 64, 69, 71, 74, 76, 82, 88, 128, 133, 134, 166, 175, 176, 185, 200, 220, 221, 245, 262, 263, 265, 273, 281, 287, 298.

4. Analysis of Network Models

List of Examples

4.1. Introduction to Network Analysis

The previous two chapters have been devoted to the construction of network models from life-cycle specifications. Now we shall move on to some of the methods available for drawing deductions from those models. Since the models discussed in this text involve the accumulations and flows of individuals in a population, the questions we can ask of these models and the deductions that we can draw from them concern the dynamics of such accumulations and flows.

Depending upon the model itself as well as upon the wishes of the modeler, analysis may proceed in any of several directions. For example, a modeler may wish, on the one hand, to deduce general trends in dynamic behavior, or, on the other hand, to deduce detailed descriptions of specific dynamics. Where detailed descriptions are desired, the modeler may wish them to be in completely literal form (i.e., with every parameter and variable represented by a letter or other symbol not implying a specific numerical value), in partially literal form (e.g., with most parameters given specific numerical values, but with time and perhaps a few key parameters represented by letters or symbols), or in completely numerical form (e.g., interval by interval values of key flows and/or accumulations). Completely literal solutions are available only for models that are very simple (very simple indeed!). Examples of such solutions can be found in many texts on population modeling or quantitative ecology, and include the generalized Malthus equation (for pure exponential or geometric growth) and the generalized solution to the Pearl-Verhulst logistic equation. An example already presented in this text is given by equation 1.19. Partially literal solutions are more generally available, but often are limited in their applicability to dynamics in the neighborhood of critical points or other specified points (see section 4.4). Examples of partially literal solutions already presented in this text are given in the answers to Examples 1.6, 1.7 and 1.8. Completely numerical solutions are the most generally available of the three, being limited by the memory capacity of available computers and by the funds available for computer rental. An example of a completely numerical solution already presented is given by the answer to Example 1.13. The more literal a solution is, the more general it is. Thus, general trends in dynamic behavior often can be deduced rather easily from a literal or partially literal solution, but seldom can be deduced from a single numerical solution.

Thus, the most specific of all deductions is that of an interval by interval numerical account of the future dynamics of a particular flow or accumulation, given the present and recent past flows and accumulations and the hypothesized or observed numerical values of the parameters of the model (e.g., the scale factors and delay durations). This kind of deduction may be carried out with a large-numbers network model, in which case the expected value of the flow or accumulation will be deduced for each interval; or it may be carried out with a stochastic network model, in which case the distribution of probabilities over the possible values of the flow or accumulation will be deduced for each interval. In either case, the model may be closed, in which case the deductions can be carried into the indefinite future; or the model may be open, with deductions drawn for the very near future and, perhaps, tentative deductions drawn for the more remote future. Open models have become especially popular in agriculture, integrated pest control, and wildlife management, where they are updated periodically with fresh field data, used for short-term predictions, and then allowed to idle until the next updating. As long as a network model can be constructed for a hypothetical or stipulated situation, purely numerical deductions in principle can be drawn. The execution of the deductive processes in some cases, however, may severely strain the memory capacity of the available computer. This is true especially of stochastic network models, where the conditional-probability arrays may have enormous numbers of elements even when the models themselves are relatively simple.

Completely numerical solutions are, by their very nature, *ad hoc* in the extreme. They apply not only to specific sets of numerical parameter values, but also to specific sets of numerical values of initial flows and accumulations. Generalization over a range of values of initial flows and accumulations or over a range of values of one or more parameters can be accomplished by multiple repetitions of the numerical analysis, with one repetition for each value or set of values selected over the range. With a partially literal solution, on the other hand, generalization over a range (sometimes limited, sometimes quite extensive) of initial flow and/or accumulation values is directly available, and repeated solution often is not necessary. On the other hand, repetitions of partially literal solutions usually are necessary for generalization over a range of values of one or more parameters. Thus, partially literal solutions are *ad hoc* with respect to parameters, but more-or-less general with respect to initial values of flows and accumulations. A completely literal solution may apply to a more-or-less limited range of dynamics and a more-or-less limited range of parameter values (depending upon the simplifying assumptions that made the literal solution possible), but over those ranges, it is general with respect to initial flow and accumulation values and with respect to parameter values.

Because of its *ad hoc* nature and the interval-by-interval accounting that is involved in its execution, completely numerical solution often is labelled *simulation*. Using the simulation label in this manner, one would say that *simulations lead to ad hoc deductions.*

4.2. Interval by Interval Accounting on a Digital Computer

Each element in the network models of this text has a well-defined operation or function that can be executed in a straightforward manner with standard digital computers. Therefore, one may view any of these network models simply as a flow chart, ready to be translated almost directly into a computer program. Scaling a variable by a constant, adding two variables, and multiplying one variable by another all are operations available directly in computer languages at many levels. Therefore, translation of the *scalor*, the *adder*, and the *multiplier* into computer instructions requires no elaboration here.

4.2.1. Digital Representation of *Time Delays* in Large-Numbers Models

Translation of the *time delay* is only slightly more complicated. Basically, it involves the assignment of a memory location to each time class (each transition interval) represented in the *time delay*. Thus if a *time delay* duration is T transition intervals, T memory locations will be required. The locations must be labelled to form an ordered sequence, as depicted in Figure 4.1. At each transition interval, the contents of each location are transferred to the next location in the sequence. Thus, in the scheme of Figure 4.1, each time the *time delay* operation is to be updated for another transition interval, the contents of location T will be transferred out of the sequence, the contents of location $T-1$ will be transferred into location T, and so on, back through the sequence until the contents of location 1 have been transferred to location 2 and appropriate new contents have been stored at location 1. It is left as an exercise for the reader to determine the rather disasterous consequences of carrying out the

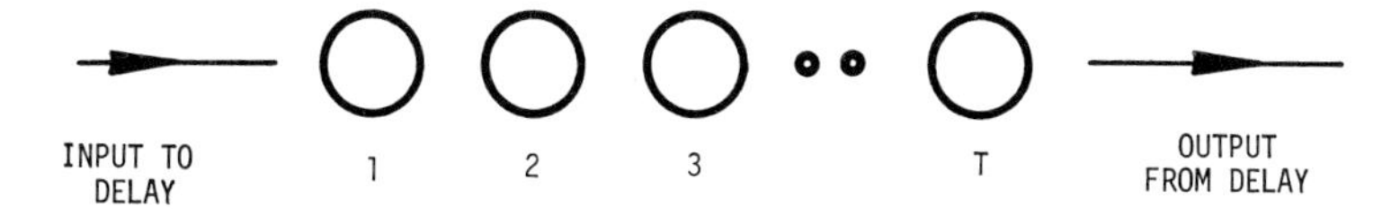

Fig. 4.1. *Time delay* represented as an ordered sequence of memory locations in a digital computer

transfers in the reverse order. If parallel processing were available, the transfers could be executed simultaneously; but until it is, the transfers must be carried out one step at a time, beginning at the output end of the delay. In programming languages that depend upon a compiler to assign memory locations, the transfer of contents along a chain of locations usually can be written simply as a manipulation of indices. Thus, in FORTRAN we might have a flow variable, X, that must pass through a delay of 25 transition intervals. We simply could consider X(I) as an array of 25 elements, which we would indicate to the compiler through a dimension statement:

```
DIMENSION X(25)
```

Then we could signal the sequential transfer of location contents by means of a simple loop

```
    ⋮
    OUTPUT=X(25)
    DO 111 J=1, 24
    I=25-J
111 X(I+1)=X(I)
    X(1)=INPUT
    ⋮
```

If the *time delay* duration to be represented varies from interval to interval, then one of the two schemes indicated in Figure 3.34 could be applied. If the duration were known *a priori*, then the input could be added directly to the updated contents of the appropriate location. In FORTRAN we could indicate this as follows:

```
    ⋮
    OUTPUT=X(T)
    DO 314 J=1, T-1
    I=T-J
314 X(I+1)=X(I)
    X(1)=0
    X(K)=X(K)+INPUT
    ⋮
```

where K is the *a priori* determined location of the point of entry for present inputs. If the duration is determined *a posteriori*, then the output could be taken as the sum of the contents of appropriate locations, before updating. In FORTRAN we could indicate this as follows:

```
          ⋮
          L=T-K
          Z=0
          DO 555 J=1, L+1
          I=T-J
          Z=Z+X(I+1)
      555 X(I+1)=0
          OUTPUT=Z
          DO 558 J=L, T-1
          I=T-J
      558 X(I+1)=X(I)
          X(1)=INPUT
          ⋮
```

Example 4.1. Binary Fission with Poisson Death. Construct a FORTRAN program for the network of Figure 4.2, which embodies the dynamics of the population of idealized protozoans of Example 3.1 with a fission interval equal to ten transition intervals.

Answer: First of all, we can assign FORTRAN symbols for floating-point variables to each of the three flows and to the survivorship parameter, and a FORTRAN integer symbol to discrete time:

$J_0(\tau)$ is X0(L)
$J_1(\tau)$ is X1(L)
$J_s(\tau)$ is XS(L)
γ is GAMMA
γ^{10} is Alpha
τ is L

Next, we need to decide which variable we wish to follow and how long we intend to follow it. Suppose that we settle upon $J_s(\tau)$ for 100 intervals; our dimension statement should include XS(100) to indicate the fact that we want 100 consecutive values for J_s (alternatively, we could ask the computer to print each value as it was generated, in which case the dimension statement XS(100)

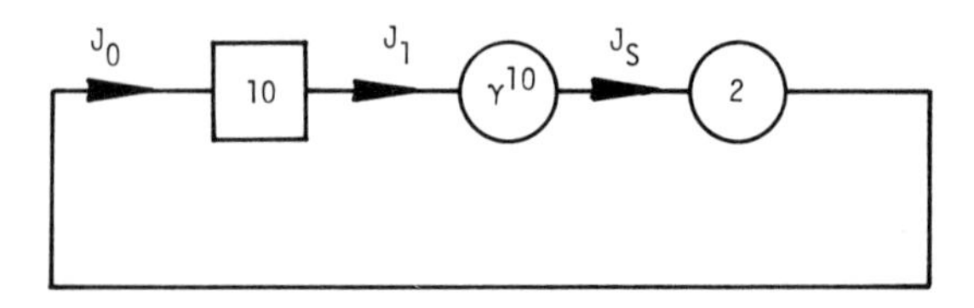

Fig. 4.2. Network model of a population undergoing binary fission

would not be required). Because J_0 enters a delay of 10 transition intervals, a dimension statement X0(10) must be made, whether or not the dynamics are printed as they are computed. J_1 may be used on an interval by interval basis and need not be retained; so no dimension statement is required for X1. The scale factor γ^{10} could be computed ahead of time on a hand calculator, or it could be computed once at the beginning of the program. Computing it over again on each program cycle of course would be a waste of processing time.

PROGRAM STATEMENT	COMMENT
DIMENSION X0 (10), XS (100)	dimension statement
GAMMA =	assign value to γ
ALPHA = GAMMA ** 10	set α equal to γ^{10}
X0(1) =	
X0(2) =	assign initial values to contents of *time delay*
X0(3) =	
⋮	
X0(10) =	
DO 20 L = 1, 100	establish time range
X1 = X0(10)	take output (J_1) from *time delay*
XS(L) = ALPHA * X1	set $J_s(\tau)$ equal to $\gamma^{10} J_1(\tau)$
DO 19 J = 1,9	
I = 10 − J	update *time delay*
19 X0(I+1) = X0(I)	
20 X0(1) = 2 * XS(L)	set *time delay* input equal to twice $J_s(\tau)$
⋮	print results: $J_s(\tau)$ for $1 \leq \tau \leq 100$

The previous example points up one very important point, namely, that before any analysis can proceed initial contents must be specified for every location (every interval, or time class) of every *time delay* in the network. This merely is another aspect of the requirements for closure.

Example 4.2. Periodic Sexual Reproduction with Age-Independent Death. Construct a FORTRAN program for a 100-year analysis of the number of nesting females from the network of Figure 4.3, which embodies the dynamics of

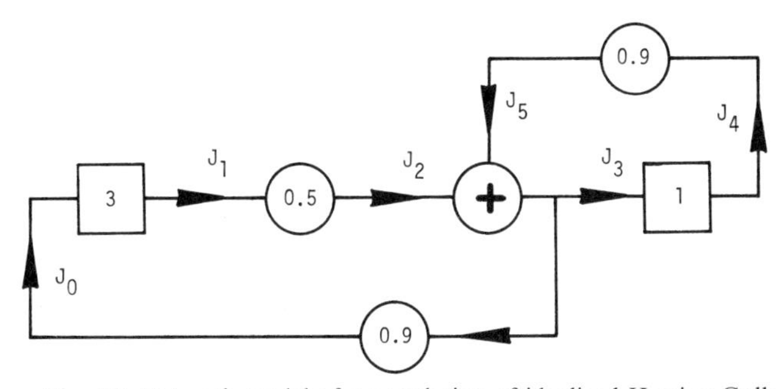

Fig. 4.3. Network model of a population of idealized Herring Gulls

the population of idealized herring gulls of Example 3.5, with a temporal resolution of one year, 0.5 of the female fledglings surviving to adulthood, and 0.9 of the adult females surviving from nesting season to nesting season.

Answer: Replacing J_0 with X0, etc., and τ with L, we have

PROGRAM STATEMENT	COMMENT
DIMENSION X0(3), X3(100)	establish duration of first *time delay* at three intervals, second *time delay* being of duration 1 requires no dimension statement, establish time range for $J_3(\tau)$.
X0(1)= X0(2)=	assign initial values to contents of first *time delay*
X0(3)= X3(1)=	assign initial contents to second *time delay*, to be consistent with those of first delay, J_3 should equal $J_0(1)/0.9$.
DO 15 L=2, 100	establish time range of 100 years
K=L−1	
X4=X3(K) X1=X0(3)	take outputs from every *time delay*
X2=.5*X1	set J_2 equal to 0.5 J_1
X5=.9*X4	set J_5 equal to 0.9 J_4
X3(L)=X2+X5	set J_3 equal to J_2+J_5 (if the second *time delay* had been of duration greater than 1, we would have updated it before computing J_3)
X0(3)=X0(2) X0(2)=X0(1)	update first *time delay*
15 X0(1)=.9*X3(L)	set J_0 equal to 0.9 J_3 and apply it as input to the first *time delay*

For large-numbers network models, *time delays* probably are the most difficult of the elements to represent in a computer program, simply because they present an excellent opportunity for bookkeeping error. Generation of specified variations of scale factors or *time delay* durations, on the other hand, generally presents no problems. Random variations can be treated in a Monte Carlo fashion, being specified either ahead of time, by *a priori* generation of the entire required random sequence, or on an interval-by-interval basis by means of random number generating routines commonly available for general-purpose computers. Periodic parameter variations can be incorporated in a program by means of simple clock routines that cycle indices through periodic sequences of values. For example, one might use the following routine to represent a ten-interval cycle:

```
   ⋮
   ALPHA=X(I)              set parameter to value
                              corresponding to phase I
   I=I+1                   update phase
   IF(10-I) 12,12,14       determine whether cycle is
                              complete
12 I=1                     If cycle complete, reset to
                              initial phase
14 CONTINUE                continue program
   ⋮
```

Parameter values that depend upon flows or accumulations in the network model often can be realized directly with functions already available in the compiler. Occasionally, the required functions are not available already, but their generation usually imposes little difficulty. Often the generation of functions, whether with routines available in the compiler or with routines specially designed by the modeler, requires a relatively large amount of time. If function generation is repeated on every program cycle, it can occupy a large part of the cycle time. Functions of discrete variables (e.g., factorials, binomial coefficients, and the like) usually span a range of values that is sufficiently small that the entire function can be stored quite easily as a table in available memory locations. In this way, it can be generated just once, before the main program is begun, and the appropriate values brought forth on each cycle of the main program. Functions of continuous variables often are sufficiently well-behaved that very simple interpolation routines can provide accurate estimates of their values from stored tables with reasonable numbers of entries, generated prior to the beginning of the main program. In this way, complicated, time-consuming function generation often can be transformed into a rather simple, much less time consuming process.

4.2.2. Digital Representations for Stochastic Network Models

The operations of the various elements of stochastic network models (Figures 3.42 and 3.43) are a bit more complicated, but they lend themselves very well to digital-computer realization. The probability vectors and conditional-probability matrices are represented quite directly as arrays of various dimensionality. Thus, $\mathbf{Pr}[X]$ can be represented in FORTRAN as an array such as PRX(I), where I is an integral variable representing the various integral values that X can assume. Similarly the array PRXY(I, J) might be used to represent the conditional-probability matrix $\mathbf{Pr}[X;Y]$, where I and J represent the various integral values that can be assumed by X and Y respectively. A *time delay* of duration greater than one transition interval effectively requires an increase by one in the dimensionality of its input vector or matrix. Thus, in FORTRAN, if we represented the various time classes in the delay by various values of the integral variable L, we could carry the vector $\mathbf{Pr}[X]$ through the delay process as a matrix array PRX(I, L) or we could carry the matrix **Pr**$[X; Y]$ through the delay as a three-dimensional array PRXY(I, J, L). Updating the *time delay* of duration T could be carried out as follows:

```
          ⋮
      DO 222 J = 1, JMAX
      DO 222 I = 1, IMAX
      OUTPUT(I, J) = PRXY(I, J, T)
      DO 220 K = 1, T - 1
      L = T - K
  220 PRXY(I, J, L + 1) = PRXY(I, J, L)
  222 PRXY(I, J, 1) = INPUT(I, J)
          ⋮
```

where IMAX is the largest value of X and JMAX is the largest value of Y.

The "scaling" of a probability vector by a conditional-probability matrix (i.e., the replacement of the *scalor* operation in the transformation from large-numbers network model to stochastic network model) is represented with extreme simplicity in most programming languages. In FORTRAN, for example, we can represent the inner product of **Pr**$[X]$ by **Pr**$[Y; X]$ as follows:

```
          ⋮
      DO 442 J = 1, JMAX
      Z = 0
      DO 440 I = 1, IMAX
  440 Z = Z + PRYX(J, I) * PRX(I)
  442 PRY(J) = Z
          ⋮
```

The operation of convolution is only slightly more complicated. If zeros were allowed as indices for arrays, then the convolution of **Pr**$[X]$ and **Pr**$[Y]$ to form **Pr**$[W]$ could be represented in FORTRAN as follows:

```
          ⋮
      WMAX = IMAX + JMAX
      DO 555 K = 0, WMAX
      Z = 0
      DO 554 I = 0, IMAX
      IF(K - I) 553, 553, 555
  553 J = K - I
  554 Z = Z + PRX(I) * PRY(J)
  555 PRW(K) = Z
          ⋮
```

This same routine, of course, could be used to represent the operation of deconditioning in Figure 3.43. It becomes only slightly more complicated when the two probabilities being operated upon are of higher dimensionality (i.e., a conditional probability being deconditioned by another conditional probability).

The operation whereby the probability of X *given* Y, $\mathbf{Pr}[X; Y]$, and the probability of Y, $\mathbf{Pr}[Y]$, are combined to yield the probability of X *and* Y, $\mathbf{Pr}[X, Y]$ also is represented very easily. In FORTRAN, for example, we could write

```
        ⋮
    DO 100 I = 1, IMAX
    DO 100 J = 1, JMAX
100 PRXAY(I, J) = PRXGY(I, J) * PRY(J)
        ⋮
```

where PRXGY(I, J) represents $\mathbf{Pr}[X; Y]$, and PRXAY(I, J) represents $\mathbf{Pr}[X, Y]$. Finally, the operation of Baye's Rule is the most complicated of all to represent, even though it is relatively simple. For example, the combination of $\mathbf{Pr}[X; Y]$ and $\mathbf{Pr}[Y]$ to form $\mathbf{Pr}[Y; X]$ can be written as follows in FORTRAN:

```
        ⋮
    DO 711 I = 1, IMAX
    Z = 0
    DO 710 J = 1, JMAX
710 Z = Z + PRXY(I, J) * PRY(J)
    PRX(I) = Z
    DO 711 J = 1, JMAX
711 PRYX(J, I) = PRXY(I, J) * PRY(J)/PRX(I)
        ⋮
```

With routines such as those presented in this section at hand, one very easily can translate a large-numbers or stochastic network model into a digital computer program. Once the resulting program has been corrected ("debugged") for the various flaws that inevitably seem to plague such schemes, it can be used for the *ad hoc*, *simulation* type of analysis described in the previous section. Repeated application of such analysis with different parameter values or different initial values of the variables may lead to generalizations that appear to transcend the details of model specification. On the other hand, repeated analysis of this kind can be quite expensive, especially in stochastic models, where the numbers of elements in high-dimension arrays may be very large. For example, if one were interested in random variables X, Y, and Z, each of which could take on 100 discrete values of interest, then the vector $\mathbf{Pr}[X]$ would have 100 elements, the matrix $\mathbf{Pr}[X; Y]$ would have 10,000 elements, and the three-dimensional array $\mathbf{Pr}[X; Y, Z]$ would have 1,000,000 elements. The numbers of operations that would be required on each program cycle would be correspondingly large; and the number of transition intervals for which dynamics could be computed for a given amount of money would be correspondingly small.

4.2.3. Examples of Digital Modeling

Figures 4.4 through 4.10 show seven network models and a digital-computer analysis corresponding to each one.

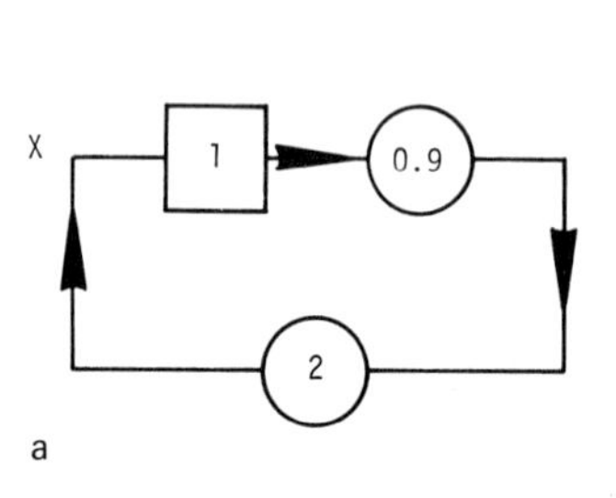

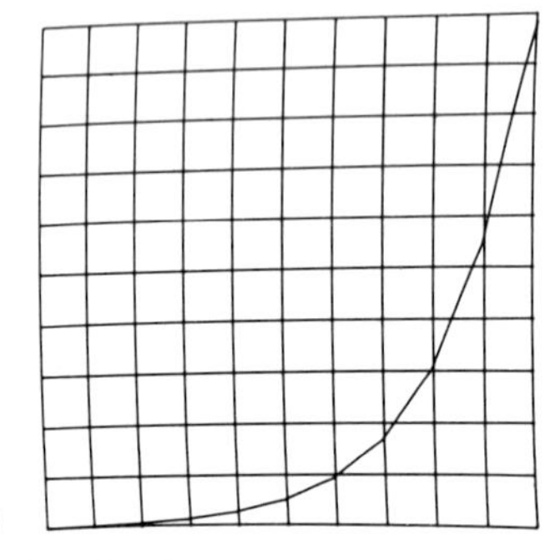

Fig. 4.4a and b. Network model of a population undergoing binary fission (a), along with a graphical display (b) of the dynamics calculated by digital computer. Maturation of the daughter cells requires one transition interval; 0.9 of the daughters survive to maturity and divide. The computer output was photographed directly from a cathode ray oscilloscope display unit connected to the computer. Horizontal grid spacings represent one transition interval each; vertical grid spacings represent twenty individuals each. At $\tau = 1$, the population was set equal to one individual. In plotting the output, the computer drew straight lines between successive data points, simulating a continuous growth curve

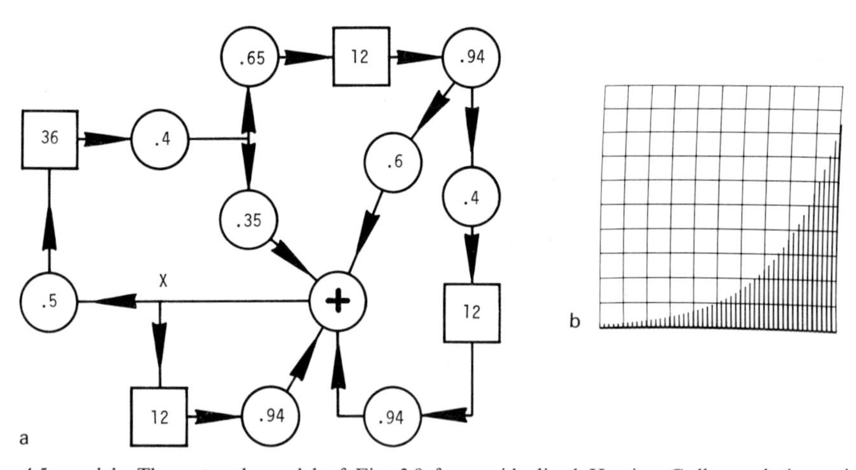

Fig. 4.5a and b. The network model of Fig. 3.9 for an idealized Herring Gull population, with numerical values inserted for each parameter (a), along with a graphical display of the dynamics calculated by a digital computer (b). Each horizontal grid spacing represents 60 months (the computer output being desplayed on a month-to-month basis); and each vertical grid spacing represents six adult females participating in nesting. The initial number of females participating in nesting was set equal to one. Since nesting occurs once every twelve months, it is displayed by the computer as a series of vertical spikes, twelve transition intervals apart

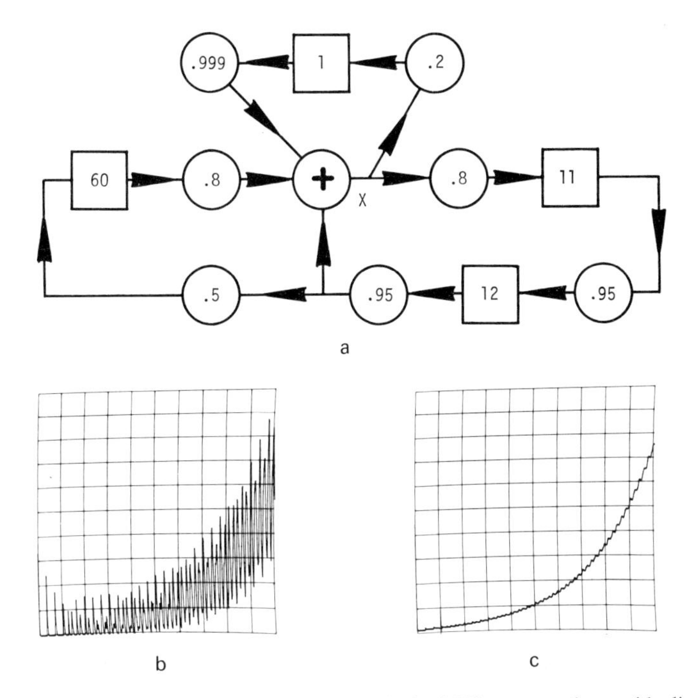

Fig. 4.6a–c. Simplified version of the network model of Fig. 2.37b representing an idealized cetacean population, with numerical values inserted for each parameter (a), along with two graphical displays (b and c) of the dynamics calculated by a digital computer. Each horizontal grid spacing represents 65 months (the output being calculated and displayed on a month to month basis). The flow of females into ovulation is displayed vertically. In (b) the modeled population was intitiated with 33 females flowing into ovulation, and each vertical grid spacing represents ten females ovulating. In (c) the modeled population was initiated with 84 females distributed uniformly over all of the time classes in the model (one in each month of the maturation process, one in each month of gestation, one in the ovulation cycle, and one in each month of lactation); and each vertical grid spacing represents five ovulating females

4.3. Graphical Analysis of One-Loop Networks with Lumpable Parameters

If a large-numbers network model comprises a single loop (as shown in Figure 4.11) with *scalors* each of whose scale factors either is constant or depends only upon the input flow to the *scalor* and with *time delays* of constant duration, then it belongs in the class of network models depicted in the title of this section. Such networks can be analyzed by means of a well-known graphical method for discrete-time systems, a method that provides general conclusions much more directly than does numerical simulation of the type described in the previous section. In such a network, the expected value of any flow variable at interval τ is uniquely determined by the value of that same flow variable at interval $\tau - T$, where T is the sum of the durations of all the *time delays* in the network. Therefore, either by simple analysis or with the aid of a computer if the *scalor* functions are complicated, we can determine and plot $J(\tau)$ as a function of

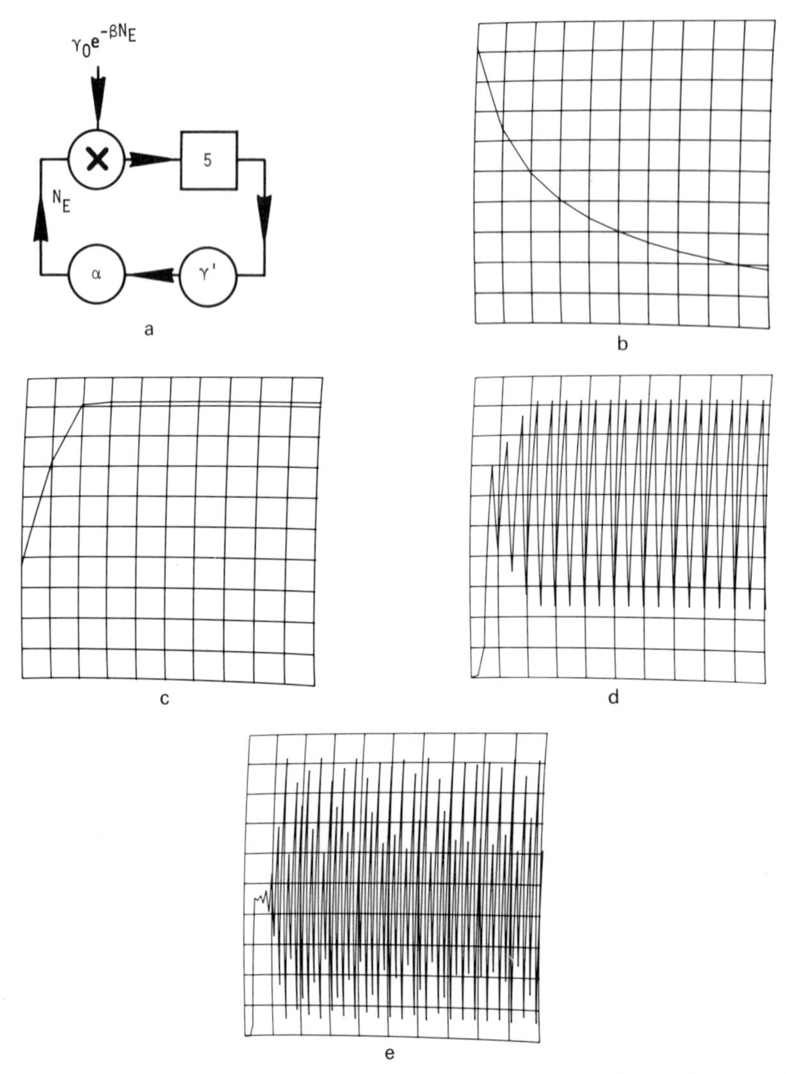

Fig. 4.7. A network version of Ricker's model (see ref. 246) of an idealized salmon population (Example 4.4), along with four graphical displays of the dynamics calculated by a digital computer. N_E is the number of fertilized eggs produced by spawners; the factor $\gamma_0 e^{-\beta N_E}$ is Ricker's estimate of the probability of survival from fertilized egg to adult; γ' is the probability of an adult surviving the gantlet of fishermen to arrive at the spawning grounds; and α is the expected number of fertilized eggs per spawner. The normalized variable βN_E is plotted vertically in the displays, discrete time horizontally. In each case, βN_E initially was set equal to one. In (b) the lumped parameter $A = \alpha\gamma'\gamma_0$ was set equal to 1.0; and each horizontal grid spacing represents five years. In (c), $A = 3.0$; and each horizontal grid spacing represents five years. In (d), $A = 10$; horizontal grid spacings represent 20 years. In (e), $A = 15$; each grid spacing represents 50 years. Note that in (b) the population declines toward zero; in (c) it increases toward a constant, non-zero level; in (d), it grows toward a stable oscillation between two non-zero levels, with the period of oscillation being ten years; and in (e), it grows toward a non-zero level about which it then fluctuates rather chaotically. The fact that this one model can lead to the first three modes of behavior has been well-known for many years; the availability of the fourth mode and its implications were first pointed out by May and Oster (see ref. 184)

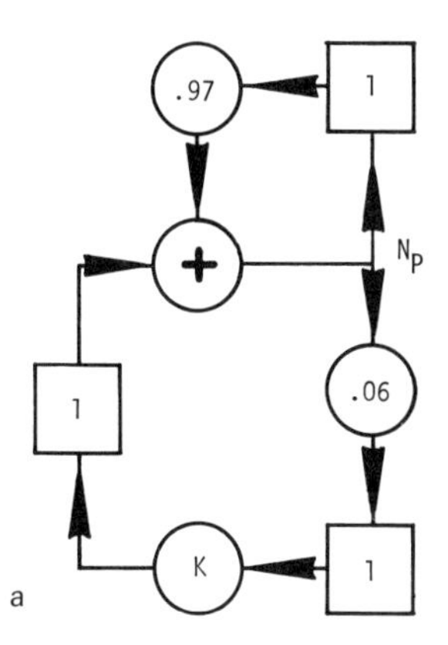

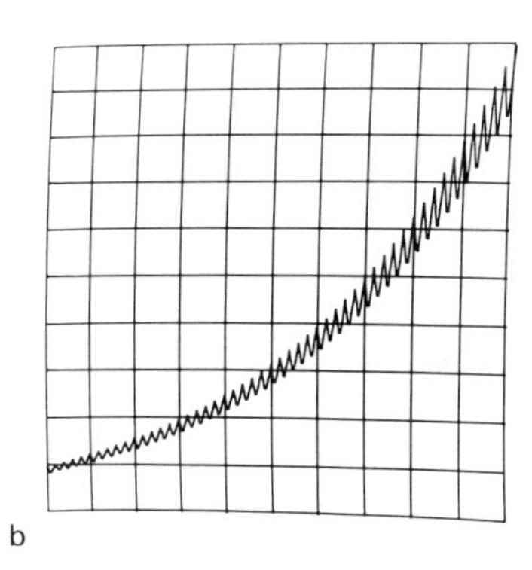

Fig. 4.8a and b. A network model of an idealized schistosome population, derived from Fig. 3.21, along with a graphical display of the dynamics calculated by a digital computer. The parameter K represents seasonal variation in the transmission of larvae through the intermediate hosts (snails). In the five months from late fall to early spring, transmission is taken to be zero, and in the seven months from mid-spring to mid-fall the transmission is taken to be one. The number of adult pairs (N_P) is displayed vertically in (b); and each horizontal grid spacing represents five years. The modeled population was initiated with one pair

$J(\tau-T)$. Then, by inspection of the resulting graph, we can draw certain, quite general conclusions concerning the dynamics of the model.

To see how this is accomplished, consider the example illustrated in Figure 4.12. In Figure 4.12a, where we simply have a graph of $J(\tau)$ as a function of $J(\tau - T)$, we can proceed on an interval-by-interval analysis as follows. Choose an initial value of J (i.e., assign a value to $J(0)$). Plot this on the horizontal axis; the vertical coordinate (the *ordinate*) of the point directly above it on the graph corresponds to the deduced expected value of J, T units of time later (i.e., to $J(T)$). Next, plot this value of $J(T)$ on the horizontal axis; the ordinate of the point above it on the graph is the expected value of $J(2T)$. Repeat the process until you have values of J for as many multiples of T as you wish. You also may repeat the process for values of J at intervals between $\tau=0$ and $\tau=T$. In other words, you may begin with $J(1)$ and from it find $J(T+1)$, then $J(2T+1)$, then $J(3T+1)\ldots$, or with $J(2)$ and then find $J(T+2)$, $J(2T+2)$, $J(3T+2)\ldots$, and so forth. Each of these sequences will depend only upon its initial value, being otherwise independent of the other sequences.

The process described in the previous paragraph can be facilitated by the presence of a reference line of slope 1, passing through the origin, as illustrated in Figure 4.12b. Beginning with the horizontal coordinate (*abscissa*) equal to $J(0)$, construct a vertical line to find the corresponding ordinate of the graph, which is equal to $J(T)$. Next, transform this ordinate value into the abscissa value for the following interval by constructing a horizontal line to the reference line. The process is repeated until values of J have been obtained for as many intervals as desired. Before proceeding into the following arguments, the reader should work through the process with Figure 4.12 until he or she is thoroughly familiar with it.

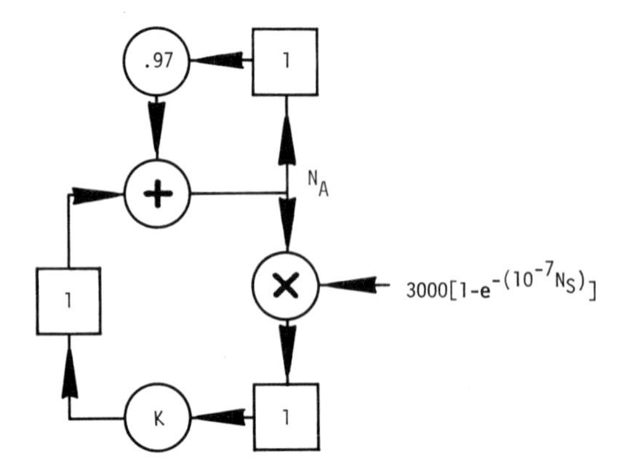

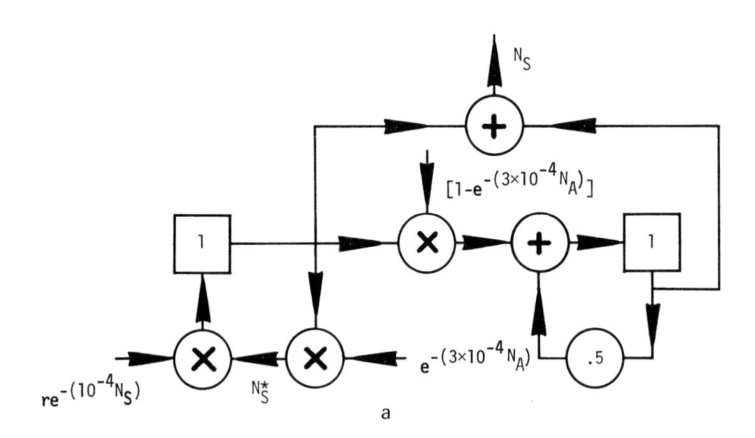

Fig. 4.9a–g. Network model of interacting schistosome and snail populations. The schistosome model is derived from that in Fig. 4.8, but with $K=1$ for all months of the year, and with the probability of a miracidium finding a snail being given by a Nicholson-Bailey relationship. The snail model is similar to that for the salmon (Fig. 4.7) in that Ricker's "stock-recruitment" relationship is embodied in the snail life cycle, providing a logistic-like saturation level for that population. In addition, the probability of survival of infected snails is taken to be reduced, and infected snails are assumed not to reproduce. N_S is the total number of snails; N_S^* is the number of uninfected snails; N_A is the number of adult pairs of schistosomes. Multiple infections of snails are allowed.

The dynamics of the model when $r=4$ are shown in the graphical displays (b) and (c). Here each horizontal grid spacing represents fifty months. The number of schistosome adult pairs is displayed in (b), where each vertical grid spacing represents 500 pairs; and the total number of snails (N_S) is displayed in (c), where each vertical grid spacing represents 2000 snails.

The dynamics of the model with $r=14$ are displayed in (d) and (e). Here each grid spacing represents 25 months. N_A is shown in (d) with each vertical grid spacing representing 1200 adult pairs; and N_S is shown in (e), with each vertical grid spacing representing 5000 snails.

For $r=20$, N_A is displayed in (f), with each vertical grid spacing representing 2500 pairs of worms; and N_S is displayed in (g), with each vertical grid spacing representing 10,000 snails. Each horizontal grid spacing represents fifty months.

The very-long-period oscillations exhibited by (d) through (g) are quite typical of complex nonlinear systems. Note that the longest *time delay* in the model represents just one month, yet the period of oscillation is nearly 50 months in (d) and (e) and approximately 80 months in (f) and (g)

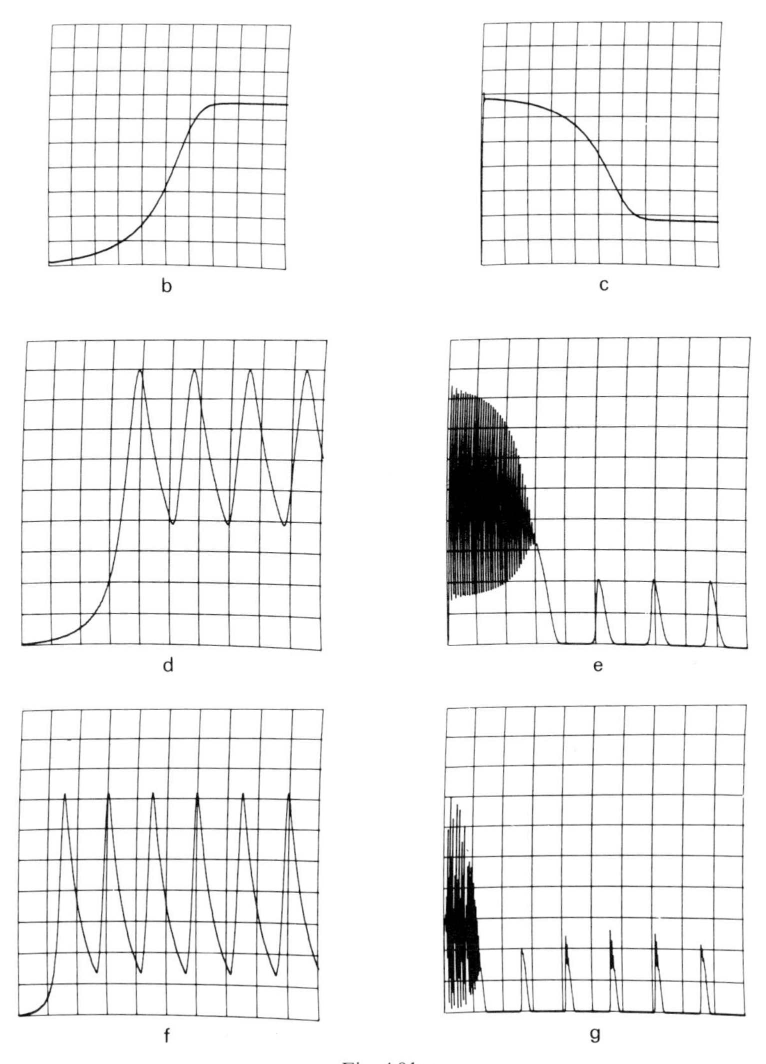

Fig. 4.9 b–g

Figure 4.13 illustrates one general feature of the dynamics of one-loop networks with lumpable parameters. Where the graph of $J(\tau)$ versus $J(\tau - T)$ lies above the reference line, the value of J will grow as time proceeds, as illustrated in Figure 4.13a. Where the graph lies below the reference line, the value of J will decline, as illustrated in Figure 4.13b. In Figures 4.13a and b, the graphs of $J(\tau)$ versus $J(\tau - T)$ happen to be straight lines passing through the origin. In such cases, J will grow or decline *geometrically* with time (i.e., as a geometric series). It is left as an exercise for the reader to demonstrate that the *common ratio* of the series is simply tan θ, where θ is the angle between the graph and the horizontal axis. In other words, the value of J grows or declines as follows:

$$J(kT) = (\tan\theta)^k J(\mathbb{O}). \qquad (4.1)$$

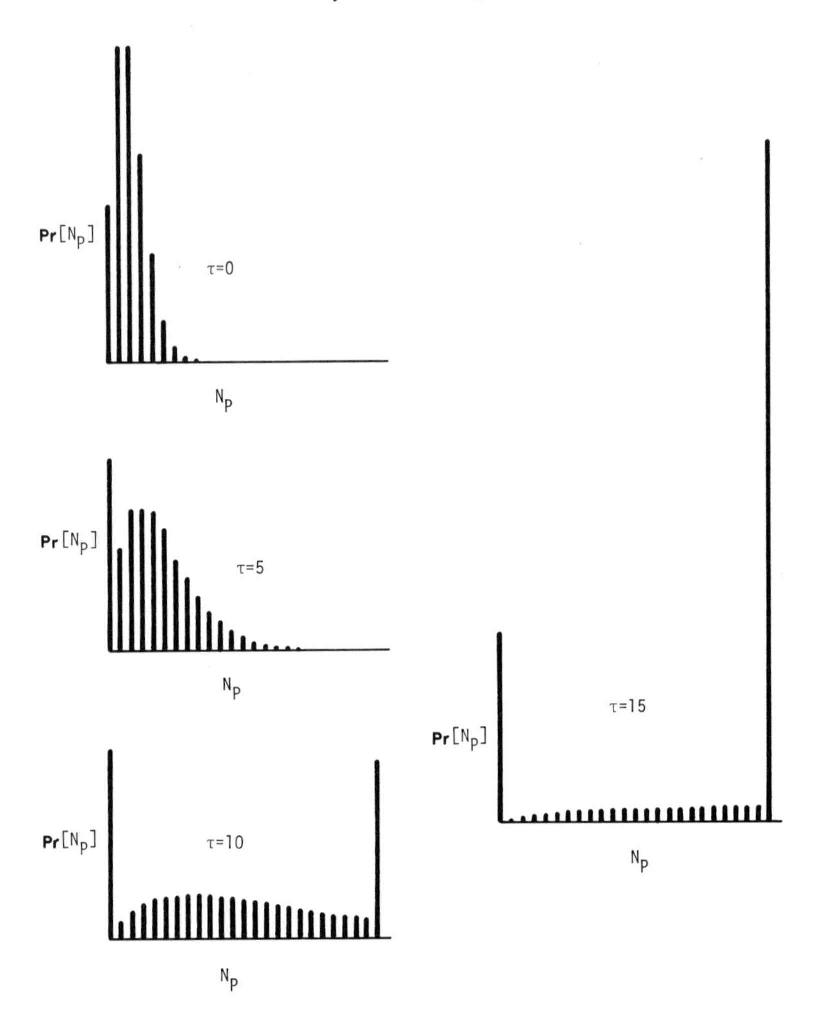

Fig. 4.10. Computer-generated response of the stochastic network model of Fig. 3.48, with specific numerical values assigned to the parameters. At $\tau=0$ the probability density function for N_P (the number of adult pairs) was established by assuming that each worm represented a Bernoulli trial with respect to whether or not it was killed by chemotherapy. The pattern in the upper left represents the expected density function for mated pairs among the worms that survived chemotherapy. After $\tau=0$, the changes in the probability density function are the result of the model, and represent the dynamics of the model. Represented along the horizontal axis are $N_P=0$, 1, 2, ..., 22, 23, and (at the far right) 24 or more. What we see here is a growth of the probability of zero pairs from 0.136 to 0.164, accompanied by a gradual growth of the probability of 24 or more (which has reached 0.597 at $\tau=15$). Thus, the probability density function exhibits a marked bifurcation as its dynamics unfold. This bifurcation corresponds to the two possible outcomes of chemotherapy: it will be successful, leading to extinction of the endemic worm population (N_P going to zero); or it will be a failure, and the endemic will be re-established (N_P going to large values, i.e., 24 or more). Evidently, the probability of success in this particular modeled example is 0.164, the asymptotic probability that $N_P=0$

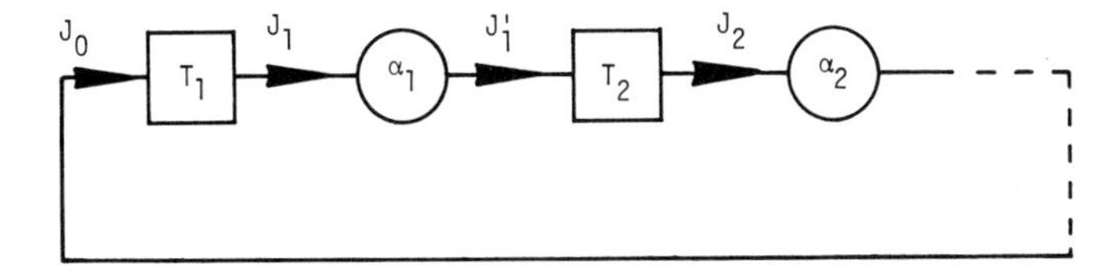

Fig. 4.11. A one-loop, lumpable network model

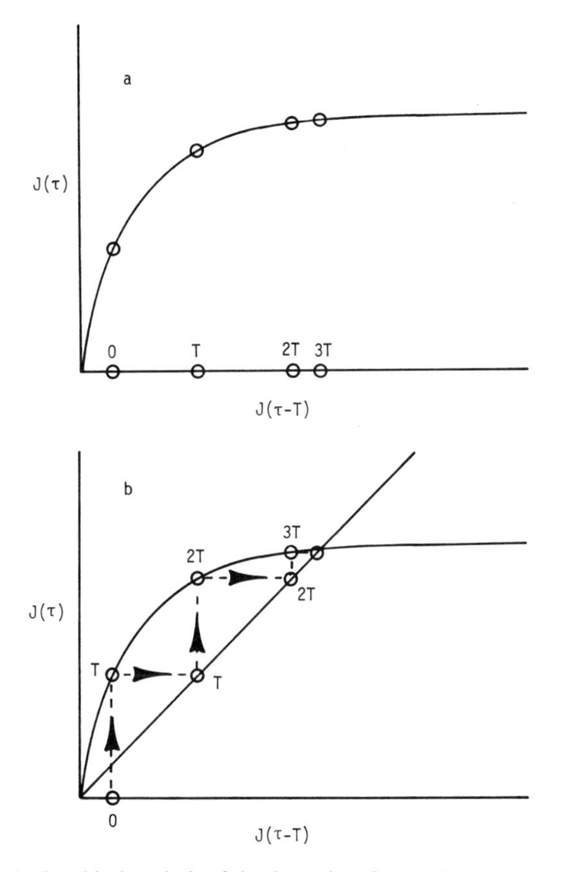

Fig. 4.12. Graphical analysis of the dynamics of a one-loop, lumpable model

Regardless of whether or not the graph is a straight line, if it is above the reference line, J grows; if it is below the reference line, J declines.

Next, we shall consider the various situations depicted in Figure 4.14, in which the graph crosses the reference line. First of all, if the graph slopes upward to the right at an angle greater than 45° as it crosses the reference line, as depicted in Figure 4.14a, then the values of J will decrease progressively in the region where the graph is below the reference line, and they will increase progressively in the region where the graph is above the reference line. In other words, J either will grow or decline, depending upon its initial value. If the

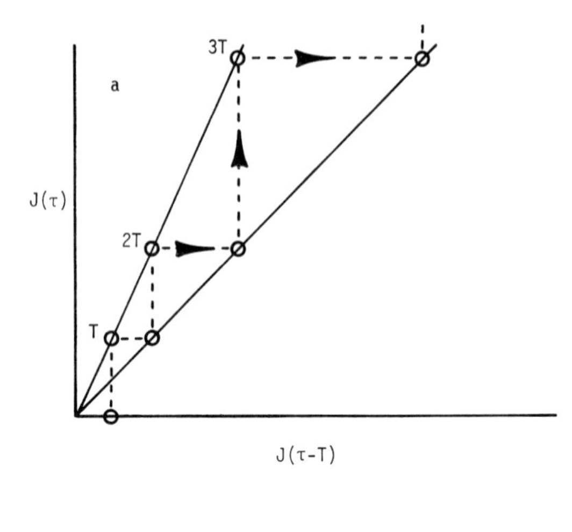

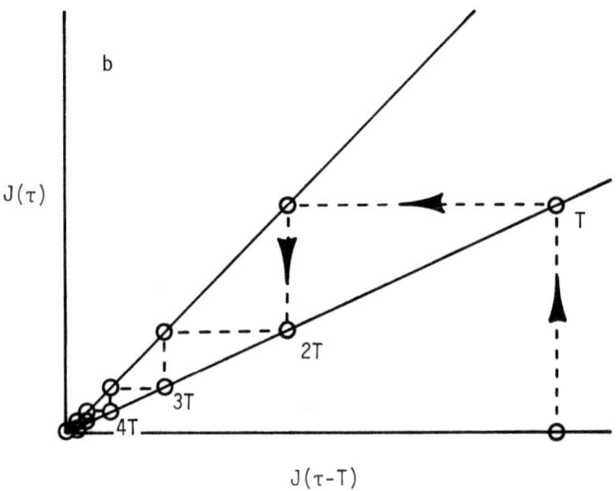

Fig. 4.13. When the graph lies above the reference line, the state variable will grow; when the graph lies below the reference line, the state variable will decline

graph is tangent to the reference line as it passes through it, as depicted in Figure 4.14b, then J will tend to remain at any initial value in the region of tangency. If the graph slopes upward to the right at an angle less than 45° as it crosses the reference line, as depicted in Figure 4.14c, then the values of J gradually will approach the ordinate (and abscissa) of the intersection as time progresses. If the value of J initially were greater than the ordinate of the intersection, then J would decrease steadily (*monotonically*) as it approaches its final value; if the value of J initially were less than the ordinate of the intersection, J would increase monotonically as it approaches its final value. If the graph is horizontal where it crosses the reference line, as in Figure 4.14d, then J will progress very quickly toward its final value (the ordinate of the intersection). If the graph slopes downward to the right at an angle of less than 45°, as depicted in Figure 4.14, then J will approach the ordinate of the intersection as time progresses, but it will not do so in a monotonically

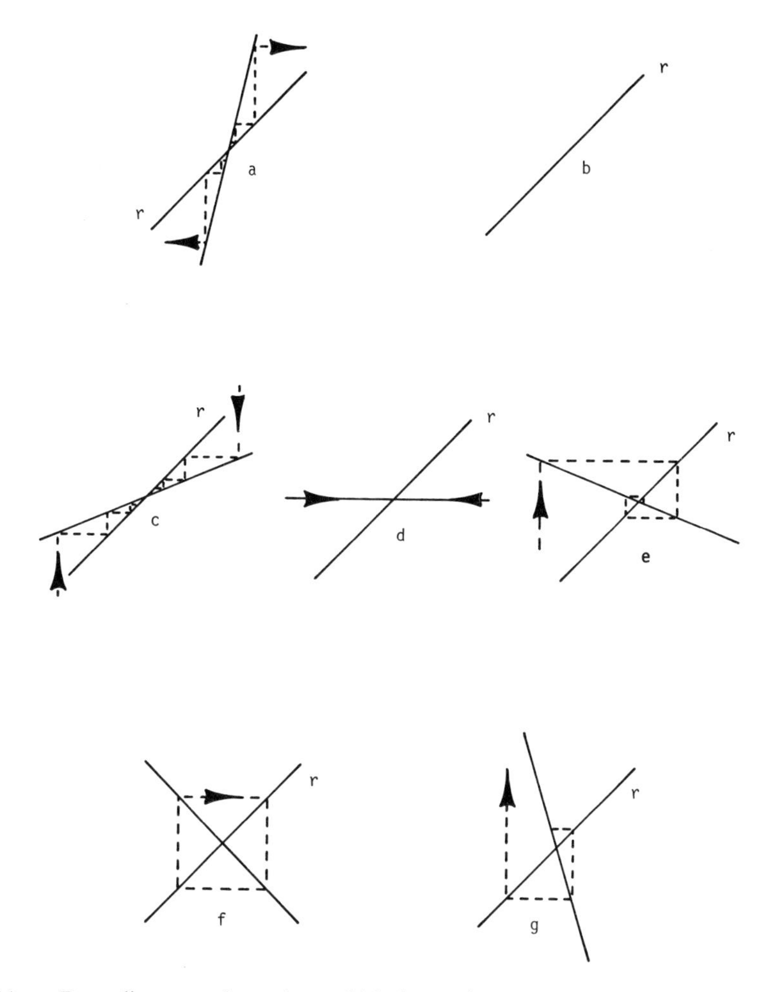

Fig. 4.14a–g. Depending upon the angle at which the graph crosses the reference line (r), the state variable may diverge from the intersection monotonically (a) or in a cyclical fashion (g); it may converge on the intersection monotonically (c) or in a cyclical fashion (e); it may cycle about the intersection on a closed path (f); it may jump immediately to the intersection (d); or it may tend not to move at all (b). Note that (b), (d), and (f) represent points in parameter space; and the probability that one of those three points is occupied is comparable to the relative size of a point, i.e., infinitesimal

increasing or monotonically decreasing manner. Instead, it will oscillate back and forth about its final value, approaching that value rapidly if the graph is nearly horizontal and increasingly slowly as the downward slope of the graph is increased. When the downward slope of the graph is precisely 45°, as depicted in Figure 4.14f, the value of J no longer will approach the ordinate of the intersection. Instead, it will oscillate back and forth between its initial value and a second value equidistant from the reference line but on the opposite side. Finally, if the graph slopes downward at an angle greater than 45°, the value of

J will move farther and farther from the ordinate of the intersection as time progresses. The rate at which it will do so increases as the downward slope of the graph increases from 45°.

The situations depicted in Figure 4.14b and f mark the boundaries between dynamics that bring about convergence of the successive values of J (toward the ordinate of the intersection of the graph and the reference line) and dynamics that bring about divergence of successive values of J. If the graphs in Figure 4.14 were straight lines, then the difference between the value of J and the ordinate of the intersection would increase or decrease as a geometric series as time progressed. Once again, it is left as a rather simple exercise for the reader to demonstrate that the common ratio of the geometric series is tan θ, where θ is the angle between the graph and the horizontal axis and is defined to be positive for slopes upward to the right, negative for slopes downward to the right.

$$J(kT)-I=(\tan\theta)^k(J(0)-I) \tag{4.2}$$

where I denotes the ordinate of the intersection.

When θ lies between 45° and 90°, the common ratio (r) of the series is a positive number, greater than 1. In such cases, r^k will increase geometrically as k increases. When θ is precisely equal to 45°, r is precisely equal to 1, in which case r^k remains unchanged as k increases. When θ lies between 0° and 45°, r lies between 0 and 1, in which case r^k will decrease geometrically as k increases. When θ is precisely equal to 0°, r is precisely equal to 0, in which case r^k remains equal to 0 as k increases. When θ lies between 0° and $-45°$, r lies between 0 and -1, in which case the magnitude of r^k decreases geometrically as k increases, and the sign of r^k oscillates between plus (for even values of k) and minus (for odd values of k). When θ is precisely equal to $-45°$, r is precisely equal to -1, in which case the magnitude of r^k remains unchanged at 1, but its sign oscillates. Finally, when θ lies between $-45°$ and $-90°$, r is a negative number of magnitude greater than 1, in which case the magnitude of r^k grows geometrically and the sign of r^k oscillates as k increases.

Thus, if the graph were a straight line intersecting the reference line, J generally would diverge from or converge on the ordinate of the intersection, depending upon the slope of the graph. In such cases, we could classify the model as being *globally* (everywhere) *divergent* or *globally convergent*. The rate of divergence or convergence also would depend upon the slope of the graph, being lesser for slopes close to $\pm 45°$ (the boundaries between divergence and convergence) and greater for slopes remote from $\pm 45°$. Of course, the graph for a model might not be a straight line, in which case its slope at an intersection with the reference line would determine whether the model were convergent or divergent for values of J close to the ordinate of the intersection, but would not determine convergence or divergence for more remote values of J. Consider, for example, the situations depicted in Figure 4.15. In Figure 4,15a, if the initial value of J corresponds to (is the abscissa or ordinate of) a point on the graph lying within the square, then the subsequent values of J will remain within the square, but will diverge monotonically from the point of intersection between the graph and the reference line. In the same figure, if the initial value of J

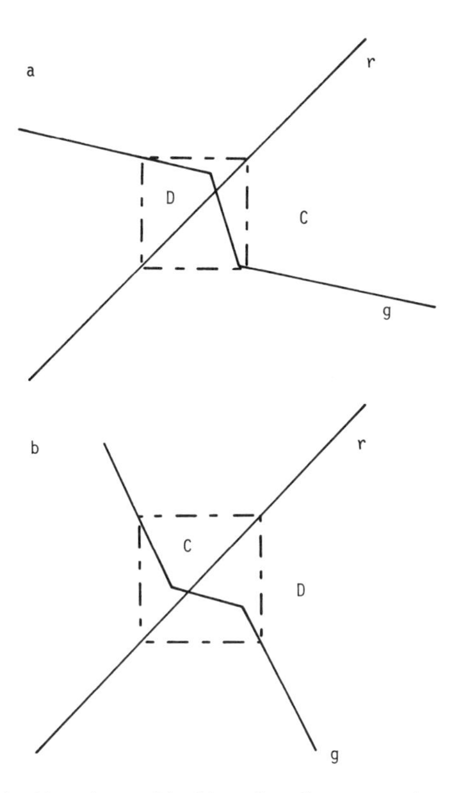

Fig. 4.15a and b. Stable (a) and unstable (b) cycles. C corresponds to region of convergence, D to region of divergence; r indicates the reference line, g the graph of the model

corresponds to a point on the graph outside of the square, the subsequent values of J will remain outside of the square but will converge monotonically toward the intersection of the graph and the reference line. If the initial value of J corresponds to a point of the graph at either corner of the square, subsequent values of J will oscillate back and forth between the two values corresponding to the two diagonally opposite corners of the square. (The reader is invited to experiment with the figure in order to be convinced of these facts.) In other words, outside of the square, the model is monotonically convergent toward a point; inside the square the model is monotonically divergent from a point. The square serves as the outer boundary of a *domain of divergence*; the model can be said to exhibit *local divergence* in the neighborhood of the intersection between graph and reference line. Similarly, in Figure 4.15b, the square forms the outer boundary of a *domain of convergence*; the model exhibits *local convergence* in the neighborhood of the intersection of graph and reference line. An important feature of each of these domains is the fact that *any value of J that is the abscissa of a point on the graph within the domain also is the ordinate of a point on the graph within the domain.*

The squares in Figures 4.15a and b were constructed graphically as follows: A straight-edge with a scale was aligned perpendicular to the reference line and, in that position, moved along the reference line until it intersected the graph at two points equidistant from the reference line. Those two points became diagonally opposite corners of the square. The other two corners lay along the reference line. It is left as an exercise for the reader to develop for himself the logical basis of this method and to convince himself of its generality.

When a *domain of divergence* is surrounded by a *domain of convergence*, then in either domain the model will tend toward the corners of the square separating the two, thence to oscillate back and forth between those corners indefinitely. Thus, in the case of the model represented in Figure 4.15a, J will tend toward the corners of the square regardless of whether it initially is in region C or region D. Since J will oscillate back and forth between the corners, the corners themselves represent alternate phases of a cycle. For rather obvious reasons, that cycle is called a *limit cycle*.

Since the dynamics of the model tend to carry J toward the limit cycle, the limit cycle itself is said to be *stable*, in that it is recovered even if extrinsic forces momentarily push the value of J away from it. Thus, in Figure 4.15a, the corners of the square represent a *stable limit cycle*. The *domain of convergence* within a stable limit cycle and the *domain of divergence* surrounding it together form the *domain of attraction* to that cycle.

When a domain of convergence is surrounded by a domain of divergence, then the corners of the square forming the boundary between the two domains form an *unstable cycle*. If the value of J coincided precisely with the ordinate or abscissa of one of the corners of the square, and there were no extrinsic forces that would shift J, then the subsequent values of J would remain on the cycle. However, if the value of J were shifted even very slightly away from the cycle, the subsequent values of J would become increasingly far from the cycle rather than returning to it.

Squares constructed in the manner of those in Figure 4.15 do not necessarily bound domains of strict convergence or divergence. Consider, for example, the situation depicted in Figure 4.16. In this situation, a given value of J can correspond both to a point on the graph inside the square and to a point on the graph outside of the square, depending upon whether J is taken to be the ordinate or the abscissa of the point. Therefore, such values of J do not lead to monotonic divergence or convergence, but to various combinations of the two; and, strictly speaking, one could not define the region within the square as a domain of divergence, even though the model clearly exhibits local divergence in the neighborhood of the intersection between graph and reference line. Similarly, the region outside of the square is not, strictly speaking, a domain of convergence, even though convergence is the general tendency in most of that region. The cycle represented by the points on the graph in the diagonally opposite corners of the square in such cases may or may not be stable; the values of J may or may not tend toward it. In fact, the sequential values of J may not form a repeated pattern at all, in which case the system gives rise to dynamics that have been given the label *chaotic* (see Figure 4.9 for an example of *chaotic dynamics*). A necessary, but not sufficient, condition for the presence of chaotic dynamics simply is the presence of part of the graph of $J(\tau)$ versus $J(\tau - T)$ lying above or below the square.

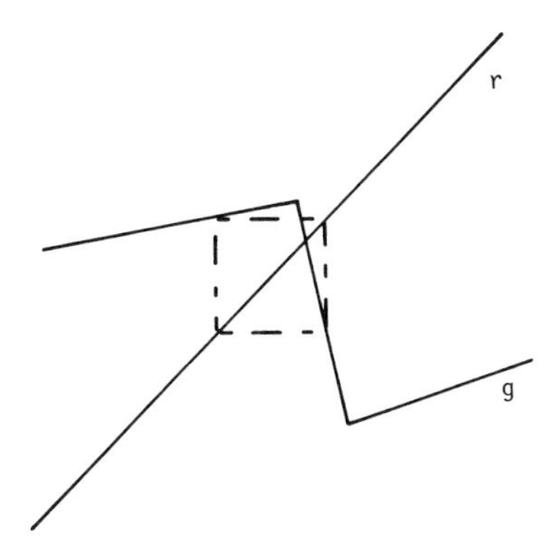

Fig. 4.16. A situation leading to chaotic dynamics about the intersection of graph (g) and reference line (r)

Example 4.3. A Simple Nonlinearity with an Inflection. Consider a population of idealized parasites with a life cycle similar to that of the schistosomes of Example 3.16, but with the survival of larvae being independent of the season, with maturation time in the intermediate host being constant, with the death rate of adults in the primary host being sufficiently high that virtually no adults are expected to survive from one interval to the next, with monogamous mating as represented by Macdonald's model (Example 3.21), and with saturation of the intermediate host, as given by the gantlet model of Example 3.23. Use graphical techniques to determine qualitatively the dynamic behavior of the model.

Answer: Because of the stipulated high death rate of adults, the individuals comprising the mated pairs present in the primary host at a given interval all will be the direct offspring of pairs present T intervals ago, where T is the duration of the maturation time from egg to adult. Using Macdonald's model in combination with the appropriate gantlet model, we can determine the general shape of the curve relating the flow of new offspring (eggs) at the present interval to the flow of offspring T intervals ago. The curve relating the flow of mated pairs to the flow of newly mature adults (according to Macdonald's model) is presented in Figure 4.17a. The curve relating the flow of larvae into intermediate hosts to the flow of eggs (according to the gantlet model) is presented in Figure 4.17b. Combining these two curves, and assuming direct proportionality between the flow of larvae into intermediate hosts and the flow of newly mature adults, and between the flow of mated pairs and the flow of eggs, one obtains a graph of the general shape of that in Figure 4.18, which relates the flow of eggs now to the flow of eggs T intervals ago.

Constructing a reference line through the graph, one immediately sees that there are two intersections. At one of these intersections, the slope of the graph clearly lies between 45° and 90°, implying that the value of the flow of eggs diverges from that point. At the second intersection, the slope of the graph lies between 0° and 45°, implying that the flow of eggs converges to that point. Therefore, the first intersection represents a *threshold*, above which the popu-

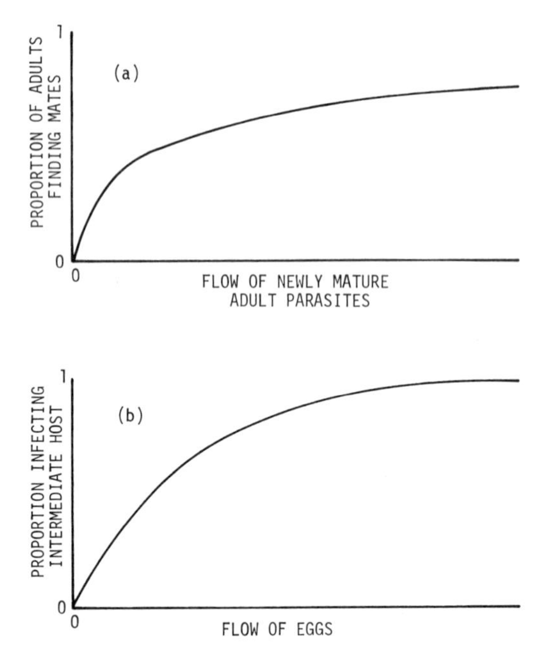

Fig. 4.17a and b. Macdonald's model (a) and the Nicholson-Bailey model (b) applied to an idealized parasite population

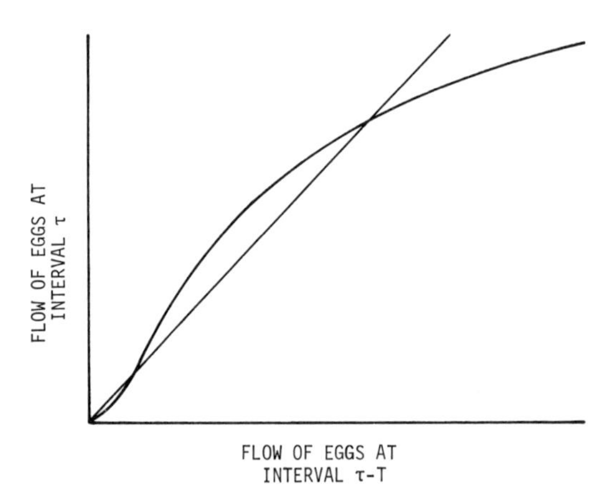

Fig. 4.18. The graph of a one-loop, lumpable schistosome model. The lower intersection of graph and reference line represents a threshold, above which an endemic is established and below which it dies away. The upper intersection represents a stable level of endemic worm population

lation of parasites is expected to grow and below which it is expected to decline to extinction. The second intersection represents a stable endemic level toward which the population will tend if it is above threshold.

It is important to note that the notion of *threshold*, like the notions of nonlinearity and linearity, is restricted to large-numbers models. With respect to stochastic models, one can ascribe a probability that a population will decline from a particular value and a probability that the population will grow from that value. Both probabilities reasonably can be expected to be finite, with no existing demarcation on one side of which the probability of decline is one and on the other side of which the probability of growth is one. Thus, the notion of threshold does not apply.

Example 4.4. Spawner-Recruit Functions. Consider a population of idealized salmon with the following life cycle: Each adult female surviving to reproduce produces α female eggs. Two years are required for a newly produced female egg to become a sexually mature adult. The probability, γ, of survival from egg to adult is dependent upon the number of eggs produced, decreasing exponentially with increasing numbers of eggs (N_e):

$$\gamma = \gamma_0 \exp[-\beta N_e]. \tag{4.3}$$

The adults (*Recruits*) run a gantlet of fisherman and other predators on their way from the sea to the spawning grounds, and each has probability γ' of surviving the gantlet to become a *spawner*. Having spawned, the adult fish dies. Use graphical techniques to determine the dynamic behavior of the population.

Answer: Let $J_S(\tau)$ be the flow of female spawners at τ, where time is resolved to one year. From the description of the life cycle, we know that

$$J_S(\tau) = \{\gamma' \gamma_0 \exp[-\beta N_e(\tau - 2)]\} N_e(\tau - 2)$$

and

$$N_e(\tau) = \alpha J_S(\tau)$$

from which we can construct the equation for the graph of the model

$$J_S(\tau) = \alpha \gamma' \gamma_0 J_S(\tau - 2) \exp[-\alpha \beta J_S(\tau - 2)]. \tag{4.4}$$

Letting $J'_S = \alpha \beta J_S$, and $A = \alpha \gamma' \gamma_0$, we can simplify this equation to the following form

$$J'_S(\tau) = A J'_S(\tau - 2) \exp[-J'_S(\tau - 2)] \tag{4.5}$$

which is plotted in Figure 4.19 for several values of A.

It can be seen that the peak of the graph occurs at $J'_S(\tau - 2) = 1$ for all values of A and that the graph intersects the reference line at the origin for all values of A and at one other point if A is sufficiently large. With elementary calculus, it is very easy to show that the slope of the graph (i.e., the tangent of the angle between the graph and the horizontal axis) is equal to A at the origin and to $1 - \log_e A$ at the second intersection when it exists. Therefore, for the intersection at the origin, equation 4.2 becomes

$$J'_S(2k) = A^k J'_S(0) \tag{4.6}$$

which describes the dynamics of the model very close to the origin (k being the number of 2-year spawning cycles since $J'_S(0)$ was specified). When it exists, the

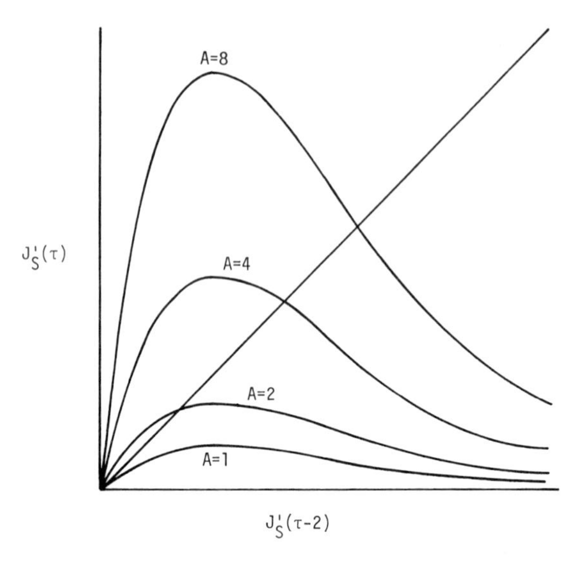

Fig. 4.19. Graphs for Ricker's spawner-recruit model with different values of the lumped parameter A. See Fig. 4.7 for computer-generated dynamics of this model

second intersection occurs at

$$J'_S = AJ'_S \exp[-J'_S] = \log_e A. \tag{4.7}$$

Therefore, for the second intersection, equation 4.2 becomes

$$J'_S(2k) = \log_e A + (1 - \log_e A)^k (J'_S(0) - \log_e A). \tag{4.8}$$

The second intersection will exist when the slope of the graph at the origin is greater than the slope (1) of the reference line, i.e., when $A > 1$. Therefore, for all values of A between zero and one, the only intersection is at the origin and the dynamics tend to converge to that intersection (i.e., A^k approaches zero as k increases). Thus, the population tends toward extinction. For all values of A greater than one, the population diverges from the intersection at the origin (i.e., A^k increases as k increases). For all values of $\log_e A$ greater than zero but less than one (i.e., values of A greater than one but less than 2.71828 ...) the factor $(1 - \log_e A)^k$ in equation 4.8 will decrease monotonically as k increases, and the dynamics converge monotonically on the second intersection. Thus, the population tends toward a steady-state level. For all values of $\log_e A$ greater than one but less than two (i.e., values of A greater than 2.71838... but less than 7.389056 ...) the factor $(1 - \log_e A)^k$ in equation 4.8 will decrease in an oscillatory manner (alternating between positive and negative values) as k increases, and the dynamics converge in an oscillatory manner on the second intersection. Thus, the population tends toward a steady-state level, but oscillates about that level as it approaches it. For all values of $\log_e A$ greater than 2, the factor $(1 - \log_e A)^k$ will grow in an oscillatory manner as k increases, and the dynamics will diverge from the second intersection.

When the dynamics diverge from both intersections (i.e., when A is greater than 7.389056 ...), then the possibility of a stable limit cycle exists. We could locate the two points of such a cycle by the graphical methods already described, or we can take another approach, setting $J'_S(\tau)=J'_S(\tau-4)$. From equation 4.5

$$J'_S(\tau)=AJ'_S(\tau-2)\exp[-J'_S(\tau-2)]$$

$$J'_S(\tau-2)=AJ'_S(\tau-4)\exp[-J'_S(\tau-4)]$$

from which

$$J'_S(\tau)=A^2 J'_S(\tau-4)\exp[-J'_S(\tau-4)]\exp[-AJ'_S(\tau-4)\exp[-J'_S(\tau-4)]] \tag{4.9}$$

and, letting $J'_S(\tau)=J'_S(\tau-4)=J'_c$,

$$1=A^2\exp[-J'_c]\exp[-AJ'_c\exp[-J'_c]]. \tag{4.10}$$

When a cycle exists, equation 4.10 should exhibit three solutions for J'_c, one at the second intersection (i.e., $J'_c=\log_e A$) and two equidistant from the second intersection. Now, if we construct a square with the latter two points at opposite corners, then the cycle definitely will be stable as long as no part of the model's graph lies above or below the square. In other words, the cycle definitely will be stable if the peak of the graph lies on or to the left of the left-hand side of the square (i.e., if the smaller of the two values of J'_c corresponding to the cycle is itself greater than or equal to one). It is rather easy to show that this will be the case for values of A that are less than 9.549 ... Thus, for values of A that are greater than 7.389 ... and less than 9.549 ..., a stable limit cycle definitely will exist, and the population will tend toward a stable oscillation with a four-year period. As the value of A is increased beyond 9.549, the dynamics eventually will become chaotic (see Figure 4.7).

4.4. Large-Numbers Models with Constant Parameters

In the previous section we introduced a graphical method that can be employed to deduce some rather general aspects of the behavior of some rather restricted systems. Specifically, the intersections of the system graph with the reference line represented critical points with respect to the system's dynamics. Depending upon the nature of the graph in the neighborhood of the intersection, a critical point could represent a stable level of flows or accumulations (i.e., the dynamics could converge to the point), it could represent a threshold level of flow or accumulation (i.e., the dynamics could diverge from it, taking distinctly different routes on the two sides of it), it could represent a focus of cyclical activity (growing, declining, or stable), or it could represent a focus of chaotic activity. Being able to deduce the dynamics of a model in the neighborhood of such critical points, one in fact can go a long way toward deducing the general dynamic properties of the model. Unfortunately, however, not many models will be of the one-loop, lumpable form.

More generally, discrete-time, large-numbers network models will have many loops. Furthermore, especially when the models have been developed and closed by hypothesis, there generally will be parameters that depend upon $\boldsymbol{n}$ and perhaps on τ. The dynamics of all such population models will possess at least one critical point (corresponding to zero population, or extinction) and possibly several more (e.g., saturation levels, threshold levels, and the like). Deduction of the dynamic behavior of the model in the neighborhood of a critical point can be carried out in a very systematic way by means of the well-developed tools of linear analysis, network analysis, and systems analysis. Once again, such deductions provide a major step toward the understanding of the general dynamic properties of the model at hand. To carry out these deductions, one modifies the network model, where necessary, to account for the coupling of variables (see section 4.13), and then substitutes parameter values appropriate to the critical point at the value of τ that is of particular interest. The parameters then are treated as being constant at the values given, and the methods of analysis of constant-parameter models are applied. If the original parameters depend upon $\boldsymbol{n}$ but not, independently, upon τ, then the results can be interpreted directly. On the other hand, if the parameters depend upon τ, then special interpretation is required. Very slow τ-dependence can be envisioned in terms of slow shifting of certain of the critical points and slow changes in the dynamics in their neighborhoods as they shift. Very rapid τ-dependence (i.e., much more rapid than the dynamics of interest in the model) may produce constant or slowly drifting parameter-value averages over periods of significance with respect to the dynamics of interest. In the former case, the deduced results are directly interpretable, and in the latter case they again can be interpreted in terms of drifting critical points. However, when the τ-dependence is neither very slow nor very fast, the dynamical behavior of the system is difficult to deduce analytically, and one often must resort to simulation methods.

Occasionally, one can employ the methods for constant-parameter models for another reason, namely that those aspects of the life cycle of interest that would lead to $\boldsymbol{n}$- or τ-dependence have been ignored or smoothed over in the gathering and processing of field data, with the result that only mean parameter values are available. Whatever the reasons for their applicability might be, the methods for constant parameter models are well established, and those aspects of the methods that are useful for discrete-time models are presented in the remaining sections of this chapter. For a specified set of constant parameter values, one can deduce with these methods whether the dynamics tend toward or away from a critical point and how rapidly those tendencies will unfold, and whether the dynamics tend to cycle about the critical point and, if so, with what frequencies.

4.5. Inputs and Outputs of Network Models

Application of the methods of network analysis to constant-parameter models generally is facilitated by the designation of one or more variables (flows or accumulations) as *inputs* to the model and one or more variables as *outputs* from it. The immediate object of the analysis then simply becomes deduction of the dynamics of the output variables in response to specified dynamics of the input variables. Depending upon what is being modeled and upon the goals of the modeler, the significances of the "input" and "output" variables may themselves be quite variable. In certain situations, the terms "input" and "output" may be quite apropos. For example, a sports fishery manager might model a trout population in a high mountain lake, selecting as his input variable the flow of newly planted trout augmenting the population and as his output the flow of trout into the creels of fishermen. In another situation, the "output" variable might represent some aspect of a population that the modeler wishes to control, but can do so only *indirectly*. In that case, the "inputs" simply would be one or more variables representing aspects of the population that are under direct control. In such cases, the "inputs" occasionally may seem more like "outputs," in that they represent a flow out of the population. For example, one might wish to model a population of Atlantic herring gulls, attempting to determine whether egg-killing would be an effective method for stemming the growth of the gull population. In that case, the outflow of killed eggs would be selected as the "input" (directly controlled) variable and the gull population would be selected as the "output" (indirectly controlled) variable. Finally, if one simply wishes to deduce the dynamics of some aspect of a population that is presumed to be operating (without human intervention), then the "output" variable simply would be one whose dynamics the modeler wishes to deduce, and the "input" variable simply can be a hypothetical initial perturbation whose role simply is to stimulate the dynamics of the "output."

Once the inputs and outputs of a network model have been specified, the network itself represents an explicit and well-defined transformation, namely the transformation of the input dynamics into the output dynamics. The input dynamics will be described by a discrete function of time. If there is only one input variable, then that function will be a *scalar function*, meaning that it will take on a single value for each transition interval. If there are several input variables, on the other hand, then the input function will be a *vector function*, with an array of values for each transition interval. Similarly, the output dynamics will be described either by a scalar function or a vector function, depending upon how many output variables have been selected. The transformation represented by the network simply converts the input function into the output function.

4.6. Linearity, Cohorts, and Superposition-Convolution

As we have mentioned in the previous section, when a network model has specified input variables and output variables, it represents a transformation: the output function is a transformation of the input function. In order to deduce the consequences of such transformations, it would be very convenient to be able to represent them in concise, manipulable mathematical form. This is especially easy to do when the network model has constant parameters. In that case, it belongs to the class of models known as *linear models* and therefore displays the following properties of linearity:

(1) For every input function there is a single, uniquely determined output function.

(2) If the input function is multiplied by a constant scale factor, then the resulting output function is multiplied by the same scale factor (i.e., if $X_{\text{out}}(\tau)$ is the output function in response to the input function $X_{\text{in}}(\tau)$, then the output function in response to $k\,X_{\text{in}}(\tau)$ will be $k\,X_{\text{out}}(\tau)$.

(3) The output function in response to two or more input functions applied together is the sum of the output functions in response to those same inputs applied separately (i.e., if $X'_{\text{out}}(\tau)$ is the output function in response to $X'_{\text{in}}(\tau)$ when the latter is applied by itself, and $X''_{\text{out}}(\tau)$ is the output function in response to $X''_{\text{in}}(\tau)$ applied by itself, then the output function in response to $X'_{\text{in}}(\tau)$ and $X''_{\text{in}}(\tau)$ applied together is $X'_{\text{out}}(\tau)$ plus $X''_{\text{out}}(\tau)$.

Property 1 often is called *uniqueness*; property 2 often is called *homogeneity*; and property 3 often is called *additivity*. Actually, if the parameters of a network model depend upon τ but not upon $\boldsymbol{n}$, then that model also is linear and displays these properties. In that case, however, the transformation represented by the network model will be time-varying.

According to property 3, if a network model is linear, then we can consider any complicated input function as being made up of several, simpler input functions applied together. The output function then can be computed as the sum of the responses to these simpler functions. Thus, for example, a vector input function can be decomposed into scalar input functions; the response to each scalar function can be deduced; and then the response to the vector function is obtained simply by addition of the scalar-function responses. Therefore, from this point on we shall consider only scalar input functions (i.e., we shall deduce responses to one input variable at a time); but our results will be easily extendable to vector input functions (i.e., to the deduction of responses to several input variables at a time).

A scalar input function itself can be decomposed in many different ways. A most convenient decomposition for discrete-time functions is separation into components each of which has a nonzero value only for one transition interval and zero values for all other intervals. Examples of such a decomposition are displayed in Figure 4.20. Quite often, discrete-time functions of this type (exhibiting only one nonzero value) are called *impulses*. However, that term is not especially appropriate for functions representing flows or accumulations of individuals in population models. Therefore, we shall use the term *cohort* in place of *impulse*. A cohort will be defined to be either

(1) all individuals occupying the same state or set of equivalent states at a given transition interval, or

(2) all individuals entering the same state or set of equivalent states at a given transition interval.

Since our inputs often will be flows of newly produced offspring, the cohort often will represent all individuals of the same age (given in number of transition intervals since production). This is a common demographic meaning of *cohort*.

It is clear that every possible scalar input function to a discrete-time population network model can be decomposed into a sequence of input cohorts. Each cohort, applied by itself, would produce a corresponding output function.

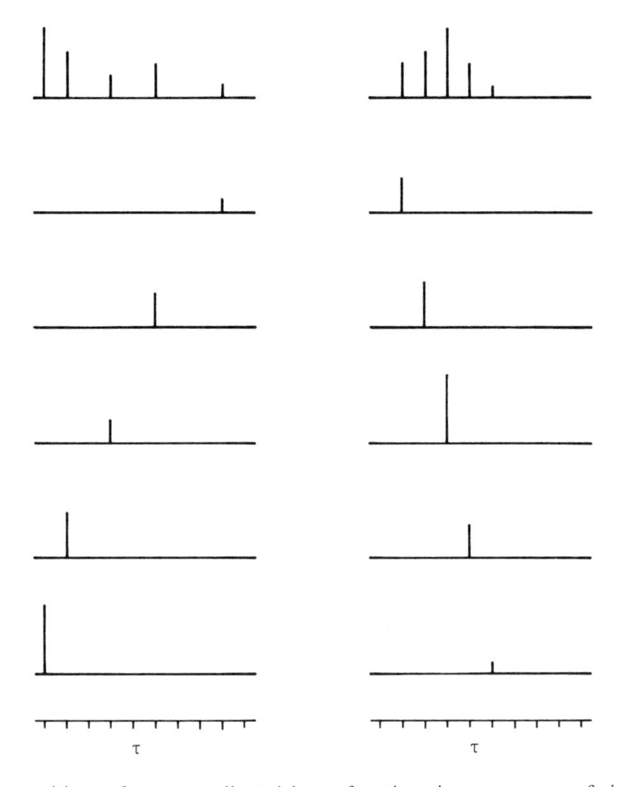

Fig. 4.20. Decompositions of two complicated input functions into sequences of simpler functions

If the network model is linear, then its output in response to the entire scalar input function simply will be the sum of the output functions in response to the individual cohorts. If the parameters of the model are time-varying, then the output response to a cohort input will vary from interval to interval. On the other hand, when the parameters of the model are constant, then its output responses to cohort inputs will be completely stereotyped; each will be a multiple of the same basic response function, merely shifted in time to correspond to the time of its cohort input. We shall call the basic response function the *unit-cohort response*.

The unit-cohort response is the output function of the network in response to a cohort of magnitude one applied at $\tau = 0$ (i.e., a cohort consisting of one individual applied at the initial transition interval). Since it is the basic building block of all output functions in response to a particular input variable (i.e., a particular input flow or accumulation), the unit-cohort response serves to characterize the dynamics of the constant-parameter network with respect to that input variable. If the output function is a vector function, then the unit-cohort response will be a vector function. On the other hand, one certainly is free to consider a vector function one element at a time. In other words, the vector unit-cohort response comprises scalar unit-cohort responses, one for each output variable; and having computed the scalar unit-cohort responses, one

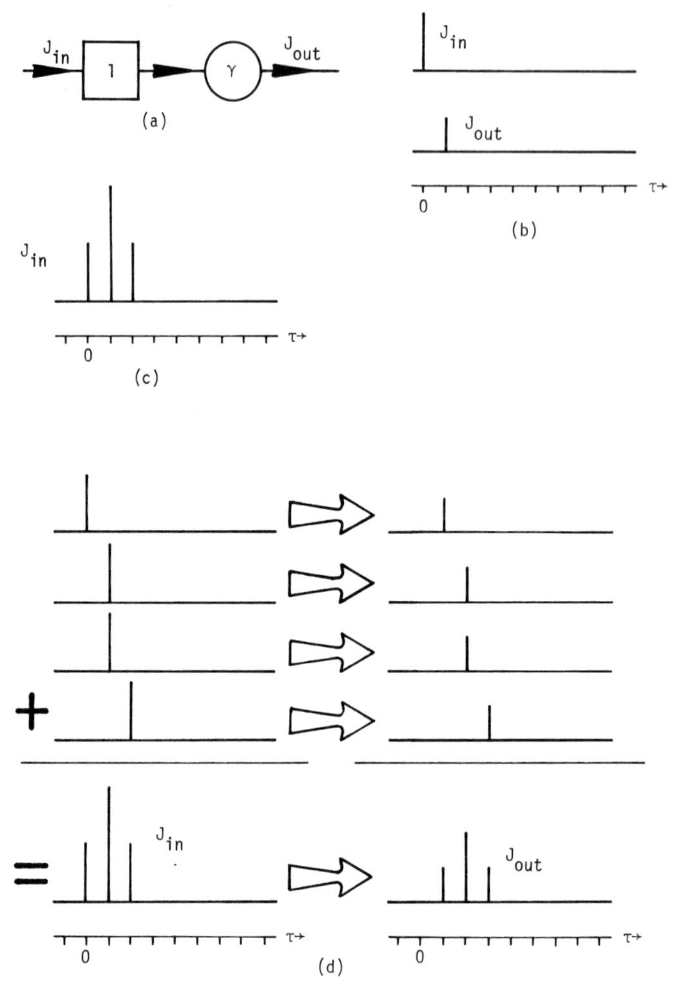

Fig. 4.21 a–d. A network model (a), its unit-cohort response (b), an input function (c), decomposition of the input function into single cohorts and superposition of the responses to the individual cohorts to generate the deduced output response (d)

then can assemble them into an array to form the vector unit-cohort response. Therefore, we lose no generality by confining our attention to scalar unit-cohort response functions. As we already have mentioned, we also lose no generality by confining our attention to scalar input functions. Therefore, the analytical tools presented in the next few sections are described in the context of scalar inputs and scalar outputs.

The process of summing individual responses to produce a total response is called superposition. It is illustrated for unit-cohort responses in Figures 4.21 through 4.24. The network of Figure 4.21 is extremely simple, as is its unit-cohort response. When a unit cohort is applied to the input at $\tau=\mathbb{0}$, the output is a single cohort of magnitude γ and occurring at $\tau=\mathbb{1}$. Now, consider the

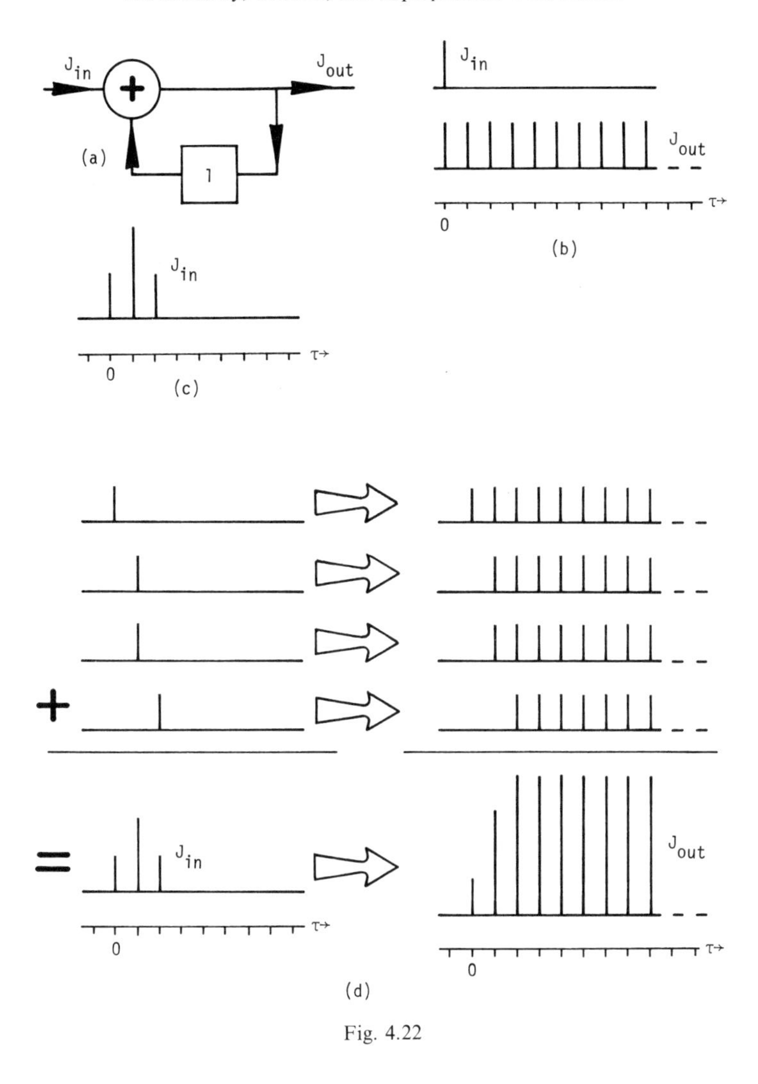

Fig. 4.22

slightly more complicated input function depicted in Figure 4.21c. This input function can be decomposed into the following components (see Figure 4.21d): a unit cohort (i.e., a cohort of one individual at $\tau = \mathbb{0}$), two unit cohorts shifted (delayed) by one transition interval (i.e., two individuals at $\tau = \mathbb{1}$), and one unit cohort shifted by two transition intervals (i.e., one individual at $\tau = \mathbb{2}$). Each of these components produces its corresponding response, as depicted in Figure 4.21d. The unit cohort produces a unit-cohort response; the two unit cohorts shifted by one interval each produce a unit-cohort response shifted by one interval; and the unit cohort shifted by two intervals produces a unit-cohort response shifted by two intervals. These component responses can be summed (superposed) to yield the total response, as depicted in Figure 4.21d.

The network of Figure 4.22a is slightly more complicated, owing to the presence of a loop. The unit-cohort response of this network is depicted in

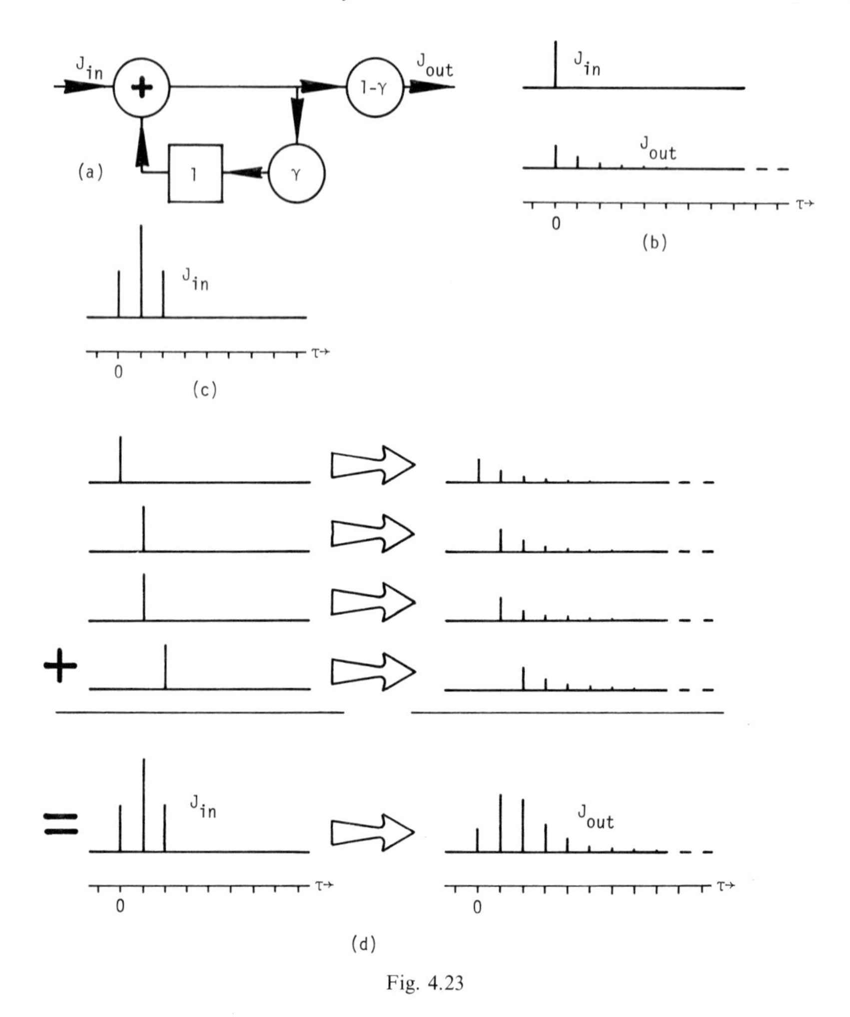

Fig. 4.23

Figure 4.22b. In Figure 4.22c is a more complicated scalar input function; and in Figure 4.22d is the decomposition of that function into unit cohorts and shifted unit cohorts. Once again, the corresponding unit-cohort responses and shifted unit-cohort responses are superposed to yield the total scalar response to the scalar input function. Figures 4.23 and 4.24 illustrate other, similar examples of the process of superposition.

With the aid of these examples, the reader should be able to develop a thorough, graphical understanding of the process of superposition. Basically, it is an operation involving two discrete functions, the input function and the unit-cohort response function. Let us restate the process in more concise form. Let $f(\tau)$ be the unit-cohort response function. Thus, for the network of Figure 4.21,

$$f(\tau)=\gamma \quad \text{when } \tau=1$$

$$f(\tau)=0 \quad \text{for all other values of } \tau$$

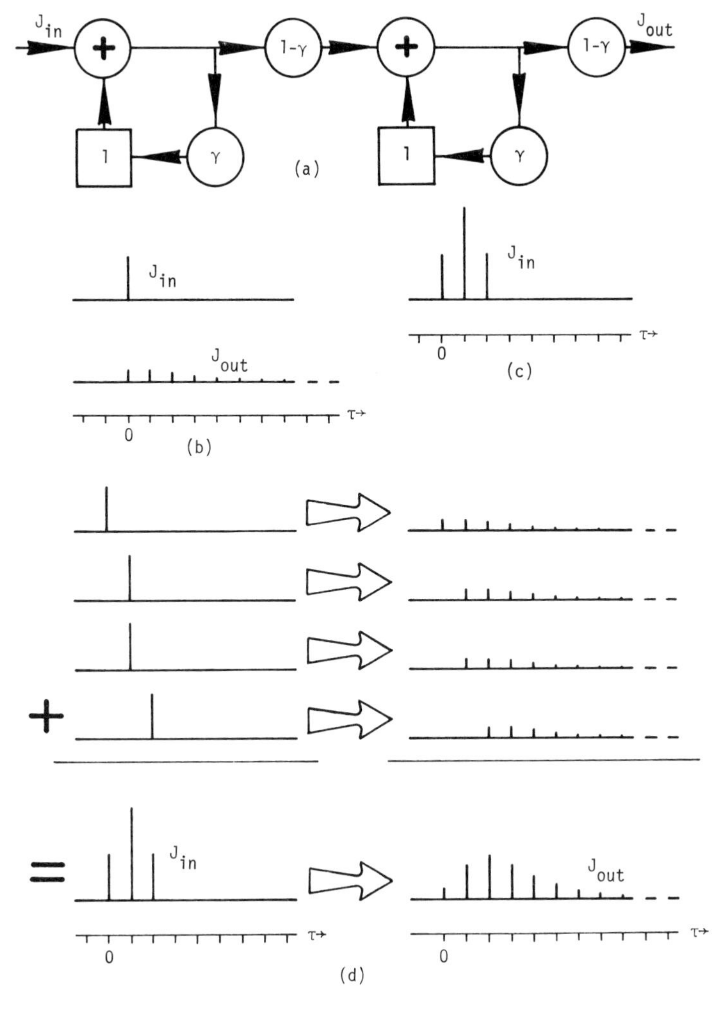

Fig. 4.24

and for the network of Figure 4.22,

$$f(\tau)=\mathbb{0} \quad \text{for all values of } \tau \text{ less than } \mathbb{1}$$
$$f(\tau)=\mathbb{1} \quad \text{for all values of } \tau \text{ greater than or equal to } \mathbb{1}.$$

With a little reflection, the reader easily should be convinced that the process of superposition is described by

$$X_{\text{out}}(\tau)=\sum_{k=-\infty}^{\tau} X_{\text{in}}(k)f(\tau-k) \tag{4.11}$$

where $X_{\text{out}}(\tau)$ is the output flow or accumulation at interval τ, and $X_{\text{in}}(k)$ is the input flow or accumulation at interval k. In order to limit the summation in equation 4.11, it is convenient to assume that $X_{\text{in}}(k)$ was zero for all intervals

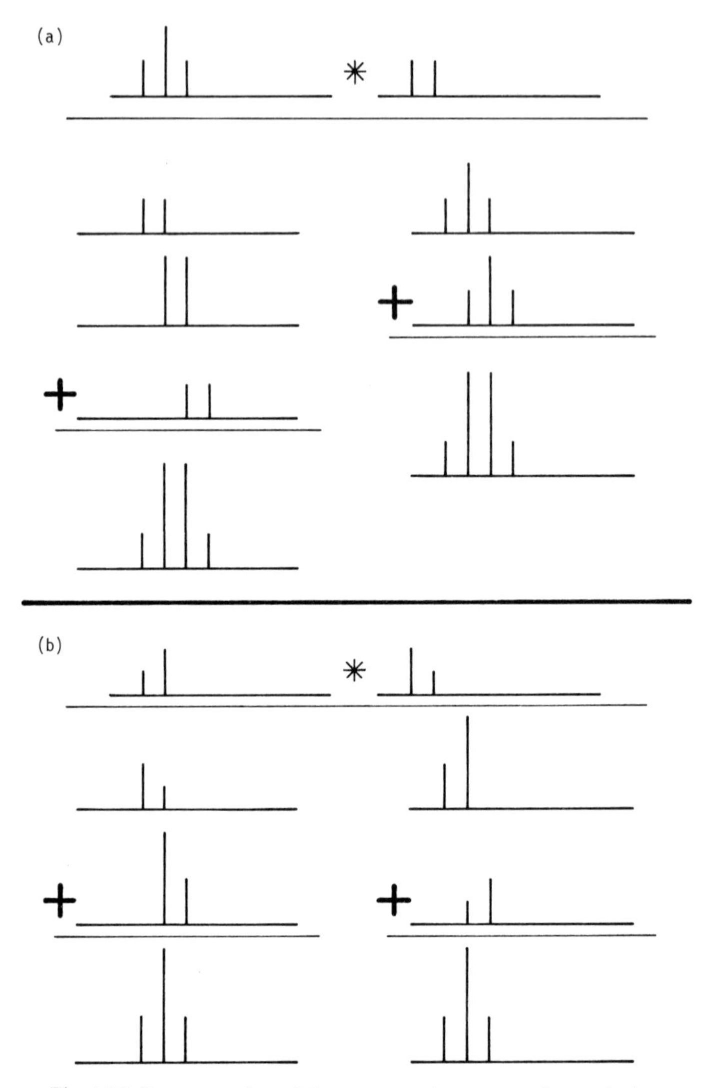

Fig. 4.25. Demonstration of the commutative nature of convolution

prior to $k=0$. In other words, we shall assume that the input begins at some interval of time and we shall designate that interval as $k=0$. In that case, the summation can begin at $k=0$, since there are no input contributions prior to that

$$X_{\text{out}}(\tau)=\sum_{k=0}^{\tau} X_{\text{in}}(k)\,f(\tau-k). \tag{4.12}$$

Comparing equations 4.12 and 3.64, one immediately can see that the process of superposition and the process of convolution are one and the same. In the case of equation 4.12 we are dealing with discrete functions of time rather than discrete probability distributions, as in equation 3.64, but the manipulations are identical. What we have done in equation 4.12 is carry out a sum of unit-cohort responses, each scaled and shifted appropriately for its corresponding cohort in the input function. Owing to the commutative nature of convolution, we equally well could have carried out a summation of input functions, each scaled and shifted to correspond to an element of the unit-cohort response. Examples of these equivalent operations are depicted graphically in Figure 4.25.

As we did at the end of Chapter 3, we shall employ an asterisk to represent the operation of convolution. Thus, we can express the input-output relationship described by equation 4.12 as follows:

$$X_{\text{out}}(\tau) = X_{\text{in}}(\tau) * f(\tau) = f(\tau) * X_{\text{in}}(\tau) \tag{4.13}$$

by which we mean precisely what is written in equation 4.12.

4.7. The z-Transform: A Shorthand Notation for Discrete Functions

In previous sections we have represented discrete functions of time graphically; it now is becoming clear that, although graphical representation may be illuminating, it will not be convenient, particularly for complicated or extensive functions and for manipulations such as convolution. What is needed is a concise representation of discrete functions that is conducive to manipulations. The linear, discrete models that we are dealing with here very often can be analyzed in terms of linear difference equations, the solutions to which usually are orderly, infinite sequences of numbers, one for each transition interval. A very powerful method of dealing with such sequences is the use of generating functions; and a natural extension of the application of generating functions was the development of the z-transform method, which now has become a well-established approach for linear, discrete networks. The z-transform is relatively simple, yet it provides an elegant shorthand notation for our discrete functions, and it meets our criterion of being conducive to manipulations. In this section, we shall introduce the basic concepts of the z-transform and discuss its manipulations.

Let us begin by considering a function comprising a single unit cohort at the kth interval, as depicted in Figure 4.26. Once our graphical symbols are understood, we very easily can represent the function in this manner, but the representation is cumbersome. We also might represent it in more formal terms, as follows:

$$f(\tau) = \mathbb{1} \quad \text{when } \tau = k$$
$$f(\tau) = \mathbb{0} \quad \text{for all other } \tau.$$

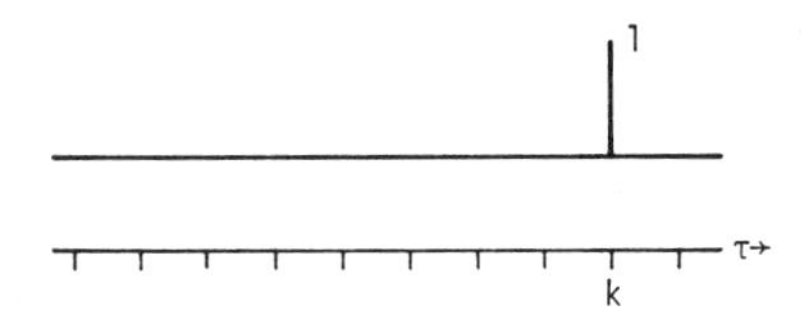

Fig. 4.26. Unit cohort at the kth interval

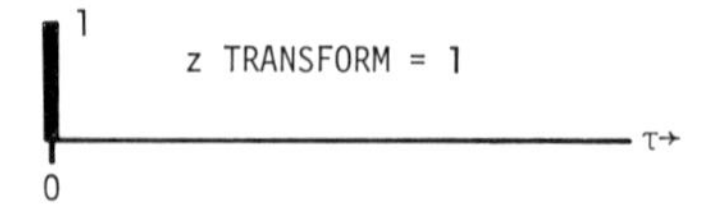

z TRANSFORM = z^{-k}
1
τ→
k

Fig. 4.27. Top: unit cohort at the 0th interval and its z transform; bottom: unit cohort at the kth interval and its z transform

This description still is rather cumbersome, and it promises to be much more so when our functions become more complicated.

Now, let us try the z-transform method. Instead of representing the function as $f(\tau)$, we shall change the independent variable and represent the function as $\mathscr{f}(z)$. In other words, we shall transform $f(\tau)$ into an equivalent, but presumably simpler function, $\mathscr{f}(z)$. To do so, we simply can begin by defining the z-transform for our present function to be

$$\mathscr{f}(z) = z^{-k}.$$

In other words, when $f(\tau)$ is the unit cohort applied at the kth interval, the corresponding function of z (*the z-transform of $f(\tau)$*) will be *defined* to be z^{-k}. This convention will prove to be especially convenient, since the standard unit cohort, applied at the $\mathbb{0}$th interval, will be represented by

$$\mathscr{f}(z) = z^{-\mathbb{0}} = \mathbb{1}.$$

These relationships are summarized in Figure 4.27.

Before we consider the transformations of more complicated functions of time, let us establish conventions of notation to be used throughout the remainder of the text: functions of time will be designated with italic letters (e.g., $f(\tau)$, $h(\tau)$, $G(\tau)$), which occasionally may bear subscripts or superscripts (e.g., $f'(\tau)$, $G_1(\tau)$); the corresponding function of z (i.e., the z-transform of the function of time) always will be designated with the script form of the same letter (same case) and will bear the same superscript or subscript, if any. Thus, $\mathscr{f}(z)$ is the z-transform of $f(\tau)$; $\mathscr{G}'(z)$ is the z-transform of $G'(\tau)$; and $\mathscr{h}_2(z)$ is the z-transform of $h_2(\tau)$.

For simplicity of manipulation, the transformation from any function of time to the corresponding function of z is taken to be *linear*. In other words, it meets precisely the same criteria that were required of a network in order to be linear: uniqueness, homogeneity, and additivity.

(1) For every discrete function of time, $f(\tau)$, there is a uniquely determined z-transform, $\mathscr{f}(z)$.

(2) If $\mathscr{f}(z)$ is the z-transform of $f(\tau)$, then $k\mathscr{f}(z)$ is the z-transform of $kf(\tau)$, where k is a constant.

(3) If $\mathscr{f}(z)$ is the z-transform of $f(\tau)$ and $\mathscr{g}(z)$ is the z-transform of $g(\tau)$, then $\mathscr{f}(z)+\mathscr{g}(z)$ is the z-transform of $f(\tau)+g(\tau)$.

Now it should be clear to the reader that, with these rules and with the basic definition depicted in Figure 4.27, we can construct the z-transform of any, arbitrary discrete function of time. According to rule 2 (homogeneity), the z-transform of a function consisting of a single cohort of size n at interval k is n times the transform for a unit cohort at interval k, or nz^{-k}. According to rule 3 (additivity), the z-transform of a function consisting of two cohorts, one of size n_1 at interval k_1 and the other of size n_2 at interval k_2, is the sum of the z-transforms of the same cohorts taken one at a time, or $n_1 z^{-k_1}+n_2 z^{-k_2}$. Extending this reasoning, we could write the general relationship

$$\mathscr{f}(z)=\sum_{\tau=-\infty}^{\infty} f(\tau)\, z^{-\tau}.$$

However, standard practice in the application of z-transforms dictates that we always consider only those discrete functions of time that are identically zero for all transition intervals prior to $\tau=0$. In other words, we once again shall assume that all dynamics begin at the 0th interval. In that case, the z-transform is defined by the following summation:

DEFINITION OF THE z-TRANSFORM

$$\mathscr{f}(z)=\sum_{\tau=0}^{\infty} f(\tau)\, z^{-\tau}. \tag{4.14}$$

So far, the z-transform has simplified our descriptions of discrete functions slightly, but perhaps not enough to justify its use. Now, however, we shall consider the conceptual and manipulatory simplifications that it brings to the process of superposition. Except for the cases of the simplest of networks combined with the simplest of input functions, this advantage of the z-transform will be compelling. We shall begin by performing an experiment. Using the usual rules of multiplication, we shall form the product of two z-transforms, $\mathscr{f}(z)$ and $\mathscr{g}(z)$. Using equation 4.14, we can write this product in the following form:

$$\mathscr{f}(z)\,\mathscr{g}(z)=\left\{\sum_{\tau=0}^{\infty} f(\tau)\, z^{-\tau}\right\}\left\{\sum_{\tau=0}^{\infty} g(\tau)\, z^{-\tau}\right\}. \tag{4.15}$$

Expanding the first of the two sums on the right-hand side, we have

$$\mathscr{f}(z)\,\mathscr{g}(z)=\{f(0)+f(1)\,z^{-1}+f(2)\,z^{-2}+f(3)\,z^{-3}+\cdots\}\,\mathscr{g}(z). \tag{4.16}$$

Considering these terms one by one, we are lead to the following interpretations: In the first term, $f(0)\,\mathscr{g}(z)$, every term of $\mathscr{g}(z)$ is present and is scaled by the factor $f(0)$; the corresponding function of time therefore is simply $g(\tau)$, with every term scaled by $f(0)$ [i.e., $f(0)\,g(\tau)$]. In the second term, $f(1)\,z^{-1}\,\mathscr{g}(z)$, every term of $\mathscr{g}(z)$ is present and is scaled by the factor $f(1)$ and multiplied by z^{-1}. The term $g(\tau)z^{-\tau}$ becomes $f(1)g(\tau)z^{-(\tau-1)}$, which represents the cohort $g(\tau)$ scaled by $f(1)$ and delayed by one additional transition interval. Thus, the function of time corresponding to $f(1)z^{-1}\mathcal{g}(z)$ is simply $g(\tau)$, with every term scaled by $f(1)$ and delayed by one transition interval [i.e., $f(1)g(\tau-1)$]. Similarly, the general term, $f(k)z^{-k}\mathcal{g}(z)$, represents $g(\tau)$ scaled by the factor $f(k)$ and delayed by k transition intervals [i.e., $f(k)g(\tau-k)$]. So the function of time that corresponds to our product is simply the sum

$$\sum_{k=0}^{\infty} f(k)g(\tau-k)$$

where

$$g(\tau-k)=0, \qquad \tau<k$$

which can be recognized as the convolution of $f(\tau)$ and $g(\tau)$. In other words, the rather complicated process of superposition, or convolution, in the time domain becomes the very simple process of multiplication in the z-transform domain.

We have introduced four properties of the z-transform. At this point we shall summarize these for easy reference.

Table 4.1. Basic properties of the z-transform

	Function of time	Corresponding function of z
(1) homogeneity	$kf(\tau)$	$k\mathcal{f}(z)$
(2) additivity	$f(\tau)+g(\tau)$	$\mathcal{f}(z)+\mathcal{g}(z)$
(3) delay by k	$f(\tau-k)$	$z^{-k}\mathcal{f}(z)$
(4) discrete convolution	$\sum_{k=0}^{\tau} f(k)g(\tau-k)$	$\mathcal{f}(z)\,\mathcal{g}(z)$

These are by far the most important properties for our purposes. However, we shall introduce other properties from time to time as they are needed; and the interested reader can consult any of the texts listed at the end of this chapter for additional properties.

4.8. The Application of z-Transforms to Linear Network Functions

According to the discussion in section 4.6, the response of a population network with constant parameters is given by the discrete convolution of the input function and the unit-cohort response function. For example, if the input function is $J_{in}(\tau)$ and the unit cohort response is $f(\tau)$, then the output function, $J_{out}(\tau)$, will be

$$J_{out}(\tau)=\sum_{k=0}^{\tau} f(k)J_{in}(\tau-k). \tag{4.17}$$

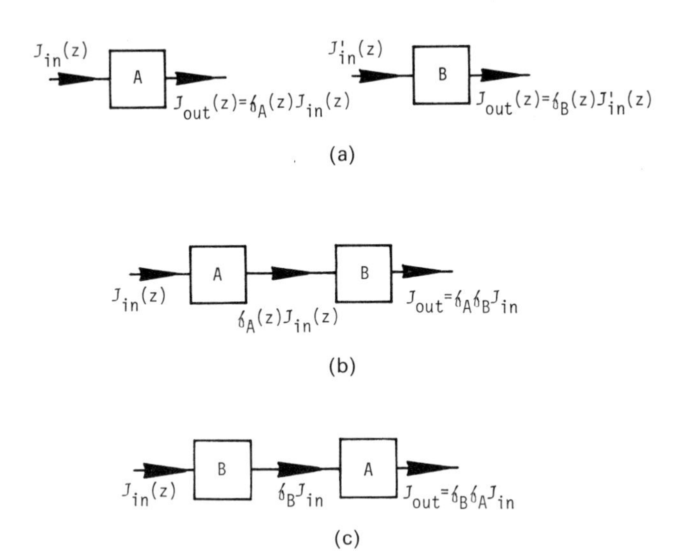

Fig. 4.28. Demonstration of the commutative nature of transformed constant-parameter network operations; the form of the output flow is independent of the order of the networks

In terms of the z-transforms of these functions, we can rewrite this relationship in the much simpler form

$$\mathscr{J}_{\text{out}}(z) = \mathscr{f}(z)\,\mathscr{J}_{\text{in}}(z). \tag{4.18}$$

In other words, in the z-transform domain, the entire operation of the network on the input function is represented by the process of multiplication by the (z-transform of the) unit-cohort response function. As we shall see, this facilitates to a compelling extent the conception of network operations.

For example, consider the two networks of Figure 4.28a. Network A has a unit-cohort response $f_{\mathsf{A}}(\tau)$, the z-transform of which is $\mathscr{f}_{\mathsf{A}}(z)$; and network B has a unit-cohort response $f_{\mathsf{B}}(\tau)$, whose z-transform is $\mathscr{f}_{\mathsf{B}}(z)$. Now, if we connect the two networks in series, as depicted in Figure 4.28b, the output of A becomes the input of B, and the overall response of the network is represented by

$$\mathscr{J}_{\text{out}}(z) = \mathscr{f}_{\mathsf{A}}(z)\,\mathscr{f}_{\mathsf{B}}(z)\,\mathscr{J}_{\text{in}}(z). \tag{4.19}$$

In other words, in the z-transform domain, the unit-cohort response of two networks in series is simply equal to the product of their individual unit-cohort responses. Since multiplication in the z-transform domain is commutative, the order of the networks is totally unimportant, as is illustrated in Figure 4.28c.

Clearly, it will be very convenient to be able to use the z-transforms of flows and accumulations directly in our network models. Let us determine how this will affect the operations represented by our standard network elements. Because of the linear nature of the z-transform (see basic properties 1 and 2 in Table 4.1 of the previous section), the operations of the *scalor* and the *adder* will be unchanged, as illustrated in Figure 4.29. According to basic property 3 of the

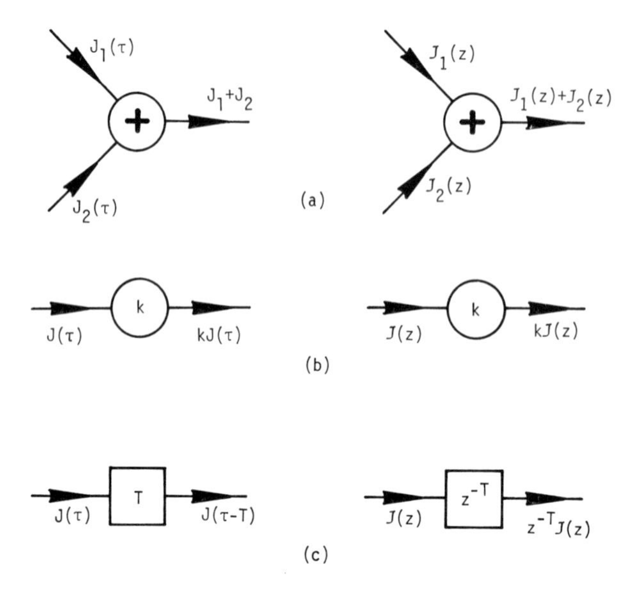

Fig. 4.29. The three common elements of constant-parameter network models (left-hand side) along with the z-transformed version of each (right-hand side)

z-transform, a *time delay* of duration k in the time domain must be represented by multiplication by z^{-k} in the z-transform domain. In other words, the transformed *time delay* becomes akin to a *scalor*. It still is a dynamic element, however, so we shall continue to designate it with a box rather than a circle. In the box we shall place z^{-k}, as illustrated in Figure 4.29c.

Now, given any linear population network model, we can redraw it in the z-transform domain simply by making the direct substitutions indicated in Figure 4.29. The next step will be to determine from the transformed network its unit-cohort response. The input function that generates the unit-cohort response, of course, is a single unit-cohort applied at the $\mathbb{0}$th transition interval. The z-transform of this function, according to the discussion of the previous section, is $\mathbb{1}$. Therefore, to determine the unit-cohort response of a network, we simply apply the transformed input function $\mathbb{1}$, follow through the network and see what emerges at the output. Perhaps the best way to illustrate the procedure is by example.

4.8.1. Straight-Chain Networks

Consider the networks of Figure 4.30. In each case there is a single path between the input and the output, with no branch points or *adders* available for signal divergence or convergence. Therefore, the networks consist entirely of *scalors* and *time delays*; and in the transformed versions, each element operates on the function entering it by a simple multiplicative operation. The transformed *scalor* multiplies its input by a constant scale factor, and the transformed *time delay* multiplies its input by an appropriate negative power of z.

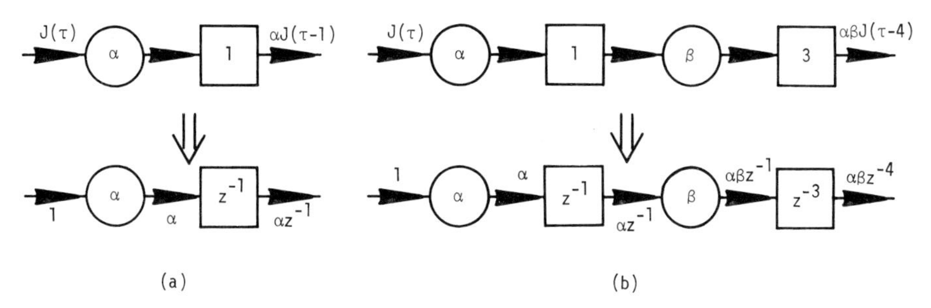

Fig. 4.30. Operation of straight-chain, constant-parameter networks. Time-domain versions (top), z-transformed versions (bottom), with flows and transformed flows indicated by the labels associated with the arrows

4.8.2. Networks with Feed Forward Loops

Consider the networks of Figure 4.31. Each contains at least one loop which begins at a branch point and ends at an adder; but these loops merely offer alternative paths between input and output, the functions cannot pass around and around them as they can in the loops of Figure 4.32. The analysis proceeds in precisely the same manner that it did for straight-chain networks, complicated only slightly by the fact that functions merge and must be summed at the adders.

4.8.3. Networks with Feedback Loops

Now we arrive at a complication that is almost guaranteed to be present in network models of populations; the networks include loops around which a function can cycle indefinitely. This one feature, all by itself, changes the basic nature of the analytical procedure. In fact, until the introduction of flow-graph methods, which are described in the following section, such loops were a major impediment to analysis.

Consider, for example, the networks of Figure 4.32. Each contains at least one such loop. The branch point associated with each loop occurs closer to the output than does the *adder* that closes the loop, leading to the designation "*feedback* loop." The network is an accumulator. The unit-cohort enters the loop and continues to pass around it indefinitely, one pass for each transition interval. On each pass, a unit cohort also flows out the output line (via the duplicating branch point). The resulting unit-cohort response is

$$f(\tau)=\mathbb{1} \qquad \text{for all intervals} \tag{4.20}$$

from which we can derive the z-transform

$$\mathscr{f}(z)=\mathbb{1}+z^{-\mathbb{1}}+z^{-\mathbb{2}}+z^{-\mathbb{3}}+\cdots. \tag{4.21}$$

However, we also can use another approach, namely consider directly the two components summing at the *adder* in the transformed network. One component,

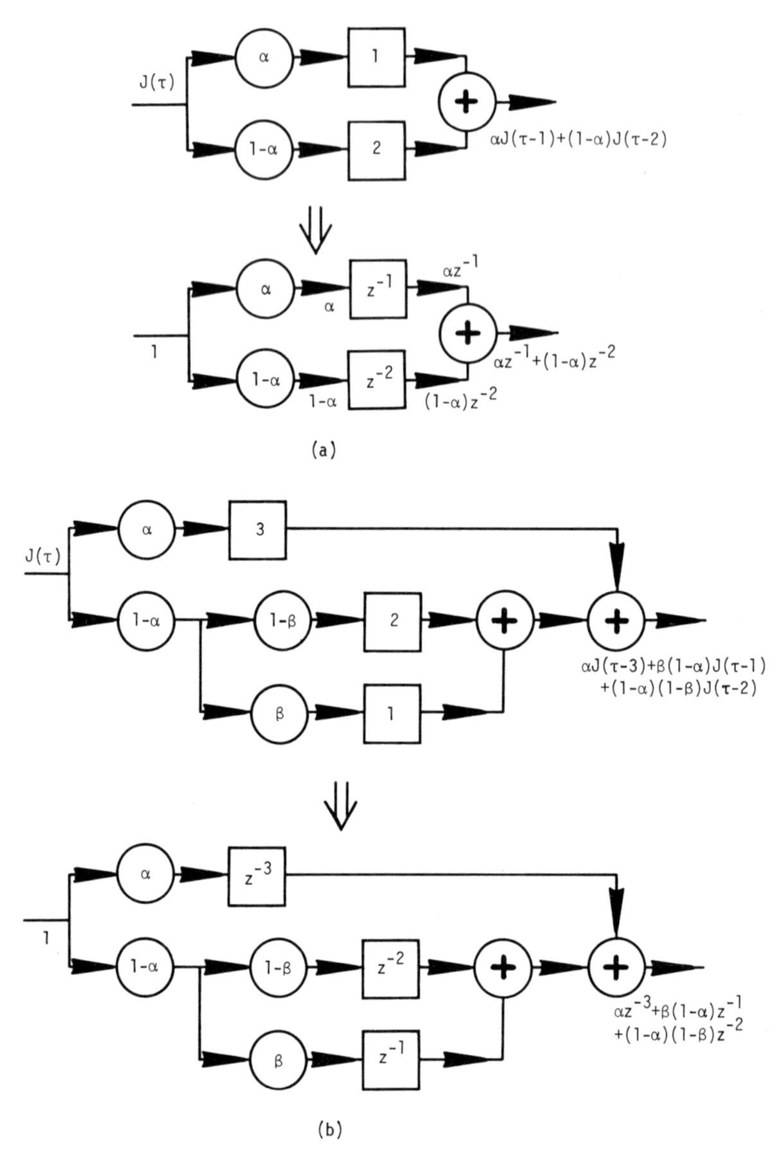

Fig. 4.31a and b. Two networks with feedforward loops (top diagram in *a* and *b*), along with the corresponding *z*-transformed networks. Flows and transformed flows are indicated by the labels associated with the arrows

of course, is the input, $\mathbb{1}$. The other is $z^{-\mathbb{1}}$ times the output, $\mathscr{J}(z)$. The sum of these two components is $\mathscr{J}(z)$ itself. Therefore, we can write

$$\mathscr{J}(z) = \mathbb{1} + z^{-\mathbb{1}}\mathscr{J}(z) \tag{4.22}$$

or

$$\mathscr{J}(z) = \frac{\mathbb{1}}{\mathbb{1} - z^{-\mathbb{1}}} = \frac{z}{z - \mathbb{1}}. \tag{4.23}$$

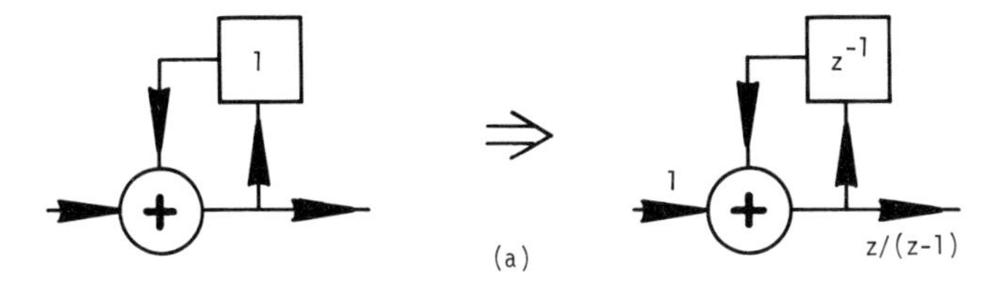

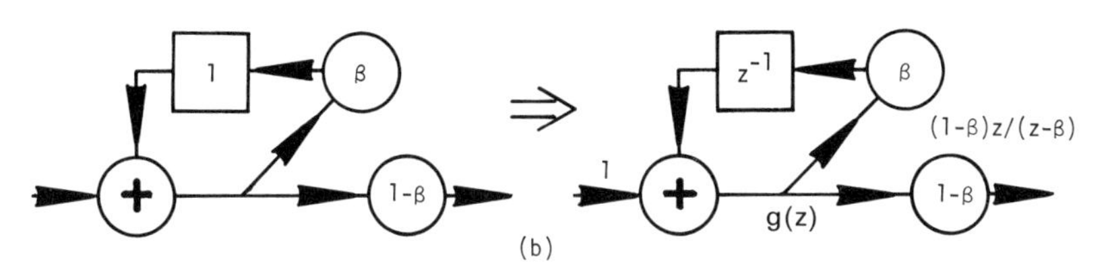

Fig. 4.32. Two networks with feedback loops (left-hand side), along with the corresponding z-transformed versions

If we carry out the division indicated by $z/(z-1)$, we can see immediately that it indeed is equal to the infinite sum on the right-hand side of equation 4.21:

$$
\begin{array}{r|l}
 & 1+z^{-1}+z^{-2}+z^{-3}+z^{-4}+\cdots \\
\hline
z-1) & z \\
 & \underline{z-1} \\
 & \quad 1 \\
 & \quad \underline{1-z^{-1}} \\
 & \qquad z^{-1} \\
 & \qquad \underline{z^{-1}-z^{-2}} \\
 & \qquad\quad z^{-2} \\
 & \qquad\quad \underline{z^{-2}-z^{-3}} \\
 & \qquad\qquad z^{-3} \\
 & \qquad\qquad \vdots
\end{array}
$$

For that reason, the function $z/(z-1)$ often is known as the *generating function* of the infinite series $1+z^{-1}+z^{-2}+\cdots$. When such a generating function is available for a series, it obviously provides a much simpler expression, which can be manipulated (e.g., added, subtracted, multiplied) much more easily than could the expanded series form. These are the reasons that generating functions are powerful tools in dealing with series. The same reasons apply to the z-transform, giving it one more, very important advantage over descriptions in which time is the independent variable. In the z-transform domain, using the basic loop analysis method of the previous paragraph, we automatically can take advantage of the powerful method of generating functions.

Let us consider some more examples before we move on. In the network of Figure 4.32b, we have the added complication of *scalors* in the feedback loop and the output line. Nonetheless, the analysis is straightforward. We consider the two components converging at the *adder*. The input is $\mathbb{1}$, the other component is $\beta z^{-\mathbb{1}} g(z)$, from which we have

$$g(z)=\mathbb{1}+\beta z^{-\mathbb{1}} g(z)$$

and

$$g(z)=\frac{\mathbb{1}}{\mathbb{1}-\beta z^{-\mathbb{1}}}=\frac{z}{z-\beta}. \tag{4.24}$$

Evidently,

$$f(z)=(\mathbb{1}-\beta)\, g(z)=(\mathbb{1}-\beta)\frac{z}{z-\beta}. \tag{4.25}$$

Now, let us use the method of long division to determine the corresponding series

$$\begin{array}{r|l} & \mathbb{1}+\beta z^{-\mathbb{1}}+\beta^{2} z^{-2}+\beta^{3} z^{-3}+\cdots \\ \hline z-\beta) & z \\ & \underline{z-\beta} \\ & \quad \beta \\ & \quad \underline{\beta-\beta^{2} z^{-\mathbb{1}}} \\ & \qquad \beta^{2} z^{-\mathbb{1}} \\ & \qquad \underline{\beta^{2} z^{-\mathbb{1}}-\beta^{3} z^{-2}} \\ & \qquad\qquad \beta^{3} z^{-2} \\ & \qquad\qquad \vdots \end{array}$$

from which

$$f(z)=(\mathbb{1}-\beta)(\mathbb{1}+\beta z^{-\mathbb{1}}+\beta^{2} z^{-2}+\beta^{3} z^{-3}+\cdots) \tag{4.26}$$

and

$$f(\tau)=(\mathbb{1}-\beta)\,\beta^{\tau} \qquad \text{for } \tau \geqq 0. \tag{4.27}$$

The network of Figure 4.33a is simply a concatenation of two networks in series, each having the form of the network of Figure 4.32b. Since we already have analyzed the latter, we can combine the result with the principle embodied in equation 4.19 to determine the unit-cohort response of the entire network. For the left-hand half of the network,

$$f'(z)=(\mathbb{1}-\beta)\frac{z}{z-\beta} \tag{4.28}$$

and for the right-hand half,

$$f''(z)=(\mathbb{1}-\gamma)\frac{z}{z-\gamma}. \tag{4.29}$$

Therefore, for the entire network

$$f(z)=f'(z)\, f''(z)=(\mathbb{1}-\beta)(\mathbb{1}-\gamma)\frac{z^{2}}{(z-\beta)(z-\gamma)}. \tag{4.30}$$

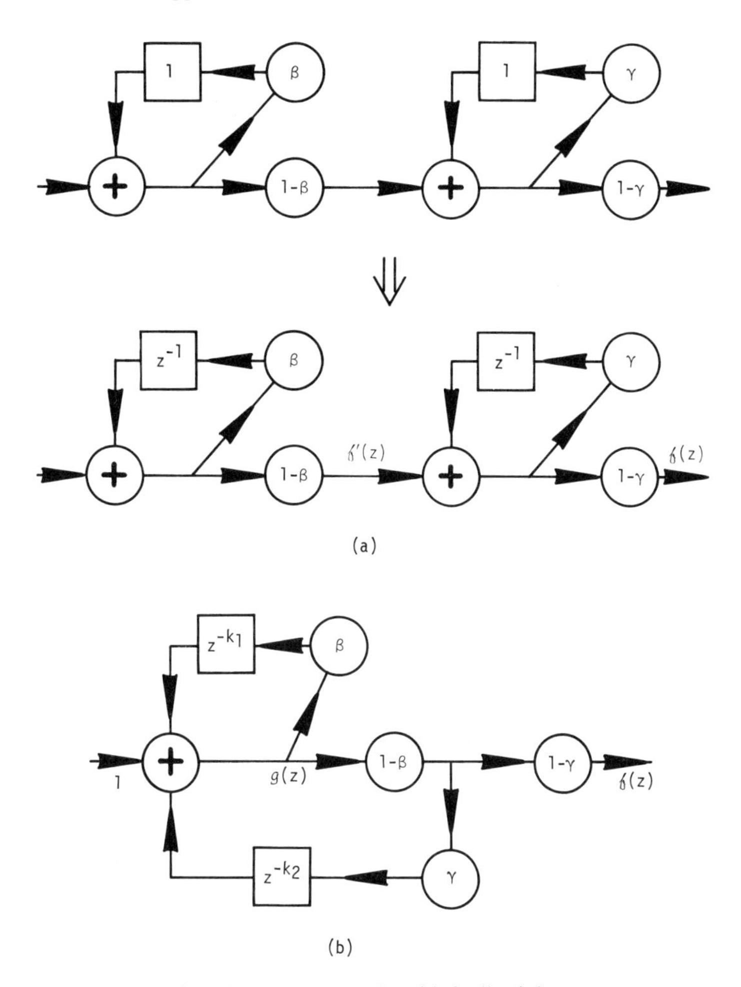

Fig. 4.33. More networks with feedback loops

The corresponding infinite series can be found by the method of long division, and from it one can determine immediately the corresponding discrete function of time. However, as we shall see in section 4.10 there is another method of converting functions of z to functions of time; and in this and many other cases, its application is far simpler.

The network of Figure 4.33b is similar to that of Figure 4.33a, but the elements are connected in a slightly different manner. Once again, we can begin by considering the functions converging at the *adder*, attempting to describe them in terms of the function emerging from the *adder*. The input function is one; the function coming in from the upper loop is $(\beta z^{-k_1})\, g(z)$; and the function coming in from the bottom loop is $(\mathbb{1}-\beta)\gamma z^{-k_2}\, g(z)$. Thus we have the equation for the combined loops:

$$g(z) = \mathbb{1} + (\beta z^{-k_1})\, g(z) + (\mathbb{1}-\beta)\,\gamma\, z^{-k_2}\, g(z) \tag{4.31}$$

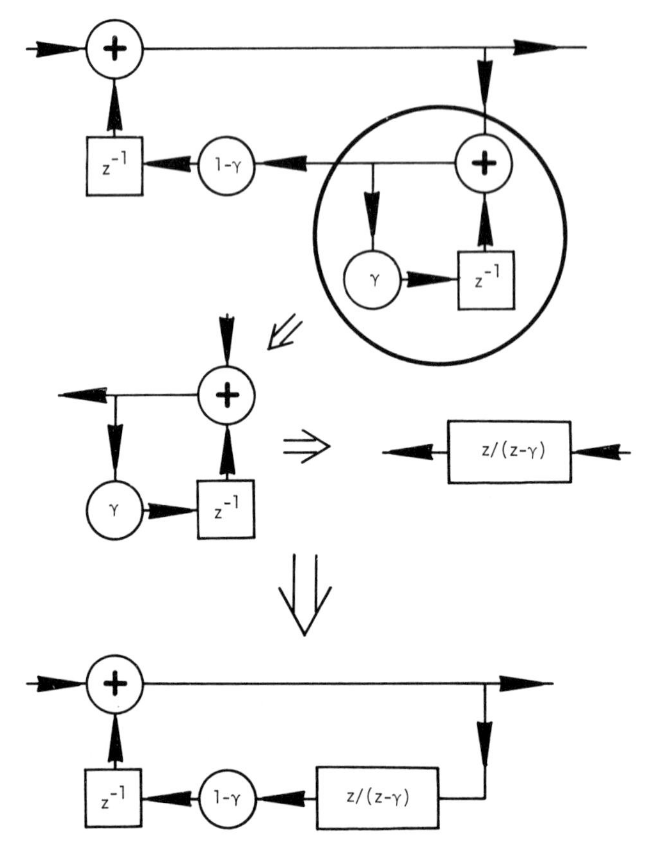

Fig. 4.34. How to deal with a feedback loop within a feedback loop

or

$$\mathscr{g}(z)=\frac{\mathbb{1}}{\mathbb{1}-\beta z^{-k_1}-(\mathbb{1}-\beta)\gamma z^{-k_2}} \tag{4.32}$$

and

$$\mathscr{f}(z)=(\mathbb{1}-\gamma)\,\mathscr{g}(z). \tag{4.33}$$

The network of Figure 4.34 exhibits a loop within a loop. In such cases, one can begin with the simple loops (i.e., those that do not contain loops within them) and replace each of them by a single element whose multiplicative parameter is the unit-cohort response of the loop being replaced. This process is illustrated in the figure.

From the last example, it should be clear to the reader that any linear network or portion thereof that has one input and one output can be replaced by a single transformed "element" whose parameter is the unit-cohort response of the network being replaced. Such complex elements are treated in precisely the same manner as the basic elements when one carries out further analyses. Whenever the parameter of such an element is a function of z, we shall represent the element with a box, since it is a dynamic element (see Figure 4.35).

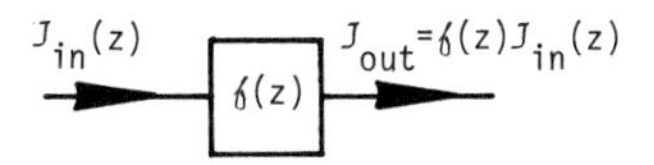

Fig. 4.35. Reduction of an entire network to a single dynamic element whose z transform is that of the unit-cohort response of the network

4.9. Linear Flow-Graph Analysis

As the networks to be analyzed become more and more complex, with multiple feedback loops and loops within loops, analysis with the methods of the previous section becomes more and more cumbersome; and as analysis becomes cumbersome, it also becomes error prone. Fortunately, for networks such as those we are dealing with, there is another analytical algorithm; and it is much simpler and much less prone to error when the networks become complex. The algorithm is based on a method known as *signal-flow-graph analysis* [see Mason and Zimmermann, 1960]. However, since we are dealing with functions to which the term *signal* is not especially appropriate, we shall change the name of the method to *linear-flow-graph analysis*. Since the method of analysis itself is quite useful, we shall introduce it first and then discuss the highly simplified algorithm that is based upon it.

Linear flow graphs usually are drawn as nodes connected by directed lines (i.e., lines with arrow heads). The nodes represent network variables and points of convergence and summation of those variables and the lines represent operations on those variables. In order to keep track of it, each node usually is assigned a number or letter, and each line is labelled with the operation that it represents (see Figure 4.36). The object of linear-flow-graph analysis is to eliminate all of the nodes but two, the one representing the input variable and the one representing the output variable. In doing so, one consolidates all of the lines into a single line connecting the input node to the output node. The operation associated with that line is the desired, unit-cohort response. Before

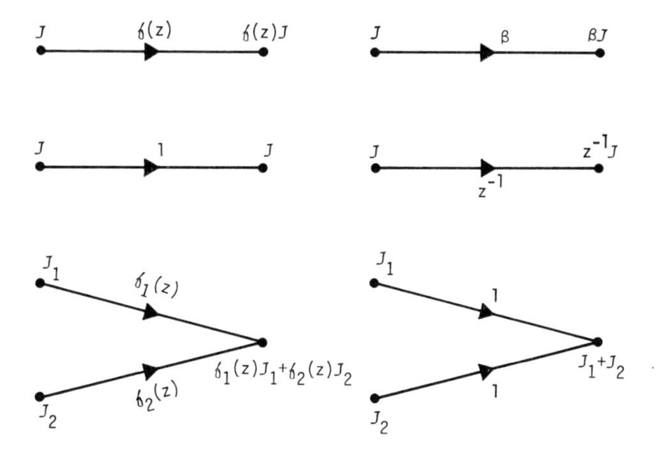

Fig. 4.36. Linear flow graph notation

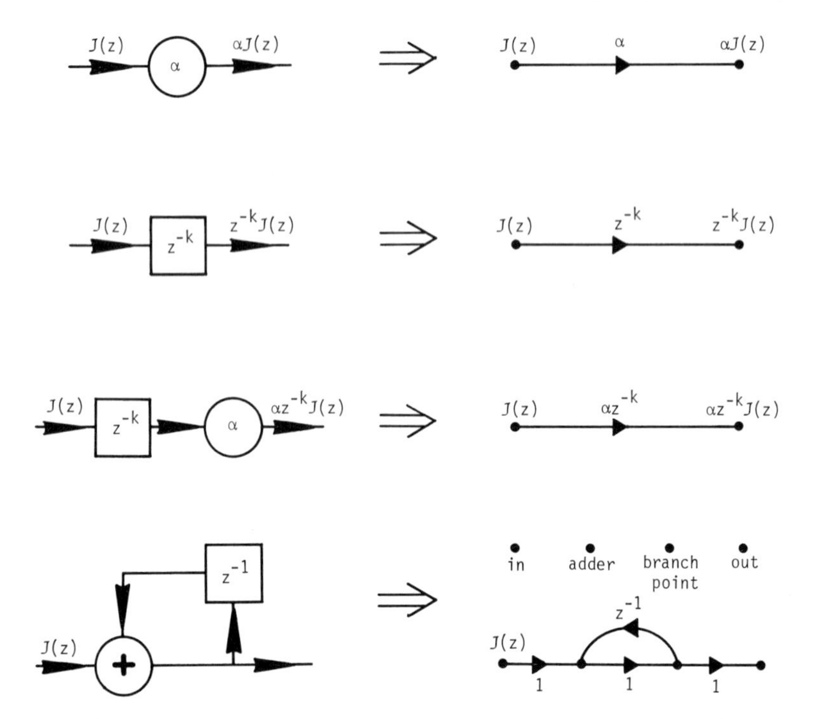

Fig. 4.37. Linear flow graph equivalents of common network elements and configurations

we can proceed with the elimination and consolidation, however, we first must convert our networks into linear flow graphs.

In order to avoid complicating our linear flow graphs unnecessarily, we can begining by converting all straight-chain segments in the network to single elements. In this way, we eliminate all variables in the network except those at branch points, those converging on *adders*, and the input and output variables. Therefore, in drawing the equivalent linear flow graph, we can begin by drawing a node for the input, a node for each branch point, a node for each *adder*, and a node for the output. We then can proceed to connect those nodes by appropriate directed lines. All of this probably is demonstrated much more clearly and succintly in pictures than it is in words. Therefore, consider Figure 4.37, in which linear-flow-graph equivalents for several basic networks are shown.

4.9.1. Graphical Reduction of Flow Graphs

For the purposes of eliminating nodes and consolidating lines in a linear flow graph, certain basic equivalences are useful. The validity of these should be apparent to the reader, so they are presented without comment in Figure 4.38. Again, since the elimination of nodes and the consolidation of lines are graphical techniques, graphical demonstration very likely is far more instructive than verbal description. Therefore, consider the following examples.

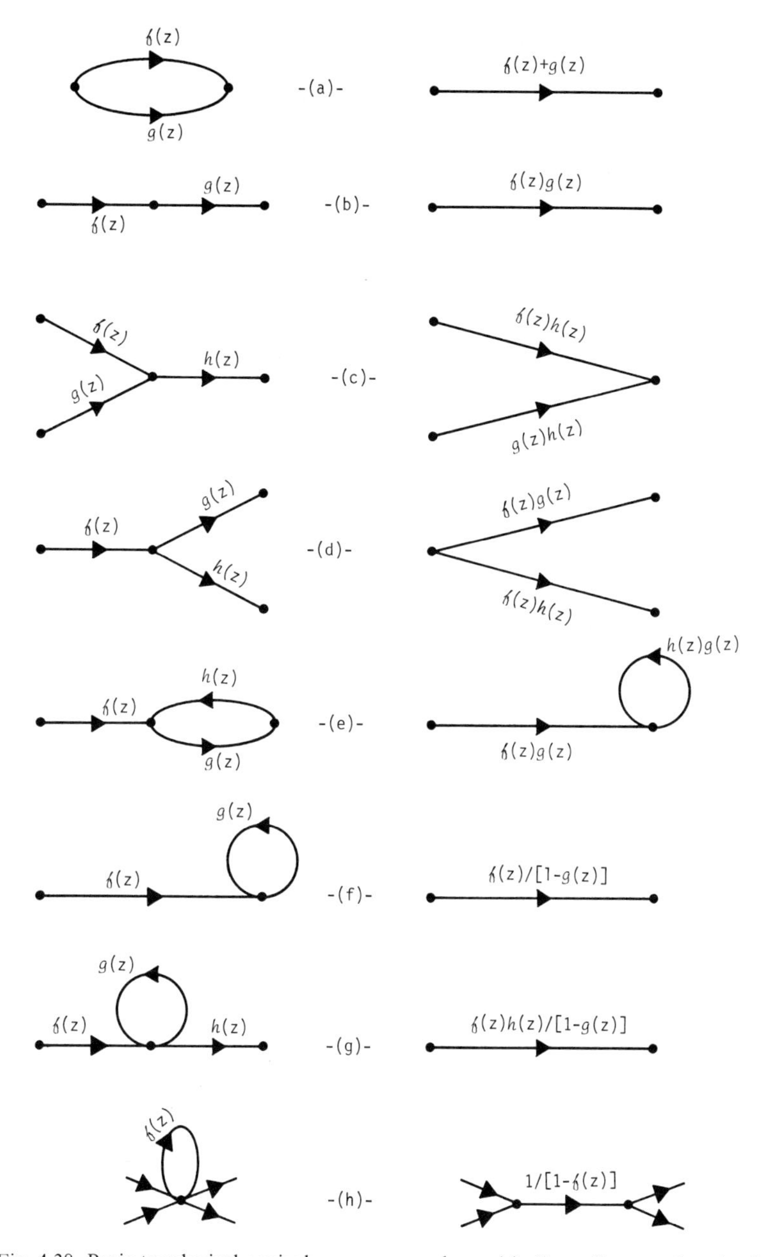

Fig. 4.38. Basic topological equivalences commonly used in linear flow graph reduction

Example 4.5. The network model of Figure 4.39 is a modified version of that in Figure 3.1. It represents the life cycle of a protozoan that undergoes binary fission with a fixed interfission interval, T_F. The input is a flow of newly produced daughter cells from an extraneous source and the output is simply our observation of the flow of newly generated daughters (we observe them but do

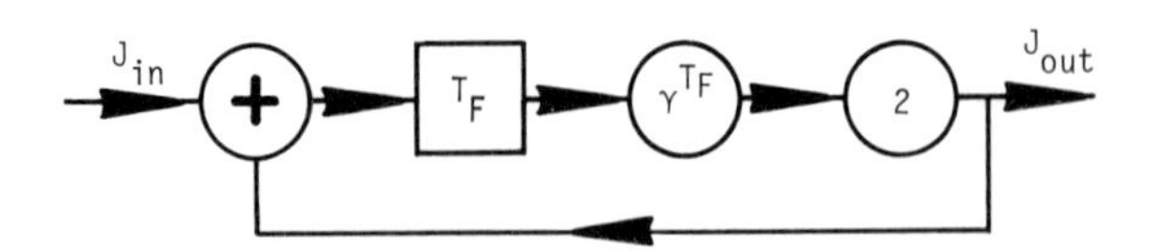

Fig. 4.39. Network model of an idealized life cycle involving binary fission

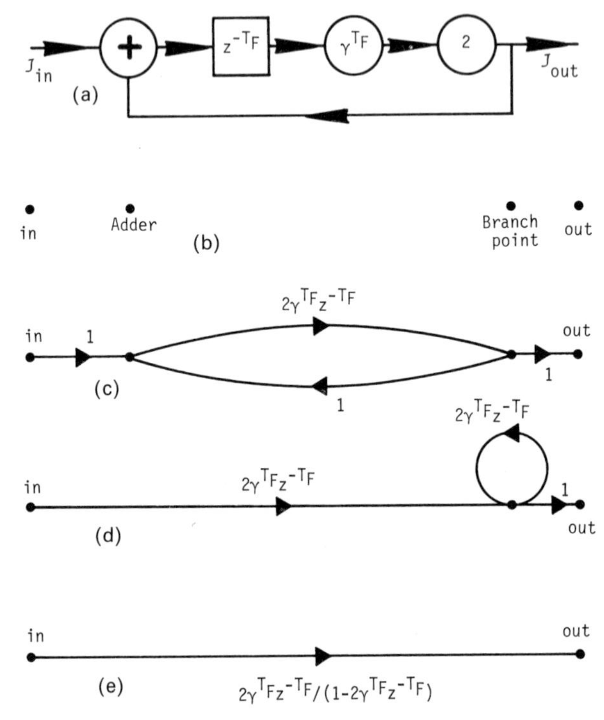

Fig. 4.40a–e. z-transformed version of the network in Fig. 4.39, along with its linear flow graph and a step-by-step linear flow graph reduction, leading to the transformed unit-cohort response (bottom)

not remove them). Draw the z-transform version of the network, then convert it to a linear flow graph and proceed to determine the unit-cohort response.

Answer: The first step brings us to the network of Figure 4.40a, whose derivation by now should be obvious to the reader. To convert this transformed network to a linear flow graph, we begin by drawing a node for the input, a node for the output, a node for the *adder*, and a node for the branch point (Figure 4.40b). The adder has two variables converging upon it, that represented by the input node and that represented by the branch-point node. We can represent the transfer of each of these variables to the *adder*'s node simply by connecting a line labelled unity (implying the transfer of an unaltered variable). Furthermore, the variable at the branch-point node and that at the output node are one and the same; therefore, we connect a line labelled unity from the branch-point node, to the output node. Finally, from the *adder*'s node to the

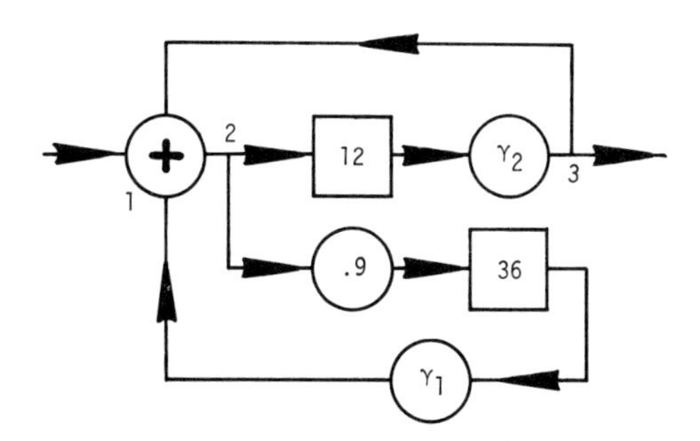

Fig. 4.41. Network model of a population of idealized Herring Gulls

branch-point node, we insert a line representing the operations on the sum that convert it to the output variable. These various lines are shown in Figure 4.40c.

To carry out our analysis from here, we simply must eliminate the *adder*'s node and the branch-point node, and leave a single line from the input node to the output node. We can remove the adder's node by employing the equivalence of Figure 4.38e, taking nodes 1, 2 and 3 to be the input node, the adder's node and the branch-point node, respectively. This leads to the configuration of Figure 4.40d. Next, employing the equivalence of Figure 4.38g, we can eliminate the branch-point node and leave a single path from the input node to the output node, as shown in Figure 4.40e. The transformed unit-cohort response appears as the parameter of the remaining line.

Example 4.6. The network of Figure 4.41 is a modified version of that in Figure 3.7. It represents the idealized life cycle of the female herring gull, with a maturation time of 36 months between fledging and the production of the first fledged brood and a nesting interval of 12 months. The input to the population is a flow of adults into it from an extraneous source (e.g., migration from another nesting colony), and the output is the observed flow of returning breeders. Construct the equivalent linear flow graph and from it determine the unit-cohort response.

Answer: In this case, we can begin with five nodes (input, output, *adder*, and two branch points), which can be connected by appropriately labelled lines. The input variable is transfered directly to the *adder*, (represented by node 1): the output of the *adder* is transfered directly to the first branch point (represented by node 2); and the variable at the second branch point (represented by node 3) is transferred directly to the output and directly back to the adder. Each of these direct transfers is represented in the linear flow graph by a line labelled unity. The feedback path from the first branch point to the *adder* is represented by the line labelled $0.9\,\gamma_1\,z^{-36}$, corresponding to the operations along that path; and the path between the two branch points is represented by a line from node 2 to node 3, labelled $\gamma_2\,z^{-12}$.

Elimination of nodes and consolidation of lines in the linear flow graph are straightforward (see Figure 4.42). First, we can eliminate node 3 by applying the

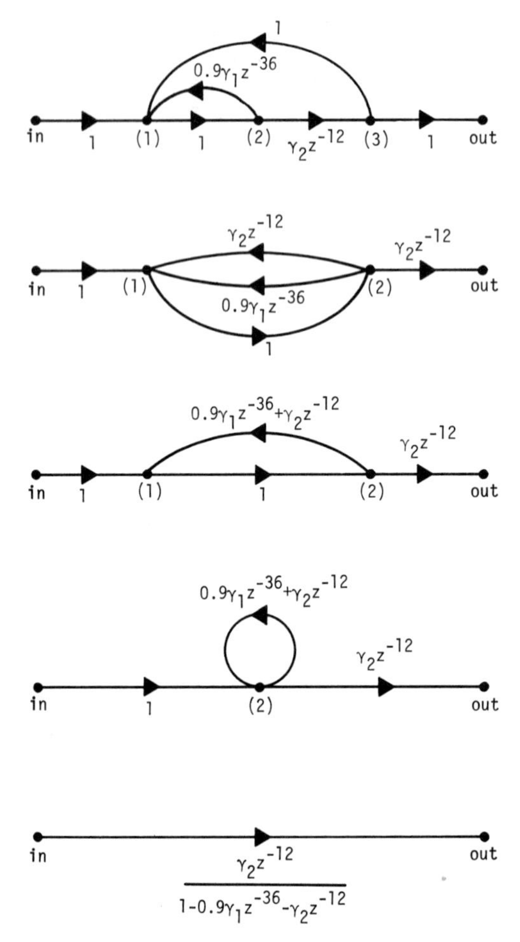

Fig. 4.42. Linear flow graph for the network of Fig. 4.41, and a step-by-step linear flow graph reduction, leading to the transformed unit-cohort response

equivalence of Figure 4.38d. Then, employing the equivalence of Figure 4.38a, we can consolidate the two parallel feedback lines from node 2 to node 1; next we can use Figure 4.38e to eliminate node 1; and finally, we can use Figure 4.38g to eliminate node 2, leaving a single line connecting the input and output nodes. The z-transform of the unit-cohort response is the parameter of that line.

Example 4.7. Figure 4.43 shows a modified version of the network of Figure 2.37b, representing the life cycle of an idealized cetacean. The *time delays* represent a 52-month interval from weaning to puberty and first ovulation, a one-month ovulation interval, an 11-month gestation period, and an 8-month nonreproductive lactation period. The various γ's are survivorships for the corresponding intervals; β is the probability of impregnation during a given ovulation; and ε is the proportion of weaned offspring that are females. The

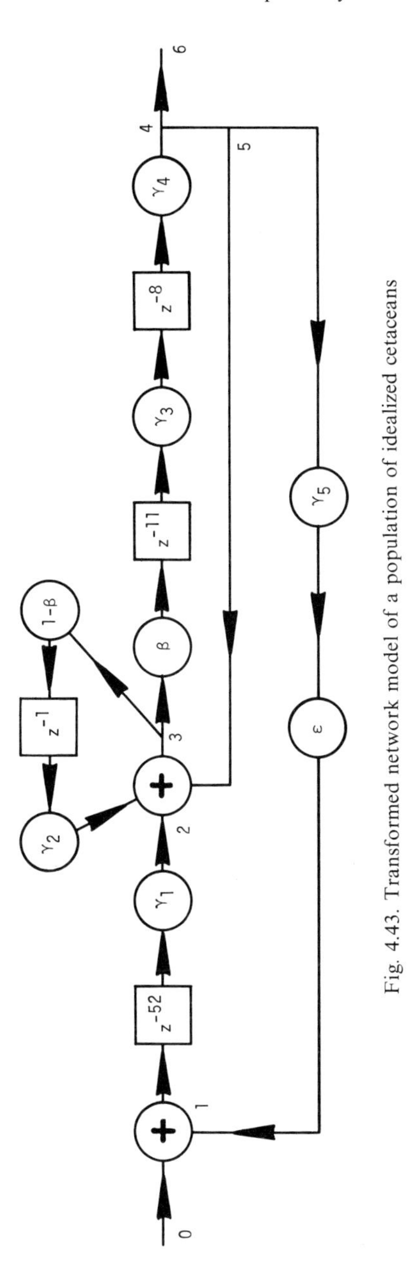

Fig. 4.43. Transformed network model of a population of idealized cetaceans

input is a flow of newly weaned females from an extraneous source; the output is the observed flow of newly weaned offspring produced by the population itself. Draw and reduce the equivalent linear flow graph to deduce the unit-cohort response of the network.

Answer: We have begun by labelling the points in the original network that will become nodes in the linear flow graph. There are six such points. The six

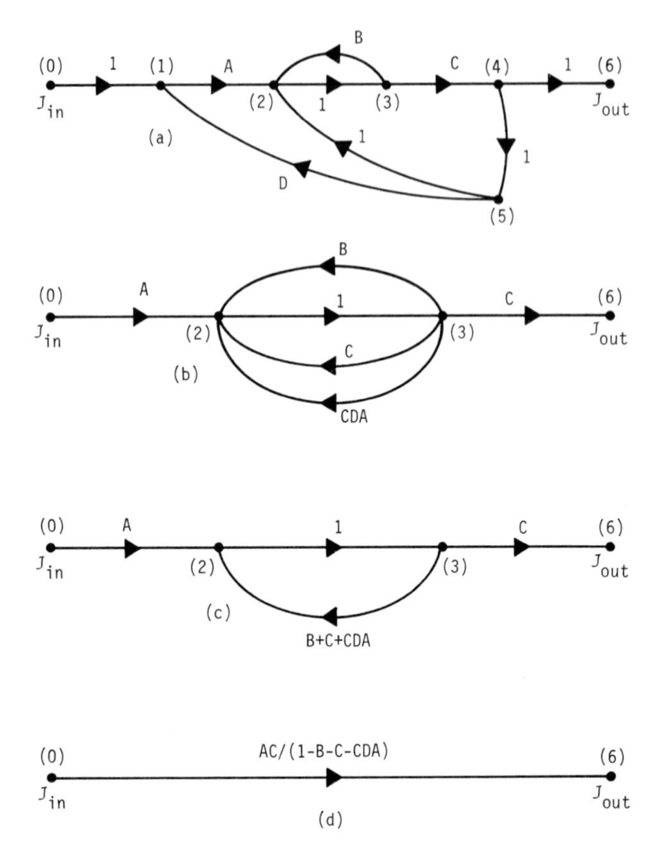

Fig. 4.44a–d. Linear flow graph and linear flow graph reduction to the transformed unit-cohort response for the network of Fig. 4.43. $A=\gamma_1 z^{-52}$; $B=\gamma_2(1-\beta)z^{-1}$; $C=\beta\gamma_3\gamma_4 z^{-19}$; $D=\varepsilon\gamma_5$

corresponding nodes appear in the flow graph of Figure 4.44a. Where the same variable appears at two nodes in the graph, the lines connecting the nodes are labelled with the multiplicative parameter 1. The other parameters in the flow graph have been abbreviated as follows:

$$A=\gamma_1 z^{-52}$$
$$B=(1-\beta)\gamma_2 z^{-1}$$
$$C=\beta\gamma_3\gamma_4 z^{-19}$$
$$D=\varepsilon\gamma_5$$

As a reduction strategy, let us attempt to bring all three feedback lines into parallel from node 3 to node 2. Inspection of the flow graph indicates that we can do this simply by application of the equivalence of Figure 4.38d at nodes 4 and 5, followed by application of Figure 4.38c at node 1. Once the lines are in parallel, their coefficients can be summed, as in the equivalence of Figure 4.38a. These steps are carried out in Figure 4.44b and c. Finally, we can apply the

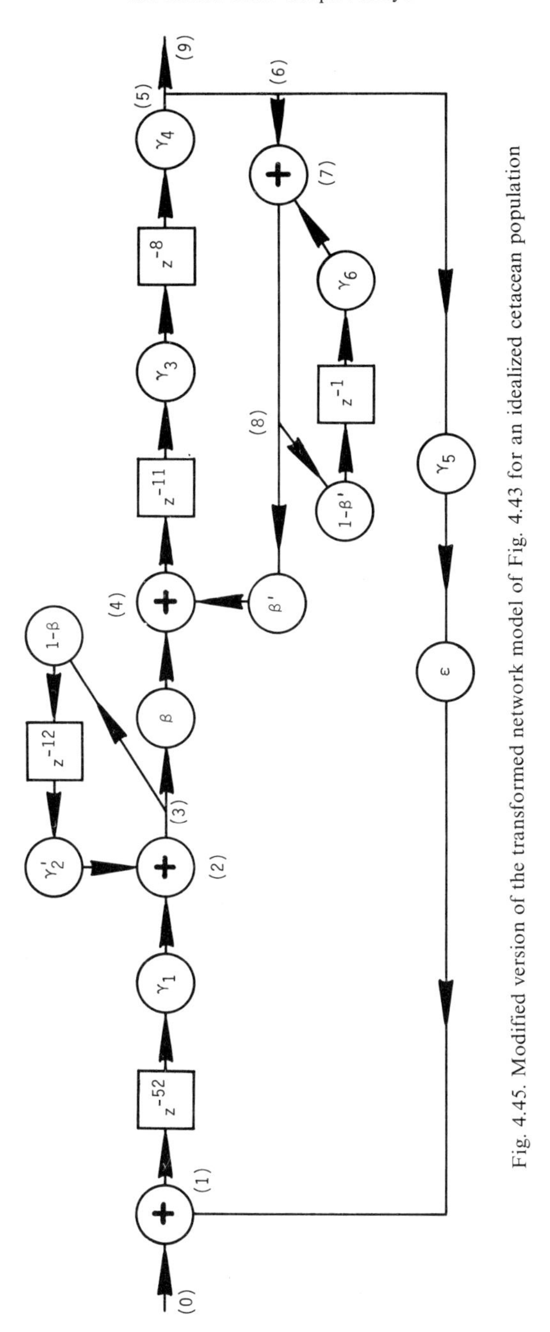

Fig. 4.45. Modified version of the transformed network model of Fig. 4.43 for an idealized cetacean population

equivalence of Figure 4.38e to remove node 2, followed by that of Figure 4.38g to obtain the reduced graph of Figure 4.44d. Now we can substitute the appropriate functions for A, B, C, and D to obtain the z-transform of the unit-cohort response:

$$f(z)=\gamma_1 z^{-52}/(1-(1-\beta)\gamma_2 z^{-1}-\beta\gamma_3 z^{-19}-\gamma_1\gamma_3\gamma_4\gamma_5\varepsilon\beta z^{-71}). \quad (4.34)$$

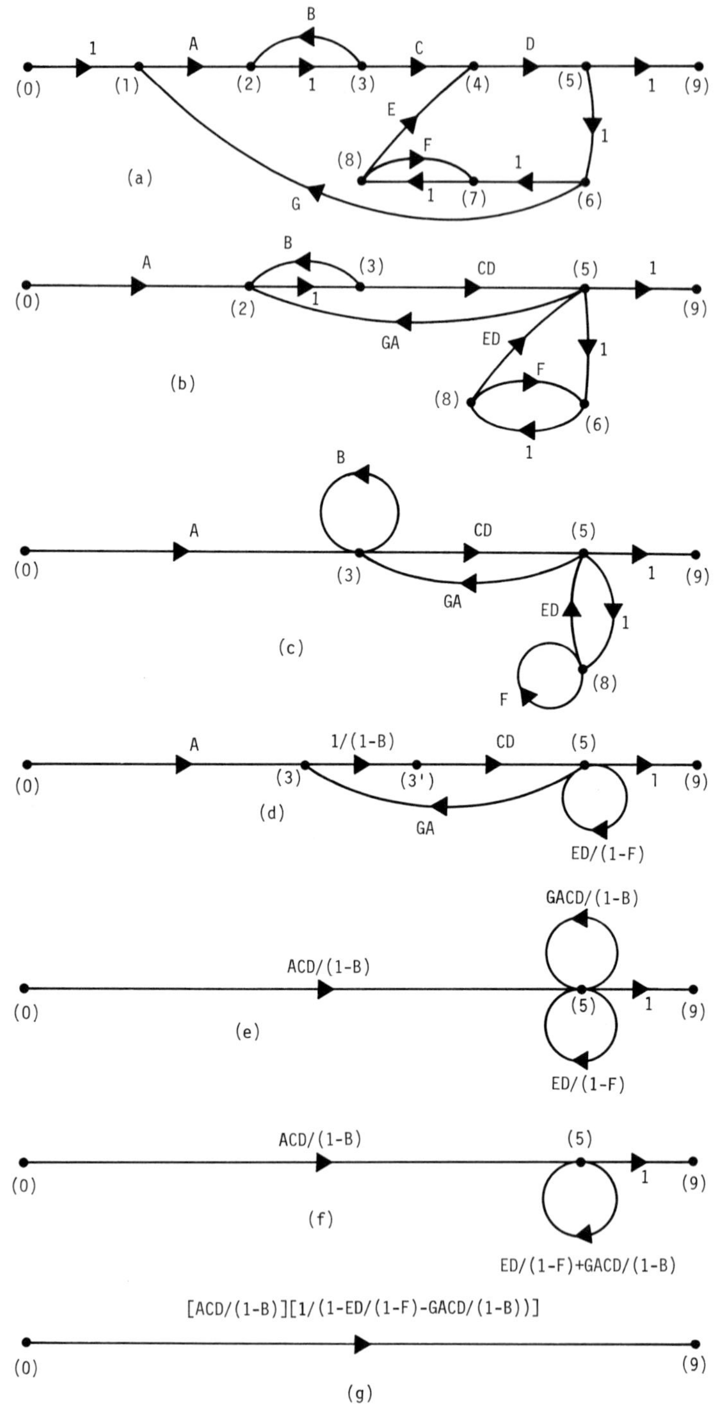

Fig. 4.46a–g. Flow graph and flow graph reduction to unit-cohort response for the network of Fig. 4.45. $A=\gamma_1 z^{-52}$; $B=\gamma_2'(1-\beta)z^{-12}$; $C=\beta$; $D=\gamma_3\gamma_4 z^{-19}$; $E=\beta'$; $F=\gamma_6(1-\beta')z^{-1}$; $G=\varepsilon\gamma_5$

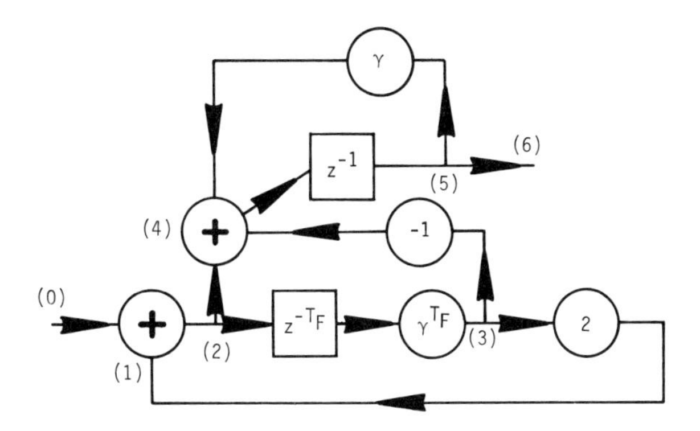

Fig. 4.47. Transformed network of an idealized population undergoing binary fission, with an accumulator providing a running total of the population

Example 4.8. The network of Figure 4.45 shows a variation on the network of Figure 4.43. It represents the same idealized cetacean life cycle, but with a different ovulation interval (12 months) for nulliparous females. The ovulation interval is one month for all females that have produced offspring. Thus, the model exhibits two loops representing ovulation. Draw and reduce the equivalent linear flow graph.

Answer: The procedure is illustrated in Figure 4.46. The initial flow graph is drawn with 10 nodes, which are labelled to correspond to the points they represent in the original network. The first step in reduction is rather arbitrary; we have chosen to apply the equivalence of Figure 4.38c to eliminate nodes 1, 4 and 7, leading to the structure in Figure 4.46b. As the next step, we apply the equivalence of Figure 4.38e to eliminate nodes 2 and 6. Then we apply Figure 4.38h to eliminate loop B at node 3, and Figure 4.38g to eliminate node 8, leading to the configuration of Figure 4.46d. Next, we apply Figures 4.38b and c to eliminate nodes 3 and 3′. This leaves us with two loops in parallel at node 5. Summing these, we have the configuration of Figure 4.46f. Finally, employing Figure 4.38g, we are led to the reduced graph of Figure 4.46g. The abbreviated form of the unit-cohort response is

$$f(z) = ACD(1-F)/(1-F-B-ED-GACD+BF+BED+FGACD) \quad (4.35)$$

which is rather complicated, but was not very difficult to deduce with the method of flow graphs.

Example 4.9. For our final example, let us return to the idealized protozoan population of Example 4.5, but for our output seek a running account of the total number of protozoa in the population. For this purpose, we simply can use a modified version (Figure 4.47) of the network model of Figure 3.2. The running total is provided by an accumulator in the network. This has led to a rather complicated network, with several loops. The linear flow graph equivalent (Figure 4.48a) is obtained in the usual way. Once again we have followed a

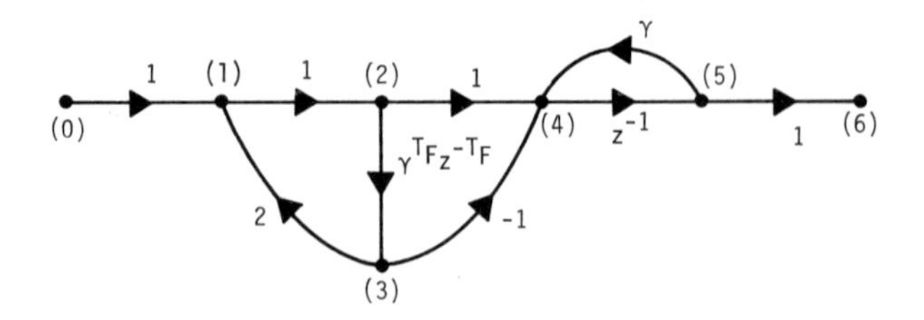

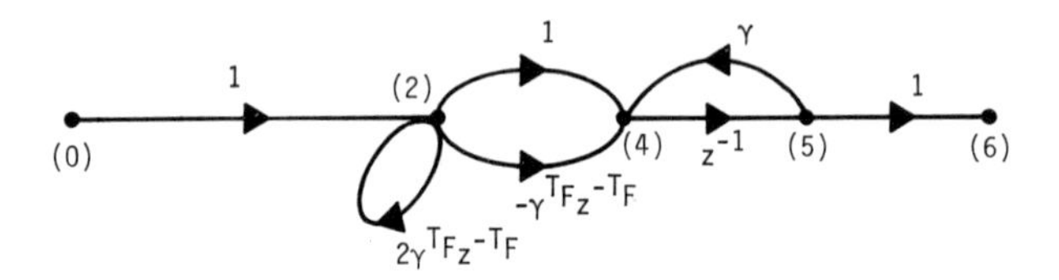

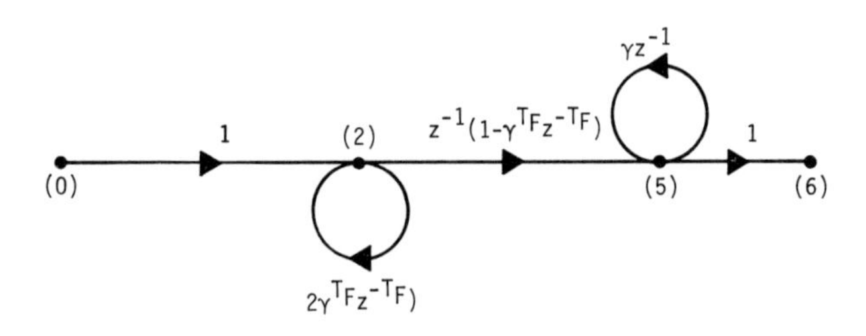

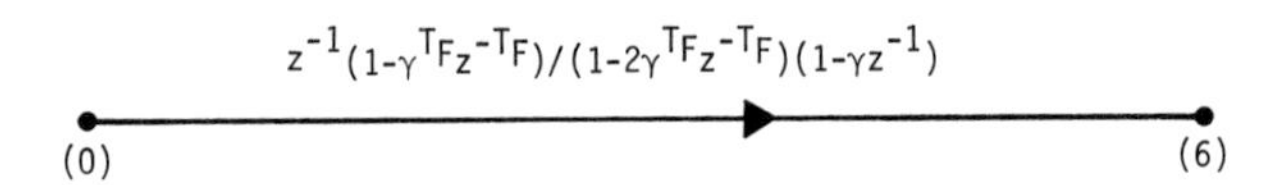

Fig. 4.48. Linear flow graph and flow graph reduction to unit-cohort response for the network of Fig. 4.47

rather arbitrary path in our reduction process. First, we have eliminated nodes 1 and 3; then we have eliminated node 4, and finally nodes 2 and 5. The z-transform of the unit-cohort response is rather complicated, but, again, not difficult to deduce by this method.

4.9.2. Mason's Rule

As we hinted at the beginning of this chapter, all of these deductive manipulations can be reduced to a rather simple algorithm. Once that algorithm and its application are understood, the z-transform of the unit-cohort response can be written by inspection from the unreduced linear flow graph. The algorithm, which was found by Mason, can be written as follows (using his notation):

$$\mathscr{f}(z) = \frac{[(P_1 + P_2 + P_3 + \cdots + P_p)(1 - L_1)(1 - L_2) \ldots (1 - L_m)]^{**}}{[(1 - L_1)(1 - L_2) \ldots (1 - L_m)]^{*}}. \quad (4.36)$$

** Drop all terms containing products of touching feedback loops or products of touching feedback loops and paths.

* Drop all terms containing products of touching feedback loops.

In this expression, a path (represented by P_i) is one possible route from the input node to the output node in which no node is encountered more than once. In other words, a path is a direct route from input to output, a route that does not pass completely around any loops. P_i is the product of multiplicative parameters of all the lines along the ith path. p is the total number of all such paths in the network, so that the sum $P_1 + P_2 + \cdots + P_p$ is taken over all possible paths. A feedback loop (represented by L_i) is a route from one node back to itself, in which no other node is encountered more than once. L_i is the product of multiplicative parameters of all the lines making up the ith feedback loop. Two or more feedback loops are said to touch if they share a common node. A path and a feedback loop touch if they share a common node.

As a demonstration of the algorithm, let us reconsider the unreduced linear flow graphs of the last five examples (Figure 4.49). The first example Figure 4.49a had one feedback loop, which is labelled L_1. The product of the parameters around that loop is

$$L_1 = 1 \times 2\gamma^{T_F} z^{-T_F}.$$

The path, P_1, and the feedback loop, L_1, share two common nodes, so they definitely do touch. Applying the algorithm of equation 4.36, we have

$$\mathscr{f}(z) = P_1(\mathbb{1} - L_1)^{**}/(\mathbb{1} - L_1)^{*} = (P_1 - P_1 L_1)^{**}/(\mathbb{1} - L_1)^{*}.$$

Dropping the terms containing products of touching paths and/or feedback loops (i.e., $P_1 L_1$) we have

$$\mathscr{f}(z) = P_1/(\mathbb{1} - L_1) = \mathbb{2}\gamma^{T_F} z^{-T_F}/(\mathbb{1} - \mathbb{2}\gamma^{T_F} z^{-T_F}) \quad (4.37)$$

which is precisely the solution we had deduced by the graph-reduction method.

The second example (Figure 4.49b) had two feedback loops, $L_1 = \mathbb{0}.\mathbb{9}\,\gamma_1 z^{-\mathbb{36}}$ and $L_2 = \gamma_2 z^{-\mathbb{12}}$, and one direct path, $P_1 = \gamma_2 z^{-\mathbb{12}}$. The path and the two feedback loops all share common nodes, so they all touch. Applying equation 4.36, we have

$$\begin{aligned}\mathscr{f}(z) &= P_1(\mathbb{1} - L_1)(\mathbb{1} - L_2)^{**}/(\mathbb{1} - L_1)(\mathbb{1} - L_2)^{*} \\ &= (P_1 - P_1 L_1 - P_1 L_2 + P_1 L_1 L_2)^{**}/(\mathbb{1} - L_1 - L_2 + L_1 L_2)^{*}.\end{aligned}$$

Since all paths and feedback loops touch, the numerator clearly will retain only one term, namely P_1, and the denominator will retain three terms, $\mathbb{1} - L_1 - L_2$.

$$\text{UNIT-COHORT RESPONSE} = P_1/(\mathbb{1} - L_1 - L_2) = \gamma_2 z^{-\mathbb{12}}/(\mathbb{1} - \mathbb{0}.\mathbb{9}\,\gamma_1 z^{-\mathbb{36}} - \gamma_2 z^{-\mathbb{12}}) \quad (4.38)$$

which again confirms our previous deductions.

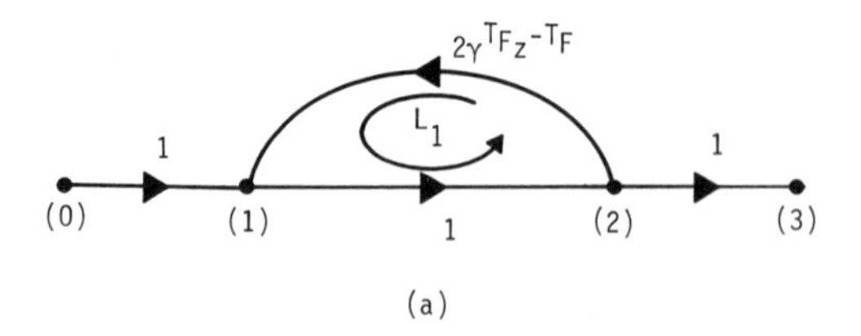

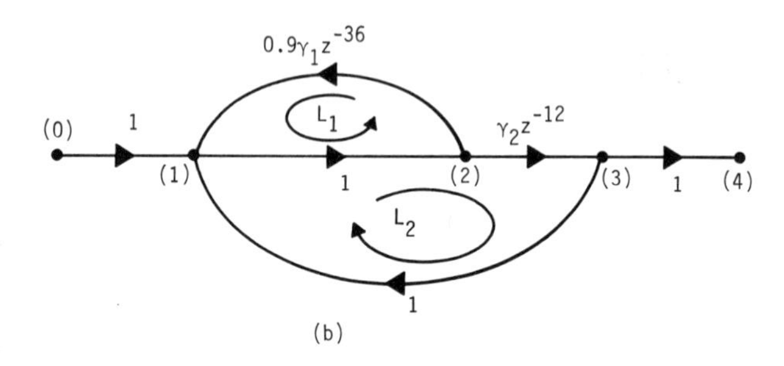

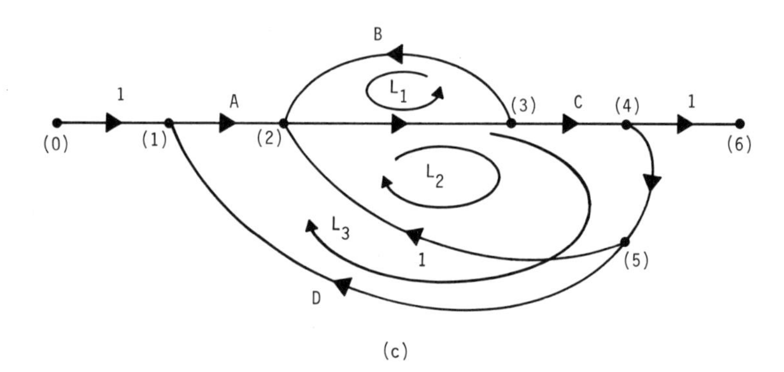

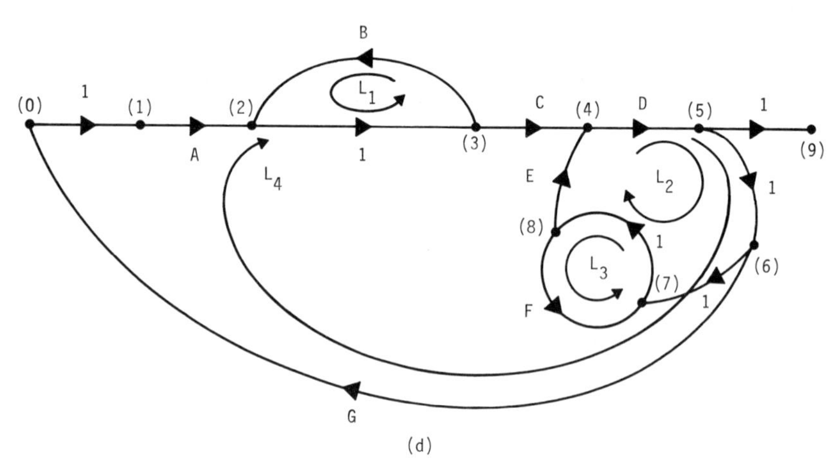

Fig. 4.49a–e. The flow graphs of Figs. 4.40, 4.42, 4.44, 4.46, and 4.48, set up to apply Mason's algorithm

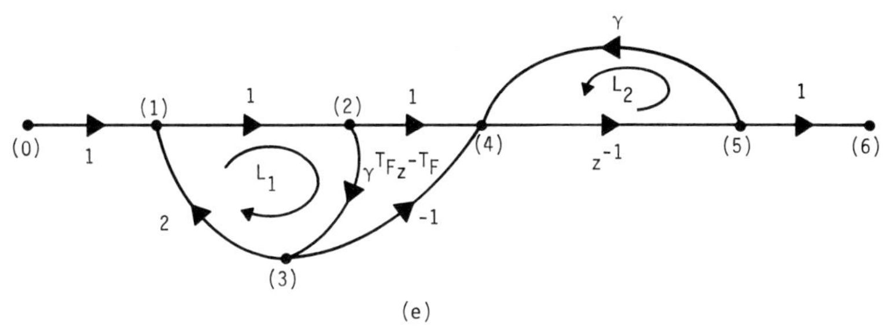

Fig. 4.49e

The third example (Figure 4.49c) had three feedback loops, $L_1 = B$, $L_2 = C$ and $L_3 = ACD$, and one direct path, $P_1 = AC$. All three feedback loops touch each other as well as the path. Therefore, when we apply equation 4.36, the numerator and denominator will contain no product terms:

$$\begin{aligned} f(z) &= P_1/(1 - L_1 - L_2 - L_3) \\ &= AC/(1 - B - C - ACD) \end{aligned} \tag{4.39}$$

which confirms our previous deduction.

In the fourth example (Figure 4.49d), there are four feedback loops:

$$L_1 = B$$
$$L_2 = DE$$
$$L_3 = F$$
$$L_4 = ACDG$$

and one path:

$$P_1 = ACD$$

The path touches loops L_1, L_2, and L_4, but does not touch L_3. Furthermore, L_3 does not share a node with either L_1 or L_4, and therefore does not touch those loops; and L_1 does not share a node with L_2. For that reason, the numerator and denominator will include the terms $P_1 L_3$, $L_1 L_3$, $L_4 L_3$, and $L_1 L_2$. All other product terms will include products of touching loops or paths and therefore are excluded.

$$\begin{aligned} f(z) &= (P_1 - P_1 L_3)/(1 - L_1 - L_2 - L_3 - L_4 + L_1 L_3 + L_4 L_3 + L_1 L_2) \\ &= (ACD - ACDF)/(1 - B - DE - F - ACDG + BF + ACDGF + BDE). \end{aligned} \tag{4.40}$$

This again confirms our previous deduction (equation 4.35). In the final example (Figure 4.49e), there are two feedback loops and two paths:

$$L_1 = 2\gamma^{T_F} z^{-T_F}$$
$$L_2 = \gamma z^{-1}$$
$$P_1 = z^{-1}$$
$$P_2 = -\gamma^{T_F} z^{-T_F} z^{-1}$$

P_1 and P_2 both touch the two loops; but the two loops do not touch one another. Carrying out the algorithm, we have

$$(P_1+P_2)[(1-L_1)(1-L_2)]^{**}/[(1-L_1)(1-L_2)]^{*}$$

but we drop all terms that contain P_1 or P_2 multiplied by L_1 or L_2. As we already have implied, one does not need to carry out the multiplication to find these terms. Clearly, the only allowable terms in the numerator are P_1 and P_2; all product terms are excluded. In the denominator the product term is not excluded. Thus we are left with

$$\begin{aligned}\text{UNIT COHORT RESPONSE} &= (P_1+P_2)/(1-L_1)(1-L_2)\\ &= (z^{-1}-\gamma^{T_F}z^{-T_F}z^{-1})/(1-2\gamma^{T_F}z^{-T_F})(1-\gamma z^{-1})\end{aligned} \tag{4.41}$$

which once again verifies our previous deduction.

By now it should be quite clear that the algorithm of equation 4.36 is an extremely powerful tool for the analysis of complicated linear networks. The method of graphical reduction can serve to provide a supplemental check on the results obtained by the algorithm.

Exercises. Construct and reduce linear flow graphs to obtain unit cohort response functions for each of the following examples.

a) Example 3.3 (Figure 3.5) with the input being the flow of new daughters from an extraneous source and the output being the flow of new daughters generated by the population itself.

Answer:

$$f(z)=2\gamma_1\beta z^{-T_m}/(1-2\gamma_1\beta z^{-T_m}-(1-\beta)\gamma_2 z^{-1}) \tag{4.42}$$

b) Example 3.4 (Figure 3.6); input: flow of new daughters from extraneous source; output: flow of new daughters generated by population itself.

Answer:

$$f(z)=2\beta\gamma z^{-T}/(1-2\beta\gamma z^{-T}-(1-\beta)\gamma z^{-T}) \tag{4.43}$$

c) Example 3.6 (Figure 3.8); input: flow of newly fledged females from extraneous source; output: annual flow of mature females into nesting.

Answer:

$$f(z)=(1-\gamma' z^{-96})\gamma_1 z^{-36}/(1-0.9\gamma_1 z^{-36}-\gamma_2 z^{-12}) \tag{4.44}$$

d) Example 3.7 (Figure 3.9); input: flow of newly fledged females from extrinsic source; output: flow of newly fledged females (including those from input).

Answer:

$$\begin{aligned}f(z)=(1-\gamma_2 z^{-12})/(1&-0.9\beta_1\gamma_1 z^{-36}-0.9(1-\beta_1)\beta_2\gamma_1\gamma_2 z^{-48}\\ &-0.9(1-\beta_1)(1-\beta_2)\gamma_1\gamma_2 z^{-60}-\gamma_2 z^{-12})\end{aligned} \tag{4.45}$$

e) Example 3.8 (Figure 3.10); same input and output as in Exercise d.

Answer:

$$\begin{aligned}\mathcal{f}(z)=&(1-\gamma_2 z^{-12})(1-\gamma_3 z^{-12})/(1+0.9\,\gamma_1\gamma_3\gamma_2' z^{-132}\\&+\gamma_1\gamma_2\gamma_2'\,0.9\,z^{-84}(1-\gamma_3 z^{-48})+0.9\,\gamma_1\gamma_3 z^{-48}(1-\gamma_2' z^{-48})\\&-0.9\,\gamma_1 z^{-36}+\gamma_2\gamma_3 z^{-24}-\gamma_2 z^{-12}-\gamma_3 z^{-12}).\end{aligned} \tag{4.46}$$

4.10. Interpretation of Unit-Cohort Response Functions: The Inverse z-Transform

Now that we have the wherewithal to generate the z-transforms of the unit-cohort responses of our network models with constant parameters, it becomes important that we have reasonably simple means of converting those transformed functions back to discrete functions of time, describing explicitly the network dynamics. In other words, given $\mathcal{f}(z)$ we want to be able to find the corresponding $f(\tau)$, and we want to be able do do so easily. The process itself, whether easy or difficult, is the inverse of the process of z-transformation, which takes us from $f(\tau)$ to the corresponding $\mathcal{f}(z)$. For that reason, $f(\tau)$, when we find it, is called the *inverse z-transform.*

When we complete our linear flow graph analysis, we have the z-transform of the unit-cohort response function in the following form:

$$\mathcal{f}(z)=R'(z^{-1})/Q'(z^{-1})=\left(\sum_{i=1}^{p} P_i\right)(1-L_1)(1-L_2)\ldots{}^{**}/(1-L_1)(1-L_2)\ldots{}^{*} \tag{4.47}$$

where P_i and L_i take the general form βz^{-k}, and R' and Q' both are polynomials in z^{-1}. Now we very well could proceed directly with these polynomials and interpret $\mathcal{f}(z)$ in terms of them. However, for several reasons, it will be easier to deal with polynomials in z rather than its reciprocal. Since we shall be dealing with finite networks, the polynomials P' and Q' both will be of finite degree (i.e., the largest exponent of z^{-1} will be finite in each case), and we can express them as follows:

$$R'(z^{-1})=a_0+a_1 z^{-1}+a_2 z^{-2}+a_3 z^{-3}+\cdots+a_{k_r} z^{-k_r} \tag{4.48}$$

$$Q'(z^{-1})=1+b_1 z^{-1}+b_2 z^{-2}+b_3 z^{-3}+\cdots+b_{k_q} z^{-k_q} \tag{4.49}$$

where k_r is the degree of the numerator; k_q is the degree of the denominator; and some of the coefficients (a_i or b_i) may be zero. To convert R' and Q' into polynomials in z rather than z^{-1}, one simply can multiply each of them by the factor z^k, where k is the larger of the two degrees, k_r or k_q:

$$\mathcal{f}(z)=z^k R'(z^{-1})/z^k Q'(z^{-1})=R(z)/Q(z). \tag{4.50}$$

Thus, if $k_r>k_q$, we have $k=k_r$ and

$$\begin{aligned}R(z)&=a_0 z^{k_r}+a_1 z^{k_r-1}+a_2 z^{k_r-2}+\cdots+a_{k_r}\\Q(z)&=z^{k_r}+b_1 z^{k_r-1}+b_2 z^{k_r-2}+\cdots+b_{k_q} z^{k_r-k_q}\end{aligned} \tag{4.51}$$

and if $k_q > k_r$, we have $k = k_q$ and

$$\begin{aligned} R(z) &= a_0 z^{k_q} + a_1 z^{k_q - 1} + a_2 z^{k_q - 2} + \cdots + a_{k_r} z^{k_q - k_r} \\ Q(z) &= z^{k_q} + b_1 z^{k_q - 1} + b_2 z^{k_q - 2} + \cdots + b_{k_r}. \end{aligned} \tag{4.52}$$

As we proceed to generate the inverse transform, there are several equally valid ways to interpret the z-transform itself. For example, in its original form, $\mathscr{f}(z) = R'(z^{-1})/Q'(z^{-1})$, we could picture it rather easily as representing a weighted sequence of repetitions of the same basic pattern. The pattern itself, $f'(\tau)$, would be determined entirely by the denominator, $Q'(z^{-1})$:

$$\mathscr{f}'(z) = 1/Q(z^{-1}). \tag{4.53}$$

The manner in which it is weighted and repeated would be determined entirely by the numerator, $R'(z^{-1})$:

$$\begin{aligned} \mathscr{f}(z) &= \frac{a_0}{Q'(z^{-1})} + \frac{a_1 z^{-1}}{Q'(z^{-1})} + \frac{a_2 z^{-2}}{Q'(z^{-1})} + \cdots + \frac{a_{k_r} z^{-k_r}}{Q'(z^{-1})} \\ &= a_0 \mathscr{f}'(z) + a_1 z^{-1} \mathscr{f}'(z) + a_2 z^{-2} \mathscr{f}'(z) + \cdots + a_{k_r} z^{-k_r} \mathscr{f}'(z). \end{aligned} \tag{4.54}$$

Recalling the basic properties of the z-transform (Table 4.1), we can see immediately that the inverse transform of $\mathscr{f}(z)$ could be expressed in the following form:

$$f(\tau) = a_0 f'(\tau) + a_1 f'(\tau - 1) + a_2 f'(\tau - 2) + \cdots + a_{k_r} f'(\tau - k_r) \tag{4.55}$$

where $f'(\tau - k)$ is defined to be zero for all values of τ less than k. In other words, the inverse transform comprises repetitions of a basic pattern, $f'(\tau)$, with each repetition weighted and shifted in time according to the dictates of a different term in $R'(z^{-1})$.

Of course, we still are left with the problem of finding the basic pattern, $f'(\tau)$, and expressing it in a compact, interpretable form. To find $f'(\tau)$, we simply can carry out the division indicated by $1/Q'(z^{-1})$, which would give it to us, term by term. Of course, if there are an indefinite number of terms in $f'(\tau)$, and we wish to express all of them, we must carry out an indefinite number of steps in our division process. We might just as well have selected the unit cohort as the repeated pattern and carried out the division directly with $R'(z^{-1})$ and $Q'(z^{-1})$ to determine the weighting factors:

$$f''(\tau) = 0 \quad \tau > 0$$

$$f''(\tau) = 1 \quad \tau = 0$$

$$\mathscr{f}''(z) = 1$$

$$\begin{array}{r|l} & a_0 + (a_1 - a_0 b_1) z^{-1} + \cdots + \\ \hline 1 + b_1 z^{-1} + b_2 z^{-2} + \cdots + b_{k_q} z^{-k_q} & a_0 + a_1 z^{-1} + a_2 z^{-2} + \cdots + a_{k_r} z^{-k_r} \\ & \underline{a_0 + a_0 b_1 z^{-1} - a_0 b_2 z^{-2} + \cdots + a_0 b_{k_q} z^{-k_q}} \\ & (a_1 - a_0 b_1) z^{-1} + (a_2 - a_0 b_2) z^{-2} + \cdots + \\ & \vdots \end{array} \tag{4.56}$$

$$\mathscr{f}(z) = a_0 \mathscr{f}''(z) + (a_1 - a_0 b_1) z^{-1} \mathscr{f}''(z) + \cdots.$$

Clearly, there is a fundamental problem here. We can express the inverse transform of $\mathscr{f}(z)$ as a repetition of basic patterns, with a particular weighting factor for each repetition. However, if we make the pattern complicated (e.g., $f'(\tau)$), then the process of determining the weighting factors is easy; but the evaluation and interpretation of the basic pattern itself are very difficult. If we make the basic pattern very simple (e.g., $f''(\tau)$), then its evaluation and interpretation are easy, but the determination of the weighting factors is very difficult. What we need here is a compromise method, presumably one involving patterns of intermediate complexity.

4.10.1. Partial-Fraction Expansion

The compromise method that conventionally is used is the decomposition of $\mathscr{f}(z)$ into sums of simple fractions of the forms

$$c_{i\kappa}\, z^{\kappa}/(z-r_i)^{\kappa} \quad \text{and} \quad c_{0\kappa}/z^{\kappa} \tag{4.57}$$

which represent the following compromise temporal patterns:

$$\text{if} \quad \mathscr{f}(z)=c_{i\kappa}\, z^{\kappa}/(z-r_i)^{\kappa} \qquad \kappa>0 \tag{4.58}$$

$$\text{then} \quad \mathscr{f}(\tau)=c_{i\kappa}\binom{\tau+\kappa-1}{\kappa-1}(r_i)^{\tau} \qquad \tau\geqq 0 \tag{4.59}$$

where

$$\binom{\tau+\kappa-1}{\kappa-1}=(\tau+\kappa-1)!/(\kappa-1)!\,(\tau)! \tag{4.60}$$

and

$$\text{if} \quad \mathscr{f}(z)=c_{0\kappa}/z^{\kappa} \qquad \kappa\geqq 0 \tag{4.61}$$

$$\begin{aligned} \text{then} \quad f(\tau)&=c_{0\kappa} && \text{when } \tau=\kappa \\ &=0 && \text{for all other values of } \tau(\tau\neq\kappa). \end{aligned} \tag{4.62}$$

To employ these compromise patterns, we somehow must be able to decompose $\mathscr{f}(z)$ into sums of terms of the forms of expression 4.57. The process of doing so is quite straightforward and is called *partial-fraction expansion.* To understand this process, consider the denominator of an arbitrary unit-cohort response function, $\mathscr{f}(z)=R(z)/Q(z)$:

$$Q(z)=z^{k}+b_1 z^{k-1}+b_2 z^{k-2}+\cdots+b_{k_q} z^{k-k_q} \tag{4.63}$$

where k is either k_q or k_r, whichever is larger. This polynomial can be expressed as the product of k_q binomial factors and a term z^{k-k_q}:

$$\begin{aligned} Q(z)&=z^{k-k_q}(z^{k_q}+b_1 z^{k_q-1}+b_2 z^{k_q-2}+\cdots+b_{k_q}) \\ &=z^{k-k_q}(z-r_1)(z-r_2)(z-r_3)(z-r_4)\ldots(z-r_{k_q}) \end{aligned} \tag{4.64}$$

where $r_1, r_2, r_3, \ldots, r_{k_q}$ are the *zeros* (often called *roots*) of the polynomial $Q(z)$.

Now, assuming that $k=k_q$ and that no two roots are the same, consider the following sum

$$S=c_{00}+c_{11}z/(z-r_1)+c_{21}z/(z-r_2)+\cdots+c_{k_q1}z/(z-r_{k_q}). \tag{4.65}$$

In the process of reconstituting this sum into a proper fraction, we obtain

$$\begin{aligned}S=\{&c_{00}Q(z)+c_{11}zQ(z)/(z-r_1)+c_{21}zQ(z)/(z-r_2)+\cdots\\&+c_{k_q1}zQ(z)/(z-r_{k_q})\}/Q(z)\end{aligned} \tag{4.66}$$

where each term $zQ(z)/(z-r_i)$ in the numerator is a simple polynomial in z, and no two polynomials are the same. By proper selections of the k_q+1 arbitrary constants, c_{i1}, therefore, we can generate in the numerator any polynomial in z of degree k_q or less. Since $R(z)$ is guaranteed to be of degree k_q or less, we certainly can choose the values of the c_{i1} such that

$$\begin{aligned}R(z)=&c_{00}Q(z)+c_{11}zQ(z)/(z-r_1)+c_{21}zQ(z)/(z-r_2)+\cdots\\&+c_{k_q1}zQ(z)/(z-r_{k_q}).\end{aligned} \tag{4.67}$$

Therefore, if the degree of $Q'(z^{-1})$ in z^{-1} is greater than that of $R'(z^{-1})$, so that $k=k_q$, and if none of the roots, r_i, of $Q(z)$ are repeated, then we can decompose $\mathscr{f}(z)$ with the following partial-fraction expansion:

$$\mathscr{f}(z)=R(z)/Q(z)=c_{00}+c_{11}z/(z-r_1)+c_{21}z/(z-r_2)+\cdots+c_{k_q1}z/(z-r_{k_q}). \tag{4.68}$$

The expansion of equation 4.68 is rather simple. The decomposition becomes slightly more complicated, however, if $k=k_r$ or if two or more of the r_i are identical. For example, if $k=k_q$, but r_1 and r_2 were identical, then the expansion of equation 4.68 would be inadequate in two respects. Since the two terms, $c_{11}z/(z-r_1)$ and $c_{21}z/(z-r_2)$ could be summed to yield a single term, $(c_{11}+c_{21})z/(z-r_1)$, the number of arbitrary constants now will be k_0 rather than k_0+1, and the degree of the reconstituted denominator will be k_q-1 rather than k_q. Thus we shall not be able to construct an arbitrary numerator of degree k_q, nor will the denominator of the reconstituted proper fraction be correct. The standard method of circumventing this difficulty is to employ in the partial-fraction expansion *all* of the terms of the form

$$c_{i\kappa}z^{\kappa}/(z-r_i)^{\kappa} \qquad 1\leqq\kappa\leqq\Phi \tag{4.69}$$

where Φ is the number of occurences of the root r_i. Thus, for example, if r_i occurs twice in the denominator of $\mathscr{f}(z)$, then the expansion would include the two terms

$$c_{i1}z/(z-r_i) \quad \text{and} \quad c_{i2}z^2/(z-r_i)^2.$$

In this manner, the number of arbitrary constants in our decomposition once again will equal k_q+1 and reconstitution once again will produce a proper fraction with the correct denominator.

When $k=k_r$, a similar problem arises. The denominator will contain a factor of the form $z^{k_r-k_q}$. If we treat this as k_r-k_q repetitions of the root $r=0$, and use expression 4.69, then we find that all of the terms so generated can be summed into a single, constant term. Therefore, once more we shall have too few arbitrary constants as well as an incorrect reconstituted denominator. The standard method of overcoming this difficulty is to use terms of the form

$$c_{0\kappa}/z^{\kappa} \qquad 1 \leqq \kappa \leqq k_r - k_q. \tag{4.70}$$

Thus, for example, if $k_r - k_q = 2$, then the partial-fraction expansion would include the two terms

$$c_{01}/z \quad \text{and} \quad c_{02}/z^2$$

which once again would introduce the needed number of arbitrary constants and would provide reconstitution of the correct denominator.

Example 4.10. Find the partial-fraction expansions of the following unit-cohort response functions:

$$\text{(a)} \ \frac{z^{-3}}{1-2z^{-1}+z^{-2}} \qquad \text{(b)} \ \frac{1}{1-z^{-1}-2z^{-2}} \qquad \text{(c)} \ \frac{1+z^{-4}}{1-3z^{-1}-10z^{-2}}.$$

Answer: The first step in each case will be to convert the numerator and denominator into polynomials in z rather than z^{-1}. To do so, we multiply both by z^k, where k is the largest power of z^{-1} in $f(z)$:

$$\text{(a)} \ f(z) = \frac{z^3}{z^3}\,\frac{z^{-3}}{1-2z^{-1}+z^{-2}} = \frac{1}{z^3-2z^2+z}$$

$$\text{(b)} \ f(z) = \frac{z^2}{z^2}\,\frac{1}{1-z^{-1}-2z^{-2}} = \frac{z^2}{z^2-z-2}$$

$$\text{(c)} \ f(z) = \frac{z^4}{z^4}\,\frac{1+z^{-4}}{1-3z^{-1}-10z^2} = \frac{z^4+1}{z^4-3z^3-10z^2}.$$

The next step is to factor the denominators into products of binomials:

(a) $Q(z) = z^3 - 2z^2 + z = z(z-1)(z-1)$

(b) $Q(z) = z^2 - z - 2 = (z-2)(z+1)$

(c) $Q(z) = z^4 - 3z^3 - 10z^2 = z^2(z-5)(z+2)$.

Then we can proceed with the decompositions:

$$\text{(a)} \ f(z) = \frac{1}{z^3-2z^2+z} = c_{00} + \frac{c_{01}}{z} + \frac{c_{11}z}{(z-1)} + \frac{c_{12}z^2}{(z-1)^2}$$

$$\text{(b)} \ f(z) = \frac{z^2}{z^2-z-2} = c_{00} + \frac{c_{11}z}{(z-2)} + \frac{c_{12}z}{(z+1)}$$

$$\text{(c)} \ f(z) = \frac{z^4+1}{z^4-3z^3-10z^2} = c_{00} + \frac{c_{01}}{z} + \frac{c_{02}}{z^2} + \frac{c_{11}}{(z-5)} + \frac{c_{12}}{(z+2)}.$$

To determine the appropriate values of the constants, $c_{i\kappa}$, we simply can reconstitute the proper practions from the expansions:

$$\text{(a) } f(z) = [c_{00} z(z-1)(z-1)^2 + c_{01}(z-1)(z-1)^2 + c_{11} z^2 (z-1)^2 + c_{12} z^3 (z-1)]/z(z-1)(z-1)^2$$

$$= [(c_{00}+c_{11}+c_{12}) z^3 + (-2c_{00}+c_{01}-c_{11}) z^2 + (c_{00}-2c_{01}) z + (c_{01})]/[z^3 - 2z^2 + z]$$

$$= R(z)/Q(z) = 1/[z^3 - 2z^2 - z]$$

and then set the coefficients of the numerator to their correct values

$$\begin{aligned} c_{00}+c_{11}+c_{12} &= 0 \\ -2c_{00}+c_{01}-c_{11} &= 0 \\ c_{00}-2c_{01} &= 0 \\ c_{01} &= 1 \end{aligned} \quad \text{from which} \quad \begin{aligned} c_{01} &= 1 \\ c_{00} &= 2 \\ c_{11} &= -3 \\ c_{12} &= 1 \end{aligned}$$

The appropriate coefficients of the partial-fraction expansions for (b) and (c) could be found in precisely the same manner. However, as we soon shall see, there is an easier way.

4.10.2. Finding the Coefficients of an Expansion

Using the simple transform pairs of equations 4.58/4.59 and 4.61/4.62 along with the partial-fraction expansion method outlined thusfar, we can find the inverse transform of any unit-cohort response function we are likely to come across in our network studies. The method is not algorithmically complete, however, because we have not yet set forth methods of finding the roots, r_i, of the denominator of $f(z)$. In addition, the method used in Example 4.10 for evaluating the constants of the partial-fraction expansion is not the most efficient procedure for that purpose. We shall take up the question of finding the roots of $Q(z)$ in the following section. In the meantime, we shall discuss an efficient method for finding the coefficients, $c_{i\kappa}$, of the partial-fraction expansion, given the roots of $Q(z)$.

Actually, the method is very simple. We can express $f(z)$ in the following, general form

$$f(z) = c_{00} + \frac{c_{11} z}{(z-r_1)} + \frac{c_{12} z^2}{(z-r_1)^2} + \cdots + \frac{c_{1\Phi_1} z^{\Phi_1}}{(z-r_1)^{\Phi_1}} + \frac{c_{21} z}{(z-r_2)} + \frac{c_{22} z^2}{(z-r_2)^2} + \cdots + \frac{c_{2\Phi_2} z^{\Phi_2}}{(z-r_2)^{\Phi_2}} + \cdots + \frac{c_{k1} z}{(z-r_k)} + \cdots + \frac{c_{k\Phi_k} z^{\Phi_k}}{(z-r)^{\Phi_k}} + \frac{c_{01}}{z} + \frac{c_{02}}{z^2} + \cdots + \frac{c_{0,k_r-k_q}}{z^{(k_r-k_q)}} \tag{4.71}$$

where only positive values of $k_r - k_q$ are used. Notice that if we multiply both sides of this equation by $(z-r_i)^{\Phi_i}$ and then set z equal to r_i, that all of the terms but one on the right-hand side will equal zero. The only nonzero term will be that with $(z-r_i)^{\Phi_i}$ in the denominator. Therefore, we immediately have the very simple expression

$$c_{i\Phi_i} = \left[\frac{(z-r_i)}{z}\right]^{\Phi_i} f(z)\bigg|_{z=r_i}. \tag{4.72}$$

In other words, to find the coefficients, $c_{i\Phi_i}$, of the partial-fraction expansion term $z^{\Phi_i}/(z-r_1)^{\Phi_i}$, where Φ_i is the number of times that the root r_i occurs in $Q(z)$, one simply multiplies $f(z)$ by $[(z-r_i)/z]^{\Phi_i}$ and then sets z equal to r_i.

To determine c_{0,k_r-k_q}, when k_r-k_q is positive, one simply multiplies $f(z)$ by $z^{k_r-k_q}$ and then sets z equal to zero in the product, in which case all of the terms but c_{0,k_r-k_q} on the right-hand side of equation 4.71 become equal to zero:

$$c_{0,k_r-k_q} = z^{(k_r-k_q)} f(z)|_{z=0} \qquad k_r \geqq k_q. \tag{4.73}$$

If k_r is less than k_q, then among the terms $c_{0\kappa}$, only c_{00} can be nonzero. To determine c_{00} in this case, one simply sets z equal to zero in $f(z)$:

$$c_{00} = f(z)|_{z=0} \qquad k_r \leqq k_q. \tag{4.74}$$

Once we have found these coefficients, we are left with the task of finding the remaining coefficients for the multiple roots (i.e., finding c_{ij} where $j<\Phi_i$). Usually, this will not be a big project; multiple roots are rather a rare occurrence in our networks. However, when they do occur, we simply begin by finding all of the coefficients that we can by the methods outlined, and then we find the few remaining coefficients by setting z in $f(z)$ equal to values other than zero or any of the r_i. This will lead to a set of linear equations, one for each unknown coefficient, which can be solved simultaneously to yield the coefficients. This procedure will be very similar to that outlined in Example 4.10(a) and usually will not be difficult if the values of z are selected judiciously.

Example 4.11. Consider Example 4.5 of the previous section. The z-transform of the unit-cohort response was found to be

$$f(z) = 2(\gamma^{T_F})(z^{-1})^{T_F}/[1 - 2(\gamma^{T_F})(z^{-1})^{T_F}].$$

The degrees of the numerator and denominator in z^{-1} are the same, namely T_F. Therefore, we can begin by multiplying both by z^{T_F} and thus converting them to polynomials in z:

$$f(z) = 2\gamma^{T_F}/z^{T_F} - 2\gamma^{T_F}.$$

Now, let us consider the particular case of $T_F = 1$ and $\gamma = 1$. In that case, the denominator has a single root:

$$r_1 = 2$$

and we have the partial-fraction expansion

$$f(z) = c_{00} + c_{11} z/(z-2).$$

To find c_{00}, we simply set z equal to zero in $f(z)$:

$$c_{00} = f(z)|_{z=0} = 2/[0-2] = -1.$$

To determine c_{11}, we simply apply the algorithm of equation 4.72:

$$c_{11} = \frac{(z-2)}{z} f(z)\bigg|_{z=2} = 1.$$

Thus

$$f(z) = -1 + z/(z-2)$$

for which the corresponding function of time is the inverse transform of -1 plus the inverse transform of $z/(z-2)$, both of which can be deduced from the transform pairs of equations 4.58/4.59 and 4.61/4.62.

$$f(\tau) = -1 + 2^{\tau} \qquad \tau = 0$$

$$f(\tau) = 2^{\tau} \qquad \tau > 0.$$

In other words, the unit-cohort response for $\tau = 0$ is zero and for all subsequent intervals it is 2^{τ}.

The reader should be able to verify this result by inspection of the network in Figure 4.39, setting γ and T_F both equal to one. A unit cohort of the flow, representing a single protozoan, is applied to the input at the 0th interval, at which time the cohort is in *time delay* T_F, and no flow has emerged yet at the output (representing production of daughters by fission). At the first interval, the original cohort has advanced through the *time delay* and divided, producing two daughters, each of which survives to recycle through the *time delay* and produce two more daughters at the second interval. Thus, at each interval after the first, the population will have doubled its value from the previous interval.

Example 4.12. Consider a network model whose unit-cohort response has the following z-transform:

$$f(z) = \frac{1}{[2 + 3z^{-1} + (z^{-1})^2]}.$$

Find the inverse transform of the unit-cohort response.

Answer: Converting the denominator into a polynomial in z leads to

$$f(z) = z^2/(2z^2 + 3z + 1).$$

The denominator has two roots

$$r_1 = -1, \qquad r_2 = -1/2$$

which lead us to the following partial-fraction expansion:

$$f(z) = c_{00} + c_{11} z/(z+1) + c_{21} z/(z+1/2).$$

Applying equations 4.72 and 4.74 to find the coefficients, we have

$$c_{00} = f(z)|_{z=0} = 0$$

$$c_{11} = [(z+1)/z] f(z)|_{z=-1} = z/(2z+1)|_{z=-1} = 1$$

$$c_{21} = [(z+1/2)/z] f(z)|_{z=-1/2} = z/(2z+2)|_{z=-1/2} = -1/2$$

which leads to

$$\mathscr{f}(z) = z/(z+1) - (1/2)\,z/(z+1/2)$$

and the corresponding inverse transform (from equations 4.58/4.59)

$$f(\tau) = (-1)^\tau - 1/2(-1/2)^\tau \qquad \tau \geqq 0.$$

Example 4.13. Find the inverse transform of

$$\mathscr{f}(z) = (1+z^{-4})/(1-3z^{-1}+2z^{-2})$$

Answer:

$$\mathscr{f}(z) = (z^4+1)/(z^4-3z^3+2z^2)$$
$$\mathscr{f}(z) = (z^4+1)/z^2(z-2)(z-1)$$
$$\mathscr{f}(z) = c_{00} + c_{01}/z + c_{02}/z^2 + c_{11}\,z/(z-2) + c_{21}\,z/(z-1)$$

$$c_{02} = z^2 \mathscr{f}(z)|_{z=0} = 1/2$$
$$c_{11} = [(z-2)/z]\,\mathscr{f}(z)|_{z=2} = 17/8$$
$$c_{21} = [(z-1)/z]\,\mathscr{f}(z)|_{z=1} = -2.$$

Up to this point, the procedure has followed the pattern of the previous examples. Now, however, we must find c_{00} and c_{01}. To do so, we can select two convenient values of z and substitute them into the expanded expression for $\mathscr{f}(z)$ to yield two equations for the two unknown coefficients. One candidate value is ∞. As z approaches infinity, $\mathscr{f}(z)$ approaches unity and the expansion of $\mathscr{f}(z)$ approaches $c_{00}+c_{11}+c_{21}$. Therefore, we have our first equation

$$c_{00} + c_{11} + c_{21} = 1.$$

Another candidate value of z is -1. When $z=-1$, $\mathscr{f}(z) = 1/3$ and

$$c_{00} - c_{01} + c_{02} + (1/3)\,c_{11} + (1/2)\,c_{21} = 1/3.$$

Substituting the coefficient values that already have been determined, we find

$$c_{00} = 7/8$$
$$c_{00} - c_{01} = 1/8; \qquad c_{01} = 3/4$$

and

$$\mathscr{f}(z) = 7/8 + 3/4\,z + 1/2\,z^2 + (17/8)\,z/(z-2) - 2\,z/(z-1)$$

from which

$$\begin{aligned} f(\tau) &= (17/8)(2)^\tau - 2(1)^\tau + 7/8 && \tau = 0 \\ &= (17/8)(2)^\tau - 2(1)^\tau + 3/4 && \tau = 1 \\ &= (17/8)(2)^\tau - 2(1)^\tau + 1/2 && \tau = 2 \\ &= (17/8)(2)^\tau - 2(1)^\tau && \tau > 2. \end{aligned}$$

Example 4.14. Find the inverse transform of

$$f(z)=1/(z-1)^2(z-2).$$

Answer:

$$f(z)=c_{00}+c_{11}z/(z-1)+c_{12}z^2/(z-1)^2+c_{21}z/(z-2)$$
$$c_{00}=f(z)|_{z=0}=-1/2$$
$$c_{21}=[(z-2)/z]f(z)|_{z=2}=1/2$$
$$c_{12}=[(z-1)^2/z^2]f(z)|_{z=1}=-1$$

let z approach infinity, $f(z)$ approaches zero

$$c_{00}+c_{11}+c_{12}+c_{21}=0$$
$$c_{11}=1$$
$$f(z)=-1/2+z/(z-1)-z^2/(z-1)^2+(1/2)\,z/(z-2)$$
$$f(\tau)=-1/2+(1)^\tau-(\tau+1)(1)^\tau+(1/2)(2)^\tau \qquad \tau=0$$
$$=(1)^\tau-(\tau+1)(1)^\tau+(1/2)(2)^\tau \qquad \tau>0.$$

4.10.3. Dealing with Multiple Roots at the Origin

Although multiple roots generally will not occur often in network models of populations, and the multiplicity of nonzero roots is unlikely to be more than two or three, the distinct possibility exists for roots of considerable multiplicity at the origin. These would correspond to the last cluster of terms in equation 4.71; and their multiplicity would be equal to k_r-k_q, the amount by which the degree of the numerator, R', is greater than the degree of the denominator, Q', in equation 4.47. As we can see from Example 4.13, one can cope rather easily with two roots at the origin. On the other hand, coping would become increasingly difficult as the multiplicity increased beyond two. Such a situation could arise in a population model if in one or more direct paths from input to output there were relatively long *time delays* that were not also in loops.

When such a situation does arise, one can deal with it quite effectively by employing the method of equations 4.54 and 4.55, having already used partial-fraction expansion to determine $f'(\tau)$. In other words, if $f(z)=R'(z^{-1})/Q'(z^{-1})$, then let $f'(z)=1/Q'(z^{-1})$, find the inverse transform, $f'(\tau)$, by partial-fraction expansion, then substitute appropriately into the right-hand side of equation 4.55. Each term in the numerator of $f(z)$ will represent $f'(\tau)$ correspondingly scaled and delayed (i.e., the term $a_k z^{-k}$ represents scaling by a_k and delay by k to yield $a_k f'(\tau-k)$). Having obtained all such terms, one simply sums them to obtain the total inverse transform. This method is especially effective when the number of terms in the numerator is small.

Example 4.15. Find the inverse transform of the following function:

$$f(z)=(1+z^{-6})/(1-5z^{-1}+6z^{-2}).$$

Answer: If we were to depend completely on the method of partial-fraction expansion, we would have

$$\begin{aligned}\mathscr{f}(z)&=(z^6+1)/z^4(z-2)(z-3)\\&=c_{00}+c_{01}/z+c_{02}/z^2+c_{03}/z^3+c_{04}/z^4+c_{11}z/(z-2)+c_{21}z/(z-3)\end{aligned}$$

in which three of the coefficients (c_{04}, c_{11}, c_{21}) can be found directly by the methods of equations 4.72 and 4.73, and the remaining four coefficients must be found by simultaneous solution of four equations, one for each distinct value selected for z. Alternatively, we could decompose $\mathscr{f}(z)$ as follows:

$$\mathscr{f}(z)=1/(1-5z^{-1}+6z^{-2})+z^{-6}/(1-5z^{-1}+6z^{-2}).$$

Choose $\mathscr{f}'(z)=1/(1-5z^{-1}+6z^{-2})$ and carry out partial-fraction expansion:

$$\begin{aligned}\mathscr{f}'(z)&=z^2/(z-2)(z-3)\\&=c_{00}+c_{11}z/(z-2)+c_{21}z/(z-3)\\c_{00}&=\mathscr{f}'(z)|_{z=0}=0\\c_{11}&=[(z-2)/z]\mathscr{f}'(z)|_{z=2}=-2\\c_{21}&=[(z-3)/z]\mathscr{f}'(z)|_{z=3}=3\\\mathscr{f}'(z)&=3z/(z-3)-2z/(z-2)\\\mathscr{f}'(\tau)&=3(3)^\tau-2(2)^\tau.\end{aligned}$$

Now, returning to our original decomposition, we have

$$\mathscr{f}(z)=\mathscr{f}'(z)+z^{-6}\mathscr{f}'(z)$$

from which

$$\begin{aligned}f(\tau)&=3(3)^\tau-2(2)^\tau & 0\leqq\tau<6\\&=3(3)^\tau-2(2)^\tau+3(3)^{\tau-6}-2(2)^{\tau-6} & \tau\geqq6.\end{aligned}$$

Note that a term $a_k f'(\tau-k)$ always is taken to be zero for values of τ less than k. This is entirely consistent with our stipulation that all functions of time, $f(\tau)$, are zero prior to $\tau=0$ (i.e., are zero for all negative arguments). Thus, in the previous example, the second term in the decomposition did not come into play until the 6th interval.

If the numerator of $\mathscr{f}(z)$ has a large number of terms, an alternative decomposition may prove to be more efficient. One simply decomposes the numerator into a sum of polynomials, each of which spans a range of degrees equal to the degree of the denominator. This leads to $\mathscr{f}(z)$ expressed as a sum of polynomial ratios, which can be written as follows:

$$\begin{aligned}\mathscr{f}(z)=P_1(z^{-1})/Q'(z^{-1})&+z^{-(k_q+1)}P_2(z^{-1})/Q'(z^{-1})\\&+z^{-2(k_q+1)}P_3(z^{-1})/Q'(z^{-1})+\cdots\end{aligned}\tag{4.75}$$

where $P_1, P_2, P_3, \ldots$ etc., are polynomials of degree equal to or less than that of Q'. Next, one can employ partial-fraction expansion to determine the inverse transforms of each polynomial ratio, $P_1/Q', P_2/Q', \ldots$, and then sum the results, each delayed by the appropriate amount (i.e., no delay for the term corresponding to P_1/Q', a delay of k_q+1 intervals for the term corresponding to P_2/Q', and so forth).

Example 4.16. Find the inverse transform of

$$\mathscr{f}(z)=(1+2z^{-1}+3z^{-2}+z^{-3}-2z^{-4})/(1-3z^{-1}).$$

Answer: We can decompose $\mathscr{f}(z)$ as follows:

$$\mathscr{f}(z)=(1+2z^{-1})/(1-3z^{-1})+z^{-2}(3+z^{-1})/(1-3z^{-1})-z^{-4}(2)/(1-3z^{-1})$$

which yields the following polynomial ratios for which inverse transforms must be found:

$$(1+2z^{-1})/(1-3z^{-1}), \qquad (3+z^{-1})/(1-3z^{-1}), \qquad 2/(1-3z^{-1}).$$

Taking them one at a time, we have

$$\mathscr{f}_1(z)=(1+2z^{-1})/(1-3z^{-1})=(z+2)/(z-3)=c_{00}+c_{11}z/(z-3)$$

$$c_{00}=-2/3 \qquad c_{11}=5/3$$

$$\begin{aligned} \mathscr{f}_1(z)&=(5/3)\,z/(z-3)-2/3 \\ f_1(\tau)&=(5/3)(3)^\tau-2/3 \qquad \tau=0 \\ &=(5/3)(3)^\tau \qquad\qquad \tau>0 \end{aligned}$$

$$\mathscr{f}_2(z)=(3+z^{-1})/(1-3z^{-1})=(3z+1)/(z-3)=c_{00}+c_{11}z/(z-3)$$

$$c_{00}=-1/3 \qquad c_{11}=10/3$$

$$\begin{aligned} \mathscr{f}_2(z)&=(10/3)\,z/(z-3)-1/3 \\ f_2(\tau)&=(10/3)(3)^\tau-1/3 \qquad \tau=0 \\ &=(10/3)(3)^\tau \qquad\qquad \tau>0 \end{aligned}$$

$$\mathscr{f}_3(z)=2/(1-3z^{-1})=2z/(z-3)$$

$$f_3(\tau)=2(3)^\tau \qquad\qquad \tau\geqq 0.$$

Summing appropriately delayed terms, we have

$$f(\tau)=f_1(\tau)+f_2(\tau-2)+f_3(\tau-4)$$

where a term is taken to be nonzero only when its argument is greater than or equal to zero.

4.10.4. Exercises

Find the inverse transforms of the following transformed unit-cohort response functions:

	Answer:
a) $f(z)=1/(1-4z^{-2})$	$f(\tau)=(1/2)(-2)^{\tau}+(1/2)(2)^{\tau}$
b) $f(z)=1/(1-z^{-1})^2$	$f(\tau)=\tau+1$
c) $f(z)=(1-z^{-8})/(1-z^{-1}-2z^{-2})$	$f(\tau)=(1/3)(-1)^{\tau}+(2/3)(2)^{\tau}$ $0\leqq\tau<8$ $f(\tau)=(2/3)(-1)^{\tau}$ $+(2/3)[(2)^{\tau}+(2)^{\tau-8}]$ $\tau\geqq 8$
d) $f(z)=(1-z^{-1}+z^{-2}-z^{-3})/(1-5z^{-1})$	$f(\tau)=1,\ \tau=0;\ \ f(\tau)=4,\ \tau=1$ $f(\tau)=21,\ \tau=2;$ $f(\tau)=(4/5)[(5)^{\tau}+(5)^{\tau-2}],\ \tau>2.$

4.11. Types of Common Ratios and Their Significances

According to the interpretation put forth in the previous section, the unit-cohort responses of our constant-parameter networks will consist of two types of components, the occasional, single cohort, represented by terms of the form

$$f(\tau)=c_{0\kappa} \quad \text{when } \tau=\kappa$$
$$f(\tau)=0 \quad \text{when } \tau\neq\kappa$$

and the modified geometric series, represented by terms of the form

$$c_{i\kappa}\binom{\tau+\kappa-1}{\kappa-1}(r_i)^{\tau}.$$

The single cohort is so simple that it warrants no further discussion. The modified geometric series, on the other hand, may be rather complicated. Furthermore, it almost always represents the most important terms in the dynamics of our linear models, so it definitely warrants further discussion. The heart of the modified geometric series is the simple geometric series itself, denoted by

$$(r_i)^{\tau}.$$

In fact, the other factor, $\binom{\tau+\kappa-1}{\kappa-1}$, very often will be simply unity, corresponding to a root that occurs only once in the denominator of $f(z)$. Therefore, a major step in the understanding of the dynamics of our linear networks will be a thorough understanding of the properties of the function $(r_i)^{\tau}$. Specifically, it is important for the reader to become familiar with the effects on this geometric series of the particular value taken by the parameter r_i, which is known as the *common ratio* of the series.

4.11.1. The General Nature of Common Ratios

The common ratios that concern us are the roots of the polynomials, $Q(z)$. As soon as we begin to consider the roots of polynomials, we encounter the so-called *imaginary number*. Consider, for example, the roots of z^2+K^2. Evidently $(z-r_1)(z-r_2)=z^2+K^2$; but

$$(z-r_1)(z-r_2)=z^2-(r_1+r_2)\,z+r_1 r_2.$$

Therefore, r_1+r_2 must be equal to zero; and $r_1 r_2$ must be equal to K^2:

$$r_1=-r_2$$
$$-r_1^2=-r_2^2=K^2.$$

It follows that

$$r_1=-r_2=\sqrt{-K^2}.$$

The numbers that one ordinarily deals with, the so-called *real numbers*, have the property that any one of them multiplied by itself produces a positive product. Clearly, on the other hand, r_1 and r_2 do not have this property and therefore are not among the real numbers. Therefore, in contradistinction to the real numbers, they usually are called *imaginary numbers*. Obviously, the imaginary numbers are closely related to the real numbers, and the various operations (such as addition, subtraction, multiplication and division) are well established by that relationship.

To begin with, we can rewrite r_1 and r_2 as follows:

$$r_1=K\sqrt{-1}$$
$$r_2=-K\sqrt{-1}.$$

If we use the standard convention, designating $\sqrt{-1}$ as $\boldsymbol{i}$, then we have

$$r_1=\boldsymbol{i}K$$
$$r_2=-\boldsymbol{i}K.$$

Now, the following relationships should be self-evident to the reader. If they are not, they should become clear with a little reflection:

$$\begin{array}{ll}
\boldsymbol{i}K_1+\boldsymbol{i}K_2=\boldsymbol{i}(K_1+K_2) & \text{sum of imaginary numbers} \\
\boldsymbol{i}K_1-\boldsymbol{i}K_2=\boldsymbol{i}(K_1-K_2) & \text{difference of imaginary numbers} \\
(\boldsymbol{i}K_1)\times(\boldsymbol{i}K_2)=-K_1K_2 & \text{product of imaginary numbers} \\
(\boldsymbol{i}K_1)/(\boldsymbol{i}K_2)=K_1/K_2 & \text{quotient of imaginary numbers} \\
K_3\times(\boldsymbol{i}K_1)=\boldsymbol{i}(K_1K_3) & \text{product of imaginary and real numbers} \\
1/(\boldsymbol{i}K_1)=-\boldsymbol{i}(1/K_1) & \text{reciprocal of imaginary number} \\
K_3\pm\boldsymbol{i}K_1=K_3\pm\boldsymbol{i}K_1 & \text{sum or difference of real and imaginary numbers.}
\end{array} \tag{4.76}$$

Notice that every operation but one leads to either a purely imaginary number (i.e., one preceded by an $\boldsymbol{i}$) or a purely real number. The one operation that does not is addition or subtraction of real and imaginary numbers. The result of such an operation is a *complex number*, having a real part and an imaginary part. The roots of polynomials, of course, may be complex as well as real or imaginary. Consider, for example, the quadratic polynomial $z^2+2K_1 z+K_2^2$. Its roots are

$$r_1=-K_1+\sqrt{K_1^2-K_2^2}$$
$$r_2=-K_1-\sqrt{K_1^2-K_2^2}.$$

If the magnitude of K_2 is greater than that of K_1, then the roots both are complex numbers.

When they are complex, the two roots r_1 and r_2 have a special relationship to one another. The real parts of the two roots are equal both in magnitude and sign; the imaginary parts are equal in magnitude but opposite in sign. Changing our notation slightly, we can make this relationship slightly more obvious:

$$r_1=a+\boldsymbol{i}b$$
$$r_2=a-\boldsymbol{i}b.$$

When two numbers are so related, they are called conjugates of one another. The conjugate of a number usually is denoted with a (*) written as a superscript. Thus r_1^* is the conjugate of r_1. Using this notation, we have

$$r_2^*=r_1$$
$$r_1^*=r_2.$$

Actually, we can think of the (*) as representing an operation which converts a number into its conjugate. Thus

$$\begin{aligned}(r^*)&=\text{the conjugate of } r\\(r^*)^*&=\text{the conjugate of } r^*=r.\end{aligned}\tag{4.77}$$

With a little reflection, the reader should easily be able to verify four more relationships:

$$\begin{aligned}&r_i^*=r_i && \text{when } r_i \text{ is purely real}\\&r_i^*=-r_1 && \text{when } r_i \text{ is purely imaginary}\\&(r_i+r_j)^*=r_i^*+r_j^* && \text{when the } r\text{'s are real, imaginary or complex}\\&(r_i\,r_j)^*=r_i^*\,r_j^* && \text{when the } r\text{'s are real, imaginary or complex.}\end{aligned}\tag{4.78}$$

Now, according to an important theorem of algebra, the set of roots of any polynomial with real, imaginary or complex coefficients is unique. In other words, any such polynomial, $A(z)$, can be factored into a unique set of binomial terms:

$$A(z)=(z-r_1)(z-r_2)(z-r_3)\ldots(z-r_k).$$

The factorization may vary only in the order of the terms, the same set of roots, $r_1, r_2, \ldots, r_k$, will appear in all possible factorizations of $A(z)$. The reader should have no difficulty in convincing himself of the validity of this theorem. However, for its rigorous proof, he should consult an elementary textbook on modern algebra. Another theorem that he will find in such a book often is called the *fundamental theorem of algebra.* Embodied in this theorem is the fact that all of the roots of $\Lambda(z)$ will be real, imaginary, or complex numbers.

There is one more theorem that concerns us here. It applies to polynomials in which the coefficients all are real numbers. Since that is the case for the polynomial, $Q(z)$, of interest to us, the theorem is applicable here. It simply states that *the nonreal roots of a polynomial with real coefficients always occur in conjugate pairs.* To prove this, simply consider the conjugate of the entire polynomial,

$$Q^*(z)=(z^*-r_1^*)(z^*-r_2^*)\dots(z^*-r_k^*) \tag{4.79}$$

(which follows from relationships 4.78). Since the coefficients of $Q(z)$ all are real, they are equal to their conjugates, and

$$Q(z)=Q^*(z^*)=(z-r_1^*)(z-r_2^*)\dots(z-r_k^*). \tag{4.80}$$

However because the roots of $Q(z)$ are unique, the set $r_1^*, r_2^*, \dots, r_k^*$ must be the same as $r_1, r_2, \dots, r_k$. This can be true only if r_i *is real, in which case* $r_i^*=r_i$, or if $r_i=r_j^*$, in which case $(r_i)^*=r_j$ and $(r_j)^*=r_i$. In other words, if r_i is imaginary or complex, it must be accompanied by another imaginary or complex root, namely its conjugate.

Thus, the common ratios that we shall be dealing with will take on one of the following forms:

r may be a positive, real number
r may be a negative, real number
or
r may be one member of a pair of conjugate complex or imaginary numbers, in which case the other member of the pair also appears as a common ratio in the unit-cohort response.

The first two cases can be dealt with directly and simply. The third situation requires a little more knowledge of the methods of complex arithmetic before it can be handled easily; so we shall postpone its consideration for the moment.

4.11.2. Real Common Ratios

Let us begin by considering the sequence $(r)^\tau$, where τ takes on integral values beginning with zero, $\tau=\mathbb{0}, \mathbb{1}, \mathbb{2}, \mathbb{3}, \dots$, and r is a positive real number. Clearly, $(r)^\tau$ always will be a positive real number. Furthermore, if r is equal to $\mathbb{1}$, $(r)^\tau$ always will be equal to $\mathbb{1}$. If r is less than $\mathbb{1}$, $(r)^\tau$ always will be less than $\mathbb{1}$, and it will diminish with increasing values of τ. If r is greater than $\mathbb{1}$, $(r)^\tau$ always will be greater than $\mathbb{1}$, and it will increase with increasing values of τ. All of these properties are displayed in Table 4.2.

Table 4.2. r^τ as a function of τ for some positive, real values of r

$\tau=0$	1	2	3	4	5	6	7
$(0.1)^\tau=1$	0.1	0.01	0.001	0.0001	0.00001	10^{-6}	10^{-7}
$(0.2)^\tau=1$	0.2	0.04	0.008	0.0016	0.0003	0.00006	0.00001
$(0.5)^\tau=1$	0.5	0.25	0.125	0.0625	0.0313	0.0156	0.0078
$(0.9)^\tau=1$	0.9	0.81	0.729	0.6561	0.5905	0.5314	0.4783
$(0.99)^\tau=1$	0.99	0.9801	0.9702	0.9606	0.9510	0.9415	0.9321
$(1.01)^\tau=1$	1.01	1.020	1.030	1.041	1.051	1.062	1.072
$(1.1)^\tau=1$	1.1	1.21	1.331	1.464	1.611	1.772	1.949
$(1.5)^\tau=1$	1.5	2.25	3.375	5.063	7.594	11.39	17.09
$(2)^\tau=1$	2.0	4.0	8.0	16.0	32.0	64.0	128.0
$(3)^\tau=1$	3.0	9.0	27.0	81.0	243.0	729.0	2,187.0
$(5)^\tau=1$	5.0	25.0	125.0	625.0	3,125.0	15.626.0	78,125.0
$(10)^\tau=1$	10.0	100.0	1,000.0	10,000.0	100,000.0	10^6	10^7

Clearly, when r is considerably greater than 1, $(r)^\tau$ grows very rapidly. As r approaches 1, the rate of growth diminishes markedly. When r becomes slightly less than 1, $(r)^\tau$ diminishes slowly; and as r approaches zero, the rate of diminution increases markedly. Seven transition intervals normally can be considered as being very few, yet the differences in the magnitudes of $(r)^\tau$ by that time are quite extreme, except for those sequences whose common ratios are very close to 1. The common ratio 1 marks a significant boundary. On one side of it we have growth, on the other we have decline. Another significant boundary is marked by the common ratio zero. Of course, when $r=0$, then $(r)^\tau$ is zero for all τ greater than zero. When r is real and greater than zero, then $(r)^\tau$ always is positive, regardless of whether it grows or diminishes. For this reason, the growth or dimunition patterns usually are called *simple* or *monotonic*.

On the other hand, when r is real and less than zero, the values of $(r)^\tau$ alternate in sign between positive (when τ is an even integer, including zero) and negative (when τ is an odd integer). Thus, if $r=-1$, we have

$$(-1)^\tau=-1 \quad \text{when } \tau \text{ is an odd integer}$$

$$(-1)^\tau=1 \quad \text{when } \tau \text{ is an even integer, including } 0.$$

If r is negative and real, but its magnitude is less than one, then the members of the sequence $(r)^\tau$ not only will have alternating signs, but their magnitudes will diminish geometrically with increasing values of τ, in exactly the manner indicated in Table 4.2. If r is a negative real number whose magnitude is greater than 1, the signs of the sequence members will alternate and their magnitudes will increase geometrically with increasing values of τ, in the manner of Table 4.2. Thus, the entries of Table 4.2 are the magnitudes of $(r)^\tau$ when r is negative.

All of these properties can be summarized by notations along a real-number line, as shown in Figure 4.50. Examples of the patterns themselves are displayed in Figure 4.51. Basically, we have seven of them. The term r^τ may (a) exhibit simple geometric growth, (b) be constant at the value one, (c) exhibit simple geometric decline, (d) be constant at the value zero, (e) exhibit alternating (or oscillating) geometric decline, with the period of alternation (i.e., the time between successive members of the same sign) being twice the transition interval,

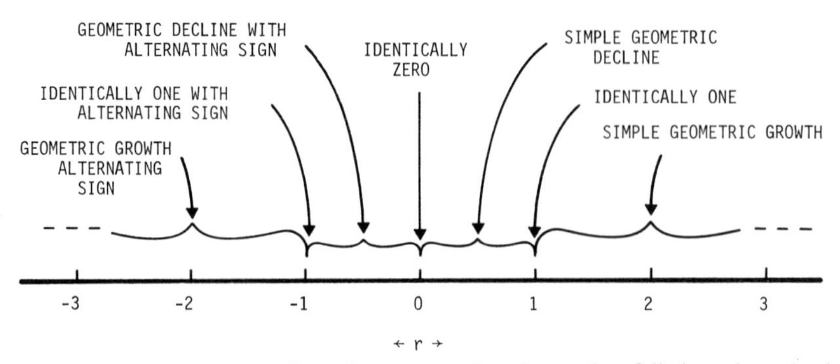

Fig. 4.50. Dynamic interpretations of common ratios whose values fall along the real axis

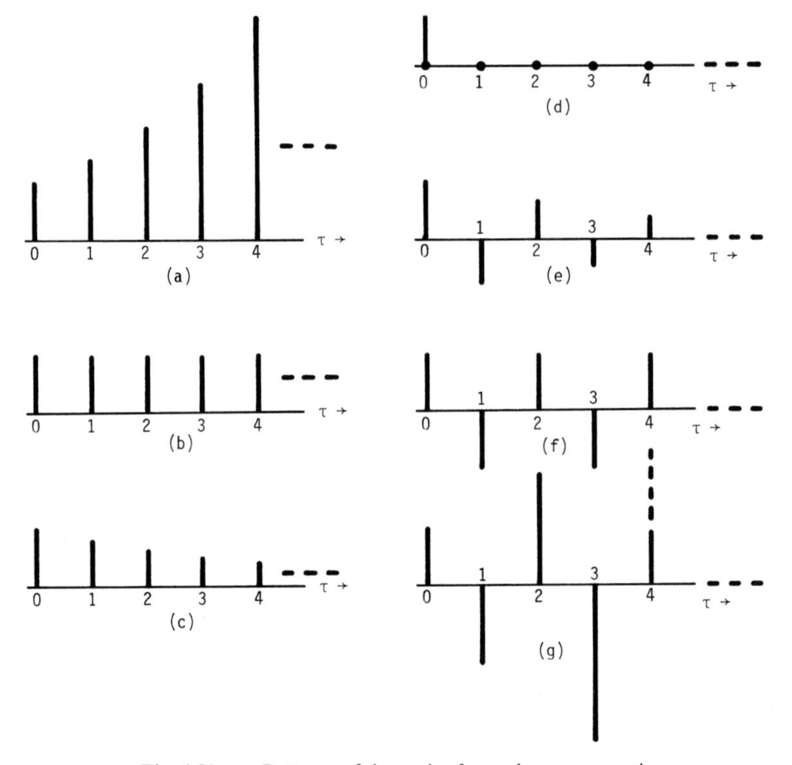

Fig. 4.51 a–g. Patterns of dynamics for real common ratios

(f) exhibit simple alternation, with a period equal to twice the transition interval, or (g) exhibit geometrically growing alternations, with the period being equal to twice the transition interval.

4.11.3. Complex Common Ratios

A glance at Figure 4.52 should convince the reader that there are at least three patterns that we have omitted. In Figure 4.52a, for example, the unit

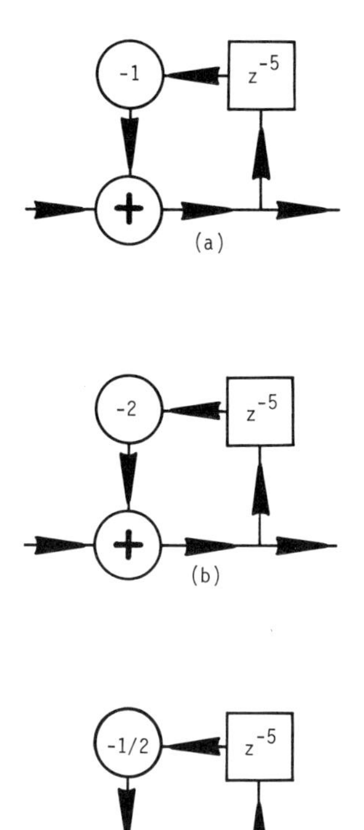

Fig. 4.52a–c. Networks that generate patterns of dynamics not included in Fig. 4.51

cohort will enter and then cycle around and around the feedback loop, changing sign on each pass and generating an output of alternating sign and of magnitude one; *but the period of alternation is greater than twice the transition interval.* As a matter of fact, it is ten transition intervals. Similarly, the unit-cohort response of the network depicted in Figure 4.52b will exhibit alternating geometric growth with a period of ten transition intervals; and that of Figure 4.52c will exhibit alternating geometric decline with a period of ten transition intervals. Thus we have the question of how alternations are represented when their periods are greater than twice the transition interval. The answer simply is that they are represented by a pair of common ratios whose values are complex conjugate numbers:

$$\begin{aligned} r_j &= a_j + \boldsymbol{i}\, b_j \\ r_j^* &= a_j - \boldsymbol{i}\, b_j. \end{aligned} \tag{4.81}$$

This introduces the interesting problem of dealing with products and powers of complex numbers. If we take the direct approach, and simply multiply complex numbers in their expanded form, the results can be quite cumbersome.

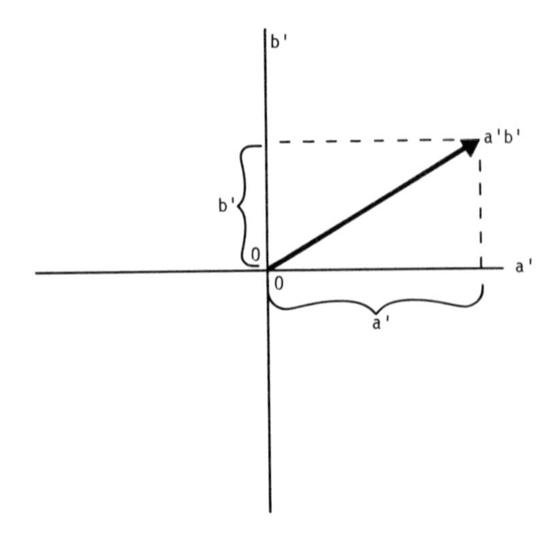

Fig. 4.53. Vector representation of a complex common ratio

Consider, for example, the lowest powers of r_j:

$$(r_j)^2=(a_j+\boldsymbol{i}\,b_j)^2=a_j^2+\boldsymbol{i}\,2\,a_j\,b_j-b_j^2=(a_j^2-b_j^2)+\boldsymbol{i}\,2\,a_j\,b_j$$
$$(r_j)^3=a_j^3+\boldsymbol{i}\,3\,a_j^2\,b_j-3\,a_j\,b_j^2-\boldsymbol{i}\,b_j^3=(a_j^3-3\,a_j\,b_j^2)+\boldsymbol{i}\,(3\,a_j^2\,b_j-b_j^3)$$
$$(r_j)^4=a_j^4+\boldsymbol{i}\,4\,a_j^3\,b_j-6\,a_j^2\,b_j^2-\boldsymbol{i}\,4\,a_j\,b_j^3+b_j^4=(a_j^4-6\,a_j^2\,b_j^2+b_j^4)+\boldsymbol{i}(4\,a_j^3\,b_j-4\,a_j\,b_j^3).$$

By induction, one can see that this process will be rather difficult as τ increases. Also, by induction, one can see that the process always leads to a complex number:

$$(r_j)^\tau=a'+\boldsymbol{i}\,b'. \tag{4.82}$$

The pair of numbers (a', b') often, for convenience, is taken to be the coordinates of a point in a plane. The a'-axis represents the real component of $(r_j)^\tau$ and is usually denoted as the *real axis*, while the b'-axis represents the imaginary component and usually is denoted as the *imaginary axis*. The entire plane usually is called the *complex plane*. The point (a', b') can be taken to denote the end of a vector, as illustrated in Figure 4.53. Viewing complex numbers in this way may add some initial complexity to their conception, but the resulting simplifications of multiplication are well worth the initial effort.

Consider two complex numbers, $X_1=a_1+\boldsymbol{i}\,b_1$ and $X_2=a_2+\boldsymbol{i}\,b_2$. X_1 and X_2 can be treated as two vectors on the complex plane, as indicated in Figure 4.54. In addition to having an imaginary and a real component, each vector also has a *magnitude* (or length) and an *angle* with respect to the positive half of the real axis. Following the usual convention, we shall denote the magnitude of X_i as $|X_i|$ and the angle of X_i as $\theta(X_i)$, or simply θ_i if it is not ambiguous. By elementary geometry, we have

$$\begin{aligned}|X_1|&=\sqrt{a_1^2+b_1^2}\\|X_2|&=\sqrt{a_2^2+b_2^2}\end{aligned} \tag{4.83}$$

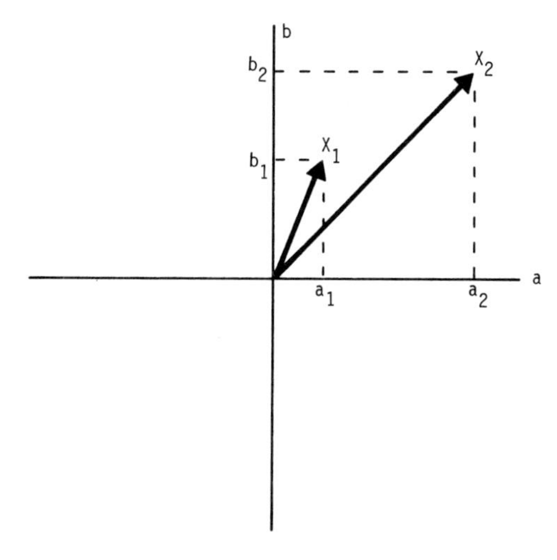

Fig. 4.54

and

$$b_1/a_1 = \tan(\theta_1); \quad \theta_1 = \tan^{-1}(b_1/a_1)$$
$$b_2/a_2 = \tan(\theta_2); \quad \theta_2 = \tan^{-1}(b_2/a_2). \tag{4.84}$$

It follows that

$$a_1 = |X_1| \cos\theta_1; \quad b_1 = |X_1| \sin\theta_1$$
$$a_2 = |X_2| \cos\theta_2; \quad b_2 = |X_2| \sin\theta_2. \tag{4.85}$$

Now, let us consider the product of X_1 and X_2:

$$X_1 \times X_2 = (a_1 + \boldsymbol{i}\, b_1)(a_2 + \boldsymbol{i}\, b_2)$$
$$= (a_1 a_2 - b_1 b_2) + \boldsymbol{i}(a_1 b_2 + a_2 b_1) = a' + \boldsymbol{i}\, b'.$$

The magnitude of the product vector is

$$|X_1 \times X_2| = \sqrt{(a')^2 + (b')^2} = \sqrt{(a_1 a_2 - b_1 b_2)^2 + (a_1 b_2 + a_2 b_1)^2}$$
$$|X_1 \times X_2| = \sqrt{a_1^2 a_2^2 - 2a_1 a_2 b_1 b_2 + b_1^2 b_2^2 + a_1^2 b_2^2 + 2a_1 b_2 a_2 b_1 + a_2^2 b_1^2}$$
$$= \sqrt{(a_1^2 + b_1^2)(a_2^2 + b_2^2)} \tag{4.86}$$
$$|X_1 \times X_2| = |X_1| \times |X_2|$$

and the angle of the product is

$$\theta(X_1 \times X_2) = \tan^{-1}(b'/a') = \tan^{-1}[(a_1 b_2 + a_2 b_1)/(a_1 a_2 - b_1 b_2)]$$
$$= \tan^{-1}[\{(b_2/a_2) + (b_1/a_1)\}/\{1 - (b_1 b_2/a_1 a_2)\}]$$
$$= \tan^{-1}[\{\tan\theta_2 + \tan\theta_1\}/\{1 - (\tan\theta_1 \tan\theta_2)\}] \tag{4.87}$$
$$= \tan^{-1}[\tan(\theta_1 + \theta_2)]$$
$$\theta(X_1 \times X_2) = \theta_1 + \theta_2.$$

If we consider complex numbers in terms of their magnitudes and angles, then equations 4.86 and 4.87 make the process of multiplication very simple. When one multiplies one complex number by another, the magnitude of the product is equal to the product of the magnitudes of the two numbers and the angle of the product is the sum of the angles of the two numbers. It follows from equations 4.85 that

$$\begin{aligned} X_1 \times X_2 &= a' + \boldsymbol{i}\, b' \\ a' &= |X_1| \times |X_2| \times \cos(\theta_1 + \theta_2) \\ b' &= |X_1| \times |X_2| \times \sin(\theta_1 + \theta_2). \end{aligned} \tag{4.88}$$

Now, we are in a position to consider the τth power of a complex root, r_j. Clearly, using the vector concept, we can write

$$\begin{aligned} |r_j^\tau| &= |r_j|^\tau \\ \theta(r_j^\tau) &= \theta_j \tau, \qquad \text{where } \theta_j \text{ is the angle of } r_j \\ r_j^\tau &= a_j' + \boldsymbol{i}\, b_j' \\ a_j' &= |r_j|^\tau \cos(\theta_j \tau) \\ b_j' &= |r_j|^\tau \sin(\theta_j \tau) \end{aligned} \tag{4.89}$$

Whenever a pair of conjugate complex roots appears in the factorization of $Q(z)$, their coefficients in the partial-fraction expansion always will be complex conjugate numbers themselves. Otherwise, the reconstituted proper fraction would have nonreal coefficients in the numerator polynomial. When the nonreal roots occur only once (i.e., r_j occurs once and r_j^* occurs once in the factorization of $Q(z)$), then one can see immediately that the corresponding coefficients of the partial fraction expansion must be conjugates:

$$c_{j1} = \frac{(z - r_j)}{z} f(z)\bigg|_{z = r_j}$$

$$c_{j1}' = \frac{(z - r_j^*)}{z} f(z)\bigg|_{z = r_j^*}.$$

Therefore,

$$c_{j1}^* = c_{j1}'$$

and the corresponding terms of the expansion become

$$\frac{c_{j1}}{(z - r_j)} + \frac{c_{j1}^*}{(z - r_j^*)}. \tag{4.90}$$

To see that these two terms combine to form a proper fraction with real coefficients in numerator and denominator, one needs only to carry out the reconstitution.

$$c_{j1} = a_{j1} + \boldsymbol{i}\, b_{j1} \qquad c_{j1}^* = a_{j1} - \boldsymbol{i}\, b_{j1}$$

$$r_j = a_j + \boldsymbol{i}\, b_j \qquad r_j^* = a_j - \boldsymbol{i}\, b_j$$

$$\frac{a_{j1} + \boldsymbol{i}\, b_{j1}}{z - a_j - \boldsymbol{i}\, b_j} + \frac{a_{j1} - \boldsymbol{i}\, b_{j1}}{z - a_j + \boldsymbol{i}\, b_j} = \frac{2 a_{j1} z - (2 a_j a_{j1} + 2 b_j b_{j1})}{z^2 - 2 a_j z + a_j^2 + b_j^2}.$$

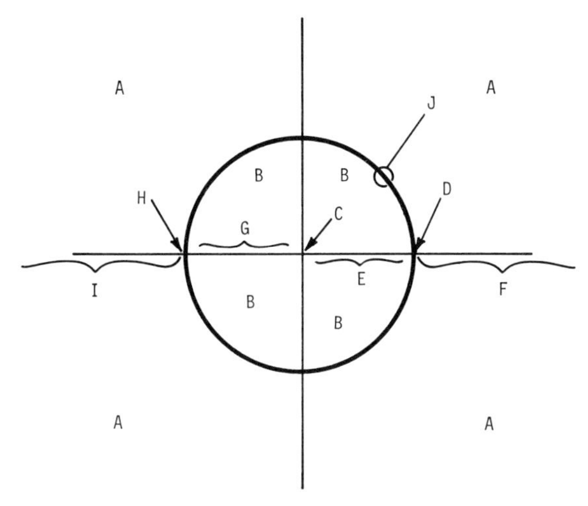

Fig. 4.55. Dynamic interpretations of common ratios over the entire complex plane. The imaginary-number axis is vertical, the real-number axis is horizontal; thus the abscissa is the real part of the common ratio, the ordinate is the imaginary part. A, the open complex plane outside the unit circle and away from the real axis corresponds to growing, oscillations with period greater than 2. *B*, the region of the complex plane inside the unit circle and away from the real axis corresponds to declining oscillations with period greater than 2. *C*, the origin corresponds to dynamics identically equal to zero for all values of τ greater than zero. *D*, the intersection of the unit circle with the positive-real axis corresponds to a constant term. *E*, the positive-real axis inside the unit circle corresponds to simple geometric decline. *F*, the positive-real axis outside the unit circle corresponds to simple geometric growth. *G*, the negative-real axis inside the unit circle corresponds to declining oscillation of period 2. *H*, the intersection of the unit circle with the negative-real axis corresponds to constant-amplitude oscillation of period 2. *I*, the negative-real axis outside the unit circle corresponds to growing oscillation of period 2. *J*, the unit circle away from the real axis corresponds to constant-amplitude oscillations of period greater than 2

The inverse transform of the pair of functions in equation 4.90 is

$$c_{j1}(r_j)^\tau + c_{j1}^*(r_j^*)^\tau. \tag{4.91}$$

Using the vector concept, we can combine these into a single, very simple term:

$$\begin{aligned}
|c_{j1}| &= \sqrt{a_{j1}^2 + b_{j1}^2} = |c_{j1}^*| \\
|r_j| &= \sqrt{a_j^2 + b_j^2} = |r_j^*| \\
\theta(c_{j1}) &= -\theta(c_{j1}^*) = \psi \\
\theta(r_j) &= -\theta(r_j^*) = \theta_j
\end{aligned}$$

$$\begin{aligned}
c_{j1}(r_j)^\tau + c_{j1}^*(r_j^*)^\tau &= |c_{j1}| \times |r_j|^\tau \cos(\theta_j \tau + \psi) + \boldsymbol{i}\, |c_{j1}| \times |r_j|^\tau \sin(\theta_j \tau + \psi) \\
&\quad + |c_{j1}| \times |r_{j1}|^\tau \cos(-\theta_j \tau - \psi) + \boldsymbol{i}\, |c_{j1}| \times |r_{j1}|^\tau \sin(-\theta_j \tau - \psi) \\
&= 2\, |c_{j1}| \times |r_j|^\tau \cos(\theta_j \tau + \psi).
\end{aligned} \tag{4.92}$$

This form not only will be easy to apply, it also is quite easy to interpret. In the first place, it represents an alternating sequence of numbers. The period, *T*, of the alternations is given (in number of transition intervals) by

$$T = 360°/\theta_j \qquad (\theta_j \text{ given in degrees}) \tag{4.93}$$

$$T = 2\pi/\theta_j \qquad (\theta_j \text{ given in radians}). \tag{4.94}$$

If $|r_j|$ is greater than one, the alternations will grow geometrically with increasing values of τ. If $|r_j|$ is equal to one, the alternations persist indefinitely with the same amplitude. If $|r_j|$ is less than one, the alternations diminish geometrically with increasing values of τ. Thus, with the common ratios that are complex conjugates, we have the possibilities of the three types of alternations that we had with negative real common ratios, but with periods greater than twice the transition interval.

Figure 4.55 summarizes the properties of the geometric sequences for common ratios over the entire complex plane.

4.11.4. Exercises

1. Evaluate the following sums:

a) $[2+i3]+[4+i7]$	*Answer:*	$6+i10$
b) $[a+ib]+[a+ib]^*$		$2a$
c) $[a+ib]-[a+ib]^*$		$i2b$

2. Evaluate the following products and powers:

a) $[4+i3][4-i3]$	*Answer:*	25
b) $[4+i3]^5$		$-124.64-i9.48$
c) $[1+i1]^4$		-4
d) $[25]/[4-i3]$		$4+i3$

3. Describe the dynamic pattern generated by each of the following common ratios:

a) $r=1/2$	*Answer:*	simple geometric decline, halving every transition interval
b) $r=2$		simple geometric growth, doubling every transition interval
c) $r_1=[4+i3]$, $r_2=[4-i3]$		oscillatory growth, fivefold every transition interval, one cycle every 9.764 ... transition intervals
d) $r=-3$		oscillatory growth, threefold every transition interval, one cycle every 2 transition intervals

4.12. The Patterns of Linear Dynamics

The unit-cohort response of a constant-parameter network model represents the expected pattern of the dynamics of the modeled population, initiated by a single individual arriving at the 0th interval. When the parameter values correspond to a total population that is close to zero, then the unit-cohort response reflects the capability of the population to recover from very low levels. On the other hand, when the parameter values correspond to a nonzero critical population level, as discussed in sections 4.3 and 4.4, then the unit-cohort response will reflect the tendency of the population to move toward or away from that critical level, or to fluctuate about it. Finally, in the special case of a model whose parameters are independent of $\boldsymbol{n}$ and τ, the unit cohort response reflects the expected pattern of dynamics for all levels of the population.

According to the discussion in sections 4.10 and 4.11, the unit-cohort response of constant-parameter models will consist of components with the following general forms:

1) single, one-time only cohorts
2) an infinite succession of cohorts exhibiting geometric growth or decline
3) an infinite succession of cohorts exhibiting oscillations superimposed upon geometric growth or decline

The infinite successions will conform to one of the following descriptions

$$c_{ij}\binom{\tau+j-1}{j-1}(r_i)^\tau \quad \text{or} \quad 2\,|c_j|\,|r_j|^\tau \cos[\theta_j+\psi]$$

where c_{ij} is a positive or negative real constant, j is a nonnegative integer, r_i is a positive or negative real number, r_j is a complex or imaginary number, c_j is either complex, imaginary or real, and θ_j and ψ are constant angles between 0 and 360 degrees (i.e., between 0 and 2π radians). Thus, a typical unit-cohort response might have the following terms:

$$\begin{aligned} f(\tau) = {} & c_{11}r_1^\tau + c_{21}(-r_2)^\tau + c_{31}r_3^\tau + c_{32}(\tau+1)\,r_3^\tau + c_{33}(\tau+2)(\tau+1)\,r_3^\tau/2 \\ & + c_4 r_4^\tau \cos[\theta_4\tau+\psi_4] + \text{one-time only cohorts} \end{aligned} \tag{4.95}$$

where we shall assume all of the parameters (r's, c's, θ_4 and ψ_4) are positive real constants.

4.12.1. Growth Patterns in Low-Level Populations and Populations with Constant Parameters

Now, let us consider the dynamics at very low population levels in models whose parameters are independent of τ but may or may not depend upon $\boldsymbol{n}$, and the dynamics at all levels in models whose parameters are independent of both τ and $\boldsymbol{n}$. In the description given by equation 4.95, two of the terms (term 21 and term 4) alternate in sign, term 21 being positive for even values of τ and negative for odd values, term 4 being negative whenever the angle $\theta_4\tau+\psi_4$ is in the second or third quadrants (i.e., between 90° and 270° or their equivalents). The four terms 11, 31, 32, and 33 always are positive. Clearly, the magnitude of each term will increase, remain constant, or decrease with time, depending on the magnitude of its common ratio. Equally clearly, if this description is to represent a real population, *the sum of all of its terms never can be less than zero.* A negative sum in this context is just as uninterpretable as an imaginary or complex sum would be. In other words, *the sum of the terms for any value of τ must be a nonnegative, real number* (which includes the possibility of zero). If our network model has been constructed properly and our analysis carried out correctly, then this constraint will be met automatically for hypothetical populations close to zero or with invariant parameters. The sum of the negative terms automatically will be counterbalanced by an equal or greater sum of positive terms.

If, in the unit-cohort response, the magnitudes of all the common ratios were identical, then, eventually, the term with the highest power of τ would grow largest and dominate. In the case of equation 4.95, for example, the ultimately dominant term would be $_{33}$, with its multiplicative factor $(\tau+2)(\tau+1)$, which contains τ raised to the second power. As τ becomes larger and larger, this term would become more and more dominant; and, eventually, $f(\tau)$ would approach $c_{33}\,\tau^2\,r_3^\tau/2$

$$\lim_{\tau\to\infty} f(\tau)=c_{33}\,\tau^2\,r_3^\tau/2.$$

It is much more usual, however, to find that the magnitudes of at least some of the common ratios are less than the magnitude of the largest. In that case, the power of τ is totally irrelevant, the term or terms with the largest common-ratio magnitude eventually will dominate the dynamics. In other words, even if $|r_a|$ is only the slightest bit greater than $|r_b|$, the following inequality always is true

$$\lim_{\tau\to\infty} |r_a|^\tau \gg \lim_{\tau\to\infty} \tau^\kappa |r_b|^\tau \qquad \text{if } |r_a|>|r_b| \tag{4.96}$$

for any finite integer, κ. In fact, for any finite κ,

$$\lim_{\tau\to\infty} \tau^\kappa |r_b|^\tau/|r_a|^\tau=0 \qquad \text{if } |r_a|>|r_b|. \tag{4.97}$$

Thus, in equation 4.95, if r_1 were larger (even slightly) than all of the other common-ratio magnitudes, then the first term eventually would dominate the dynamics completely:

$$\lim_{\tau\to\infty} f(\tau)=c_{11}\,r_1^\tau \qquad \text{for equation 4.95 if } r_1 \text{ is largest.}$$

Clearly, *if $f(\tau)$ is to remain positive for all values of τ, an oscillating term never can dominate.* Therefore, for populations close to zero or with invariant parameters, *the magnitudes of the common ratios of oscillating terms always must be less than or equal to the common-ratio magnitude of at least one positive, nonalternating term.*

4.12.2. Time Required for Establishment of Dominance

One question that we have not answered yet is just how many intervals will be required to pass before a particular term becomes dominant over another. Before we answer that question, of course, we must establish a criterion of dominance. Let us state simply that one term will be considered dominant over another beyond a certain transition interval if for all subsequent intervals the ratio of the magnitude of the dominated term to that of the dominant term falls below a certain value, ε. In other words, term $a\kappa''$ will be considered dominant over term $b\kappa'$ after the kth interval if

$$|c_{b\kappa'}|\binom{\tau+\kappa'-\mathbb{1}}{\kappa'-\mathbb{1}}|r_b|^\tau \Big/ |c_{a\kappa''}|\binom{\tau+\kappa''-\mathbb{1}}{\kappa''-\mathbb{1}}|r_a|^\tau<\varepsilon \tag{4.98}$$

for all values of τ greater than k. As we already have mentioned, this will be possible if and only if $|r_a|>|r_b|$. Employing the expansion of $\binom{m}{n}$ (see equation B.3), one can restate this inequality as follows:

$$\frac{|c_{b\kappa'}|(\kappa''-\mathbb{1})!\,(\tau+\kappa'-\mathbb{1})!}{|c_{a\kappa''}|(\kappa'-\mathbb{1})!\,(\tau+\kappa''-\mathbb{1})!}\left|\frac{r_b}{r_a}\right|^\tau<\varepsilon. \tag{4.99}$$

Taking the logarithm of both sides and rearranging terms slightly, one obtains

$$\begin{aligned}\tau>&\log[|c_{b\kappa'}|/\varepsilon|c_{a\kappa''}|]/\log|r_a/r_b|\\&+\log\left[\frac{(\tau+\kappa'-\mathbb{1})!\,(\kappa''-\mathbb{1})!}{(\tau+\kappa''-\mathbb{1})!\,(\kappa'-\mathbb{1})!}\right]\Big/\log|r_a/r_b|.\end{aligned} \tag{4.100}$$

Thus, we have an implicit expression for the time required for the establishment of dominance. In the case of a transformed unit-cohort response with no repeated roots in the denominator, κ' and κ'' both equal one and inequality 4.100 becomes explicit with respect to discrete time.

$$\tau>\log[|c_{b\kappa'}|/\varepsilon|c_{a\kappa''}|]/\log[|r_a/r_b|]. \tag{4.101}$$

If the transformed unit-cohort response has repeated roots in the denominator and κ'' is greater than κ', then the second term on the right-hand side of inequality 4.100 will be negative; and dominance will be established earlier than predicted by 4.101. One can determine by trial and error the minimum value of τ satisfying inequality 4.100 in this case. Since the upper limit of the possible minimum values is given by the first term on the right-hand side and the lower limit is zero, the range of the trial and error search is limited. On the other hand, if κ' were greater than κ'', then the second term on the right-hand side of inequality 4.100 would be positive and the upper limit of the search range would not be specified. In that case, the search range in fact might be very large. Nevertheless, there is a very efficient algorithm for converging on the minimum time. One simply evaluates the first term on the right-hand side,

$$A=\log[|c_{b\kappa'}|/\varepsilon|c_{a\kappa''}|]/\log|r_a/r_b| \tag{4.102}$$

then substitutes the result, A, for τ in the second term,

$$B=\log\left[\frac{(A+\kappa'-\mathbb{1})!\,(\kappa''-\mathbb{1})!}{(A+\kappa''-\mathbb{1})!\,(\kappa'-\mathbb{1})!}\right]\Big/\log|r_a/r_b| \tag{4.103}$$

and then sums the two numbers

$$C = A + B. \tag{4.104}$$

From this point on the algorithm becomes repetitive, with the values of B and C being updated on each repetition:

$$\text{Set } B = \log\left[\frac{(C+\kappa'-1)!\,(\kappa''-1)!}{(C+\kappa''-1)!\,(\kappa'-1)!}\right]\Big/\log|r_a/r_b|. \tag{4.105}$$

$$\text{Then} \quad \text{Set } C = A + B.$$

Repeat until successive values of C differ by less than one. The minimum time satisfying inequality 4.100 will be extremely close to this final value of C; and trial and error will be very efficient from this point on.

Example 4.17. Consider the two terms $10(0.5)^\tau$ and $1000(0.3)^\tau$. Which of these terms eventually will dominate? If the criterion for domination at interval τ is the dominant term being at least 100 times as large as the dominated term at interval τ and all subsequent intervals, when does domination first occur?

Answer: The answer to the first question is determined strictly by the relative magnitudes of the common ratios, the term with the larger common ratio always being ultimately dominant. In this case that term is $10(0.5)^\tau$. To determine when domination (according to our criterion) first takes place, one simply finds the minimum value of τ that satisfies inequality 4.101 with $\varepsilon = 0.01$.

$$\tau > \log[1000/(0.01)\,10]/\log[0.5/0.3] = 18.030\ldots.$$

In other words, dominance according to our criterion is established at the 19th transition interval.

Example 4.18. Consider the two terms $2(0.9)^\tau$ and $8(-0.5)^\tau$. If the criterion for domination at interval τ is a ratio of at least 1000 to 1 between dominant and dominated terms at τ and all subsequent intervals, which term ultimately will dominate and when will it first achieve dominance?

Answer: One again, to answer the first part we simply consider the *magnitudes* of the common ratios. The term with the larger common ratio magnitude, and therefore the one that ultimately will dominate, is $2(0.9)^\tau$. The interval at which domination is achieved is given by inequality 4.101:

$$\tau > \log[8/(0.0001)\,2]/\log[0.9/0.5] = 14.110\ldots$$

from which we can see that domination is achieved in the 15th transition interval. In other words, in only fifteen transition intervals, the dominated term is diminished to 1/1000th of the dominant term and never again rises above that relative level.

Example 4.19. Consider the two terms $5(\tau+1)(2)^\tau$, and $7(1.414)^\tau \cos[\pi\tau/8]$. Which eventually will dominate? If the criterion for domination is a ratio of $1/100$, when will domination be achieved?

Answer: The first term will dominate since its common ration (2) is of greater magnitude than that (1.414) of the second. The time of domination can be determined from inequality 4.100:

$$\tau > \log[7/(0.01)5] \log[2/1.414] + \log[1/(\tau+1)]/\log[2/1.414].$$

Reducing the expressions on the right-hand side, we have

$$\tau > 14.2523\ldots - (6.6409\ldots)\log_{10}\tau$$

from which it is clear that the upper limit of the time for achievement of domination is 15 intervals (the next greater integer to the first term on the right). Inspection also clearly indicates that 15 intervals is more than adequate, however, so we might try a smaller candidate, such as 5:

$$5 \overset{?}{>} 14.2523\ldots - (6.6409\ldots)\log_{10} 5 = 9.6105\ldots.$$

Since the right-hand side of the inequality decreases in value as τ increases, it now is clear that the appropriate value of τ will lie between 5 and 9.6105, probably closer to the latter. Therefore, for our next trial, let us try 9:

$$9 \overset{?}{>} 14.2523\ldots - (6.6409\ldots)\log_{10} 9 = 7.9152\ldots.$$

The only possibility remaining now is 8; and one very quickly can determine that it does not satisfy the inequality. Therefore, domination is achieved at the ninth transition interval.

Example 4.20. Consider the two terms $(0.91)^\tau$ and $\binom{\tau+10}{10}(0.9)^\tau$. Clearly the term (0.91) has the greater common-ratio magnitude and therefore eventually will dominate. Determine the interval at which domination is achieved if our criterion of dominance is a ratio of $100:1$.

Answer: In this case, we can apply inequality 4.100 and the algorithm of equations 4.102 through 4.105.

$$\tau > \log[1/(0.01)1]/\log[0.91/0.90] + \log\left[\binom{\tau+10}{10}\right]/\log[0.91/0.90].$$

Because the second term on the right-hand side is positive, the first term does not provide the upper bound on the time of domination; so we must resort to the algorithm:

(1) $A = 416.763\ldots$

(2) $B = \log\left[\binom{427}{10}\right]/\log[0.91/0.90] = 4{,}104.789\ldots$

(3) $C = A + B = 4{,}521.552\ldots$

(4) $B = \log\left[\binom{4532}{10}\right] / \log[0.91/0.90] = 6{,}251.209\ldots$

(5) $C = A + B = 6{,}667.972\ldots$

(6) $B = \log\left[\binom{6678}{10}\right] / \log[0.91/0.90] = 6{,}602.322\ldots$

(7) $C = A + B = 7{,}019.085\ldots$

(8) $B = 6{,}648.841\ldots$

(9) $C = 7{,}065.604\ldots$

(10) $B = 6{,}654.747\ldots$

(11) $C = 7{,}071.510\ldots$

(12) $B = 6{,}655.514\ldots$

(13) $C = 7{,}072.277\ldots.$

At this point C has changed by less than one (in going from step 11 to step 13). Therefore, we are extremely close to our answer and can proceed by trial and error. Trying $\tau = 7{,}073$, we find

$$7073 \overset{?}{>} 416.763 + \log\left[\binom{7083}{10}\right] / \log[0.91/0.90] = 7072.405\ldots.$$

Clearly, 7073 satisfies the inequality. Next we can try 7072 to verify that it does not satisfy the inequality:

$$7072 \overset{?}{>} 416.763 + \log\left[\binom{7082}{10}\right] / \log[0.91/0.90] = 7072.277\ldots.$$

Thus, in fifteen steps, we have determined that domination, according to our criterion, is achieved at the 7073rd transition interval. In many modeling situations the dynamics would not be evaluated over nearly that many intervals, in which case eventual domination by $(0.91)^\tau$ is irrelevant as far as analysis is concerned.

In the first three examples, domination occurs in a very few transition intervals, a situation typical of many linear population models. In fact, the situation of Example 4, where domination required thousands of transition intervals is very unlikely to occur. A much more likely prospect is the occurrence of two or more terms with common ratios of the same magnitude, with no one of them becoming dominant no matter how many intervals go by. Thus, if domination is to occur, it usually takes place early in the growth pattern. Once it has occurred, the pattern itself usually is greatly simplified.

4.12.3. Geometric Patterns of Growth and the Biotic Potential

For the population at low levels or with invariant parameters the ultimately dominant terms will be of the form

$$J(\tau) = c_{ij} \binom{\tau + j - 1}{j - 1} (r_i)^{\tau} \tag{4.106}$$

where r_i is a positive real number. As time progresses, the expression on the right-hand side will approach closer and closer to a pure geometric progression. Thus, for large values of τ,

$$J(\tau + T)/J(\tau) \simeq (r_i)^T. \tag{4.107}$$

Since $\binom{\tau - j - 1}{j - 1}$ approaches $\tau^{(j-1)}/(j-1)!$ as τ becomes large (compared to j), the criterion for the validity of approximation 4.107 can be deduced from the binomial expansion of $(\tau + T)^{(j-1)}$ to be

$$\tau \gg (j - 1)\, T. \tag{4.108}$$

Usually, j will be equal to one, in which case inequality 4.108 always is valid, and approximation 4.107 becomes an equality. The single term on the right-hand side of 4.107 has a special significance, since it represents the ultimate rate of growth or decline of the modeled population, given the parameter values specified.

This ultimate rate of growth very often is abstracted as a single number or index so that it can be compared easily from population to population. Once dominance has been achieved and, in addition, approximation 4.107 has become valid for the dominant term, then in a given number (T) of intervals the population increases or decreases by very nearly the same proportion (i.e., r_i^T-fold). There are two basic indices that one might extract from such a pattern: (1) the mean time required for a certain, standard amount of proportional growth or decline, and (2) the mean amount of proportional growth or decline in a certain, standard amount of time. Neither of these indices necessarily has to represent an integral number of transition intervals. For example, if a population is undergoing a geometric pattern of growth in which it doubles three times (i.e., increases eightfold) in twenty transition intervals, one perfectly well might characterize it as having a mean doubling time of 20/3 transition intervals, in spite of the fact that 20/3 intervals is, in principle, unobservable in the model. Once 4.107 has become valid for the dominant term, the mean time, T_Λ, required for a Λ-fold growth or decline is given by

$$T_\Lambda \simeq \log[\Lambda]/\log[r_i] \tag{4.109}$$

in numbers of transition intervals. The mean proportional growth, Λ_s, in a standard amount, T_s, of time is given directly by 4.107:

$$\Lambda_s \simeq (r_i)^{T_s} \tag{4.110}$$

where T_s is given in numbers of transition intervals.

Example 4.21. Consider a unit-cohort response in which the term $15(2.7)^{\tau}$ ultimately dominates. If the transition interval is twenty days, what is the mean doubling time of the modeled population once dominance is established?

Answer: Using 4.109, one can see immediately that the mean doubling time will be

$$T_2 \simeq \log[2]/\log[2.7] \quad \text{transition intervals}$$
$$= 13.957\ldots \quad \text{days.}$$

Example 4.22. Consider a unit-cohort response dominated by $(1.4)^{\tau}$, where the transition interval is 365 days. Determine the amount of proportional growth expected in one day.

Answer: Employing 4.110 with $T_s = 1/365$, we have

$$\Lambda_s \simeq (1.4)^{(1/365)} = 1.000922\ldots \quad \text{fold per day.}$$

Example 4.23. Repeat Example 4.22, but let T_s be one year and the transition interval be one day.

Answer:

$$\Lambda_s \simeq (1.4)^{(365)} = (2.17136\ldots) \times 10^{53} \quad \text{fold per year.}$$

Both of these indices of growth are quite descriptive. The Λ-folding time is a more compact description in the sense of its values spanning a far more restricted range of magnitudes. However, an increased potential for growth is reflected in a decrease of this index. Some population biologists prefer to convert it to a similarly compact description, but one whose value increases as the potential for growth increase. The standard means of doing this is simply to take its reciprocal, which would be the number of Λ-foldings per transition interval. However, when $\Lambda = e$, where e is the base of the Naperian logarithm

$$e = 2.718281828\ldots$$

and when the parameter values correspond to hypothetically optimal conditions for growth of the modeled population, then the index often is called the *biotic potential* of the modeled population:

$$\begin{aligned} \alpha &= \text{Biotic Potential} = \log[r_i]/\log[e] \\ &= \log_e[r_i] \quad e\text{-foldings per transition interval.} \end{aligned} \tag{4.111}$$

Although it is compact and it increases with increasing potential for growth, the biotic potential perhaps is not as suggestive or immediately descriptive as the other two indices. However, with a little experience, one very quickly can develop an intuitive sense of its meaning. The biotic potential varies markedly from one idealized species to another. At one extreme, we find the very slow reproducers, such as the California condor, and some of the large mammals. The condor apparently produces only one egg every two years. If we assume absolutely no mortality, and we assume that half of the chicks are female, then the population could be expected to double every four years:

$$N_{\text{condor}} = N_0(2)^{\tau}$$

where the transition interval is four years. The corresponding biotic potential is

$$\alpha_{\text{condor}} = \frac{\log_e(2)}{4} = 0.1743 \quad e\text{-foldings per year.}$$

At the other extreme, we find the rapid reproducers such as the pomace flies (*Drosophila*) which, according to Borror and DeLong (1966), may produce 25 generations per year, with each female laying up to 100 eggs per generation; approximately half of which are female. With no mortality, a population of such flies might increase fiftyfold in each twenty-fifth of a year:

$$N_{\text{Drosophila}} = N_0(50)^{\tau}$$

where the transition interval is $1/25$ year. The corresponding biotic potential is

$$\alpha_{\text{Drosophila}} = \frac{\log_e[50]}{1/25} = 97.8 \ldots \quad e\text{-foldings per year.}$$

Thus, we probably nearly have spanned the range of biotic potentials for multicellular organisms. Most such biotic potentials very likely will fall between 0.1 and 100. The effective difference between one end of this range and the other is enormous. Translated into mean proportional growth in one year's time, we have

$$e^{0.1743\ldots} = 1.1904 \ldots \quad \text{fold}$$

for the condor and

$$e^{97.8\ldots} = 2.978 \ldots \times 10^{42} \quad \text{fold}$$

for the pomace fly. According to Borror and DeLong, if $2.978\ldots \times 10^{42}$ pomace flies were packed tightly together, 1000 to a cubic inch, they could form a ball extending nearly all the way from the sun to the earth.

4.12.4. Patterns of Dynamics Near Nonzero Critical Levels

In the subsequent section, we shall consider the modifications of a network that are necessary to convert it to a constant-parameter model appropriate to a nonzero critical level. Even before we do so, however, we can predict confidently the types of dynamic behaviors that we can expect. Clearly, the dynamics ultimately may converge on the critical level or they ultimately may diverge from it. In converging or diverging, the dynamics may progress monotonically (with ever increasing or ever decreasing values of the variable of interest) or they may progress in on oscillatory fashion (moving from one side of the critical level to the other).

By its very nature, a critical level is one from which a modeled population will move only if it is perturbed. The unit-cohort input represents a test perturbation of one additional individual. Usually, it is most convenient to consider the resulting dynamics in terms of deviations from the critical level. When the critical level is zero, such deviations always must be positive; and the dynamics and the ultimately dominant terms thus are constrained. When the critical level is nonzero, on the other hand, the net deviations may be positive, negative, or oscillating between the two. Thus, the unit-cohort responses are not constrained as they are near the zero level.

4.12.5. Exercises

1. Find the ultimately dominant terms of each of the following transformed unit-cohort responses:

	Answer:
a) $(1+2z^{-1})/(1-2z^{-1}+z^{-2})$	$3(\tau+1)$
b) $1/(z-2)(z^2+2)$	$(1/12)(2)^\tau$
c) $1/(z-2)(z^2+4)$	$(1/16)(2)^\tau+(0.044419\ldots)(2)^\tau\cos[\pi\tau/4+\pi/4]$ ($\pi/4$ rad. $=45°$)
d) $1/(z-2)(z^2+9)$	$(0.0154\ldots)(3)^\tau\cos[(0.98\ldots)\tau+(0.98\ldots)]$ ($0.98\ldots$ rad. $=56.3\ldots$ deg.)

4.13. Constant-Parameter Models for Nonzero Critical Levels

Let $\boldsymbol{JN}(\tau)$ be a vector whose elements are the values of all of the accumulation and flow variables in a network model at interval τ; and let $\boldsymbol{JN}_c$ be that same vector when the values of all flows and accumulations correspond precisely to a nonzero critical level. Evidently, by the very nature of a critical level, if $\boldsymbol{JN}(\tau)=\boldsymbol{JN}_c$ and there are no perturbations at interval τ, then $\boldsymbol{JN}(\tau+1)$ also will equal $\boldsymbol{JN}_c$. Next, consider a network model constructed in the manner described in section 3.2, but with *scalor* and *time-delay* parameters that vary in the manner described in sections 3.3 and 3.4. Thus, for the ith *scalor* in the network, we have the parameter $\gamma_i(\boldsymbol{JN},\tau)$, and for the jth *time delay* we have the parameter $T_j(\boldsymbol{JN},\tau)$. Our object will be to restructure the network for a particular value of τ (i.e., to remove the time-dependence of all parameters) and for $\boldsymbol{JN}=\boldsymbol{JN}_c$. Therefore, let

$$\begin{aligned}\gamma_i'(\boldsymbol{JN})&=\gamma_i(\boldsymbol{JN},\tau)\big|_{\text{evaluated at the interval of interest}}\\ T_j'(\boldsymbol{JN})&=T_j(\boldsymbol{JN},\tau)\big|_{\text{evaluated at the interval of interest}}\,.\end{aligned}\tag{4.112}$$

4.13.1. Replacement for the *Scalor*

Now, let us take these elements one at a time. First, consider the *scalor*, with its parameter $\gamma_i'(\boldsymbol{JN})$. If the flow into it, J_i', is at the critical level, J_{ic}', and all other flows and accumulations also are at their critical levels, so that $\boldsymbol{JN} = \boldsymbol{JN}_c$, then the flow, J_i, out of the *scalor* simply is

$$J_i = \gamma'_i(\boldsymbol{JN}_c)\, J'_{ic} = \gamma'_{ic}\, J'_{ic} \tag{4.113}$$

where γ'_{ic} is the value of γ'_i at the critical level and $\gamma'_{ic} J'_{ic}$ must be the critical level for J_i. Now, let us consider the deviation of J_i from its critical level in response to a unit-cohort deviation of J'_i. Evidently, that deviation, ΔJ_{ii}, is given by

$$\begin{aligned} \Delta J_{ii} &= (\gamma'_{ic} + \Delta\gamma'_{ii})(J'_{ic} + \mathbb{1}) - \gamma'_{ic} J'_{ic} \\ &= \gamma'_{ic} + \Delta\gamma'_{ii}(J'_{ic} + \mathbb{1}) \end{aligned} \tag{4.114}$$

where $\Delta\gamma'_{ii}$ is the deviation of γ'_i owing to the increase by one of J'_i. Now, if J'_{ic} is very large compared to one, which generally must be presumed for all variables in a large-numbers model, then

$$\Delta J_{ii} \simeq \gamma'_{ic} + \Delta\gamma'_{ii} J'_{ic}. \tag{4.115}$$

Furthermore, if we make the usual assumption for linearization about a critical level, namely that the deviation of γ'_i is directly proportional to the deviation of J'_i for small deviations of the latter (i.e., that $\Delta\gamma'_i = h\,\Delta\gamma'_{ii}$ when $J'_i = J'_{ic} + h$ and all other flows and accumulations are at their critical levels), then we can relate deviations in J_i to small deviations in J'_i as follows:

$$\Delta J_{ii} \simeq (\gamma'_{ic} + \Delta\gamma'_{ii} J'_{ic})\, \Delta J'_i. \tag{4.116}$$

Therefore, for small deviations of J'_i about its critical level, we can substitute the value $(\gamma'_{ic} + \Delta\gamma'_{ii} J'_{ic})$ for the parameter of the ith *scalor* in our network.

However, we are not through yet. We have not considered the effect on flow J_i of small deviations in flow J_j or accumulation N_k. Although neither of these variables is the input to the ith *scalor*, each affects the output of that *scalor* by changing its parameter, $\gamma'_i(\boldsymbol{JN})$. Let $\Delta\gamma'_{ik}$ be the deviation of γ'_i owing to the increase by one of the kth variable over its critical level (i.e., in response to $N_k = N_{kc} + \mathbb{1}$). Evidently

$$\Delta J_{ik} = (\gamma'_{ic} + \Delta\gamma'_{ik})(J'_{ic} + \Delta J'_i) - \gamma'_{ic} J'_{ic} \tag{4.117}$$

where ΔJ_{ik} is the deviation of J_i owing to the deviation in N_k; and $\Delta J'_i$ is any deviation that happens to the present in J'_i. Invoking the assumption of large numbers ($J'_{ic} \gg \Delta J'_i$) and linearity ($\Delta\gamma'_i = h\,\Delta\gamma_{ik}$ when $N_k = N_k + h$ and all other flows and accumulations are at their critical levels), we can simplify equation 4.117 to the following form:

$$\Delta J_{ik} \simeq (\Delta\gamma'_{ik} J'_{ic})\, \Delta N_k. \tag{4.118}$$

If X_j is the jth variable (either a flow or accumulation), then the total deviation of J_i is given by the following equation:

$$\Delta J_i = (\gamma'_{ic} + \Delta\gamma'_{ii} J'_{ic})\, \Delta J'_i + J'_{ic} \sum_j \Delta\gamma'_{ij}\, \Delta X_j \tag{4.119}$$

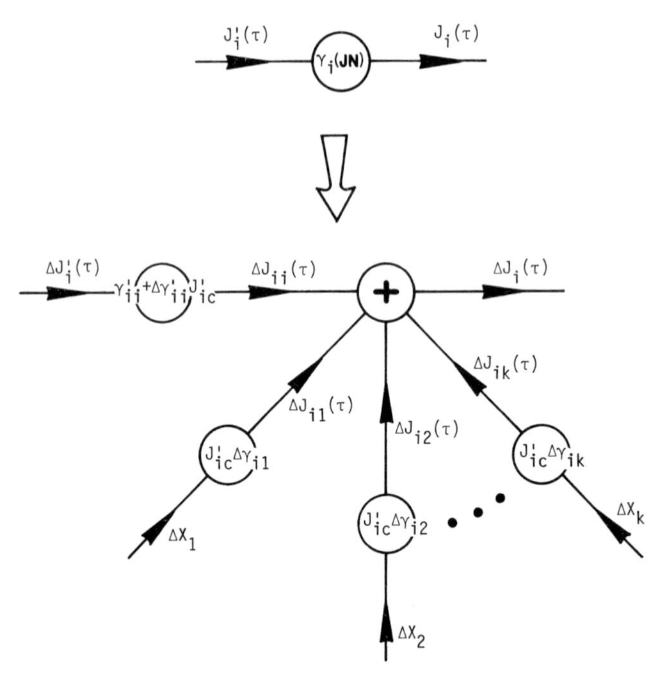

Fig. 4.56. A nonlinear *scalor* (top) and its constant-parameter replacement for small deviations of flows and accumulations about their critical levels

where the sum on the right-hand side is carried out over all variables other than J_i'. A network representation of this relationship is shown in Figure 4.56. Here, the summation is carried out at an *adder* connected to every variable in the network through an appropriate *scalor*. Thus, we have the required network replacement for the *scalor* $\gamma_i(\boldsymbol{JN})$ linearized about $\boldsymbol{JN}_c$. Although the replacement seems rather complicated, it usually will be simplified considerably by the lack of postulated dependence of γ_i on many of the network variables, in which case many of the paths into the *adder* will have *scalors* with parameters equal to zero and therefore may be eliminated completely.

Example 4.24. For the idealized salmon population of Example 4.4, construct a network model with a variable *scalor*. Then, by the methods of this section, find a replacement for the *scalor* that will convert the network into one with constant parameters and valid close to a nonzero critical level.

Answer: A network model embodying the general dynamics of the idealized population is shown in Figure 4.57. Here N_e is the seasonal egg production, J_e is the flow of surviving eggs, and J_s is the flow of surviving two-year-old fish into spawning. According to equation 4.7, the only nonzero critical level is given by

$$J_{sc} = (\mathbb{1}/\alpha\,\beta)\log_e(\alpha\,\gamma_0\,\gamma'). \tag{4.120}$$

Now, our variable *scalor* depends only upon one network variable, namely N_e. From equation 4.120 and the configuration of Figure 4.57, it follows that the

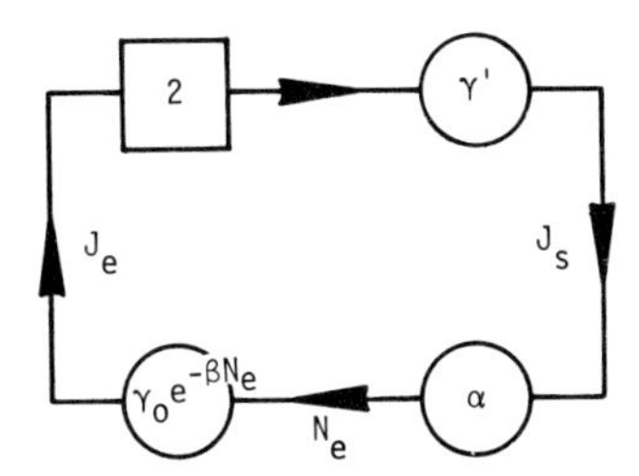

Fig. 4.57. Network model of an idealized salmon population

critical level of N_e is

$$N_{ec}=(1/\beta)\log_e(\alpha\gamma_0\gamma'). \tag{4.121}$$

Knowing that

$$\gamma=\gamma_0\exp[-\beta N_e] \tag{4.122}$$

we have

$$\gamma(N_{ec})=\gamma_c=1/\alpha\gamma' \tag{4.123}$$

and we easily can determine that in response to a unit cohort added to N_{ec},

$$\Delta\gamma=(\exp[-\beta]-1)/\alpha\gamma'. \tag{4.124}$$

From Figure 4.56, we see that the replacement scale factor is

$$\gamma_c+\Delta\gamma N_{ec}=1/\alpha\gamma'+\log_e[\alpha\gamma_0\gamma'](\exp[-\beta]-1)/\alpha\beta\gamma' \tag{4.125}$$

which leads to the modified network model of Figure 4.58, applicable for small excursions from the critical level. Generally, one would expect the magnitude of β to be very small, otherwise γ would vary quite drastically even for very small changes in N_e (i.e., changes of the order of one or two individuals); and this is inconsistent with a large-numbers model, where N_e necessarily is very large. In other words, if β were not small, the function $\gamma(N_e)$ would have disproportionalety large jumps (see section 1.20). If, as we should expect, β is very small, then $(\exp[-\beta]-1)/\beta$ is very nearly equal to -1.0; and the replacement scale factor becomes

$$(1/\alpha\gamma')(1-\log_e[\alpha\gamma_0\gamma']). \tag{4.126}$$

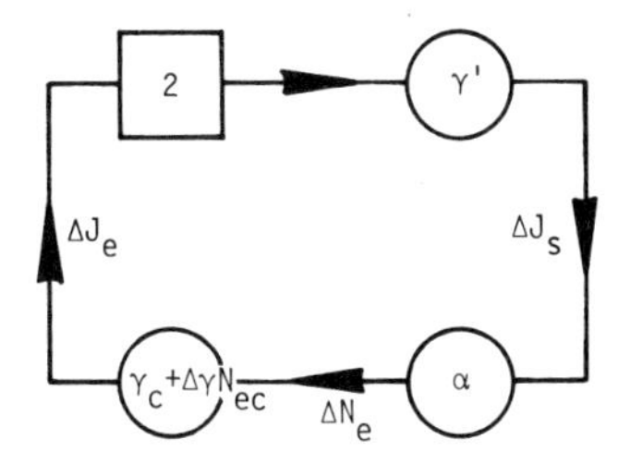

Fig. 4.58. Constant-parameter version of the network of Fig. 4.57, linearized for small deviations about the critical level

4.13.2. Modification of the *Time Delay*

In discrete-time network models modified to correspond to nonzero critical levels, the time delays remain as they are, but with their parameters (durations) set to the values corresponding to $\boldsymbol{JN}_c$. This represents an approximation that is consistent with the temporal resolution of the model (i.e., one transition interval) and is valid as long as a unit cohort added anywhere leads to a duration deviation, $\Delta T'$, of much less than one transition interval; in other words,

$$|\Delta T'(\boldsymbol{JN})| \ll \mathbb{1} \tag{4.127}$$

when a unit cohort is added to any component of $\boldsymbol{JN}$. Again, we should expect this to be true in large-numbers models with no disproportionately large jumps in their parameters.

If inequality 4.127 is untrue to the extent that

$$|\Delta T'(\boldsymbol{JN})| \geqq \mathbb{1}, \tag{4.128}$$

then an interesting thing happens when ΔT is positive (i.e., T is increasing). In that case, the *delay* duration will be increasing more rapidly than individuals are progressing through the *delay* (the latter rate of course being one transition interval per transition interval), and the flow out of the *delay* therefore will be zero. Thus, for one or more added unit cohorts, the output of the *time delay* would be zero; but, if unit cohorts were subtracted rather than added, the output of the delay would not be zero. In fact it would increase with increasing numbers of subtracted unit cohorts. As those cohorts were added back again, however, the output flow once again would stop. The result of all this would be an essential and profound nonlinearity that could not be eliminated from the model. Therefore, if inequality 4.128 is true at the critical point, our model cannot be linearized at that point.

4.13.3. Dynamics Close to a Nonzero Critical Level

Now we have a more general alternative to the graphical method of section 4.3. One simply restructures the network model to be valid for small excursions from the critical level, obtaining by this process a model with constant parameters. The next step is to transform the network into a linear flow graph, then apply Mason's rule to determine the denominator of the transformed unit-cohort response, then find the roots of the denominator (see section 4.14), and then determine which root or roots have the largest magnitude. Those roots, of course, will represent the ultimately dominant terms of the unit-cohort response. If the largest magnitude is less than $\mathbb{1}.\mathbb{0}$, then the dynamics will *converge* to the critical level; if it is greater than $\mathbb{1}.\mathbb{0}$, then the dynamics will *diverge* from the critical level; and if (by some very remote chance) it is precisely equal to $\mathbb{1}.\mathbb{0}$, then one must consider the nonlinearities in order to deduce the ultimate outcome. Convergence and divergence may be single-sided (i.e., with the dynamics approaching or moving away from the critical level from one side only), or they may be oscillatory (moving from one side of the critical level to the other as they approach it or move away from it). When divergence is single-sided, it may progress steadily upward, to values greater than the critical level, or it may progress steadily downward. The direction that is taken depends upon the perturbation initiating the divergence and upon the nondominant terms in the unit-cohort response. One can determine the direction for a unit-cohort perturbation simply by determining the sign of the coefficient corresponding to the dominant term (i.e., the coefficient for that term in the partial-fraction expansion). If it is negative, we have downward divergence; if it is positive, we have upward divergence.

Oscillatory divergence or convergence is signalled by a dominant term with a common ratio that is negative and real, complex, or imaginary.

Example 4.25. Taking $(\exp[-\beta]-1)/\beta$ to be equal to -1.0, deduce the general characteristics of the dynamics of the model of Figure 4.58 for small excursions from the nonzero critical level.

Answer: The parameter of the replaced scalor is given by expression 4.126. The network has a single loop, and the product of transformed terms around that loop is

$$(1-\log_e[\alpha\gamma_0\gamma'])z^{-2}.$$

Therefore, by Mason's rule, the denominator of the transformed unit-cohort response is

$$1-(1-\log_e[\alpha\gamma_0\gamma'])z^{-2}$$

from which

$$r_1, r_2 = \pm\sqrt{1-\log_e[\alpha\gamma_0\gamma']}. \tag{4.129}$$

The situation in this case is complicated slightly by the fact that $|r_1|=|r_2|$. When $\alpha\gamma_0\gamma'<1$, then the magnitude of r_1 and r_2 is greater than 1.0 and divergence occurs. The form of the unit-cohort response in that case will be

$$\begin{aligned} \Delta f'(\tau) &= c_1(|r_1|)^\tau + c_2(-|r_1|)^\tau + c_0 \qquad & \tau=0 \\ &= c_1(|r_1|)^\tau + c_2(-|r_1|)^\tau & \tau>0. \end{aligned} \tag{4.130}$$

To determine c_0, c_1 and c_2, one must select input and output variables and find the numerator of the corresponding transformed unit-cohort response. For all combinations in the network of Figure 4.58, c_1 will be positive and $|c_1|\geqq|c_2|$. Therefore, the dynamics will be single-sided, progressing steadily upward. If the perturbation were a single individual removed rather than added, then c_1 would be negative and $|c_1|\geqq|c_2|$, and the dynamics would be single-sided in the opposite direction, progressing steadily downward.

When $1<\alpha\gamma_0\gamma'<2.71828\ldots$, r_1 and r_2 will be real with magnitude less than 1.0. The unit-cohort response again will be given by equations 4.130; and c_1 again will be positive with $|c_1|>|c_2|$, leading to single-sided convergence from above. When $2.71828\ldots<\alpha\gamma_0\gamma'<5.43656\ldots$, r_1 and r_2 will be imaginary with magnitude less than 1.0, in which case oscillatory convergence will occur.

Finally, when $\alpha\gamma_0\gamma' > 5.43656\ldots$, r_1 and r_2 will be imaginary with magnitude greater than 1.0, in which case oscillatory divergence will occur. It is left as an exercise for the reader to compare these conclusions with those reached by graphical analysis in Example 4.4.

4.14. Finding the Roots of $Q(z)$

Now we come to the heart of the linear analysis problem. How does one go about finding the various nonzero roots of the denominator of the z-transform of the unit-cohort response function? Except for the simplest cases, the usual answer to this question is completely numerically, *by trial and error.* Fortunately, there are efficient computer programs readily available for carrying out the numerical trial-and-error procedure. Occasionally, however, we may wish to find roots by hand, perhaps with the aid of a small calculator; or we may wish to program a microcomputer to carry out the root-finding process. In either case, it helps to know some of the algorithms that can be used. In this section, we discuss some of the principles and methods of root-finding that can be applied to $Q(z)$.

In the first place, let us review the nature of $Q(z)$ and its roots. We know, for example, that $Q(z)$ was formed from another polynomial, $Q'(z^{-1})$, which had the following form:

$$Q'(z^{-1}) = (1 - L_1)(1 - L_2)(1 - L_3)\ldots(1 - L_m)^* \tag{4.131}$$

where L_i is the product of all the scale factors and transformed *time delays* around loop i; and the (*) indicates that all terms containing products of touching loops are dropped. Each L_i can be expressed in the following form:

$$L_i = b'_{k_i} z^{-k_i}.$$

Therefore, in its expanded form,

$$\begin{aligned}
Q'(z^{-1}) = {} & 1 - b'_1 z^{-k_1} - b'_2 z^{-k_2} - \cdots \\
& - b'_m z^{-k_m} + b'_1 b'_2 z^{-(k_1+k_2)} + b'_1 b'_3 z^{-(k_1+k_3)} + \cdots \\
& + b'_1 b'_m z^{-(k_1+k_m)} + b'_2 b'_3 z^{-(k_2+k_3)} + \cdots + b'_2 b'_m z^{-(k_2+k_m)} + \cdots \\
& + b'_{m-1} b_m z^{-(k_{m-1}+k_m)} - b'_1 b'_2 b'_3 z^{-(k_1+k_2+k_3)} - \cdots
\end{aligned}$$

(all distinct products of three L's) +
(all distinct products of four L's) −
(all distinct products of five L's) ...
with all terms containing products of touching L's being dropped. (4.132)

The degree, k_q, of $Q'(z^{-1})$ will be equal to or less than the sum of all of the *time delays* in the model being analyzed. In forming $Q(z)$, we simply multiply every term in $Q'(z^{-1})$ by z^{-k}, where k is the degree, k_r, of the numerator of $\mathscr{f}(z)$ or the degree, k_q, of $Q'(z^{-1})$, whichever is greatest. If the degree of the numerator is greater, then $Q(z)$ will have $k_r - k_q$ roots equal to zero. These will be quite obvious and require no further discussion here. The number of nonzero roots will be k_q, and these also will be the roots of the polynomial

$$Q_1(z) = z^{k_q} Q'(z^{-1}). \tag{4.133}$$

Therefore, it is this polynomial upon which we should focus our attention. Equivalently, we could deal directly with $Q'(z^{-1})$, finding its roots and then taking their reciprocals, which are the roots of $Q(z)$.

There is one situation in which the latter approach is most efficient and obvious. If *none of the loops in the linear flow graph of our model touches one another*, then none of the terms on the right-hand side of equation 4.131 will be dropped out. In that case, we immediately have a partial factorization of $Q'(z^{-1})$:

$$Q'(z^{-1}) = (1 - b'_1 z^{-k_1})(1 - b'_2 z^{-k_2}) \dots (1 - b'_3 z^{-k_m}). \tag{4.134}$$

To complete the factorization, we simply can consider the existing factors one by one:

$$(1 - b'_j z^{-k_j}) = (1 - r_1 z^{-1})(1 - r_2 z^{-1})(1 - r_3 z^{-1}) \dots (1 - r_{k_j} z^{-1}) \tag{4.135}$$

and notice that when z^{-1} is equal to the reciprocal of any of the r's in the factorization, then $(1 - b'_j z^{-k_j})$ must be zero. These are the roots (or zeros) with respect to z^{-1}. To determine the roots with respect to z (i.e., the corresponding roots in $Q_1(z)$), we simply take their reciprocals. In other words, the r's themselves are the corresponding roots of $Q_1(z)$.

The values of the r's in this special case are very easy to find. They simply are the k_j values (real, imaginary and complex) of

$$r_j = (b'_j)^{1/k_j}. \tag{4.136}$$

Recalling the vector concept of multiplication of complex numbers, we can write

$$\begin{aligned} b'_j &= (r_j)^{k_j} \\ &= |r_j|^{k_j} \cos(\theta_j k_j) + \boldsymbol{i}\, |r_j|^{k_j} \sin(\theta_j k_j). \end{aligned} \tag{4.137}$$

Since b'_j is a real number (positive or negative), $\sin(\theta_j k_j)$ must be equal to zero. Therefore,

$$k_j \theta_j = n \times 180° \quad \text{or} \quad n \times \pi \quad \text{radians}$$

where n is an integer; and

$$\theta_j = \frac{n \times 180°}{k_j} \quad \text{or} \quad \frac{n \times \pi}{k_j}. \tag{4.138}$$

Since $\cos(n\pi)$ is equal to one for all even values of the integer n and equal to minus one for all odd values of n, we can reduce equation 4.137 to the following form:

$$\begin{aligned} b'_j &= +|r_j|^{k_j} \quad \text{when } n \text{ is even} \\ b'_j &= -|r_j|^{k_j} \quad \text{when } n \text{ is odd.} \end{aligned} \tag{4.139}$$

Thus, if b_j is positive, n must be an even integer, and if b_j is negative, n must be an odd integer. In either case, we have

$$|r_j|^{k_j} = |b_j|$$

or

$$|r_j| = |(b_j)^{1/k_j}| \tag{4.140}$$

which is the positive, real value of the k_jth root of b_j. The k_j values of r_j therefore can be expressed as follows:

If no loops are touching

$$r_{j,n} = |(b_j)^{1/k_j}| \cos\left(\frac{n\pi}{k_j}\right) + i\,|(b_j)^{1/k_j}| \sin\left(\frac{n\pi}{k_j}\right) \tag{4.141}$$

where n is any of the first k_j odd integers if b_j is negative, and any of the first k_j even integers (including 0) if b_j is positive.

In general, however, we can expect some touching among the loops in our linear flow graph. In that case, the first thing to do is remove any factors $(1 - L_i)$ in $Q'(z^{-1})$ that represent loops that touch no others. From that point on, it usually is best to deal with the polynomial $Q_1(z)$, since its roots are much more confined than are those of $Q'(z^{-1})$; and we very likely will be using numerical trial and error methods. The logical first step in the solution of any polynomial is an attempt to simplify it by reducing its degree, since roots are easier to locate for polynomials of lower degree. The classic example of this is the fourth-degree polynomial in which the terms of odd degree are missing:

$$P(y) = y^4 + a\,y^2 + b. \tag{4.142}$$

The standard method of approach is to find the roots with respect to y^2 first, then proceed to find the roots with respect to y. Let $y^2 = x$, which leads to the simple quadratic polynomial

$$P(x) = x^2 + a\,x + b \tag{4.143}$$

the roots of which can be determined very easily:

$$P(x) = (x^2 + a\,x + b) = (x - r_1')(x - r_2'). \tag{4.144}$$

Replacing y^2 for x, we now have a partial factorization of the original polynomial:

$$P(y) = (y^2 - r_1')(y^2 - r_2') \tag{4.145}$$

and we can complete the factorization with the method of the previous paragraph. Thus, by reducing the degree of the polynomial with a change of variable, we were able to accomplish a partial factorization, in which our original polynomial is expressed as a product of polynomials of lower degree in the original variable. The roots of each of these polynomial factors of course are roots of the original polynomial, and they can be found separately and more easily.

4.14.1. Euclid's Algorithm

We can generalize this procedure as follows: To accomplish a partial factorization of the polynomial $Q_1(z)$, substitute $x=z^\zeta$, where ζ is the *greatest common divisor* of the degrees of the various terms of $Q_1(z)$, the greatest common divisor being the largest integer that divides into each of the degrees an integral number of times. In the example of the previous paragraph, the degrees of the terms were 4, 2, and 0. We can ignore the term of degree 0, since any integer divides into it 0 times and thus can be considered a divisor of it. This leaves 4 and 2, the greatest common divisor of which is 2, which goes once into 2 and twice into 4. Quite often, one can determine the greatest common divisor of a set of degrees by inspection, just as we did here. On the other hand, this occasionally may be difficult. For example, consider the two integers 219 and 365. The fact that the greatest common divisor is 73 certainly is not immediately obvious. In such cases, we can employ an ancient procedure, known as Euclid's algorithm, to determine systematically the greatest common divisor of any pair of numbers.

To understand this algorithm, consider two integers whose greatest common divisor is ζ:

$$A=n\zeta, \qquad B=m\zeta.$$

Clearly, n and m are relatively prime integers (integers whose greatest common divisor is one). Otherwise, A and B would have a common divisor greater than ζ. Furthermore, if n is larger than m, then we can write

$$n=am+b$$

where a and b are integers, with b and m being relatively prime (otherwise n and m would not be relatively prime) and b is less than m. Thus, we have

$$A=am\zeta+b\zeta \quad \text{and} \quad B=m\zeta.$$

Now, if we divide B (the smaller of the two integers) into A (the larger), we have

$$\begin{array}{r@{\,}l} & a \\ \cline{2-2} m\zeta) & am\zeta+b\zeta \\ & am\zeta \\ \cline{2-2} & b\zeta \end{array}$$

where the remainder is $b\zeta$. It is clear that the greatest common divisor of $B(=m\zeta)$ and the remainder $(=b\zeta)$ is the same as the greatest common divisor of A and B. At this point, let us generalize the procedure. If we divide the smaller of two integers into the larger, the greatest common divisor of the remainder and the smaller integer is the same as the greatest common divisor of the two original integers. Since the remainder is smaller than either of the two original integers, the problem of finding that greatest common divisor is simplified somewhat. We can simplify it further by dividing the remainder into the smaller integer, producing a second remainder. The greatest common divisor of the second remainder and the first is still the one that we are seeking, so we can continue this division process until we finally produce a remainder that is zero, at which point we can conclude that the immediately preceding, nonzero remainder is the sought-after greatest common divisor.

Example 4.26. Find the greatest common divisor of the two integers 219 and 365.

Answer: Applying Euclid's algorithm, we have

Step 1.

$$\begin{array}{r} 1 \\ 219\overline{)\,365} \\ \underline{219} \\ \text{remainder } = 146 \end{array}$$

Step 2.

$$\begin{array}{r} 1 \\ 146\overline{)\,219} \\ \underline{146} \\ \text{remainder } = 73 \end{array}$$

Step 3.

$$\begin{array}{r} 2 \\ 73\overline{)\,146} \\ \underline{146} \\ \text{remainder } = 0 \end{array}$$

Step 4.
the immediately preceding nonzero remainder is 73,
which therefore is the greatest common divisor of 219 and 365.

There is another way in which Euclid's algorithm can be used to simplify our polynomial. If $Q_1(z)$ has repeated roots, Euclid's algorithm will enable us to eliminate them. To understand the procedure, consider the factorized form of $Q_1(z)$:

$$Q_1(z)=(z-r_1)(z-r_2)(z-r_3)\ldots(z-r_j)^m\ldots(z-r_k) \tag{4.146}$$

where r_j is a root repeated m times. The first derivative of $Q_1(z)$ with respect to z is

$$\begin{aligned}\frac{dQ_1(z)}{dz}&=(z-r_2)(z-r_3)\ldots(z-r_j)^m\ldots(z-r_k)+(z-r_1)(z-r_3)\ldots(z-r_j)^m\ldots(z-r_k)\\&\quad+(z-r_1)(z-r_2)\ldots(z-r_j)^m\ldots(z-r_k)+(z-r_1)(z-r_2)(z-r_3)\ldots m(z-r_j)^{m-1}\\&\quad\ldots(z-r_k)+\cdots.\end{aligned} \tag{4.147}$$

Clearly, the additive terms in the first derivative will have in common only those factors, such as $(z-r_j)$ that occur multiply in $Q_1(z)$. Thus, the binomial factors that appear multiply in $Q_1(z)$ also appear in its first derivative, and they are the only binomial factors that do so. If we can determine the binomials that are common divisors of $Q_1(z)$ and its first derivative, therefore, we have determined the multiple roots of $Q_1(z)$. The product of all such binomials is a polynomial that can be considered to be the greatest common divisor of $Q_1(z)$ and its first derivative with respect to z. Clearly, however, when we say that one polynomial divides another, our concept is somewhat different from that when we say one integer divides another. In the abstract, the two concepts merge, however. In the case of integers, we say that one divides another if the quotient is an integer. Likewise, in the case of polynomials, we say that one divides another if the quotient is a polynomial.

Since Euclid's algorithm is much easier to demonstrate than to describe, let us proceed with an example. Consider the polynomial

$$Q_1(z)=z^4-5z^3+9z^2-7z+2.$$

Its first derivative with respect to z is

$$\frac{dQ_1(z)}{dz}=4z^3-15z^2+18z-7.$$

We wish to determine the greatest common divisor of these two polynomials. To do so we begin by dividing the polynomial of lower degree into that of higher degree:

$$\begin{array}{rl}
 & (1/4z)-(5/16) \\
\hline
4z^3-15z^2+18z-7) & z^4-5z^3+9z^2-7z+2 \\
 & z^4-\frac{15}{4}z^3+\frac{18}{4}z^2-\frac{7}{4}z \\
\hline
 & -\frac{5}{4}z^3+\frac{18}{4}z^2-\frac{21}{4}z+2 \\
 & -\frac{5}{4}z^3+\frac{75}{16}z^2-\frac{90}{16}z+\frac{35}{16} \\
\hline
\text{remainder } = & -\frac{3}{16}z^2+\frac{6}{16}z-\frac{3}{16}.
\end{array}$$

Now, by analogy with our previous discussion, it is clear that the greatest common divisor of our original polynomials is also the greatest common divisor of the remainder and the divisor. Therefore, we can continue the process by dividing the remainder into the original divisor. We can simplify the procedure in two ways. First of all, we are interested only in the roots of the greatest common divisor. Therefore, we can consider all polynomials with the same roots as being equivalent. Clearly, if we multiply every coefficient in a polynomial by the same constant, we generate a polynomial that has the same roots and therefore is equivalent. Therefore, we can replace the remainder with its simpler equivalent

$$z^2-2z+1$$

which we obtained by multiplying every coefficient by $-16/3$. Secondly, we can use the *method of detached coefficients* in carrying out our division. The procedure should be self-evident to the reader. We wish to divide the polynomial z^2-2z+1 into the polynomial $4z^3-15z^2+18z-7$. We shall do so as follows:

$$
\begin{array}{rl}
 & 4-7 \\
\cline{2-2}
1-2+1) & 4-15+18-7 \\
 & 4-\ \ 8+\ 4 \\
\cline{2-2}
 & \quad -\ 7+14-7 \\
 & \quad -\ 7+14-7 \\
\cline{2-2}
\text{remainder} = & \qquad 0
\end{array}
$$

In other words, we carry along the coefficients *in their proper positions*, but to avoid clutter, we omit the powers of z, carrying them along only by implication. Since the remainder in this last division was zero, we conclude that the immediately preceding remainder (z^2-2z+1) is the greatest common divisor that we seek. This polynomial can be factorized by inspection:

$$z^2-2z+1=(z-1)(z-1)=(z-1)^2.$$

From the expansion of equation 4.147 one can see that if $(z-1)^2$ is a common divisor of $Q_1(z)$ and its first derivative, then $Q_1(z)$ must contain the factor $(z-1)^3$. Similarly, if the greatest common divisor had been $(z-r_1)(z-r_2)$, where r_1 and r_2 were different roots, then $Q_1(z)$ would include the factors $(z-r_1)^2$ and $(z-r_2)^2$. In other words, $Q_1(z)$ contains each of the binomial factors contained in the greatest common divisor, but the degree of each of these factors is greater by one in $Q_1(z)$.

By carrying out the simple procedure of Euclid's algorithm, we have determined that $Q_1(z)=z^4-5z^3+9z^2-7z+2$ contains the multiple binomial factor $(z-1)^3$. To complete our simplification process, therefore, we simply must divide the factor into the original polynomial:

$$(z-1)^3=z^3-3z^2+3z-1$$

$$
\begin{array}{rl}
 & 1-2 \\
\cline{2-2}
1-3+3-1) & 1-5+9-7+2 \\
 & 1-3+3-1 \\
\cline{2-2}
 & \quad -2+6-6+2 \\
 & \quad -2+6-6+2 \\
\cline{2-2}
 & \qquad 0
\end{array}
$$

Therefore,

$$z^4-5z^3+9z^2-7z+2=(z-1)^3(z-2).$$

In many polynomials, $Q_1(z)$, the degrees of the various terms will have a greatest common divisor equal to 1, in which case the polynomials are not susceptible to partial factorization by the substitution $x=z^{\zeta}$; and many polynomials will not have repeated roots, in which case they cannot be simplified by elimination of those roots. Nonetheless, one should determine whether or not such simplifications are possible with a particular $Q_1(z)$ before proceeding with the task of finding the roots by trial and error.

4.14.2. Descartes' Rule and Sturm Sequences

From the discussions in the previous sections, we know the following facts about the k_q nonzero roots of $Q_1(z)$:

1) All of the roots lie somewhere in the complex plane.
2) All of the nonreal roots occur in conjugate pairs.
3) If the parameters of the model are invariant or if they have been fixed at values corresponding to very low levels (i.e., the zero critical point), then one positive, real root will have a magnitude equal to or greater than that of any other root.

When (3) applies, we can draw a circle on the complex plane that encloses or includes all of the roots of $Q_1(z)$. The center of the circle is at the origin and its radius is the magnitude of the largest positive, real root (i.e., the magnitude of the common ratio of the dominant term). Once we have drawn such a circle, our search will be reasonably well confined (see Figure 4.59). In such cases, a reasonable first step, therefore, is to locate the roots along the positive, real axis. Even if (3) does not apply, location of the positive, real roots (if any) will allow us to simplify the polynomial (by reducing its degree), and therefore still is a logical first step.

To obtain a rough estimate of just how many positive, real roots $Q_1(z)$ has, we can write it in the *canonical form* and apply *Descartes*' rule. The canonical form simply has each term placed to the left of every term of lower degree, and the coefficient of the term of highest degree equal to 1.0. Thus, the term of highest

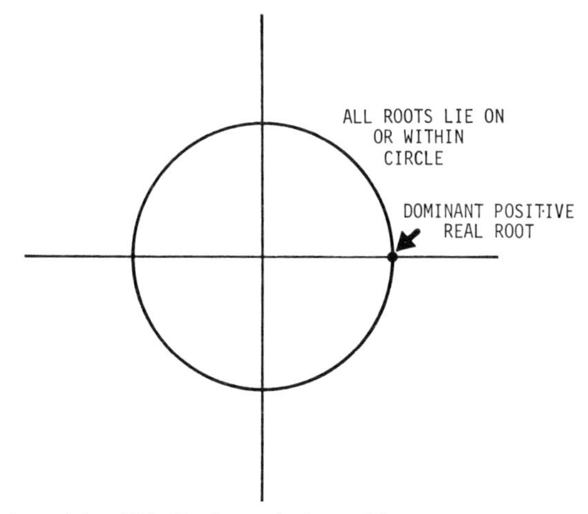

Fig. 4.59. For network models of idealized populations with constant parameters, and for models linearized for small deviations from the zero critical level, the root of largest magnitude will be the dominant positive real root; and all other roots will lie on or within the circle centered at the origin of the complex plane and passing through the dominant positive real root

degree in $Q_1(z)$, i.e., the term z^{k_q}, will be on the far left, followed immediately by the term of next highest degree, and so forth:

$Q_1(z)$ IN CANONICAL FORM

$$Q_1(z)=z^{k_q}+b_1 z^{k_q-1}+b_2 z^{k_q-2}+\cdots+b_{k_q-1} z+b_{k_q} \tag{4.148}$$

where the coefficients (the b_i's) are positive or negative real numbers (positive or negative sums and products of scale factors in the model being analyzed). To apply Descartes' rule, we begin at the left-hand side of the polynomial in canonical form and proceed to count the number of changes of sign of the terms. According to Descartes' rule, *the number of positive, real roots is either equal to the number of sign changes, or less than it by an even integer.* For example, in the polynomial

$$Q_1(z)=z^6-5z^5-2z^4+7z^3+15z^2.$$

The signs of the terms are $+, -, -, +, +$. There are two sign changes, one from $+$ to $-$ in going from the first term on the left to the second term, and one from $-$ to $+$ in going from the third term to the fourth term. Therefore, the number of positive, real roots of $Q_1(z)$ is either two or none. If we were dealing with the zero critical level or a model with invariant parameters, there would have to be at least one positive, real root, and we could conclude that there were in fact two positive, real roots.

If the number of sign changes in the terms of the canonical form, is three or more, then the conclusions from Descartes' rule will be ambiguous. In that case, we can apply Sturm's theorem, which allows us to obtain the precise number of roots lying on any segment of the real axis. To apply Sturm's theorem, we first must form a *Sturm sequence* of polynomials, the algorithm for which is quite straightforward. The first member (Ξ_1) of the sequence is the original polynomial itself, $Q_1(z)$. The second member (Ξ_2) is the derivative of $Q_1(z)$ with respect to z:

$$\Xi_1(z)=Q_1(z)=z^{k_q}+b_1 z^{k_q-1}+b_2 z^{k_q-2}+\cdots+b_{k_q-1} z+b_{k_q} \tag{4.149}$$

$$\Xi_2(z)=\frac{dQ_1(z)}{dz}=k_q z^{k_q-1}+b_1(k_q-1) z^{k_q-2}+b_2(k_q-2) z^{k_q-3}+\cdots+b_{k_q-1}. \tag{4.150}$$

To obtain the remaining members, one simply carries out Euclid's algorithm with Ξ_1 and Ξ_2, setting the successive members of the sequence equal to the negative remainders. If $\Xi_1(z)$ and $\Xi_2(z)$ both are divisible by a common polynomial, then the last remainder will be zero and the one immediately preceding it will be the common polynomial. Neither the zero nor the common polynomial should be included as members of the Sturm sequence. If $\Xi_1(z)$ and $\Xi_2(z)$ are not divisible by a common polynomial, then the last remainder will be a constant; and it should be included as a member of the Sturm sequence. If one has used Euclid's algorithm in an attempt to find the repeated roots of $Q_1(z)$, then all of the members of the Sturm sequence already are available.

Now, according to the Sturm theorem, *the number of nonrepeated real roots lying between any two real numbers, c_1 and c_2, is the difference between the number of sign changes in the Sturm sequence with $z=c_1$ and the number of sign changes with $z=c_2$.*

THE STURM SEQUENCE

$$\Xi_1(z), \Xi_2(z), \Xi_3(z), \Xi_4(z), \ldots, \Xi_k(z)$$

where k is the number of nonrepeated roots of all kinds in $Q_1(z)$.

Example 4.27. Determine the number of positive, real roots in each of the following polynomials:

(a) $Q_1(z) = z^3 - 8z^2 + 3z - 1$

(b) $Q_1(z) = z^4 - 7z^3 + 17z^2 - 17z + 6$

(c) $Q_1(z) = z^7 - 2z^6 - z^4 - z + 1$ for zero critical level.

Answer: (a) Here we have three changes of sign in the canonical form of the polynomial. Therefore, by Descartes' rule, there must be either three or one positive, real roots. To determine which number is correct, we can form a Sturm sequence, as follows:

$$\Xi_1(z) = z^3 - 8z^2 + 3z - 1, \qquad \Xi_2(z) = 3z^2 - 16z + 3.$$

Applying Euclid's algorithm, we have

Step 1.

$$\begin{array}{r|l} & 1/3 - 8/9 \\ \hline 3 - 16 + 3) & 1 - 8 + 3 - 1 \\ & 1 - \frac{16}{3} + 1 \\ \hline & -\frac{8}{3} + 2 - 1 \\ & -\frac{8}{3} + \frac{128}{9} - \frac{8}{3} \\ \hline & -\frac{110}{9} + \frac{5}{3} \end{array}$$

remainder $= -\frac{110}{9}z + \frac{5}{3}$ or, equivalently, $-22z + 3$

Step 2.

$$\begin{array}{r|l} & -3/22 + 343/484 \\ \hline -22 + 3) & 3 - 16 + 3 \\ & 3 - \frac{9}{22} \\ \hline & -\frac{343}{22} + 3 \\ & -\frac{343}{22} + \frac{1029}{484} \\ \hline \end{array}$$

remainder $= +\frac{423}{484}$

Thus, our last remainder is a constant. The Sturm sequence comprises $\Xi_1(z)$, $\Xi_2(z)$, and the negatives of the successive remainders

$$\Xi_3(z)=22z-3, \qquad \Xi_4(z)=-\frac{423}{484}.$$

Now, applying the Sturm theorem, we simply establish the limiting vaues, c_1 and c_2, between which we want to know the number of real roots. Since we are interested in positive, real roots, we can set c_1 equal to zero (one end of the positive, real axis) and c_2 equal to an indefinitely large positive value. When we do this, we have

$$\Xi_1(0)=-1, \qquad \Xi_2(0)=+3, \qquad \Xi_3(0)=-3, \qquad \Xi_4(0)=-\frac{423}{484}$$

$$\Xi_1(c_2)\to +c_2^3, \quad \Xi_2(c_2)\to +3c_2^2, \quad \Xi_3(c_2)\to +22c_2, \quad \Xi_4(c_2)=-\frac{423}{484}.$$

The number of sign changes in the sequence for $c_1=0$ is two; the number of changes in the sequence for c_2 (very large) is one. The difference between two and one is one, which therefore is the actual number of distinct positive, real roots.

(b) $Q_1(z)=z^4-7z^3+17z^2-17z+6.$

By Descartes' rule, there are either four positive real roots, two positive real roots, or no positive real roots. Again we must employ a Sturm sequence to clarify the situation:

$$\Xi_1(z)=z^4-7z^3+17z^2-17z+6$$

$$\Xi_2(z)=4z^3-21z^2+34z-17$$

$$\begin{array}{r|l}
 & 1/4-7/16 \\
\hline
4-21+34-17) & 1-7+17-17+6 \\
 & 1-\frac{21}{4}+\frac{34}{4}-\frac{17}{4} \\
\hline
 & -\frac{7}{4}+\frac{34}{4}-\frac{51}{4}+6 \\
 & -\frac{7}{4}+\frac{147}{16}-\frac{238}{16}+\frac{119}{16} \\
\hline
 & -\frac{11}{16}+\frac{34}{16}-\frac{23}{16}
\end{array}$$

$\Xi_3(z)=11z^2-34z+23$

$$\begin{array}{r l}
 & \frac{4}{11}-\frac{95}{121} \\
\hline
11-34+23\,) & 4-21+34-17 \\
 & 4-\frac{136}{11}+\frac{92}{11} \\
\hline
 & -\frac{95}{11}+\frac{282}{11}-17 \\
 & -\frac{95}{11}+\frac{3230}{121}-\frac{2185}{121} \\
\hline
 & -\frac{128}{121}+\frac{128}{121}
\end{array}$$

$\Xi_4(z)=z-1$

$$\begin{array}{r l}
 & 11-23 \\
\hline
1-1\,) & 11-34+23 \\
 & 11-11 \\
\hline
 & -23+23 \\
 & -23+23 \\
\hline
 & 0
\end{array}$$

Since the last remainder is zero, we discard the previous remainder, which in this case was $\Xi_4(z)$. This leaves a Sturm sequence with three members. Letting c_1 be zero and c_2 be very large, we have

$$\Xi_1(0)=+6, \qquad \Xi_2(0)=-17, \qquad \Xi_3(0)=+23$$
$$\Xi_1(c_2)\to +c_2^4, \qquad \Xi_2(c_2)\to +4c_2^3, \qquad \Xi_3(c_2)\to +11c_2^2.$$

The number of sign changes with $c_1=0$ is two; the number with c_2 very large is zero. Therefore, according to the Sturm theorem, the number of *nonrepeated* positive real roots is two. Since $z-1$ is a common divisor of Ξ_1 and Ξ_2, its roots are the repeated roots of $Q_1(z)$. Therefore, in addition to the two nonrepeated roots, we have the root $+1$ which is repeated. Since Descartes's rule tells us that there must be four, two or zero positive real roots, the root $+1$ must occur twice in $Q_1(z)$.

(c) In the canonical form of $Q_1(z)$ in this case, there are two changes of sign. Therefore, according to Descartes' rule, there are two positive, real roots or none. Since the polynomial was derived for a zero critical level, there must be at least one; therefore we conclude that there in fact are two.

4.14.3. Locating the Real Roots

Once we know how many positive, real roots there are, we can proceed systematically to find them. Usually, they will be transcendental numbers, so we can only approximate them. To do so, we simply can carry out an iterative, trial-and-error method of one sort of another (the most commonly applied being the *secant method*, which is not treated here, and *Newton's method*, which is more efficient for our purposes and is treated here). Basically, we shall search for positive, real values of r_j such that

$$Q_1(r_j)=0. \tag{4.151}$$

Employing a Taylor's series expansion of $Q_1(z)$, we have

$$Q_1(r_j)=Q_1(a_1)+(r_j-a_1)\left[\frac{dQ_1}{dz}\bigg|_{z=a_1}\right]+(1/2)(r_j-a_1)^2\left[\frac{d^2Q_1}{dz^2}\bigg|_{z=a_1}\right] + \cdots = 0 \tag{4.152}$$

which is a polynomial equation in r_j-a_1, and can be rewritten as follows:

$$P(r_j-a_1)=C_1(r_j-a_1)+C_2(r_j-a_1)^2+C_3(r_j-a_1)^3+\cdots=-Q_1(a_1). \tag{4.153}$$

Now, we can proceed by letting a_1 be our best initial guess of the value of a positive, real root of $Q_1(z)$. Substituting this guess for r_j in $Q_1(r_j)$, we immediately have $Q_1(a_1)$, the far right-hand side of equation 4.153. If $Q_1(a_1)$ equals zero, then of course our initial guess was correct. In the very likely event that it was not equal to zero, then our guess was in error; and the amount (ε) of that error is given by the smallest, real solution to equation 4.154:

$$C_1(a_1)\,\varepsilon+C_2(a_1)\,\varepsilon^2+C_3(a_1)\,\varepsilon^3+\cdots+=-Q_1(a_1) \tag{4.154}$$

where $C_i(a_i)$ is the value of C_i corresponding to our guess, a_i. Of course solving this equation for the error will be nearly as difficult as solving equation 4.151 itself. However, if the magnitude of the error is sufficiently small (i.e., considerably less than 1.0), then the terms of degree greater than one or two will be negligible in comparison to the first or second term. Newton's method, in fact, employs only the first term:

$$C_1(a_1)\,\varepsilon\simeq -Q_1(a_1). \tag{4.155}$$

Using this approximation, we can estimate the error in our initial guess, then correct our guess by adding the estimated error to it, and then substitute our new guess for r_j in $Q_1(r_j)$ and dQ_1/dz to yield new values of Q_1 and C_1. This process can be repeated until we have a corrected guess that meets our established criterion of accuracy for the roots of $Q_1(z)$.

Of course, this iterative procedure will work only if our initial error is sufficiently small to make approximation 4.155 reasonably close. Otherwise, we are in danger of diverging from the desired root rather than converging upon it. In fact, iterative numerical processes such as this simply are nonlinear, discrete-time systems, perfectly analogous to the models discussed in section 4.3. The desired root is a critical level, toward which the dynamics of the iterative process may converge, from which they may diverge, or about which they may oscillate or be chaotic. In the case of Newton's method, the process is analogous to the dynamics of a one-loop, lumpable model with

$$a_k = a_{k-1} - Q_1(a_{k-1})/C_1(a_{k-1}) \tag{4.156}$$

where a_k is the kth estimate of the root, and a_0 (the initial estimate) is specified. The criteria for convergence, divergence, oscillation and chaos are precisely those given in section 4.3. Fortunately, we know that if we initiate the iterative process close enough to the desired root, convergence will be guaranteed.

Very often, the largest positive, real root of $Q_1(z)$ will be very close to 1.0; and setting a_0 equal to 1.0 is likely to lead to convergence of Newton's method. If this fails, then one can fall back on the Sturm sequence to narrow down the location of real roots systematically and efficiently. In this case, the process is analogous to an artillery exercise. To begin with, we presumably have placed two widely separated "shots," one at $z=c_1$ and the other at $z=c_2$, and from sign-change differences between the two corresponding Sturm sequences we know that we have bracketed one or more target roots. By judicious selection of subsequent "shots" and observation of any shifts in sign-change patterns, we can bracket and close on single target roots within this sector, halving our error each time. Once we are sufficiently close to a target root to apply Newton's method, we can shift to it and from that point on converge much more rapidly.

Example 4.28. Find the positive, real root of $z^5 + z^4 + z^2 - 2$.

Answer: Using Newton's method (equation 4.156),

$$a_{k+1} = a_k - (a_k^5 + a_k^4 + a_k^2 - 2)/(5\,a_k^4 + 4\,a_k^3 + 2\,a_k) \tag{4.157}$$

in which we have made the following substitutions:

$$Q_1(a_k) = a_k^5 + a_k^4 + a_k^2 - 2$$

$$C_1(a_k) = \left.\frac{dQ_1}{dz}\right|_{z=a_k} = 5\,a_k^4 + 4\,a_k^3 + 2\,a_k\,.$$

Let our first guess be $a_0 = 1.0$. Substituting into equation 4.157, we obtain our second guess

$$a_1 = 1.0 - 1/11 = 0.909090909\ldots.$$

Continuing the iteration, we find

$$a_2 = 0.909090909\ldots - 0.0158258091\ldots = 0.8932651\ldots.$$

Noting that the estimated error has decreased markedly from the first to the second iteration, we now can be reasonably confident that we are on a path to convergence. Moving on, we find

$$a_3 = 0.8932651\ldots - 0.0004254\ldots = 0.8928396\ldots$$

$$a_4 = 0.8928396\ldots - 0.0000002303\ldots = 0.8928393697\ldots$$

$$a_5 = 0.8928393697\ldots - 0.0000000000\ldots = 0.8928393697\ldots.$$

Thus, in five steps, we have approximated the target root to ten decimal places. Notice that the number of correct significant figures in our estimation doubled on every iteration. This phenomenon often is called *quadratic convergence.*

To see if we can converge even more rapidly, let us repeat the process with the quadratic term of the Taylor's series as well as the linear term:

$$C_1(a_k)\,\varepsilon + C_2(a_k)\,\varepsilon^2 \simeq -Q_1(a_k) \tag{4.158}$$

$$\varepsilon_k \simeq \left[-C_1(a_k) \pm \sqrt{C_1^2(a_k) - 4\,Q_1(a_k)\,C_2(a_k)}\right]/2\,C_2(a_k) \tag{4.159}$$

$$a_{k+1} = a_k + \varepsilon_k \tag{4.160}$$

$$a_0 = 1.$$

Always choosing the smaller value of ε_k, we continue,

$$a_1 = 1 - 0.1090408\ldots = 0.8905914$$
$$a_2 = 0.8905914\ldots - 0.02264829\ldots = 0.892856229\ldots$$
$$a_3 = 0.892856229\ldots - 0.0001685939\ldots = 0.8928393697\ldots.$$

Thus, in three steps we attain the accuracy that required four steps in Newton's method. However, as one easily can determine from comparison of equations 4.159 and 4.156, the steps themselves in Newton's method are considerably simpler, requiring much less computational effort. In fact, in terms of computational expense, we lose more than we gain by including the quadratic term.

Example 4.29. Find the positive, real root of $z^3 - 3z^2 + 2z - 3$.

Answer: Using Newton's method and starting with $a_0 = 1$, we have

$$a_{k+1} = a_k - (a_k^3 - 3a_k^2 + 2a_k - 3)/(3a_k^2 - 6a_k + 2) \tag{4.161}$$
$$a_0 = 1.0$$
$$a_1 = -2.0$$
$$a_2 = -0.8846153385\ldots$$
$$a_3 = -0.0758270383\ldots$$
$$a_4 = 1.20615848\ldots$$
$$a_5 = -2.45849594\ldots$$
$$a_6 = -1.28575958\ldots$$
$$a_7 = -1.34519534\ldots$$
$$a_8 = -0.4707827733\ldots$$

and so forth. In other words, we are not converging on the target root. Therefore, let us turn to the Sturm sequence:

$$\Xi_1 = z^3 - 3z^2 + 2z - 3$$
$$\Xi_2 = 3z^2 - 6z + 2$$
$$\Xi_3 = -2z + 7$$
$$\Xi_4 = 71/4.$$

Since we are looking for a positive, real root, let us use $z=0$ as our first trial shot. In that case, we find the first term, Ξ_1, is negative and the rest are positive; in other words, we have one sign change. Next, let us see if we can bracket the root with a second shot at $z=10$. In that case, $\Xi_1=717$, $\Xi_2=242$, $\Xi_3=-13$, $\Xi_4=71/4$; so we now have *two* sign changes and therefore have indeed bracketed the target root. Next, we can halve the sector containing the root, by setting $z=5$: $\Xi_1=57$, $\Xi_2=47$, $\Xi_3=-3$, $\Xi_4=71/4$; we still have two sign changes, so the target lies between $z=0$ and $z=5$. Halving the sector again, we set $z=2.5$: $\Xi_1=-1.1125$, $\Xi_2=5.75$, $\Xi_3=2$, $\Xi_4=71/4$; we now have one sign change, so the target lies between $z=2.5$ and $z=5$. Perhaps we now are close enough to converge with Newton's method. Letting $a_0=2.5$, we generate the following sequence with equation 4.161.

$$
\begin{aligned}
a_0 &= 2.5 \\
a_1 &= 2.69565218\ldots \\
a_2 &= 2.67208073\ldots \\
a_3 &= 2.67169998\ldots \\
a_4 &= 2.67169988\ldots
\end{aligned}
$$

which clearly has converged very close to the target.

4.14.4. Locating Imaginary and Complex Roots

Once all of the real roots have been found (by means of the Sturm sequence and Newton's method), $Q_1(z)$ can be divided by the corresponding binomial factors; and the remainder polynomial, $Q_2(z)$, will have only imaginary and complex roots. The degree of $Q_2(z)$ will be even. Repeated roots can be removed from it by application of Euclid's algorithm, as discussed in section 4.14.1. Therefore, we shall be left with a polynomial with nonrepeated, complex or imaginary roots, occurring in conjugate pairs. If the degree of $Q_2(z)$ is 2 or 4, the roots can be found directly by the analytical methods for quadratic or quartic equations. If the degree of $Q(z)$ is greater than 4, then numerical methods will be required.

The process of carrying out a numerical search for roots of $Q_2(z)$ over the complex plane can be envisioned topologically. Imagine a surface lying over the complex plane, with each point on the surface having an elevation $|Q_2(z)|$ above the plane. Clearly, this surface will have a depression corresponding to each root (zero) of $Q_2(z)$, and the lowest point in such a depression will have elevation zero and correspond to the root itself. The strategy of the numerical search is to locate each depression, then move into it along the path of steepest descent to its lowest point. Once a depression has been located, the process of descent can be accomplished by means of Newton's method (with complex numbers now rather than real numbers) or by other, similar iterative methods. Perhaps the most common of these is Bairstow's method, which is carried out with real numbers and which, like Newton's method, is quadratically convergent when initiated sufficiently close to the target root. Bairstow's method is carried out with *quadratic factors*, which incorporate conjugate pairs of candidate complex roots:

$$r = a_r + i\, b_r$$
$$(z - r)(z - r^*) = z^2 - 2a_r z + (a_r^2 + b_r^2). \tag{4.162}$$

Letting

$$m = 2a_r \quad \text{and} \quad -p = a_r^2 + b_r^2 \tag{4.163}$$

we have the simplified quadratic factor

$$g(z) = z^2 - mz - p. \tag{4.164}$$

The object now is to correct iteratively the coefficients m and p until $g(z)$ becomes a divisor of $Q_2(z)$ (i.e., division of $Q_2(z)$ by $g(z)$ leaves no remainder). Of course we only can approximate that situation, but we can do so with arbitrary accuracy. Bairstow's method of approximation proceeds through the following steps:

1) Divide $Q_2(z) = z^n + a_1 z^{n-1} + \cdots + a_n$ by the candidate quadratic factor, $g(z) = z^2 - mz - p$, generating the quotient polynomial $q(z) = q_0 z^{n-2} + q_1 z^{n-3} + \cdots + q_{n-2}$ and the binomial remainder $q_{n-1} z + q_{nn}$. Numerically, this process is carried out quite easily in the following sequence:

$$\begin{aligned}
q_0 &= 1\\
q_1 &= mq_0 + a_1\\
q_2 &= mq_1 + pq_0 + a_2\\
&\vdots\\
q_j &= mq_{j-1} + pq_{j-2} + a_j\\
&\vdots\\
q_{n-2} &= mq_{n-3} + pq_{n-4} + a_{n-2}\\
q_{n-1} &= mq_{n-2} + pq_{n-3} + a_{n-1}\\
q_n &= mq_{n-1} + pq_{n-2} + a_n = q_{nn} + mq_{n-1}.
\end{aligned}$$

2) Divide the quotient, $q(z)$, by $g(z)$, generating a second binomial remainder, $q'_{n-1} z + q'_{nn}$. Again, the process can be executed numerically through the sequence

$$\begin{aligned}
q'_2 &= 1\\
q'_3 &= mq'_2 + q_1\\
q'_4 &= mq'_3 + pq'_2 + q_2\\
&\vdots\\
q'_j &= mq'_{j-1} + pq'_{j-2} + q_{j-2}\\
&\vdots\\
q'_{n-1} &= mq'_{n-2} + pq'_{n-3} + q_{n-3}\\
q'_n &= mq'_{n-1} + pq'_{n-2} + q_{n-2} = q'_{nn} + mq'_{n-1}.
\end{aligned}$$

3) Set $R = pq'_{n-1} + mq'_n$ and set $D = (q'_n)^2 - Rq'_{n-1}$.
4) The estimated errors in m and p respectively are

$$\varepsilon_m = (q_{n-1}\, q'_n - q_n\, q'_{n-1})/D$$
$$\varepsilon_p = (q'_n\, q_n - R\, q_{n-1})/D.$$

5) Correct the estimates of m and p by subtracting the corresponding error estimates from each. Then, with the corrected candidate quadratic factors, go back to step (1) and repeat the cycle until the error estimates fall within your established criteria of accuracy for the target root.

Example 4.30. Find the quadratic factor of the polynomial $z^4 + z^2 + 2z + 6$ using Bairstow's method.

Answer: Assuming that by trial and error we have found that the quadratic factor $z^2 + 2.1z + 2.1$ is a good candidate, we have $m = -2.1$, $p = -2.1$ and

	$q_0 = 1$	$q'_2 = 1$	$R = -12.243$	$\varepsilon_m = -0.0945258\ldots$
$a_1 = 0$	$q_1 = -2.1$	$q'_3 = -4.2$	$D = 49.1803$	$\varepsilon_p = -0.0969272\ldots$
$a_2 = 1$	$q_2 = 3.31$	$q'_4 = 10.03$		
$a_3 = 2$	$q_3 = -0.541$			
$a_4 = 6$	$q_4 = 0.1851$.			

Correcting m and p, we have $m = -2.0054742\ldots$, $p = -2.0030728\ldots$ and

$q_0 = 1$	$q'_2 = 1$	$R = -10.134642\ldots$	$\varepsilon_m = -0.005459182\ldots$
$q_1 = -2.0054742\ldots$	$q'_3 = -4.0109484\ldots$	$D = 41.427454\ldots$	$\varepsilon_p = -0.003075406\ldots$
$q_2 = 3.01885397\ldots$	$q'_4 = 9.05963470\ldots$		
$q_3 = -0.0371229\ldots$			
$q_4 = 0.02746480\ldots$			

Correcting again, we have $m = -2.0000150\ldots$, $p = -1.9999973\ldots$. These approximations are very close to the coefficients of the actual quadratic factor, which in this case happens to be $z^2 + 2z + 2$. Another iteration of Bairstow's method would double again the number of correct significant figures, which at this point seems to be quite unnecessary.

The chief problem with Bairstow's method is the fact that the initial estimates of m and p usually must be quite close to the target in order for quadratic convergence to occur. Until a reasonably accurate estimate has been obtained, other methods, including simple trial and error, are likely to be more efficient. Once such an estimate is available, Bairstow's method can be used to sharpen it. One iterative method that does not require an accurate initial estimate is that of Friedman, which is executed as follows:

1) With both polynomials arranged from left to right in order of descending powers of z, divide

$$Q_2(z) = z^n + a_1 z^{n-1} + \cdots + a_n$$

by the candidate quadratic factor

$$g(z)=z^2-mz-p$$

to obtain the quotient

$$q(z)=q_0 z^{n-2}+q_1 z^{n-3}+\cdots+q_{n-2}$$

whose coefficients are given by

$$q_0=1 \qquad q_1=m q_0+a_1$$

$$q_j=m q_{j-1}+p q_{j-2}+a_j.$$

2) Divide each term in the quotient by q_{n-2} and arrange in order of ascending powers to obtain the equivalent polynomial

$$q'(z)=1+q'_1 z+q'_2 z^2+\cdots+q'_{n-2} z^{n-2}$$

where

$$q'_j=q_{n-2-j}/q_{n-2}.$$

3) With $Q_2(z)$ arranged in order of ascending powers, divide it by $q'(z)$ to obtain the new quotient

$$g'(z)=g'_0+g'_1 z+g'_2 z^2$$

where

$$g'_0=a_n \qquad g'_1=a_{n-1}-q'_1 a_n \qquad g'_2=a_{n-2}-q'_2 a_n-g'_1 q'_1.$$

4) Divide each term of $g'(z)$ by g'_2 and arrange in order of descending powers to obtain the corrected candidate quadratic factor

$$g(z)=z^2-mz-p$$

where

$$-m=g'_1/g'_2 \qquad -p=g'_0/g'_2.$$

5) Return to step (1) and repeat until corrections in $g(z)$ fall within the criterion of accuracy.

Example 4.31. Find a quadratic factor of the polynomial z^4+z^2+2z+6 using Friedman's method.

Answer: Let us begin arbitrarily with the candidate quadratic factor $g(z)=z^2+z+1$. We have $m=-1$, $p=-1$ and

	$q_0=1$	$q'_1=-1$	$g'_0=6$
$a_1=0$	$q_1=-1$	$q'_2=1$	$g'_1=8$
$a_2=1$	$q_2=1$		$g'_2=3$
$a_3=2$			
$a_4=6$			

from which the corrected coefficients of $g(z)$ are $m=-2.6666\ldots$, $p=-2.0$ and

$$\begin{array}{lll} q_0=1 & q_1'=-0.43636\ldots & g_0'=6 \\ q_1=-2.66666\ldots & q_2'=0.163636\ldots & g_1'=4.61816\ldots \\ q_2=6.11111\ldots & & g_2'=2.03336\ldots \end{array}$$

from which $m=-2.27119\ldots$, $p=-2.95078\ldots$ and

$$\begin{array}{lll} q_0=1 & q_1'=-0.70808199\ldots & g_0'=6 \\ q_1=-2.27119\ldots & q_2'=0.311176695\ldots & g_1'=6.2484919\ldots \\ q_3=3.207524\ldots & & g_2'=3.5538428\ldots \end{array}$$

from which $m=-1.75823\ldots$, $p=-1.68831\ldots$ and

$$\begin{array}{lll} q_0=1 & q_1'=-0.731166213\ldots & g_0'=6 \\ q_1=-1.75823\ldots & q_2'=0.416135620\ldots & g_1'=6.38997278\ldots \\ q_2=2.403062 & & g_2'=3.17848738\ldots \end{array}$$

from which $m=-2.0103816\ldots$, $p=-1.8876903\ldots$ and

$$\begin{array}{lll} q_0=1 & q_1'=-0.637418317\ldots & g_0'=6 \\ q_1=-2.0103816\ldots & q_2'=0.317063346\ldots & g_1'=5.82450990 \\ q_2=3.153943 88\ldots & & g_2'=2.81026922 \end{array}$$

from which $m=-2.07258075\ldots$, $p=-2.13502676\ldots$.

Since our successive estimates of m and p now are reasonably close to one another, we must be fairly close to the target. At this point we could switch to Bairstow's method and hope to establish quadratic convergence.

Often, Friedman's method converges more rapidly than it did in Example 4.31. Its convergence becomes slower as the magnitudes of the complex root pairs become comparable; in fact, when the magnitudes are nearly equal, Friedman's method may not converge at all. In Example 4.31, the magnitudes of the two root pairs were 1.414 ... and 1.732 ... respectively, and convergence was slow and oscillatory. An alternative to this rigidly systematic search routine is a simple, heuristic trial-and-error search. One could divide $Q_2(z)$ by a candidate quadratic factor, inspect the remainder, then use his or her own judgement to correct the candidate factor and repeat the division. The object, of course, is to reduce the coefficients of the remainder to small values. By observing the effects of successive corrections on these coefficients, one can make an educated guess at the subsequent correction. Once the remainder coefficients are sufficiently small, Bairstow's method can be used to sharpen the final approximation.

4.14.5. Estimating the Magnitude of the Dominant Common Ratio

Of course, if one merely wishes to know whether the dynamics ultimately will converge on or diverge from a particular critical point, then he or she need not locate all of the roots of $Q_1(z)$. Only the root or roots of largest magnitude will affect this conclusion; and one need not even locate those roots precisely. In fact, all he or she requires is to know whether the magnitude of the dominant root or roots is greater than 1.0 or less than 1.0. If it is greater, divergence occurs; if it is less, convergence occurs.

In the case of the critical level at zero, the dominant root must be positive and real. One can apply the Sturm sequence to determine whether or not there are real roots between $z=1$ and z indefinitely large and positive. Thus, for the zero critical level, convergence occurs if there are no differences between the number of sign changes in the Sturm sequence for $z=1$ and the number of sign changes in the Sturm sequence for z indefinitely large and positive.

Example 4.32. Determine whether $Q_1(z)=z^4+3.75z^3-4.25z^2-3z+1$, which was derived for a model operating close to the zero critical level, has a root whose magnitude is greater than 1.0.

Answer: If the model were constructed correctly, the root of greatest magnitude must be positive and real. Applying a Sturm sequence to locate that root, we have

$$\begin{aligned}
\Xi_1 &= z^4+3.75z^3-4.25z^2-3z+1 = -1.5\;(z=1); &&= +z^4 &&(z \text{ indefinitely large})\\
\Xi_2 &= 4z^3+11.25z^2-8.5z-3 = 3.75\;(z=1); &&= +4z^3 &&(z \text{ indefinitely large})\\
\Xi_3 &= -4.8z^2-0.3z+1.7 = -3.4\;(z=1); &&= -4.8z^2 &&(z \text{ indefinitely large})\\
\Xi_4 &= -7.7z+0.9 = -6.8\;(z=1); &&= -7.7z &&(z \text{ indefinitely large}).
\end{aligned}$$

Even without calculating Ξ_5, we can see that there will be a difference of one sign change between the sequence evaluated at $z=1$ and that evaluated at indefinitely large, positive values of z (i.e., at $z=1$, there is a change of sign from Ξ_1 to Ξ_2, another from Ξ_2 to Ξ_3; at z indefinitely large and positive, there is a change from Ξ_2 to Ξ_3). Therefore, we conclude that one root lies on the real axis to the right of 1.0; and the dynamics diverge. In other words, at very low levels this modeled population will grow rather than decline.

Under certain circumstances, one can deduce ultimate divergence from a critical level without going through the effort of generating a Sturm sequence. A necessary condition for

$$Q_2(z)=z^n+a_1z^{n-1}+\cdots+a_n \tag{4.165}$$

to have no roots of magnitude greater than or equal to 1.0 is that

$$|a_j|<\binom{n}{j} \tag{4.166}$$

for all values between 1 and n of the integer j. If this test fails for any value of j, then the magnitude of at least one root is greater than or equal to 1.0; and the dynamics ultimately will not converge on the critical level. On the other hand, if the test does not fail (i.e., inequality 4.166 is true for all values of j), then no conclusions can be drawn; the dynamics may diverge or they may converge.

Although it does not always lead to conclusive results, the test of inequality 4.166 is more generally applicable than the Sturm-sequence test, applying to both zero and nonzero critical levels.

When the dynamics about a nonzero critical level are being considered, and the test of inequality 4.166 is not conclusive, then one cannot revert to the Sturm sequence, since the roots of largest magnitude may be complex or imaginary. What is needed is a test that is both conclusive and applicable to all types of roots (real, imaginary, and complex). Such a test is provided by the Routh-Hurwitz algorithms. These algorithms originally were designed for continuous-time models, in which the key question is whether or not any roots of the denominator polynomial have nonnegative real parts. For discrete-time models, the key question is whether or not any roots of the denominator have magnitudes greater than or equal to 1.0. The simple transformation

$$z=(s+1)/(s-1); \qquad s=(z+1)/(z-1) \tag{4.167}$$

carries every (complex, real, or imaginary) value of z with magnitude greater than 1.0 into a value of s with positive real part, every (complex, real, or imaginary) value of z with magnitude less than 1.0 into a value of s with negative real part, and every (complex, real, or imaginary) value of z with magnitude precisely equal to 1.0 into a value of s with real part precisely equal to zero (i.e., z into a purely imaginary value of s). To see that this is so, simply substitute $a+\boldsymbol{i}b$ for z in the right-hand expression:

$$s=[a+\boldsymbol{i}b+1]/[a+\boldsymbol{i}b-1]$$

which, through simply algebraic manipulation, leads to

$$s=[a^2+b^2-1-2\boldsymbol{i}b]/[a^2+b^2-2a+1].$$

The magnitude of z is simply $\sqrt{a^2+b^2}$ while that of $z-1$ is $\sqrt{a^2+b^2+1-2a}$, therefore

$$s=[|z|^2-1-\boldsymbol{i}b]/[|z-1|^2]. \tag{4.168}$$

Thus, when $|z|$ is greater than 1.0, the real part of s is positive and so forth.

If $(s+1)/(s-1)$ is substituted for z in $Q_1(z)$ to yield the function $F(s)$, then where $Q_1(z)$ is zero for certain values of z whose magnitudes are greater than or equal to 1.0, $F(s)$ will be zero for corresponding values of s whose real parts are nonnegative. $F(s)$ can be converted into a proper fraction of the form

$$\begin{aligned}F(s)=[(s+1)^n+a_1(s+1)^{n-1}(s-1)+\cdots+a_{n-1}(s+1)(s-1)^{n-1}\\+a_n(s-1)^n]/(s-1)^n.\end{aligned} \tag{4.169}$$

Clearly, $F(s)$ is zero when and only when its numerator polynomial, $P_1(s)$, is zero. Therefore, the Routh-Hurwitz algorithms can be used directly on $P_1(s)$ to determine whether or not $Q_1(z)$ has any roots of magnitude greater than or equal to 1.0 (i.e., by determining whether or not $P_1(s)$ has any roots with nonnegative real parts). Probably the most direct and easily remembered form of the Routh-Hurwitz algorithms involves a process similar to, but significantly different from that of Euclid's algorithm. The polynomial

$$P_1(s)=b_n s^n+b_{n-1} s^{n-1}+b_{n-2} s^{n-2}+\cdots+b_1 s+b_0 \tag{4.170}$$

is decomposed into a polynomial comprising all terms of odd degree in $P_1(s)$,

$$P_0(s)=b_k s^k+b_{k-2} s^{k-2}+\cdots+b_3 s^3+b_1 s \tag{4.171}$$

and a polynomial comprising all terms of even degree in $P_1(s)$,

$$P_e(s)=b_j s^j+b_{j-2} s^{j-2}+\cdots+b_4 s^4+b_2 s^2+b_0. \tag{4.172}$$

Next, we divide the polynomial (P_e or P_0) if smaller degree into the other one to yield a single quotient term, $\alpha_1 s$, and a remainder. Next, the remainder becomes the divisor and the previous divisor becomes the dividend, to yield a new, single-term quotient, $\alpha_2 s$, and a new remainder. Again the remainder becomes the divisor and the previous divisor becomes the dividend to yield yet another single-term quotient and another remainder. This process (known as *continued-fraction expansion*) is repeated until the final remainder is zero.

The roots of $P_1(s)$ all will have negative real parts if and only if all of the coefficients, $\alpha_1, \alpha_2, \ldots, \alpha_n$, of the quotients are positive. Therefore, the roots of $Q_1(z)$ all will have magnitudes less than 1.0 if and only if $\alpha_1, \alpha_2, \ldots, \alpha_n$ all are positive.

Example 4.33. Determine whether or not the roots of the following polynomials have magnitudes less than 1.0:

a) $Q_1(z)=z^5+3z^4+4z^3+2z^2+z+1$

b) $Q_1(z)=z^5+3z^4/2+2z^3+z^2+z/2+1/2$.

Answer: First of all, we can apply the rather quick test of inequality 4.166.

a) $|a_1|=3 \overset{?}{<} \binom{5}{1}=5 \quad |a_2|=4 \overset{?}{<} \binom{5}{2}=10 \quad |a_3|=2 \overset{?}{<} \binom{5}{3}=10$

$|a_4|=1 \overset{?}{<} \binom{5}{4}=5 \quad |a_5|=1 \overset{?}{<} \binom{5}{5}=1.$

Since 1.0 is not less than 1.0, the last of the test inequalities is not true; therefore, the magnitudes of the roots of $Q_2(z)$ are not all less than 1.0.

b) Since a_5 now is 1/2 rather than 1.0, all of the test inequalities are true, so the test is inconclusive and we must proceed to the Routh-Hurwitz algorithm. Substituting $(s+1)/(s-1)$ for z in $Q_1(z)$, we have

$$\begin{aligned} F(s)&=(s+1)^5/(s-1)^5+(3/2)(s+1)^4/(s-1)^4+2(s+1)^3/(s-1)^3\\ &\quad+(s+1)^2/(s-1)^2+(1/2)(s+1)/(s-1)+1/2\\ &=(13s^5+13s^4+26s^3+2s^2+9s+1)/(s-1)^5 \end{aligned}$$

$$P_1(s)=13s^5+13s^4+26s^3+2s^2+9s+1$$

$$P_0(s)=13s^5+26s^3+9s$$

$$P_e(s)=13s^4+2s^2+1.$$

Dividing with detached coefficients, we have

$$
\begin{array}{rrrl}
 & & 1 & \\
\cline{2-3}
13 \quad 2 \quad 1) & 13 \quad 26 & 9 & \qquad \alpha_1 = 1 \\
 & 13 \quad 2 & 1 & \\
\cline{2-3}
 & 24 & 8 &
\end{array}
$$

$$
\begin{array}{rlll}
 & 0.541666\ldots & & \\
\cline{2-3}
24 \quad 8) & 13 \qquad 2 & 1 & \qquad \alpha_2 = 0.541666\ldots \\
 & 13 \qquad 4.3333\ldots & & \\
\cline{2-3}
 & \qquad -2.3333\ldots & 1 & \\
 & & & \\
 & \qquad -10.2857\ldots & & \\
\cline{2-3}
-2.333\ldots \quad 1) & 24 \qquad 8 & & \qquad \alpha_3 = -10.2857\ldots
\end{array}
$$

Since α_3 is negative we need go no further. $P_1(s)$ definitely has at least one root with nonnegative real parts; and, therefore, $Q_1(z)$ has at least one root whose magnitude is not less than 1.0. In other words, the dynamics represented by the transformed unit-cohort response with $Q_1(z)$ as its denominator will not converge on the critical level in this case.

Two computationally convenient alternatives to the Routh-Hurwitz algorithms have been developed by Jury and his colleagues. The more recent of these (labeled the *Inners approach*) involves the construction of a succession of simple matrices from the coefficients of $Q_1(z)$ and the evaluation of the determinant of each matrix. The other alternative, often called the *Jury test*, involves the construction of a simple table from the coefficients of $Q_1(z)$ and inspection of the relative magnitudes of the entries in that table. Without resorting to tabular form, we can describe the second method as follows:

Given

$$Q_1(z) = z^n + a_1 z^{n-1} + \cdots + a_{n-1} z + a_n$$

let,

$$b_n = 1 - a_n^2,\, b_{n-1} = a_1 - a_n a_{n-1}, \ldots, b_{n-j} = a_j - a_n a_{n-j}, \ldots, b_1 = a_{n-1} - a_n a_1$$

and

$$c_n = b_n^2 - b_1^2,\, c_{n-1} = b_n b_{n-1} - b_1 b_2, \ldots, c_{n-j} = b_n b_{n-j} - b_1 b_{j+1}, \ldots,$$

$$c_2 = b_n b_2 - b_1 b_{n-1}$$

and

$$d_n = c_n^2 - c_2^2, \ldots, d_{n-j} = c_n c_{n-j} - c_2 c_{j+2}, \ldots, d_3 = c_n c_3 - c_2 c_{n-1}$$

and so on, until only three elements are generated:

$$s_n = r_n^2 - r_{n-3}^2, \qquad s_{n-1} = r_n r_{n-1} - r_{n-3} r_{n-2}, \qquad s_{n-2} = r_n r_{n-2} - r_{n-3} r_{n-1}.$$

The roots of $Q_1(z)$ all will have magnitudes less than 1.0 if and only if

$$Q_1(z=1)>0 \tag{4.173}$$

$$(-1)^n Q_1(z=-1)>0 \tag{4.174}$$

$$\begin{aligned} &|a_n|<1\\ &|b_n|>|b_1|\\ &|c_n|>|c_2|\\ &|d_n|>|d_3|\\ &\quad\vdots\\ &|s_n|>|s_{n-2}|. \end{aligned} \tag{4.175}$$

Example 4.34. Apply the Jury test to $Q_1(z)=z^5+(3/2)z^4+2z^3+z^2+(1/2)z+1/2$.

Answer: Appling tests 4.173 and 4.174 first, we have

$$Q_1(z=1) \quad =13/2 \overset{?}{>} 0$$
$$(-1)^5 Q_1(z=-1)=1/2 \overset{?}{>} 0$$

both of which pass, and therefore are inconclusive. Continuing, we have

$$\begin{array}{llllll} a_1=3/2, & a_2=2, & a_3=1, & a_4=1/2, & a_5=1/2 & |a_5|<1\\ b_5=3/4, & b_4=5/4, & b_3=3/2, & b_2=0, & b_1=-1/4 & |b_5|>|b_1|\\ c_5=1/2, & c_4=15/16, & c_3=3/2, & c_2=5/16 & & |c_5|>|c_2|\\ d_5=39/256, & d_4=0, & d_3=117/256 & & & |d_5|\not>|d_3|. \end{array}$$

Thus, the test finally fails at d_5 and shows conclusively that $Q_1(z)$ has at least one root whose magnitude is not less than 1.0.

Finally, if one wishes to estimate the actual magnitude of the dominant root (as opposed simply to determining whether or not it is less than 1.0), and, perhaps, the magnitudes of some of the other roots, then Graeffe's method of *root squaring* is applicable. Given the polynomial

$$Q_1(z)=z^n+a_1 z^{n-1}+\cdots+a_n \tag{4.176}$$

with roots $r_1, r_2, \ldots, r_n$, such that

$$|r_1|>|r_2|>|r_3|>\cdots>|r_n| \tag{4.177}$$

one can derive a new polynomial

$$F(-z^2)=(-z^2)^n+b_1(-z^2)^{n-1}+\cdots+b_n \tag{4.178}$$

where

$$
\begin{aligned}
b_1 &= a_1^2 - 2a_2. \\
b_2 &= a_2^2 - 2a_3a_1 + 2a_4 \\
b_3 &= a_3^2 - 2a_4a_2 + 2a_5a_1 - 2a_6 \\
&\vdots \\
b_{n-3} &= a_{n-3}^2 - 2a_{n-2}a_{n-4} + 2a_{n-1}a_{n-5} - 2a_na_{n-6} \\
b_{n-2} &= a_{n-2}^2 - 2a_{n-1}a_{n-3} + 2a_na_{n-4} \\
b_{n-1} &= a_{n-1}^2 - 2a_na_{n-2} \\
b_n &= a_n^2.
\end{aligned}
\tag{4.179}
$$

The roots of $F(-z^2)$ are $-r_1^2, -r_2^2, \ldots, -r_n^2$. Repeating the process with $F(-z^2)$, we obtain $F(-z^4)$. Continuing to repeat it, we eventually will have $F(-z^k)$, where k is sufficiently large that

$$|r_1^k| \gg |r_2^k| \gg |r_3^k| \gg \cdots \gg |r_n^k|. \tag{4.180}$$

At this point, the polynomial, F, will be factorable by inspection:

$$F(-z^k) = (-z^k)^n + b_1(-z^k)^{n-1} + \cdots + b_{n-1}(-z^k) + b_n \tag{4.181}$$

$$r_j^k \simeq b_j/b_{j-1} \tag{4.182}$$

for real roots, and

$$|r_j|^{2k} \simeq b_{j+1}/b_{j-1} \tag{4.183}$$

for complex or imaginary roots. To determine which coefficients correspond to real roots and which correspond to complex or imaginary roots, one simply compares $F(-z^k)$ with the previous polynomial, $F(-z^{k/2})$:

$$F(-z^{k/2}) = (-z^{k/2})^n + b_1'(-z^{k/2})^{n-1} + \cdots + b_{n-1}'(-z^{k/2}) + b_n'. \tag{4.184}$$

Where

$$b_{j-1} \simeq (b_{j-1}')^2 \qquad b_j \simeq (b_j')^2 \tag{4.185}$$

the coefficient b_j corresponds to a real root whose magnitude is given by equation 4.182. Where

$$b_{j+1} \simeq (b_{j+1}')^2 \qquad b_j \not\simeq (b_j')^2 \qquad b_{j-1} \simeq (b_{j-1}')^2 \tag{4.186}$$

the coefficients b_j and b_{j+1} correspond to a pair of conjugate complex or imaginary roots, whose magnitudes are given by equation 4.183. When several "nonsquares" are enclosed between two "squares," then several pairs of conjugate roots of approximately the same magnitude are represented.

Example 4.35. Use Graeffe's method to determine the actual magnitudes of the roots of the polynomial $Q_1(z) = z^5 + 3z^4 + 4z^3 + 2z^2 + z + 1$.

Answer: Following the algorithm of equations 4.179, we have the coefficients of

$Q_1(z)$: $a_1=3,\ a_2=4,\ a_3=2,\ a_4=1,\ a_5=1\ (n=5)$

$F(-z^2)$: $b_1=9-8=1,\ b_2=16-12+2=6,\ b_3=4+36+2=42,$
$b_4=1-4=-3,\ b_5=1$

$F(-z^4)$: $b_1=1-12=-11,\ b_2=36-4-6=26,\ b_3=4+36+2=42,$
$b_4=9-4=5,\ b_5=1$

$F(-z^8)$: $b_1=121-52=69,\ b_2=676+924+10=1610,$
$b_3=1764-260-22=1482,\ b_4=25-84=-59,\ b_5=1$

$F(-z^{16})$: $b_1=1541,\ b_2=2.38746\times 10^6,\ b_3=2.38644\times 10^6,\ b_4=517,\ b_5=1$

$F(-z^{32})$: $b_1=-2.400\ldots\times 10^6,\ b_2=5.69263\ldots\times 10^{12},$
$b_3=5.69263\ldots\times 10^{12},\ b_4=4.505\ldots\times 10^6,\ b_5=1.$

At this point, we can see that

$$\begin{aligned}
b_0 &= 1 = (b_0')^2\\
b_1 &= -2.4\ldots\times 10^6 \not\simeq 1541^2 = (b_1')^2\\
b_2 &= 5.69263\ldots\times 10^{12} \simeq (2.3874\ldots\times 10^6)^2 = (b_2')^2\\
b_3 &= 5.69263\ldots\times 10^{12} \simeq (2.3864\ldots\times 10^6)^2 = (b_3')^2\\
b_4 &= 4.505\ldots\times 10^6 \not\simeq (517)^2 = (b_4')^2\\
b_5 &= 1 = (b_5')^2.
\end{aligned}$$

Therefore, b_1 and b_2 correspond to a pair of conjugate roots, r_1 and r_2, whose magnitude is given by

$$|r_1|=|r_2|\simeq(5.69263\ldots\times 10^{12})^{1/64}=1.58234\ldots$$

b_3 corresponds to a real root, r_3, whose magnitude is given by

$$|r_3|\simeq(1)^{1/32}=1$$

and b_4 and b_5 correspond to a pair of conjugate roots, r_4 and r_5, whose magnitude is given by

$$|r_5|=|r_4|\simeq(1/5.69263\ldots\times 10^{12})^{1/64}=0.63197\ldots.$$

4.15. Network Responses to More Complicated Input Patterns

Consider a hypothetical flow variable whose temporal pattern is given by $J(\tau)$, and which is injected at a designated input point in a network. According to equation 4.12, the corresponding response of a particular output variable of that network simply is the convolution of the input pattern with the response of the same output variable to a unit cohort applied at the designated input point (i.e., convolution of the input pattern with the *appropriate* unit-cohort response function, $f(\tau)$):

$$X_{\text{out}}(\tau) = J(\tau) * f(\tau). \tag{4.187}$$

Of course, one of the major advantages of the z-transform is its simplification of the convolution process:

$$\mathscr{X}_{\text{out}}(z) = \mathscr{J}(z)\,\mathscr{f}(z). \tag{4.188}$$

In general, both $\mathscr{J}(z)$ and $\mathscr{f}(z)$ can be expressed as ratios of polynomials in z:

$$\mathscr{f}(z) = R(z)/Q(z) \tag{4.189}$$

$$\mathscr{J}(z) = A(z)/B(z) \tag{4.190}$$

and

$$\mathscr{X}_{\text{out}}(z) = A(z)\,R(z)/B(z)\,Q(z). \tag{4.191}$$

The temporal pattern of the output, $X_{\text{out}}(\tau)$, can be found by the methods of sections 4.14 and 4.10: First the roots of the product polynomial $B(z)\,Q(z)$ are found, and then the inverse transform of $\mathscr{X}_{\text{out}}(z)$ is found by partial-fraction expansion. Clearly, the roots of $B(z)\,Q(z)$ will be the roots of $B(z)$ and the roots of $Q(z)$. Therefore, in general, $\mathscr{X}_{\text{out}}(z)$ can be expressed as follows:

$$\begin{aligned}\mathscr{X}_{\text{out}}(z) = c_{00} &+ c_{11}\,z/(z-r_{b1}) + c_{21}\,z/(z-r_{b2}) + \cdots + c_{n1}\,z/(z-r_{bn})\\ &+ c'_{11}\,z/(z-r_{q1}) + c'_{21}\,z/(z-r_{q2}) + \cdots + c'_{m1}\,z/(z-r_{qm})\\ &+ \text{terms representing multiplicities of roots in } B(z)Q(z)\end{aligned} \tag{4.192}$$

where r_{bi} is the ith distinct root of $B(z)$ and r_{qi} is the ith distinct root of $Q(z)$.

4.15.1. The Natural Frequencies of a Constant-Parameter Network Model

Clearly, the roots of $B(z)$ depend completely upon the specific temporal pattern assumed by the input flow. The roots of $Q(z)$, on the other hand, reflect inherent dynamic properties of the network itself and therefore have a very general significance that transcends the specifics of input pattern. The roots of $Q(z)$ thus provide an essentially universal characterization of a constant-parameter network. In continuous-time models, the analogous entities are called the "natural frequencies" of the network model, the typical unit of imaginary roots and imaginary parts of complex roots being one radian per second (one radian per second equals $1/2\pi$ cycles per second) and the typical unit of the real roots and the real parts of complex roots being one *neper* per second (one neper per second equals one e-folding per second). The term "natural frequency" applies equally well to a root of $Q(z)$ for discrete-time networks, but the units are modified slightly. The roots of $Q(z)$ become the common ratios of terms in $X_{\text{out}}(\tau)$. The *magnitude* (sometimes called the "modulus" or "absolute value") of a root is the amplification (or attenuation) per transition interval of the corresponding term in $X_{\text{out}}(\tau)$, and the *angle* (sometimes called the "argument" or "amplitude") of a root is the amount by which the angle of the corresponding term in $X_{\text{out}}(\tau)$ changes per transition interval. The term in $X_{\text{out}}(\tau)$ will have undergone a complete cycle when its angle has changed by 360 deg (i.e., 2π rad).

The frequency of complete cycles (given in cycles per transition interval) therefore is the angle of the root divided by 360 (or 2π). Thus, the magnitude of the root of $Q(z)$ corresponds to the real part of a natural frequency in continuous-time models; and the angle of the root of $Q(z)$ corresponds to the imaginary part of a continuous-time natural frequency. The unit of the root's angle is *one degree per transition interval* or *one radian per transition interval*; and the unit of the root's magnitude can be conceived as *one folding per transition interval* (thus a magnitude m represents m-folding per transition interval).

The natural frequencies of a constant-parameter model also go by other names, such as "the poles" of the model, "the eigenvalues" of the model, and "the roots (or zeros) of the characteristic equation (or polynomial)" of the model. The last concept already was introduced in section 1.14. Clearly, for a discrete-time, constant-parameter network, the characteristic polynomial is $Q(z)$, and the characteristic equation is $Q(z)=0$. Mason's rule provides a simple means by which the characteristic polynomial may be derived:

CHARACTERISTIC POLYNOMIAL OF A CONSTANT-PARAMETER NETWORK

$$\begin{aligned} Q'(z^{-1}) &= (1-L_1)(1-L_2)\dots(1-L_m)^* \\ Q(z) &= z^k q(1-L_1)(1-L_2)\dots(1-L_m)^* \end{aligned} \tag{4.193}$$

where all terms containing products of touching loops are dropped. This polynomial embodies the quintessence of all of the inherent dynamic properties of the constant-parameter network for which it was derived.

4.15.2. Exciting and Observing the Natural Frequencies of a Network

Although the natural frequencies of a network represent the essence of its inherent dynamic behavior, it nevertheless is quite possible that some or all of those natural frequencies will be absent from a particular output response. In other words, the roots of $Q(z)$ may or may not appear as common ratios in a particular response. A question of concern to network and systems modelers is, "Under what circumstances do the natural frequencies of a network appear in a response, and under what circumstances do they not?" To answer this question, one simply can inspect equation 4.191:

$$\mathscr{X}_{\text{out}}(z) = A(z)R(z)/B(z)Q(z).$$

Clearly, if there are no polynomial factors common to the numerator, $A(z)R(z)$ and the denominator, $B(z)Q(z)$, i.e., if the greatest common divisor of the numerator and the denominator is simply a pure number, then all of the roots of $B(z)$ *and* $Q(z)$ will be represented by terms in the partial-fraction expansion of $\mathscr{X}_{\text{out}}(z)$ and therefore will appear as common ratios in $X_{\text{out}}(\tau)$. Furthermore, if all polynomial factors common to the numerator and denominator are contained in $B(z)$, then all of the roots of $Q(z)$ will appear as common ratios in the response.

Thus our question can be abstracted as follows: "Under what circumstances might the numerator and denominator of $\mathscr{X}_{\text{out}}(z)$ contain common polynomial

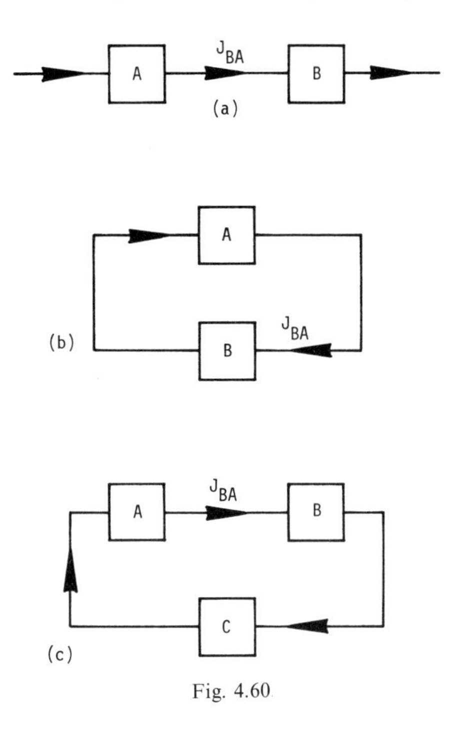

Fig. 4.60

factors that are not contained in $B(z)$?" There are three such circumstances, two related to the structure of the network model and one related to the values of the parameters of the network model and the input flow pattern. Let us begin by considering a structural circumstance related to the input that is being used to produce X_{out}. Inputs to a constant-parameter, large-numbers network model fall into two importantly different categories, namely (1) those inputs that *do not* alter the structure of the network, and (2) those inputs that effectively do alter the network's structure. Category 1 comprises all inputs that can be represented as injection into the network of a flow whose temporal pattern is completely independent of any variables internal to the network. The second category comprises all other inputs, namely those that can be represented as injection of flow patterns *determined by one or more variables internal to the network.* The unit-cohort input belongs to category 1. In fact any predetermined pattern of input flow belongs to category 1. Similarly, any pattern of flow generated by one network and injected as input into a second will belong to category 1 as long as no flow from the second is injected directly or indirectly back into the first. Thus, for example, the flow J_{BA} in Figure 4.60a is a category-1 input to network B, whereas J_{BA} in Figures 4.60b and c is a category-2 input to network B. In general, the numerator, $A(z)$, of a category-2 input shares polynomial factors in common with $Q(z)$ and not found in $B(z)$. Therefore, a category-2 input generally fails to excite some or all of the natural frequencies of a network. For this reason, category-1 inputs usually are more interesting from a scientist's point of view, since they tend to excite rather than mask the inherent dynamic properties of a

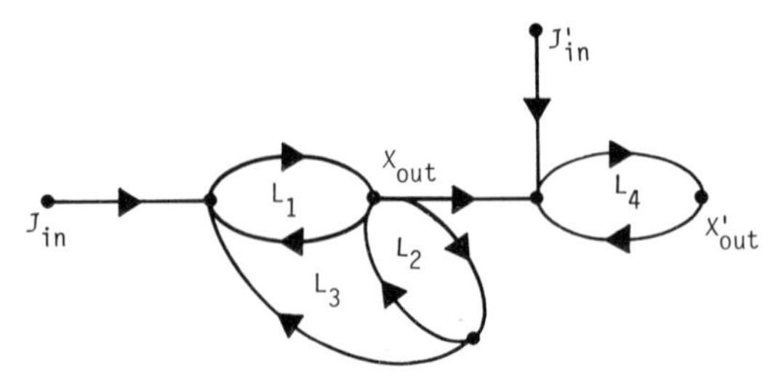

Fig. 4.61. Reducible network comprising two irreducible subnetworks ($L_1L_2L_3$ and L_4)

system. From the point of view of an engineer, a resource manager, or the like, on the other hand, category-2 inputs usually are the more interesting, since they often represent the best means whereby a system may be controlled or managed.

The second structural circumstance under which natural frequencies may be absent from a response can be derived rather easily from Mason's rule:

$$R(z)/Q(z)=[(P_1+P_2+\cdots+P_p)(\mathbb{1}-L_1)(\mathbb{1}-L_2)\ldots(\mathbb{1}-L_m)]^{**}/[(\mathbb{1}-L_1)(\mathbb{1}-L_2)\ldots(\mathbb{1}-L_m)]^{*} \tag{4.194}$$

where numerator terms are dropped if they contain products of loops that touch or the product of a path with a loop that touches it, and denominator terms are dropped if they contain products of loops that touch. The touching of loops with other loops is a structural property of a network itself, and is completely independent of the choice of input and output variables (i.e., where input flow is injected and which variable is observed in response). If every member of a set of loops touches at least one other member of the same set, then that set of loops represents an irreducible Markov chain (i.e., a collection of Markovian states, every one of which can be reached directly or indirectly from every other one, see section 1.9). Therefore, a network comprising such a set of loops is designated an *irreducible network* or an *irreducible subnetwork*. An actual network model may consist of a concatenation of several irreducible subnetworks. For example, the model of Figure 4.61 is made up of two irreducible subnetworks, one comprising loops L_1, L_2 and L_3, and the other comprising L_4.

If a network model is not irreducible, then it always is possible to select an input flow and an output variable such that one or more natural frequencies of the network will be absent from the output response. One simply selects the input and output variables in such a way that one or more of the irreducible subnetworks is not touched by any path. That subnetwork will be represented by the same polynomial factor in $R(z)$ and $Q(z)$, and that factor therefore will be cancelled by division. Thus, for example, in Figure 4.61 the irreducible subnetwork comprising L_4 does not touch the path from J_{in} to X_{out}. Therefore, the factor $(\mathbb{1}-L_4)$ appears both in the numerator and the denominator of the appropriate unit-cohort response; and the corresponding natural frequencies will be absent from $X_{out}(\tau)$. Their absence is quite fitting and proper, since the effects of L_4 are hidden from X_{out} and L_4 therefore is *structurally isolated* from the portion of the network generating X_{out}. Similarly, the path from J'_{in} to X'_{out}

does not touch the subnetwork containing L_1, L_2 and L_3; and the natural frequencies corresponding to those loops are absent from X'_{out} in response to J'_{in}. In this case J'_{in} is *structurally isolated* from L_1, L_2 and L_3 and is incapable of exciting those loops and their natural frequencies. *Structural isolation* leads to the absence of natural frequencies in responses of networks that are not irreducible.

Generally, if a network model is irreducible and is stimulated by category-1 inputs, then all of its natural frequencies can be expected to appear in its response. However, one or more of the natural frequencies may in fact be absent, owing to very specific relationships (perhaps fortuitous, perhaps by design) among the parameters of the network or between the parameters of the network and those of the input flow pattern. Thus, for example, one might specify independently an input pattern that just happened to match a category-2 pattern for the network, in which case one or more natural frequencies would be absent from the response. Similarly, one might select the parameters of an irreducible network model in such a way that $R(z)$ and $Q(z)$ shared common, polynomial factors. In doing so, he or she, either deliberately or accidently, effectively would isolate some portion of the network with respect to the input or the output. Euclid's algorithm applied to $R(z)$ and $Q(z)$ provides a simple test for such isolation. Since $R(z)$ depends specifically on the choices of input and output variables, this form of isolation also depends upon those choices.

A network is called *controllable* with respect to a particular input variable if all of its natural frequencies (i.e., all of its loops) can be excited by a proper choice of category-1 pattern (i.e., one that does not fortuitously happen to match a category-2 pattern); and it is called *observable* with respect to a particular output variable if all of its natural frequencies, when excited, can be seen in the dynamics of that variable. If Euclid's algorithm fails to uncover any polynomial factors in common to $R(z)$ and $Q(z)$, then the network is both observable and controllable with respect to the selected input and output variables. In the network of Figure 4.61, input J'_{in} is incapable of exciting loops L_1, L_2 and L_3; so the network is not controllable with respect to J'_{in}. Furthermore, the natural frequencies elicited by excitation of loop L_4 are not seen in the dynamics of X_{out}; so the network is not observable with respect to X_{out}. Whether the network is controllable with respect to J_{in} and observable with respect to X'_{out} depends upon its parameter values and can be determined by Euclid's algorithm.

If a network is controllable with respect to a particular input, observable with respect to a particular output, and the input pattern belongs to category 1 (and does not fortuitously match a category-2 pattern), then all of the network's natural frequencies will appear in the dynamics of the output.

4.15.3. Responses to Category-1 Inputs

In general, one can expect the responses of any network to a category-1 input to exhibit some or all of the network's natural frequencies *plus* any additional frequencies present in the input itself. The calculation of the complete response, with all its components, can be visualized as follows: The input flow is decomposed into single cohorts, one for each transition interval. Each input cohort elicits a response pattern, which becomes part of the total response,

$X_{out}(\tau)$. To obtain $X_{out}(\tau)$, one simply sums (*superposes*) the individual cohort responses, each of which simply is the unit-cohort response scaled by the actual size of the corresponding input cohort and delayed to correspond to the time of arrival of that input cohort (see section 4.6). Although the process of superposition seems rather complicated in the time domain, it becomes quite simple in the z-transform domain, where it is executed by ordinary multiplication of the transformed input pattern and the unit-cohort response (equation 4.188). To deduce the response dynamics, one simply transforms the input pattern, carries out the multiplication, and then finds the inverse transform by the methods already described in sections 4.10 and 4.14. Occasionally, the input pattern will be the unit-cohort response of another network, in which case the transform of the pattern is available directly from the analysis (e.g., by Mason's rule) of that network. On the other hand, the pattern may be postulated independently by the modeler, in which case he or she will need to be able to find the pattern's transform.

For this purpose, equation 4.14 may be used, in which case the pattern tacitly is taken to be a sequence of single cohorts. On the other hand, the input pattern actually may comprise sums or differences of geometric sequences of the form given in the right-hand side of equation 4.59. In that case, the transformation of equation 4.14 is not the most convenient form; it can be replaced by terms of the form of the right-hand side of equation 4.58 (i.e., by the *generating functions* of the component sequences in the input pattern). Of course, recognition of complicated sums and differences of geometric sequences may not be an easy task. On the other hand, if the modeler is free to postulate the input pattern, then he or she should be free to design it to conform to a known combination of sequences, with known z transform. A list of standard transforms, such as that in Table 4.3, often is useful in the selection or design of postulated input patterns.

Example 4.36. A sports fisheries manager plans to stock a small pond. Taking into account the estimated fishing pressure and other mortality factors, she models the population with a temporal resolution of one month. Linearized in the neighborhood of the zero critical level (i.e., for low density population), the model reduces to a constant-parameter network with the following transformed unit-cohort response:

$$\mathcal{N}(z) = 1/(z - 0.9)$$

where N is the fish population and the transition interval is one month. Because the root (0.9) is less than 1.0, the manager deduces that the stocked population will die away geometrically with time unless it is supplemented from time to time with planted fish. Therefore, she proposes a periodic planting scheme, whereby 10,000 fish are introduced in a single batch every four months. Assuming the original stocking is 20,000 fish, and that the supplementary planting begins four months later, deduce the dynamics of the modeled population under the proposed scheme.

Answer: Letting $\mathcal{N}'(z)$ be the z transform of the fish population and $\mathcal{I}_{in}(z)$ be the transformed input pattern (comprising initial stock plus periodic plantings),

Table 4.3. z transforms of standard input patterns

Pattern description	Time function	Transformed function
1. Single cohort of size N at interval k	$J(\tau)=N \quad \tau=k$ $J(\tau)=0 \quad \tau\neq k$	$\mathcal{J}(z)=Nz^{-k}$
2. Constant input flow, N per interval, beginning at interval k	$J(\tau)=0 \quad \tau<k$ $J(\tau)=N \quad \tau\geqq k$	$\mathcal{J}(z)=Nz^{-k+1}/(z-1)$
3. Ramp (staircase) input flow, slope A individuals per interval per interval, beginning at interval k	$J(\tau)=0 \quad \tau<1$ $J(\tau)=A(\tau-k) \quad \tau\geqq k$	$\mathcal{J}(z)=Az^{-k+1}/(z-1)^2$
4. Pulse of N individuals per interval beginning at interval k_1 and ending at interval k_2	$J(\tau)=0 \quad \tau<k_1$ $J(\tau)=N \quad k_1\leqq\tau\leqq k_2$ $J(\tau)=0 \quad \tau>k_2$	$\mathcal{J}(z)=Nz(z^{-k_1}-z^{-k_2})/(z-1)$
5. Cohort of size N applied periodically, every kth interval, beginning with interval k_1	$J(\tau)=0 \quad \tau<k_1$ $J(\tau)=N \quad \tau-k_1=nk$, $n=$integer $J(\tau)=0$ otherwise	$\mathcal{J}(z)=Nz^{k-k_1}/(z^k-1)$
6. Cosinusoidally varying input flow, amplitude A, period K intervals, beginning at interval k with phase ψ	$J(\tau)=0 \quad \tau<k$ $J(\tau)=A\cos(\omega\tau+\psi) \quad \tau\geqq k$ $\omega=2\pi/K$ radians/interval	$\mathcal{J}(z)=2Az^{-k+1}\dfrac{z\cos\psi-\cos(\omega-\psi)}{z^2-2z\cos\omega+1}$

we have

$$\mathcal{N}'(z)=[1/(z-0.9)]\,\mathcal{J}_{\text{in}}(z).$$

Evidently, $J_{\text{in}}(\tau)$ comprises a single initial cohort of 20,000 (i.e., pattern 1 in Table 4.3, with $N=20{,}000$ and $k=0$),

$$\mathcal{J}'(z)=20{,}000$$

and a cohort of size 10,000 applied every 4th interval beginning with the 4th interval (i.e., pattern 5 in Table 4.3, with $N=10{,}000$, $k=4$, and $k_1=4$),

$$\mathcal{J}''(z)=10{,}000/(z^4-1).$$

Thus,

$$\mathcal{J}_{\text{in}}(z)=\mathcal{J}'(z)+\mathcal{J}''(z)=20{,}000+10{,}000/(z^4-1)$$

and

$$\mathcal{N}'(z)=[1/(z-0.9)][20{,}000+10{,}000/(z^4-1)]$$

from which

$$\mathcal{N}'(z)=[20{,}000\,z^4-10{,}000]/[(z-0.9)(z^4-1)].$$

Since the roots of $z^4 - 1$ are 1, -1, $\boldsymbol{i}$, and $-\boldsymbol{i}$, we have the following partial-fraction expansion of $N'(z)$:

$$\begin{aligned}\mathcal{N}'(z) = {} & c_{00} + c_{11} z/(z - 0.9) + c_{21} z/(z - 1) + c_{31} z/(z + 1) \\ & + c_{41} z/(z + \boldsymbol{i}) + c_{41}^{*} z/(z - \boldsymbol{i})\end{aligned}$$

where

$$\begin{array}{lll} c_{00} = -11{,}111.11\ldots & c_{11} = -10{,}086.91\ldots & c_{21} = 25{,}000 \\ c_{31} = -1{,}315.78\ldots & c_{41} = -1{,}234.09\ldots - \boldsymbol{i}(1{,}381.21\ldots) & \\ c_{41}^{*} = -1{,}243.09\ldots + \boldsymbol{i}(1{,}381.21\ldots). & & \end{array}$$

Therefore,

$$\begin{aligned} N'(0) &= 0 \\ N'(\tau) &= 25{,}000 - 1{,}315.78\ldots(-1)^{\tau} - (3{,}716.45\ldots)\cos[\pi\tau/2 + 0.8379\ldots \text{ rad}] \\ &\quad - 10{,}086.91\ldots(0.9)^{\tau} \qquad \tau > 0 \end{aligned}$$

from which

$$\begin{array}{lll} N'(1) = 20{,}000 & N'(2) = 18{,}000 & N'(3) = 16{,}200 \\ N'(4) = 14{,}580 & N'(5) = 23{,}122 & N'(6) = 20{,}809.8\ldots \\ N'(7) = 18{,}728.8\ldots & N'(8) = 16{,}855.9\ldots & N'(9) = 25{,}170.3\ldots \\ N'(10) = 22{,}653.3\ldots & N'(11) = 20{,}387.9\ldots & N'(12) = 18{,}349.1\ldots \\ N'(13) = 26{,}514.2\ldots & N'(14) = 23{,}862.8\ldots & N'(15) = 21{,}476.5\ldots \\ N'(16) = 19{,}328.8\ldots & N'(17) = 27{,}396.0\ldots & \ldots. \end{array}$$

Ultimately, the term $-10{,}086.91\ldots(0.9)^{\tau}$ will die away, and the four-month cycle will approach

$$\begin{array}{ll} N'(4n+1) = 29{,}078.2\ldots & N'(4n+2) = 26{,}170.4\ldots \\ N'(4n+3) = 23{,}553.3\ldots & N'(4n) = 21{,}198.0\ldots. \end{array}$$

Up to this point, our z-transform analysis of constant-parameter networks has been based on the explicit assumption that the deviations from the critical level of all of the accumulations and flows, other than the input flow, initially are zero. Thus, for example, if a model has been linearized in the neighborhood of the zero critical level, then the tacit assumption becomes that of zero initial flow and accumulation values. Dynamics based on the assumption of zero initial deviations very often are labelled *zero-state dynamics* by systems theorists. Fortunately, the assumption is not a particularly restrictive one. In fact, in order to convert zero-state dynamics into the dynamics appropriate to specific, nonzero initial deviations from the critical level, one simply postulates input flows that provide the necessary initial deviations. As long as we are free to postulate input flow patterns injected anywhere in the network, any initial distribution of nonzero deviations can be established.

Example 4.37. Suppose that the pond of the previous example had an indigenous population of fish, and that the model of this population (with one-month transition interval) exhibited the unit-cohort response

$$\mathcal{N}(z)=1/(z-0.9)$$

indicating that the population would die away if not augmented by planted fish. The manager decides to plant 10,000 fish every four months, with the first planting taking place when the indigenous population is 20,000. Deduce the resulting dynamics of the modeled population.

Answer: The analysis is carried out in the same manner as it was in Example 4.36. The input flow once again has two components, one representing the periodic planting beginning at the initial interval

$$\mathcal{J}''(z)=10{,}000\ z^4/(z^4-1)$$

and the second designed to provide an initial population of 20,000. Now we have a small problem. By virtue of the way we have defined flows and accumulations in discrete time (see section 2.8), a flow applied at the initial interval ($\tau=0$) will not produce an accumulation until the subsequent interval ($\tau=1$). What we want, however, is to establish an initial accumulation of 20,000 at interval $\tau=0$. To do so, we must apply a component of input flow of 20,000 at interval $\tau=-1$. With very little reflection concerning the definition of the z-transform, the reader should be able to see that the transform of this component of the input flow must be

$$\mathcal{J}'(z)=20{,}000\ z.$$

Thus,

$$\mathcal{J}_{\text{in}}(z)=\mathcal{J}'(z)+\mathcal{J}''(z)=20{,}000\ z+10{,}000\ z^4/(z^4-1)$$

and

$$\mathcal{N}'(z)=[20{,}000\ z^5+10{,}000\ z^4-20{,}000\ z]/(z-0.9)(z^4-1)$$

from which

$$N'(\tau)=25{,}000-1{,}315.78\ldots(-1)^\tau-(3{,}716.45\ldots)\cos[\pi\tau/2+0.8379\ldots\text{ rad}]$$
$$-1{,}198.02(0.9)^\tau$$

$$N'(0)=20{,}000\quad N'(1)=28{,}000\quad N'(2)=25{,}200$$
$$N'(3)=22{,}680\quad N'(4)=20{,}412\quad N'(5)=28{,}371..$$
$$N'(6)=25{,}533..\quad N'(7)=22{,}980..\quad N'(8)=20{,}682..$$
$$N'(9)=28{,}614..\quad N'(10)=25{,}752..\quad N'(11)=23{,}177..$$
$$\vdots$$
$$N'(4n)=21{,}198\quad N'(4n+1)=29{,}078\quad N'(4n+2)=26{,}170$$
$$N'(4n+3)=23{,}553.$$

Example 4.38. In Example 4.25, the transformed unit-cohort response for small excursions from the nonzero critical level was

$$\Delta\mathcal{J}(z)=z^2/(z^2-A^2)$$

where

$$A^2=1-\log_e[\alpha\gamma_0\gamma'].$$

Deduce the dynamics of the model for various values of A, when the initial perturbation is

$$\Delta J(0)=-1$$

(i.e., one individual less than the critical level at $\tau=0$).

Answer: To establish a perturbation of flow at $\tau=0$, we simply must inject the appropriate flow as input at that *same* interval. In this case, the input flow should be -1 at $\tau=0$:

$$\mathcal{J}_{\text{in}}(z)=-1.$$

Applying this, we have

$$\Delta\mathcal{J}(z)=-z^2/(z^2-A^2)$$

from which

$$\Delta\mathcal{J}(z)=-(1/2)\,z/(z-A)-(1/2)\,z/(z+A)$$

and

$$\Delta J(\tau)=-(1/2)\,A^{\tau}-(1/2)(-A)^{\tau}$$

or

$$\Delta J(0)=-1,\quad \Delta J(1)=0,\quad \Delta J(2)=-A^2,\quad \Delta J(3)=0,\quad \Delta J(4)=-A^4,$$
$$\Delta J(5)=0,\quad \Delta J(6)=-A^6,\quad \Delta J(7)=0,\quad \Delta J(8)=-A^8,\quad \Delta J(9)=0,\ldots.$$

Clearly, when the magnitude of A is less than 1.0, the dynamics converge on the critical level; when the magnitude of A is greater than 1.0, the dynamics diverge from it. When A is real, convergence or divergence is single-sided, below the critical level. When A is imaginary, convergence or divergence is oscillatory, with the dynamics alternating from above to below the critical level.

4.15.4. The Initial- and Final-Value Theorems, Steady-State Analysis

There is a very simple test that can be applied to a z-transform function to determine the value of its inverse transform at $\tau=0$. One can employ this test on any network variable to determine whether or not the selected transformed input flow has, in fact, established the desired initial value. The test itself is known as the *Initial-Value Theorem*, and can be stated quite simply as follows:

INITIAL-VALUE THEOREM

$$f(\tau=0)=\lim_{z\to\infty}\mathcal{f}(z). \qquad (4.195)$$

In other words, the value of $f(\tau)$ at $\tau=0$ is equal to the value approached by $\mathcal{f}(z)$ as z becomes indefinitely large.

Example 4.39. Test the initial values of the transformed response functions of Examples 4.37 and 4.38:

$$\mathcal{N}'(z) = [20{,}000\, z^5 + 10{,}000\, z^4 - 20{,}000\, z]/(z-0.9)(z^4-1)$$
$$\Delta\mathcal{J}(z) = -z^2/(z^2 - A^2).$$

Answer: As z becomes indefinitely large, the terms of highest degree in z in the numerator and denominator polynomials of each transformed response function become overwhelmingly dominant. Thus

$$\mathcal{N}'(z) \text{ approaches } 20{,}000\, z^5/z^5 = 20{,}000$$

and

$$\Delta\mathcal{J}(z) \text{ approaches } -z^2/z^2 = -1.$$

Therefore, the initial values of the time functions are

$$N'(0) = 20{,}000$$
$$\Delta J(0) = -1.$$

When the unit-cohort response, $\mathcal{f}(z)$, of a network exhibits no roots of magnitude greater than or equal to 1.0 in its denominator, then the corresponding function of time, $f(\tau)$, eventually will decline to zero. By the same token, the corresponding output function,

$$X_{\text{out}}(\tau) = f(\tau) * J_{\text{in}}(\tau)$$

also eventually will decline to zero unless J_{in} provides continual augmentation. Thus, for example, $f(\tau)$ might be the unit-cohort response of the fish population of Examples 4.36 and 4.37; unless the flow of planted fish into the pond continually augments the population, the population eventually will decline to zero. If the input flow takes the form of pattern 2, pattern 3, pattern 5, or pattern 6 in Table 4.3, then, eventually, the output variable will assume the same pattern. This follows simply from

$$X_{\text{out}}(\tau) = c_{00} + c_{11}(r_{b1})^\tau + c_{21}(r_{b2})^\tau + \cdots + c'_{11}(r_{q1})^\tau + c'_{21}(r_{q2})^\tau + \cdots \quad (4.196)$$

which is the inverse transform of the right-hand side of equation 4.192, where r_{bi} is the ith common ratio of the input function and r_{qi} is the ith natural frequency of the network model. Since, for the situation under discussion here,

$$|r_{qi}| < 1.0$$

and for patterns 2, 3, 5 and 6,

$$|r_{bi}| = 1.0$$

the terms in r_{bi} eventually will dominate.

Now, in some situations the modeler may be interested in the ultimate dynamics represented by equation 4.196, and have no interest in the transient phenomena taking place as those ultimate dynamics are being established. When the common ratios of the ultimate dynamics have magnitudes equal to 1.0, then those dynamics themselves often are referred to as the "steady-state" response. The dynamics prior to establishment of steady state often are referred to simply as "transients." To deduce the steady-state response of a network model, one merely must determine the common ratios of magnitude 1.0 in the input function and the coefficients corresponding to each of them in the partial-fraction expansion. This can be done without knowledge of the natural frequencies of the network model (i.e., without going through the tedious procedure of finding the roots of $Q(z)$), *as long as one is certain that none of those natural frequencies has magnitude equal to or greater than* 1.0 (which, of course, can be determined by application of the Routh-Hurwitz algorithms or the Jury test). The procedures for finding the coefficient are simply those of section 4.10.2:

$$c_{jk}=[(z-r_{bj})/z]^k \mathscr{J}_{\text{in}}(z)\mathscr{f}(z)|_{z=r_{bj}} \tag{4.197}$$

and so forth.

When J_{in} happens to conform to pattern 2 of Table 4.3, then the steady-state response has a single term

$$c_{11}(1)^\tau = c_{11}$$

where

$$c_{11}=[(z-1)/z]\,\mathscr{J}_{\text{in}}(z)\,\mathscr{f}(z)|_{z=1} \tag{4.198}$$

which is especially easy to evaluate. In this case, c_{11} is the ultimate (final) value assumed by the inverse transform of $\mathscr{J}_{\text{in}}(z)\mathscr{f}(z)$. Equation 4.198 often is cited under the label "Final-Value Theorem," as follows: If $G(\tau)$ approaches a unique, finite value, c_{11}, as time becomes indefinitely large, then that value is given by

FINAL-VALUE

$$\lim_{\tau\to\infty} G(\tau)=c_{11}=[(z-1)/z]\,\mathscr{G}(z)|_{z=1}. \tag{4.199}$$

Example 4.40. Carry out a steady-state analysis for the situation depicted in Example 3.47.

Answer: The input in this case is pattern 5 with $k=4$, which has four common ratios of magnitude 1.0: 1, -1, i, $-i$. The corresponding coefficients, determined by equation 4.197, are 25,000, $-1{,}315.78\ldots$, $-1{,}243.09\ldots - i(1{,}381.21\ldots)$, and $-1{,}243.09\ldots + i(1{,}381.21\ldots)$ respectively. Combining the terms representing the conjugate pair of common ratios, we have the following steady-state response:

$$N'(\tau)|_{ss}=25{,}000-1{,}315.78\ldots(-1)^\tau-(3{,}716.45\ldots)\cos[\pi\tau/2+0.8379\ldots]$$

or, as in Example 4.37,

$$N'(4n) = 21{,}198, \quad N'(4n+1)=29{,}078,$$
$$N'(4n+2)=26{,}170, \quad N'(4n+3)=23{,}553.$$

Example 4.41. Find the steady-state response of the system of Example 4.38, with $A=0.25$, driven by a cosinusoidal input of amplitude B and frequency ω radians per interval.

Answer: In this case, the z transform of the input has the form

$$\mathscr{I}(z)=2Bz(z-\cos\omega)/(z^2-2z\cos\omega+1)$$

the roots of whose denominator are

$$r=\cos\omega+i\sin\omega \quad \text{and} \quad r^*=\cos\omega-i\sin\omega.$$

The coefficient corresponding to r in the partial-fraction expansion is given by

$$c=[(z-r)/z]\,\mathscr{I}(z)f(z)|_{z=r} \tag{4.200}$$

where $f(z)$ is the transformed unit-cohort response of the model. Since r is a root of the denominator of $\mathscr{I}(z)$, we can write equation 4.200 as follows:

$$c=\{[(z-r)/z]\,\mathscr{I}(z)|_{z=r}\}f(r).$$

But, in this case,

$$[(z-r)/z]\,\mathscr{I}(z)=B.$$

Therefore,

$$c=Bf(r)=Bf(r^*) \tag{4.201}$$

or,

$$c=Bf(\cos\omega+i\sin\omega). \tag{4.202}$$

In other words, to find the steady-state response to a pure consinusoidal input, one simply substitutes $\cos\omega+i\sin\omega$ for z in the transformed unit-cohort response and multiplies the resulting *steady-state transfer function* by the amplitude of the input cosine wave.

In this particular case, we have

$$f(z)=z^2/(z^2-0.25)$$

and

$$f(r)=f(\cos\omega+i\sin\omega)=(\cos\omega+i\sin\omega)^2/[(\cos\omega+i\sin\omega)^2-0.25].$$

With a little manipulation, we find that

$$|f(r)|=1.0$$

$$\psi_{f(r)}=\tan^{-1}[\sin 2\omega/(4-\cos 2\omega)].$$

Indicating that the amplitude of the steady-state reponse is the same as the amplitude of the input cosine wave:

$$|c|=B|f(r)|=B$$

but the *phase* of the steady-state response is shifted by $\psi_{f(r)}$ with respect to that of the input.

4.16. Elements of Dynamic Control of Networks

Suppose that in the immediate neighborhood of a particular critical level, the behavior (i.e., divergence, convergence, oscillation) deduced from a network model is not that desired by the modeler. In other words, the hypothetical system does not behave in a manner that is satisfactory to the modeler. How might she or he alter the model's behavior so as to make it correspond to the desired behavior? There are two basic alternatives: (1) alter the parameters of the model, or (2) provide appropriate category-2 inputs to the model. Clearly, in the case of population models, such procedures represent candidate management or control strategies that might be applied to the actual population being modeled. In such cases, the parameters are those related to the life cycle (e.g., fecundity, rate of maturation, gestation period, and the like), which may be rather difficult to alter in the actual population in a systematic way. On the other hand, one in principle should be able to apply category-2 inputs of one sort or another to the actual population. The range of available inputs may or may not encompass the desired modifications.

Regardless of the method selected, the goal of the procedure can be expressed in terms of the transformed unit-cohort response of the network model linearized about the critical level in question:

$$\mathscr{f}(z)=R(z)/Q(z). \tag{4.203}$$

One or more of the roots of the polynomial $Q(z)$ are inappropriate; therefore the structure of the network must be modified so as to generate a new unit-cohort response:

$$\mathscr{f}'(z)=R'(z)/Q'(z)$$

in which all of the roots of $Q'(z)$ are appropriate.

4.16.1. Linear Control with Category-2 Inputs

Suppose that we observe a variable, $X(\tau)$, of a constant parameter network; and from $X(\tau)$ we form an input flow variable, $J_{\text{in}}(\tau)$ to be injected into the network so that it merges with existing flow, $J(\tau)$. Since $J_{\text{in}}(\tau)$ is dependent on a variable of the network, namely $X(\tau)$, it clearly is a category-2 input. Now, if $J_{\text{in}}(\tau)$ is a linear, dynamic function of $X(\tau)$,

$$J_{\text{in}}(\tau)=a_0X(\tau)+a_1X(\tau-\mathbb{1})+a_2X(\tau-\mathbb{2})+\cdots+a_nX(\mathbb{0}) \tag{4.204}$$

then its z transform simply is

$$\mathscr{J}_{\text{in}}(z)=\mathscr{A}(z)\mathscr{X}(z) \tag{4.205}$$

where

$$\mathscr{A}(z)=a_0+a_1z^{-\mathbb{1}}+a_2z^{-\mathbb{2}}+\cdots. \tag{4.206}$$

It can be generated by insertion of a secondary, constant-parameter network with transformed unit-cohort response $\mathscr{A}(z)$, providing feedback from $X(\tau)$ to $J(\tau)$, as depicted in Figure 4.62. When this is the case, J_{in} provides linear control on the original network.

By application of this linear, category-2 input, we clearly have altered the basic structure of the model. According to Mason's rule, the transform of the

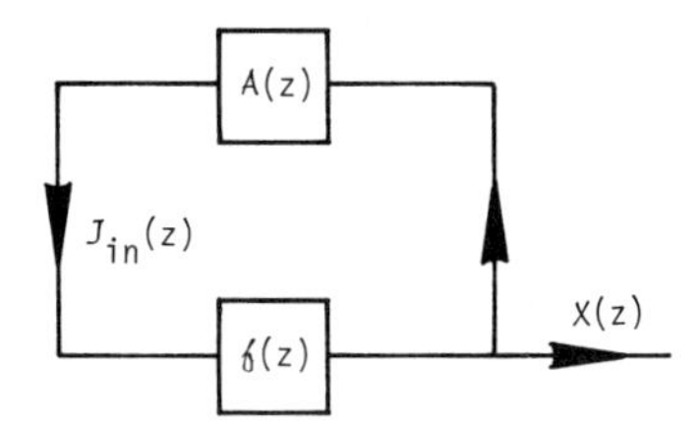

Fig. 4.62. Typical pattern for feedback control of a dynamic process. $\mathscr{f}(z)$ represents the process being controlled, and $\mathscr{A}(z)$ the feedback network providing the control

modified response of $X(\tau)$ to a unit cohort applied at J_{in} will be

$$\mathscr{f}'(z)=R'(z)/Q'(z)=\mathscr{f}(z)/[1-\mathscr{f}(z)\mathscr{A}(z)]. \tag{4.207}$$

Substituting the right-hand of equation 4.203 for $\mathscr{f}(z)$ and the right-hand side of

$$\mathscr{A}(z)=B(z)/C(z)$$

for $\mathscr{A}(z)$, we have

$$\mathscr{f}'(z)=R'(z)/Q'(z)=R(z)\,C(z)/[Q(z)\,C(z)-R(z)B(z)]. \tag{4.208}$$

The natural frequencies of the modified network are the roots of the polynomial $Q'(z)$

$$Q'(z)=Q(z)\,C(z)-R(z)B(z). \tag{4.209}$$

Therefore, in order to provide appropriate control with a linear, category-2 signal generated by feedback from $X(\tau)$ to $J(\tau)$, one simply must design a network that will produce the appropriate roots in $Q'(z)$. Thus, for example, if convergence on the critical level is desired, one would attempt to design the feedback network so that all of the roots of $Q'(z)$ have magnitudes less than 1.0. One approach is to select a specific $Q'(z)$ that has satisfactory roots, then from knowledge of $Q(z)$ and $R(z)$ design appropriate polynomials $B(z)$ and $C(z)$. The only fundamental constraint on the design procedure is that the degree of $C(z)$ must be at least as great as the degree of $B(z)$, otherwise predictive elements (*time delays* with negative durations) will be required in the feedback network, making it *unrealizable in principle.*

Example 4.42. A network model linearized about the nonzero critical level of interest has the following transformed unit-cohort response:

$$\mathscr{f}(z)=1/(z^2+5z+6)$$

whose denominator has the roots $r_1=-3$, $r_2=-2$. The deduced oscillatory divergence is undesirable. Therefore, design a feedback network to provide linear dynamic control such that the modified network exhibits the two natural frequencies $r_1=0.5$, $r_2=0.8$ and thus converges on the critical level.

Answer: The desired denominator polynomial is

$$Q'(z)=(z-0.5)(z-0.8)=z^2-1.3z+0.4.$$

From the original transformed unit-cohort response, we have

$$R(z)=1$$
$$Q(z)=z^2+5z+6.$$

Substituting into equation 4.209, we find

$$Q'(z)=z^2-1.3z+0.4=(z^2+5z+6)C(z)-B(z).$$

With a little experimentation, the reader soon will become convinced that there is no solution to this equation that conforms to the fundamental constraint (i.e., that the degree of C be at least as great as that of B). However, we can find conforming solutions to

$$Q'(z)=z^3-1.3z^2+0.4z=(z^2+5z+6)C(z)-B(z). \tag{4.210}$$

Here, we simply have added a root $r=0$, which will alter the dynamics only during the initial interval and have no effects otherwise. Now, letting

$$C(z)=z+c_0$$
$$B(z)=b_1z+b_0$$

and matching the right- and left-hand sides of equation 4.210 term by term, we find

$$\begin{aligned} 6c_0-b_0&=0\\ 6+5c_0-b_1&=0.4\\ 5+c_0&=-1.3 \end{aligned} \tag{4.211}$$

from which

$$c_0=-6.3$$
$$b_0=-37.8$$
$$b_1=-25.9$$

and

$$\mathscr{A}(z)=(-25.9z-37.8)/(z-6.3).$$

Dividing numerator and denominator by z, we have

$$\mathscr{A}(z)=(-25.9-37.8z^{-1})/(1-6.3z^{-1})$$

which can be realized as a network with

$$P_1=-25.9, \quad P_2=-37.8z^{-1}, \quad L_1=6.3z^{-1}$$

such as that depicted in Figure 4.63.

Examining equations 4.211 and their solutions, the reader will notice that two of the coefficients of the desired polynomial, $Q'(z)$, are obtained through the feedback process as *small differences of large parameters* (e.g., 6, 5×6.3, and 25.9 combining to yield 0.4). When this is the case, the coefficients of $Q'(z)$, and thus the dynamic properties of the modified network, are very sensitive to pro-

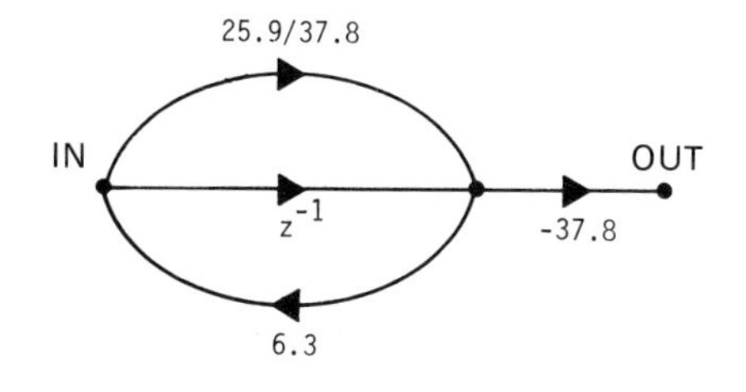

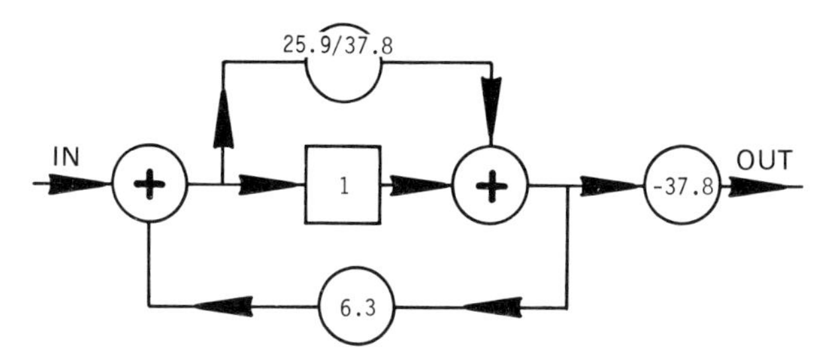

Fig. 4.63. Realization of a control network for the problem of Example 4.42

portionately small errors in the parameters either of the feedback network or of the original network. Needless to say, this may be undesirable. If the sensitivity is unacceptably high, one can try other candidate points of connection for the feedback network or other candidate roots in $Q'(z)$.

When category-2 input flow is used to modify a network in the neighborhood of a nonzero critical point, that flow generally must be expected to take on both positive and negative values. If such a flow represents a candidate control strategy to be applied to an actual population, then it must be possible to provide both positive and negative inputs at the appropriate point in that population. In other words, individuals of the appropriate state must be removable, and individuals of that same state must be available for augmentation. Clearly, the practical constraints of realization imposed by this requirement may be just as severe as the fundamental constraint of no anticipation. These difficulties may not exist for the problem of control about the zero critical level. In that case, the candidate category-2 input flow may be of one sign only (e.g., augmentation alone or removal alone).

Example 4.43. A network model linearized for excursions from the zero critical level exhibits the following transformed unit-cohort response:

$$f(z) = z^2/(z^2 - 2z - 4).$$

By application of the test of inequality 4.166 to the denominator of this function, we know that the network model exhibits natural frequencies of magnitude greater than 1.0. The resulting divergence from zero is undesirable and must be checked. Therefore, design a feedback network to provide linear dynamic control such that the modified network exhibits a single dominant natural frequency equal to 1.0 (i.e., with the magnitudes of all other natural frequencies being less than 1.0).

Answer: Applying equation 4.209, we have

$$Q'(z)=(z^2-2z-4)\,C(z)-z^2 B(z).$$

To minimize the complexity in the feedback network, we should attempt to minimize the degree of our modified polynomial. Therefore, we first should attempt a solution with $C(z)$ and $B(z)$ both equal to constants, which would lead to a $Q'(z)$ of degree two:

$$Q'(z)=a_0 z^2-a_1 z-a_2.$$

According to inequality 4.166, $|a_1/a_0|$ must be less than or equal to 2.0 and $|a_2/a_0|$ must be less than or equal to 1.0 in order for the roots of $Q'(z)$ to have magnitudes less than or equal to 1.0. We can accomplish this by setting $C(z)=1$, leading to $a_1=2$, $a_2=4$, and setting $B(z)=-K$, leading to $a_0=1+K$. By scaling K up, we can scale the magnitudes of a_1/a_0 and a_2/a_0 down:

$$Q'(z)=(1+K)\,z^2-2z-4.$$

The roots of $Q'(z)$ are

$$r_1, r_2=1/(1+K)\pm\sqrt{5}/(1+K)$$

setting $r_1=1$, we have

$$K=\pm\sqrt{5}.$$

Applying the further constaint that $|r_2|<1$, we have

$$K=\sqrt{5}$$

and

$$r_1=1, \quad r_2=-0.381966\ldots$$

which definitely meet our specifications. The feedback network in this case is extremely simple, comprising a scalor with parameter $-K$, as depicted in Figure 4.64.

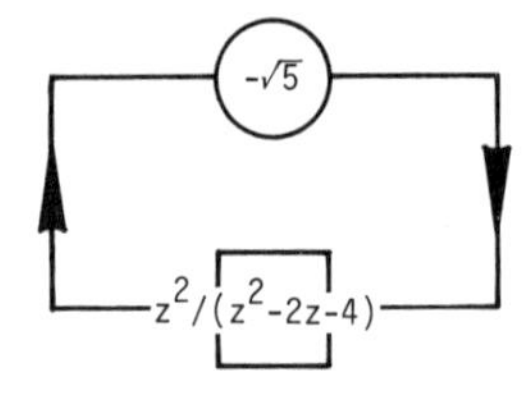

Fig. 4.64. Realization of a feedback control network for the problem of Example 4.43

The feedback networks of the previous two examples were designed to alter the natural frequencies of the original networks, and thus to *control* their dynamic behavior. If these networks represent candidate control schemes to be applied to an actual population, then one simply may view them as *control algorithms* (perhaps to be executed with pencil and paper, perhaps with the aid of a computer). The input to the feedback network (i.e., X_{out}) represents data input to the control algorithm. Obviously, the two examples presented here were very simple, so that the selection of appropriate feedback networks did not require much sophistication. On the other hand, as the original network dynamics become more complex and/or the goals of control more specific and rigid, then the design of appropriate feedback networks becomes increasingly difficult. Factors such as sensitivity to parameter variation and cost of control implementation often must be taken into account; and the design goal becomes one of finding an optimum control scheme rather than one that merely provides appropriate natural frequencies. The selected control scheme very well may involve category-2 flows derived from several variables of the original network and injected at several locations in that network. Further information on the art of linear dynamic control can be found in any of numerous texts on the subject.

4.16.2. Comments on Nonlinear Control and Regulation

Occasionally it is the location of the critical level rather than the dynamic behavior around it that the modeler wishes to alter. In fact, a modeler may wish to create a nonzero critical level when none at all existed before. The creation of a critical level toward which dynamics converge very often is given the name *regulation* (i.e., the system is *regulated* in such a way that it tends to be held at the created critical level, which often is called a *set point*). Shifting an existing critical level involves alteration of existing nonlinearities in a model; creating a critical level where none existed before involves either alteration of existing nonlinearities or introduction of new nonlinearities. In either case, the process can be given the general label of *nonlinear control.*

The deduction of a strategy for obtaining the maximum sustainable yield or harvest from an exploited population falls into the category of a nonlinear-control problem. The solution to the problem is very well known for one-loop networks with lumpable parameters (section 4.3). One simply selects the network variable, $J(\tau)$, that represents organisms in the state appropriate for harvesting, then plots $J(\tau+T)$ vs. $J(\tau)$ where T is the total delay around the single loop, as was done in Figure 4.12. A reference line, $J(\tau+T)=J(\tau)$ is drawn and the maximum yield occurs at the level where the graph of $J(\tau+T)$ vs. $J(\tau)$ lies above that line by the greatest distance. That distance, in fact, translates directly into the yield. The graphical procedure is illustrated in Figure 4.65. Notice that the maximum yield does not, in general, occur at the peak of the graph $J(\tau+T)$ vs. $J(\tau)$. In fact, unless the slope of that graph is discontinuous, the maximum yield will occur to the left of the peak, where the slope of the graph is 1.0 (i.e., parallel to the reference line).

A one-loop model has lumpable parameters only if no interaction between resolved age classes is represented in it. For example, if the model is designed to

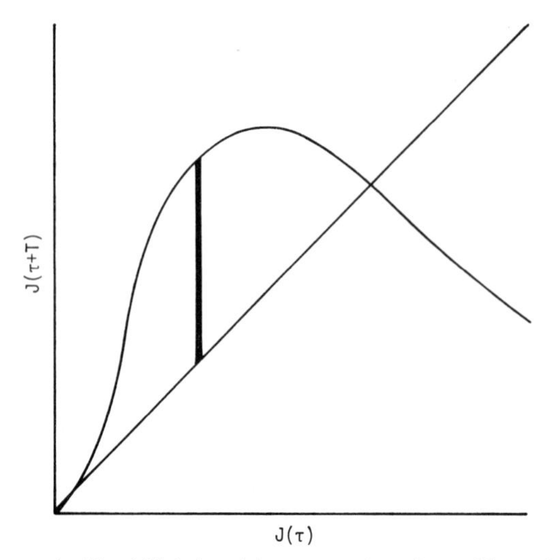

Fig. 4.65. Maximum sustainable yield deduced from a one-loop lumpable model. The heavy vertical line spans the greatest distance between the reference line and the model's graph, where the graph is above the reference line. Note that the slope of the graph at its intersection with the heavy line is one. If the population of one generation is the value directly beneath the heavy line, then the population of offspring will equal $J(\tau + T)$ at the top of the heavy line; and one can return the modeled population to its previous state by harvesting a number of offspring equal to the length of the heavy line. Thus, every T units of time, one can harvested that same number. Since the heavy line is the longest of its type that can be drawn, it represents the maximum yield that can be taken repeatedly from the modeled population

include mortality of one age class being dependent upon the number of individuals in another age class, then the parameters of that model are not lumpable (see section 4.3). In that case, the method of Figure 4.65 may approximate the maximum sustainable yield when the dependence is weak, but will not be applicable when the dependence is strong. Thus, if we included strong age-class interaction in a fisheries model, for example, we would be forced to use an alternative method of deducing the harvesting strategy.

Probably the most common methods of design of nonlinear control strategies involve the use of iterative algorithms that seek out maxima or minima within the stipulated constraints of the problem. A major problem in nonlinear control, therefore, is the design of algorithms that converge efficiently. As we already have seen, this is precisely the problem faced by the modeler wishing to find the natural frequencies of a linear model whose characteristic polynomial is of degree five or greater.

4.17. Dynamics of Constant-Parameter Models with Stochastic *Time Delays*

Up to this point, our methods of constant-parameter network analysis have been applied to models in which the durations of key life-cycle processes (e.g., maturation, gestation, regression, lactation, and the like) were represented as being fixed. Although the durations of many such processes are very close to being constant, there usually is some variability, so that the processes exhibit a stochastic nature. A modeler may wish to include an estimate of this stochastic nature in his model, representing the process not as a single *time delay*, but as an array of *time delays*, each with a potential duration of the process and an associated probability. This possibility already was mentioned briefly at the end of section 3.4; but with the availability of the z transform, we now are in a position to examine it efficiently and in much more detail.

4.17.1. z-Transforms of Stochastic *Time Delays*

Consider the flow, $J_{\text{in}}(\tau)$, of individuals into a particular process, such as maturation, gestation, or the like. Let the duration, T, of the process be a random variable with probability density function $p(T)$. The expected corresponding flow, $J_{\text{out}}(\tau)$, out of the process is simply the convolution of $J_{\text{in}}(\tau)$ and $p(T)$:

$$J_{\text{out}}(\tau) = \sum_{T=0}^{\tau} J_{\text{in}}(\tau - T)\, p(T) \tag{4.212}$$

where $J(\tau)$ is taken to be zero for all values of τ less than zero. Of course, the process of discrete convolution is most conveniently carried out in terms of z transforms:

$$\mathscr{J}_{\text{out}}(z) = \mathscr{J}_{\text{in}}(z)\, \not{p}(z) \tag{4.213}$$

where

$$\not{p}(z) = \sum_{T=0}^{\infty} p(T) z^{-T} \tag{4.214}$$

and $\mathscr{J}(z)$ is defined in the usual manner.

If $p(k)$ is the probability that the process duration is k intervals for an individual *entering* that process, then any mortality assumed to occur during the process will prevent $p(k)$ from being conservative:

$$\sum_{k=0}^{\infty} p(k) < 1.$$

In which case the mean and the various moments of the duration distribution will be infinite (since some of those that enter the process never will leave it). On the other hand, if $p(k)$ is the probability that an individual *emerging* from the process required k intervals to complete it, then the effects of mortality are set aside and $p(k)$ is conservative:

$$\sum_{k=0}^{\infty} p(k) = 1. \tag{4.215}$$

If the summation on the left-hand side of equation 4.215 were carried out over finite limits, it could be viewed as discrete convolution of $p(T)$ and $f(T)$, where

$$f(T) = 1, \qquad T \geq 0$$

$$G(T) = \sum_{k=0}^{T} f(T-k)\, p(k) = \sum_{k=0}^{T} p(k). \tag{4.216}$$

It follows that since $f(z)=z/(z-1)$,

$$\mathcal{G}(z)=[z/(z-1)]\,p(z). \tag{4.217}$$

In order to evaluate the limit of $G(T)$ as T becomes indefinitely large (i.e., as the limit of the summation approaches infinity), one simply can invoke the final-value theorem (equation 4.199), which applies here because $G(T)$ is strictly bounded to values less than or equal to 1.0:

$$\lim_{T\to\infty} G(T)=[(z-1)/z]\mathcal{G}(z)|_{z=1}=p(z)|_{z=1}. \tag{4.218}$$

Therefore, in the z-transform domain, equation 4.215 becomes

CONSERVATION CONDITION

$$p(z)|_{z=1}=1. \tag{4.219}$$

Equation 4.219 provides a simple test of conservation for a candidate transformed probability function.

4.17.2. Moments of Stochastic *Time Delay* Distributions

The mean, variance, and other moments of the distribution of delay durations can be computed directly from $p(z)$ as long as $p(T)$ is conservative. It follows from

$$p(z)=p(0)+p(1)z^{-1}+p(2)z^{-2}\ldots$$

that

$$d\,p(z)/dz=-p(1)z^{-2}-2\,p(2)z^{-3}-3\,p(3)z^{-4}-\cdots$$

and, therefore, $-z[d\,p(z)/dz]$ is the transform of $Tp(T)$. The kth moment, μ_k', about the origin (i.e., the kth *initial moment*) of a probability distribution is defined as follows:

$$\mu_k'=\sum_{T=0}^{\infty} T^k p(T). \tag{4.220}$$

Employing the previous result, the transform pair of equations 4.216 and 4.217, and the final-value theorem, we can evaluate μ_k' from $p(z)$:

$$\mu_k'=[-z(d/dz)]^k\,p(z)|_{z=1} \quad (\text{intervals}^k) \tag{4.221}$$

where $[-z(d/dz)]^k$ indicates k repetitions of the process (operation) described within the brackets. For example

$$\begin{aligned}[-z(d/dz)]^2\,p(z)&=-z\{d/dz[-z\,d\,p(z)/dz]\}\\&=-z[-d\,p(z)/dz-zd^2\,p(z)/dz^2].\end{aligned}$$

As indicated on the right-hand side of equation 4.221, the unit of the kth moment is the transition interval to the kth power (e.g., μ_1' is given in intervals, μ_2' is given in intervals squared).

The central moments of the duration distribution can be derived directly from the moments about the origin:

MEAN DURATION

$$T_m = \mu_1' = -z[d\,p(z)/dz]|_{z=1} = -d\,p(z)/dz|_{z=1} \tag{4.222}$$

VARIANCE

$$\sigma_T^2 = \mu_2' - T_m^2 = \{d\,p(z)/dz + d^2\,p(z)/dz^2 - [d\,p(z)/dz]^2\}|_{z=1} \tag{4.223}$$

COEFFICIENT OF SKEWNESS (OBLIQUITY)

$$\gamma_1(T) = [\mu_3' - 3\,T_m\sigma_T^2 - T_m^3]/\sigma_T^3. \tag{4.224}$$

4.17.3. Examples of Stochastic *Time Delays* and Their Transforms

Anyone familiar with probability theory will recognize immediately that what we have presented as the z-transform of our probability function is in fact identical to the *generating function* of that probability function. The expressions for the various moments, therefore, are precisely those employed for generating functions. Thus, according to equation 4.213, the z-transform of the flow out of a stochastic *time delay* simply is the product of the transformed input flow and the generating function of the duration probabilities.

Three discrete probability distributions commonly encountered are the *binominal distribution*, the *negative binomial distribution* (sometimes called the *Pascal distribution*), and the *uniform distribution*. The properties of these distributions including their probability generating functions, are summarized in Table 4.4. The negative binomial distribution always is skewed toward delay durations greater than the mean (it therefore exhibits a positive coefficient of skewness). The uniform distribution is symmetric about the mean (its coefficient of skewness therefore is zero); and the binomial distribution may be skewed toward longer durations or toward shorter durations, or it may be symmetric (coefficient of skewness may be positive, negative, or zero). Add to these three the possibility of offsetting with a *time delay* of fixed duration, and we have a reasonably large and versatile repertoire of stochastic *time delays*.

4.17.4. Effects of *Time Delay* Distributions on Dynamics in a One-Loop Model

Consider the simple cycle depicted in Figure 4.66. Newly formed organisms enter a maturation process of duration T. At the end of the process, each emerging organism immediately produces some number of offspring and then becomes reproductively inactive or dies. The parameter b is the product of the expected proportion of offspring that will survive to maturity and the expected number of offspring produced by a surviving mature individual. Thus b represents the *reproductive potential* of a newly produced offspring. We shall explore three situations, namely (1) T is fixed and equal to T_0, (2) T is a random variable

Table 4.4. Stochastic time delays

Name	$p(T)$		Generating function (z transform)	Mean (intervals)	Standard deviation (intervals)	Skewness
Fixed duration (T_0)	$p(T)=1 \quad T=T_0$ $p(T)=0 \quad T \neq T_0$		z^{-T_0}	T_0	0	0
Uniform from T_1 to T_2 inclusive	$p(T)=1/(T_2-T_1+1)$ $p(T)=0 \quad T<T_1,\ T>T_2$	$T_1 \leqq T \leqq T_2$	$\dfrac{z^{-T_1}-z^{-T_2}}{(1-z^{-1})(T_2-T_1+1)}$	$(T_1+T_2)/2$	$(T_2-T_1)/2\sqrt{3}$	0
Negative binomial (Pascal)[a]	$\binom{T+k-1}{k-1}(1-\beta)^k \beta^T$	$T \geqq 0$	$[(1-\beta)z/(z-\beta)]^k$	$k\beta/(1-\beta)$	$\sqrt{k\beta}/(1-\beta)$	$(1+\beta)/\sqrt{k\beta}$
Binomial[a]	$\binom{k}{T}(1-\beta)^{k-T}\beta^T$	$T \geqq 0$	$(\beta z^{-1}+1-\beta)^k$	$k\beta$	$\sqrt{k\beta(1-\beta)}$	$(1-2\beta)/\sqrt{k\beta(1-\beta)}$

[a] $0 \leqq \beta \leqq 1 \quad k=$ positive integer.

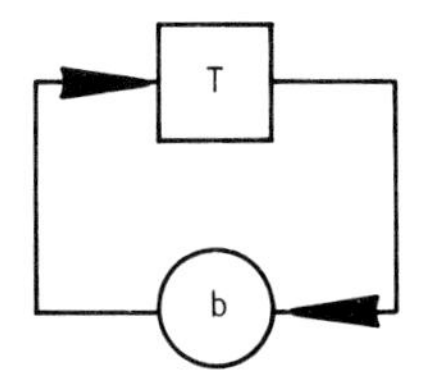

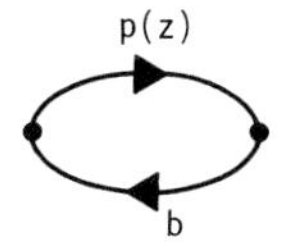

Fig. 4.66. Network model embodying simple natality with a stochastic *time delay* (distribution of times to offspring production)

with a negative binomial distribution, and (3) T is a random variable with a binomial distribution.

In each of these three cases, the denominator of the unit-cohort response is simply

$$Q(z) = 1 - b\,p(z). \tag{4.225}$$

In the case of a fixed duration, T_0, we have

$$Q(z) = 1 - b z^{-T_0} \tag{4.226}$$

the roots of which are the various T_0th roots of b:

$$r = b^{1/T_0}$$

the jth such root being

$$r_j = |b^{1/T_0}|\{\cos[2\pi j/T_0] + i \sin[2\pi j/T_0]\}. \tag{4.227}$$

In the case of durations distributed according to the negative binomial, we have

$$Q(z) = 1 - b[(1-\beta)z/(z-\beta)]^k \tag{4.228}$$

the jth root of which is

$$r_j = \beta[1 - a\cos(2\pi j/k) + i\,a\sin(2\pi j/k)]/[1 - 2a\cos(2\pi j/k) + a^2] \tag{4.229}$$

where

$$a = (1-\beta)|b^{1/k}|.$$

Finally, in the case of binomially distributed durations,

$$Q(z) = 1 - b(\beta z^{-1} + 1 - \beta)^k \tag{4.230}$$

the jth root of which is

$$r_j = \beta|b^{1/k}|[\cos(2\pi j/k) - a - i\sin(2\pi j/k)]/[1 - 2a\cos(2\pi j/k) + a^2] \tag{4.231}$$

where, once again,

$$a=(1-\beta)|b^{1/k}|.$$

In all three cases, the roots corresponding to $j=0$ are real and positive, and therefore represent nonoscillatory convergence on or divergence from the zero level (i.e., pure geometric growth or decline). All of the remaining roots represent oscillatory dynamics. In the case of the fixed *time delay*, all of the roots have precisely the same magnitude. Therefore, for that distribution, the ratio of the magnitude of the oscillation represented by the jth root to the magnitude of the nonoscillatory pattern represented by the 0th root will remain constant for all time. In other words, in the case of the fixed *time delay*, any oscillatory activities present in the system (such as those that might be triggered by transitory stimuli) will persist indefinitely. This is a well-known property of such a system.

However, in the cases of the other two distributions, the magnitudes of the positive real roots are greater than the magnitudes of the other roots, so that the nonoscillatory pattern eventually will dominate and oscillatory activities gradually will fade relative to the dominant pattern. An interesting question that arises is, "How does the time required for dominance of the nonoscillatory pattern (i.e., the time required for the oscillations to fade away) depend upon the parameters of the stochastic *time delay* distributions?"

4.17.5. An Elementary Sensitivity Analysis

In answering this question, we shall be determining the *sensitivity* of the persistence of oscillations to variations in the parameters of the stochastic *time delay* distribution. In effect, we shall be examining the propensity of the models toward sustained wave activity. The significant information for the two distributions is carried by the following parameters:

(1) For the negative binomial distribution

$$\omega_j=\tan^{-1}\{[(1-\beta)|b^{1/k}|\sin(2\pi j/k)]/[1-(1-\beta)|b^{1/k}|\cos(2\pi j/k)]\} \quad \text{radians per interval.} \tag{4.232}$$

(2) For the binomial distribution

$$\omega_j=\tan^{-1}\{\sin(2\pi j/k)/[1-(1-\beta)|b^{1/k}|\cos(2\pi j/k)]\} \quad \text{radians per interval.} \tag{4.233}$$

(3) For both distributions

$$\begin{aligned}\alpha_j&=[1-(1-\beta)|b^{1/k}|]/\sqrt{1-2(1-\beta)|b^{1/k}|\cos(2\pi j/k)+(1-\beta)^2|b^{1/k}|^2}\\&=|r_j|/r_0 \quad \text{foldings per interval.}\end{aligned} \tag{4.234}$$

A transient disturbance of the model at τ_0 will trigger waves whose frequencies are given by ω_j. As time progresses, the ratio of the magnitude of the jth wave to the magnitude of the nonoscillatory pattern will diminish geometrically as

$$\alpha_j^{\tau-\tau_0}.$$

Examination of equation 4.234 reveals that the more persistent waves are those for which j is small, and the most persistent wave is that corresponding to $j=1$. From equations 4.232 and 4.233, it is clear that these also are the waves of lowest frequency.

Equations 4.232, 4.233, and 4.234 express ω_j and α_j in terms of the principal parameters of the binomial and negative binomial distributions, namely β and k. The sensitivities of ω_j and α_j to changes in these parameters are given by $\partial\omega_j/\partial\beta$, $\partial\alpha_j/\partial\beta$, $\omega_j(k)/\omega_j(k+1)$, and $\alpha_j(k)/\alpha_j(k+1)$. Clearly, these sensitivities will vary with β and k. Although these sensitivities may be interesting in their own right, a more directly interpretable approach would be to express α_j and ω_j in terms of the central moments of the distributions and examine their sensitivities with respect to variations of those moments (e.g., to variations of the mean, T_m, or the standard deviation, σ_T). As they stand, equations 4.232, 4.233, and 4.234 are not particularly amenable to this approach. However, if we focus our attention on the more persistent waves (i.e., low values of j), if k is moderate to large, and if b is moderate to small, then those equations can be replaced by reasonably accurate, considerably simplified approximations.

The approximations are based upon the following, well-known series:

$$|b^{1/k}|=1+(1/k)\log_e(b)+(1/k)^2[\log_e(b)]^2/2!+(1/k)^3[\log_e(b)]^3/3!+\cdots \qquad (4.235)$$

$$\sin(2\pi j/k)=2\pi j/k-(2\pi j/k)^3/3!+(2\pi j/k)^5/5!-\cdots \qquad (4.236)$$

$$\cos(2\pi j/k)=1-(2\pi j/k)^2/2!+(2\pi j/k)^4/4!-(2\pi j/k)^6/6!+\cdots. \qquad (4.237)$$

If k is moderate to large and b is moderate to small (how moderate in each case is left for the reader to decide), the following approximation is reasonably accurate:

$$|b^{1/k}|\simeq 1+(1/k)\log_e(b). \qquad (4.238)$$

If j is small and k is moderate to large, the following approximations also are reasonably accurate:

$$\sin(2\pi j/k)\simeq 2\pi j/k \qquad (4.239)$$

$$\cos(2\pi j/k)\simeq 1-(2\pi j/k)^2/2. \qquad (4.240)$$

Inserting approximations 4.238, 4.239 and 4.240 into equations 4.232, 4.233, and 4.234, we have

$$\omega_j\simeq 2\pi j(1-\beta)/k\beta=2\pi j/T_m \quad \text{rad/interval} \qquad (4.241)$$

for the negative binomial,

$$\omega_j\simeq 2\pi j/k\beta=2\pi j/T_m \quad \text{rad/interval} \qquad (4.242)$$

for the binomial, and

$$\alpha_j\simeq 1-2\pi^2 j^2(1-\beta)/k^2\beta^2=1-2\pi^2 j^2\sigma_T^2/T_m^3 \qquad (4.243)$$

for both distributions.

Table 4.5. Dominant growth patterns in a simple natality cycle with stochastic time delays

Value of k	Binomial distribution						Negative binomial distribution					
	$\beta=0.1$		$\beta=0.5$		$\beta=0.9$		$\beta=0.1$		$\beta=0.5$		$\beta=0.9$	
	T_b/T_m	σ_T/T_m	T_b/T_m	σ_T/T_m	T_b/T_m	σ_T/T_m	T_b/T_m	σ_T/T_m	T_b/T_m	σ_T/T_m	T_b/T_m	σ_T/T_m
1	–	–	–	–	0.654	1.054	–		–	–	0.950	0.333
2	–	–	0.648	1.000	0.817	0.745	–		0.786	0.707	0.978	0.236
4	–	–	0.826	0.707	0.906	0.527	–		0.905	0.500	0.990	0.167
8	0.463	1.118	0.913	0.500	0.953	0.373	0.489	1.061	0.955	0.354	0.995	0.118
16	0.767	0.791	0.957	0.354	0.976	0.264	0.785	0.750	0.978	0.250	0.998	0.083
32	0.888	0.559	0.978	0.250	0.988	0.186	0.898	0.530	0.989	0.177	0.999	0.059
64	0.945	0.395	0.989	0.177	0.994	0.132	0.950	0.375	0.995	0.125	0.999	0.042
128	0.973	0.280	0.995	0.125	0.997	0.093	0.975	0.265	0.997	0.088	1.000	0.029
256	0.986	0.198	0.997	0.088	0.998	0.066	0.988	0.188	0.999	0.063	1.000	0.021
512	0.993	0.140	0.999	0.063	0.999	0.047	0.994	0.133	0.999	0.044	1.000	0.015

The frequency of the jth wave for the fixed *time delay* (i.e., the angle in radians of the jth root) as given by equation 4.227 is $2\pi j/T_0$. Thus we can see that for moderate to small b, small j, moderate to large k, the frequency of the jth wave for both the binomial and the negative binomial distributions of durations is approximately the same as the frequency of the jth wave for a time delay whose duration is fixed at T_m, the mean of the distribution. That frequency is inversely proportional to T_m, with the period of the lowest-frequency wave ($j = 1$) being approximately equal to T_m.

For moderate to small b, small j, and moderate to large k, the ratio α_j depends as follows on the mean and variance of the duration distribution: For fixed mean, α_j decreases in direct proportion to the variance (i.e., as the square of the standard deviation). For fixed variance, α_j is diminished as the inverse cube of the mean duration. Therefore, for both distributions, the propensity of the network toward sustained waves *increases* rather markedly *with increasing mean duration* if the standard deviation is fixed, and *decreases* rather markedly *with increasing standard deviation* if the mean is fixed. Although we have not demonstrated it here, the propensity toward sustained waves also decreases with increasing coefficient of skewness in these distributions.

Finally, it might be interesting to compare the dominant, nonoscillatory dynamic patterns in the cases of the two stochastic *time-delay* models with that deduced for the model with a fixed *time delay*. In the case of the fixed *time delay*, the dominant pattern simply is one of b-fold change (increase or decrease) every T_0 intervals. Thus, in that special case, the b-folding time is precisely equal to the mean maturation time. For the two stochastic *time-delay* models, we can calculate directly from r_0 the ratio of the b-folding time, T_b, to the mean maturation time, T_m, and compare the results. This has been done for $b=2$ (see Fig. 4.66) over ten values of k and three values of β, and the results are displayed in Table 4.5. Along with these results are displayed the corresponding values of σ_T/T_m. Notice that for all moderate values of standard deviation, the b-folding time approximates the mean maturation time, the former in each case being slightly less than the latter. For all six distributions represented, the b-folding time is within ten percent of the mean maturation time for all values of standard deviation less than or equal to half the mean maturation time. Thus, even if the distribution of delay durations is not especially narrow, as far as the dominant pattern of dynamics is concerned the stochastic *time delay* appears to behave very much like a fixed time delay of duration T_m.

4.17.6. When the Minimum Latency is Not Finite

Introduction of the stochastic *time delay* leads to the possibility of instantaneous feedback around a loop in a network model. This would occur if $p(0) \neq 0$; in other words, a nonzero probability of instantaneous passage through the *time delay* in the loop. This will create a problem only if the product of $p(0)$ and the *scalar* parameters around the loop is too large. This product, the *instantaneous loop gain* must be less than 1.0. Otherwise, the dynamics of the model will be overwhelmed totally by an instantaneous explosion of the variables in the loop (it is left as an exercise for the reader to show that this is so). In the cases of the binomial and negative binomial distributions, the value of $p(0)$ simply is

$$p(0)=(1-\beta)^k. \tag{4.244}$$

In the case of the uniform distribution, $p(0)$ is zero unless $T_1=0$, in which case

$$p(0)=1/(T_2+1). \tag{4.245}$$

Many processes to be represented by *time delays* in our network models exhibit apparently finite minimum latencies. In other words, the probability of completing the process prior to some finite, minimum time appears to be zero. In that case, the diligent modeler should incorporate that minimum latency or a very good approximation to it in any stochastic *time delays* in his model. Thus should the problem of instantaneous explosion be avoided at the outset. Finite minimum latency can be incorporated in a uniformly distributed *delay*, simply by making $T_1>1$. In the case of binomial or negative binomial *delay*, it can be incorporated in the form of a fixed *time delay* inserted in series with the stochastic *delay*, or it can be approximated either by making β sufficiently large (i.e., sufficiently close to 1.0) or k sufficiently large. The effectiveness of the latter method is demonstrated in a continuous-time model of binary fission proposed by Kendall (1948). He selected a gamma distribution (the continuous-time analog of the negative binomial) for the times between fissions in his model; and with $k=20$, he found that the distribution of delays was very closely matched to the fission intervals observed by Kelly & Rahn (1932) in bacteria, approximating very well the finite minimum latency.

4.18. The Inverse Problem: Model Synthesis

A problem that arises in many areas of modeling, both of living and of nonliving systems, is that of inference of the (presently unobservable) mechanisms underlying the operation of the system, given a limited set of observed dynamic behavior. Typically, input-output relationships involving an incomplete subset of the state variables of the system (i.e., inputs applied at one or more, but not all, state variables and outputs observed at one or more, but not all, state variables) are observed, then an attempt is made to synthesize a model that generates those same input-output relationships; and, if such a model is found, inferences are drawn from it with respect to *all* of the state variables and their interactions. Although this process does not seem likely to be common in population modeling, nevertheless a brief discussion of it and its pitfalls seems to be in order.

In its simplest form, the process involves the synthesis of constant-parameter models on the basis of the observed, linear (or very nearly linear) input-output relationship between a single pair of state variables. In terms of our discrete-time population models, this would be analogous to observation of the response of one output variable $X_{\text{out}}(\tau)$ to a unit cohort applied at one input site of a natural population, followed by synthesis of a network model generating the same unit-cohort response, followed in turn by inference of the life-history

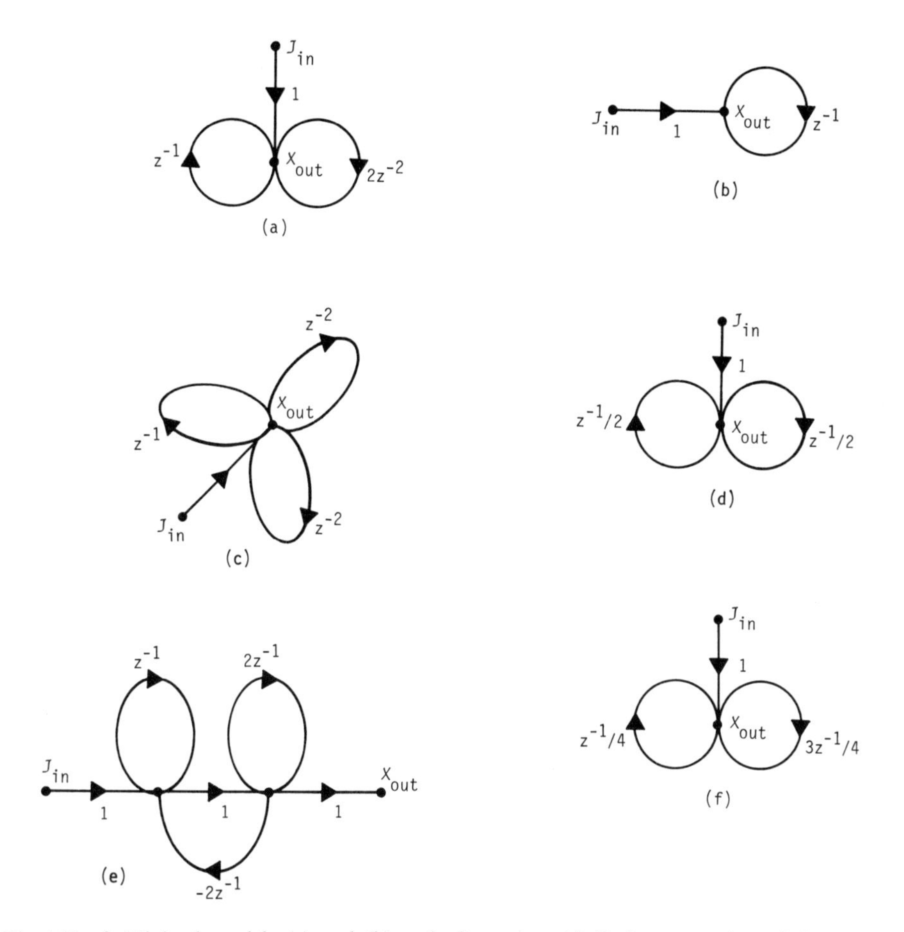

Fig. 4.67 a–f. Minimal models (a) and (b) and alternatives (c)–(f) for generation of the output functions given by equations 4.246 and 4.247

details of the members of the population from the topography and parameter values of the model. Terms that often appear in conjunction with such exercise, apparently to add credence to the inferences, are "unique model" and "minimal model." Although it may be minimal with respect to one criterion or another (see Kalman, 1968), a network model based on the dynamic relationship between two state variables simply cannot be unique. Furthermore, the fact that a model is minimal with respect to some criterion does not necessarily bear on its structural similarity to the system being modeled.

Suppose that by diligent, repeated observation we were able to conclude with reasonable confidence that two natural populations exhibited the following expected dynamic behavior in response to unit-cohort inputs:

(1) $$X_{\mathrm{out}}(\tau) = (2/3)(2)^{\tau} + (1/3)(-1)^{\tau} \qquad (4.246)$$

(2) $$X_{\mathrm{out}}(\tau) = 1. \qquad (4.247)$$

The corresponding z-transforms are

$$\text{(1)} \qquad \mathscr{f}(z) = (2/3)z/(z-2) + (1/3)z/(z+1) = z^2/(z^2 - z - 2) = 1/(1 - z^{-1} - 2z^{-2}) \tag{4.248}$$

$$\text{(2)} \qquad \mathscr{f}(z) = z/(z-1) = 1/(1 - z^{-1}). \tag{4.249}$$

Next, suppose that we stipulate that the corresponding network with the minimum number of loops is the minimal model. Applying Mason's rule in reverse, we immediately would conclude that for population (1) we have a two-loop minimal model, and for population (2) we have a one-loop minimal model, as depicted in Figure 4.67a and b. On the other hand, in each case there is an *infinite number of alternative models*, none of which can be eliminated by the available data. Some of these alternatives are shown in Figure 4.67c through *f*. Without independent information about the life history of the organism involved, we would have no clue as to which topography to select or how to relate the loops in an arbitrarily selected topography to hypothetical phases in the organisms life history. In other words, without further information, the simple input-output response, no matter how diligently obtained, would be infinitely ambiguous with respect to underlying life-history details.

4.19. Application of Constant-Parameter Network Analysis to More General Homogeneous Markov Chains

So far, our constant-parameter network models represent a special class of Markov process, namely one in which exactly one transition is stipulated to occur for each resolvable interval of time. Thus, the number of transitions that have occurred since the initial interval is precisely equal to the number of units of temporal resolution that have passed since the initial interval; and the number of transitions is a linear measure of time. More generally, one might dissociate transitions from time, keeping track of the inter- and intrastate transitions in their order of occurrence but ignoring altogether the associated time. Thus, for example, in a population with noninterbreeding generations, one might keep a running count of the transitions from generation to generation without noting the corresponding times. In this way, descriptions of the dynamics of the progression from generation to generation can transcend any variations in generation intervals that might occur. The dynamics themselves will be given as functions of k, the generation number, rather than as functions of τ. Nevertheless, all of the methods described in this chapter for constant-parameter networks will be directly applicable to the analysis of these dynamic functions as long as the Markov processes are homogeneous (i.e., have constant transition probabilities).

Consider, for example, simple binary fission described by the following equation:

$$N(k) = 2N(k-1) \tag{4.250}$$

where k is the generation number, and $N(k)$ is the corresponding expected state (size) of the population. Taking the z transform of the functions on either side of the equation, we have

$$\mathcal{N}(z)=2z^{-1}\mathcal{N}(z) \quad (4.251)$$

from which

$$1=2z^{-1}$$

$$r=2$$

and

$$N(k)=r^k=2^k. \quad (4.252)$$

Of course we knew the solution to equation 4.250 and need not have applied the z transform. On the other hand, not many Markov processes will be reduced to such simple descriptions.

Example 4.44. The Brother-Sister Mating Problem. Consider the following classic, highly-idealized example: A single locus with two panmictic alleles, A_1 and A_2, occurs in a diploid organism. A male and a female are selected at random from each generation to produce the entire subsequent generation. Given that the first such pair comprises genotypes A_1A_1 and A_2A_2, deduce the dynamics (on a generation by generation basis) of the genetic drift of the population with respect to this locus.

Answer: Defining the state of a generation, with respect to this locus, in terms of the genotypes of the two individuals selected from it to produce the subsequent generation, we have the following Markovian states:

$$A_1A_1/A_1A_1, \quad A_1A_2/A_1A_1, \quad A_1A_2/A_1A_2,$$
$$A_1A_2/A_2A_2, \quad A_1A_1/A_2A_2, \quad A_2A_2/A_2A_2$$

plus others which are equivalent owing to the panmictic nature of the alleles. The transition probabilities from state to state can be calculated very easily. Thus, for example, in state A_1A_2/A_1A_1 we have the following genotype probabilities for the gametes produced by the mated pair:

$\Pr[A_1]=0.5$	$\Pr[A_1]=1$
$\Pr[A_2]=0.5$	$\Pr[A_2]=0$
for one sex	for the other sex

leading to the following genotype probabilities for offspring

$$\Pr[A_1A_1]=0.5$$
$$\Pr[A_1A_2]=0.5$$
$$\Pr[A_2A_2]=0$$

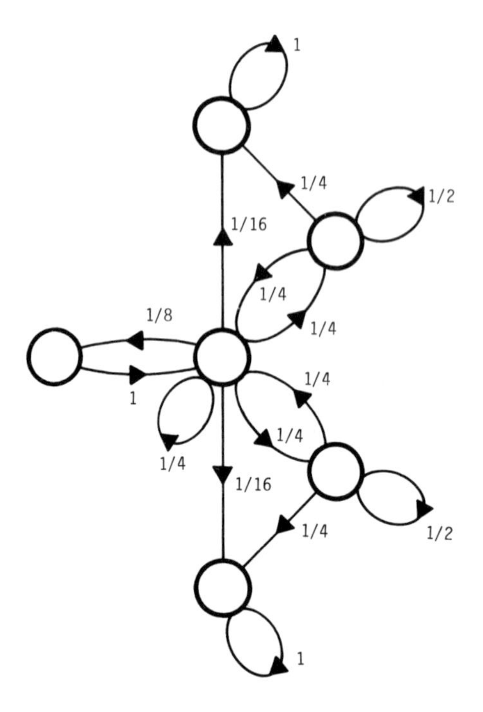

Fig. 4.68. Brother-sister mating problem represented as a network of Markovian states

and thence to the following probabilities for the subsequently selected pair

$$\Pr[A_1 A_1/A_1 A_1] = 0.25$$
$$\Pr[A_1 A_2/A_1 A_1] = 0.5 \quad \text{(two ways, male } A_1 A_1 \text{ or female } A_1 A_1\text{)}$$
$$\Pr[A_1 A_2/A_1 A_2] = 0.25$$
$$\Pr[A_1 A_2/A_2 A_2] = 0$$
$$\Pr[A_1 A_1/A_2 A_2] = 0$$
$$\Pr[A_2 A_2/A_2 A_2] = 0$$

which are the transition probabilities from $A_1 A_2/A_1 A_1$ to the various states.

Now, in the manner of Chapter 1, we can construct a network representation of the entire Markov process, as illustrated in Figure 4.68. Since each transition from one state to another represents the emergence of a new generation, which will be signified by a unit incrementation of the variable k, we can think of the passage along a transition path as being analogous to delay by one generation. The z-transformed network, therefore, is formed simply by multiplying each transition probability by z^{-1}, as shown in Figure 4.69. Remembering that the variables associated with the nodes in this network are the probabilities of occupancy of the corresponding state, we can proceed to analyze the transformed network in the usual manner, beginning with application of Mason's rule.

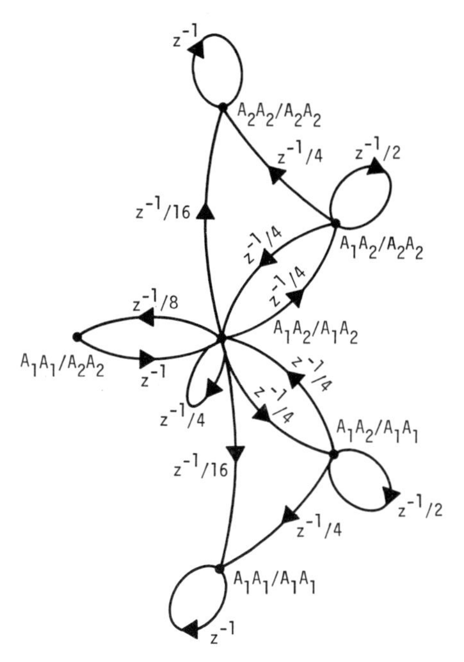

Fig. 4.69. Linear flow graph for the brother-sister mating problem

There are eight loops to consider:

$$L_1=(1/8)\,z^{-2},\quad L_2=(1/16)\,z^{-2},\quad L_3=(1/16)\,z^{-2},\quad L_4=(1/4)\,z^{-1},$$
$$L_5=(1/2)\,z^{-1},\quad L_6=(1/2)\,z^{-1},\quad L_7=z^{-1},\quad L_8=z^{-1}.$$

According to the specification of the problem, the state designated as input is $A_1\,A_1/A_2\,A_2$, which has an initial probability of occupancy equal to 1. We shall consider the generation-dependent probability of occupancy of each state, one at a time, as the output variable. Beginning with $A_1\,A_1/A_2\,A_2$ itself, we have

$$\mathscr{P}\imath\,[A_1\,A_1/A_2\,A_2]=\Pr[A_1\,A_1/A_2\,A_2(0)]\,[(1-L_2)(1-L_3)(1-L_4)(1-L_5)(1-L_6)$$
$$\cdot(1-L_7)(1-L_8)^*]/[(1-L_1)(1-L_2)(1-L_3)\ldots(1-L_8)^{**}].$$

Since L_7 and L_8 do not touch any other loops, the factors $(1-L_7)$ and $(1-L_8)$ will remain intact (no terms dropped) in the denominator and the numerator, and therefore will be cancelled by division. Dropping the terms involving touching loops from the denominator, we find

$$\begin{aligned}Q'(z^{-1})&=1-(L_1+L_2+\cdots+L_6)+L_1\,L_5+L_1\,L_6+L_4\,L_5+L_4\,L_6+L_2\,L_6\\&\quad+L_3\,L_5+L_5\,L_6-L_1\,L_5\,L_6-L_4\,L_5\,L_6\\&=1-(5/4)\,z^{-1}+(1/4)\,z^{-2}+(1/8)\,z^{-3}-(1/32)\,z^{-4}\end{aligned}$$

from which

$$Q(z)=z^4-(5/4)\,z^3+(1/4)\,z^2+(1/8)\,z-(1/32).$$

The roots of $Q(z)$ all happen to be real and can be found quite quickly by trial and error:

$$r_1 = 1/4, \quad r_2 = 1/2, \quad r_3 = 0.80902\ldots, \quad r_4 = -0.30902\ldots.$$

Taking $\Pr[A_1 A_1/A_2 A_2(0)]$ to be 1, as indicated in the problem statement, we find that the numerator of $\mathscr{P}r[A_1 A_1/A_2 A_2]$ is

$$\begin{aligned} R'(z^{-1}) &= 1 - (L_2 + L_3 + \cdots + L_6) + L_4 L_5 + L_4 L_6 + L_2 L_6 \\ &\quad + L_3 L_5 + L_5 L_6 - L_4 L_5 L_6 \\ &= 1 - (5/4)\, z^{-1} + (7/16)\, z^{-2} - (1/16)\, z^{-3} \end{aligned}$$

from which

$$R(z) = z^4 - (5/4)\, z^3 + (7/16)\, z^2 - (1/16)\, z.$$

Partial-fraction expansion leads to

$$\begin{aligned} \mathscr{P}r[A_1 A_1/A_2 A_2] &= c_0 + c_1\, z/(z - 1/4) + c_2\, z/(z - 1/2) \\ &\quad + c_3\, z/(z - 0.80902) + c_4\, z/(z + 0.30902) \end{aligned}$$

$$\begin{aligned} c_0 &= [R(z)/Q(z)]|_{z=0} = 0 \\ c_1 &= [(z - 1/4)/z][R(z)/Q(z)]|_{z=1/4} = -1/5 \\ c_2 &= [(z - 1/2)/z][R(z)/Q(z)]|_{z=1/2} = 1/2 \\ c_3 &= [(z - 0.80902)/z][R(z)/Q(z)]|_{z=0.80902} = 0.0146\ldots \\ c_4 &= [(z + 0.30902)/z][R(z)/Q(z)]|_{z=-0.30902} = 0.6854\ldots \end{aligned}$$

from which

$$\begin{aligned} \Pr[A_1 A_1/A_2 A_2] &= -(1/5)(1/4)^k + (1/2)(1/2)^k \\ &\quad + (0.0146\ldots)(0.80902)^k + (0.6854\ldots)(-0.30902)^k \\ &= 1\,(k=0), \quad = 0\,(k=1), \quad = 0.1875\ldots(k=2) \\ &= 0.0469\ldots(k=3), \quad = 0.0398\ldots(k=4), \quad = 0.0186\ldots(k=5) \\ &= 0.0123\ldots(k=6), \quad = 0.0070\ldots(k=7), \quad = 0.0047\ldots(k=8) \\ &= 0.0031\ldots(k=9), \quad = 0.0022\ldots(k=10), \quad = 0.0017\ldots(k=11) \\ &= 0.0006\ldots(k=15), \quad = 0.0002\ldots(k=20), \quad = 0.00007\ldots(k=25). \end{aligned}$$

Eventually, the term $0.0146(0.80902)^k$ dominates, and the probability of genotype $A_1 A_1/A_2 A_2$ diminishes by 0.80902 per generation.

To complete the solution, we must calculate the dynamics for each of the other genotypes:

$$\begin{aligned} \mathscr{P}r[A_1 A_2/A_1 A_2] &= z^{-1}[1 - L_5 - L_6 + L_5 L_6]/Q'(z^{-1}) \\ &= z^{-1}[1 - z^{-1} + (1/4)\, z^{-2}]/Q'(z^{-1}) \\ &= [z^3 - z^2 + (1/4)\, z]/Q(z) \\ &= 0.8\, z/(z - 1/4) + 0.4944\ldots\, z/(z - 0.80902) - 1.2944\ldots\, z/(z + 0.30902) \end{aligned}$$

$$\begin{aligned}\Pr[A_1 A_2/A_1 A_2] &= (0.8)(1/4)^k + (0.4944\ldots)(0.80902)^k - (1.2944\ldots)(-0.30902)^k \\ &= 0\,(k=0), \quad = 1\,(k=1), \quad = 0.25\,(k=2) \\ &= 0.3125\,(k=3), \quad = 0.2031\ldots(k=4), \quad = 0.1758\ldots(k=5) \\ &= 0.1377\ldots(k=6), \quad = 0.1125\ldots(k=7), \quad = 0.0906\ldots(k=8).\end{aligned}$$

Eventually, the term $0.4944(0.80902)^k$ dominates and the probability of genotype $A_1 A_2/A_1 A_2$ diminishes by 0.80902 per generation.

$$\begin{aligned}\mathscr{P}r[A_1 A_2/A_2 A_2] &= \mathscr{P}r[A_1 A_1/A_1 A_2] = (1/4)\, z^{-2}[1-(1/2)\, z^{-1}]/Q'(z^{-1}) \\ &= (1/4)[z^2-(1/2)\, z]/Q(z) \\ &= -0.8\, z/(z-1/4) + 0.4\, z/(z-0.80902) + 0.4\, z/(z+0.30902)\end{aligned}$$

$$\begin{aligned}\Pr[A_1 A_2/A_2 A_2] &= -0.8(1/4)^k + 0.4(0.80902)^k + 0.4(-0.30902)^k \\ &= 0\,(k=0), \quad = 0\,(k=1), \quad = 0.25\,(k=2) \\ &= 0.1875\,(k=3), \quad = 0.1718\ldots(k=4), \quad = 0.1367\ldots(k=5) \\ &= 0.1123\ldots(k=6), \quad = 0.0896\ldots(k=7), \quad = 0.0734\ldots(k=8)\end{aligned}$$

and, once again, the term in $(0.80902)^k$ will dominate and the probabilities of genotypes $A_1 A_2/A_2 A_2$ and $A_1 A_1/A_1 A_2$ will diminish by 0.80902 per generation. The probabilities of genotypes A_1A_1/A_1A_1 and A_2A_2/A_2A_2 may be calculated in the same manner; or they can be obtained by noting that they are equal to each other and that their sum must equal

$$1 - \Pr[A_1 A_1/A_2 A_2] - \Pr[A_1 A_2/A_1 A_2] - 2\Pr[A_1 A_1/A_1 A_2]$$

from which

$$\begin{aligned}\Pr[A_1 A_1/A_1 A_1] &= 0.5 + 0.1(1/4)^k - 0.25(1/2)^k - (0.4545\ldots)(0.80902)^k \\ &\quad + (0.1045\ldots)(-0.30902)^k \\ &= 0\,(k=0), \quad = 0\,(k=1), \quad = 0.1562\ldots(k=2) \\ &= 0.2265\ldots(k=3), \quad = 0.2910\ldots(k=4), \quad = 0.3391\ldots(k=5) \\ &= 0.3687\ldots(k=6), \quad = 0.3949\ldots(k=7), \quad = 0.4156\ldots(k=8).\end{aligned}$$

Eventually, the terms $0.5-(0.4545\ldots)(0.80902)^k$ dominate and the difference between $\Pr[A_1 A_1/A_1 A_1]$ and 0.5 diminishes by 0.80902 per generation.

4.20. Some References for Chapter 4

The numbers in the following lists refer to entries in the Bibliography at the end of the text. The lists themselves are designed to guide the interested reader into some of the literature relevant to particular topic areas treated in Chapter 4.

On Computer Simulation, Algorithmic Models

24, 31, 36, 67, 81, 91, 100, 107, 109, 119, 144, 147, 186, 218, 219, 223, 224, 236, 242, 248, 261, 292, 293, 294.

On Graphical Analysis, Critical Points, Limit Cycles, Chaos

65, 77, 83, 84, 100, 109, 111, 121, 126, 136, 152, 156, 163, 169, 170, 171, 181, 182, 183, 184, 186, 219, 229, 233, 251, 252, 266, 267, 274, 278, 282, 285, 286, 296.

On Constant-Parameter Network Analysis, *z* Transforms, Root Finding

4, 32, 43, 87, 89, 121, 122, 123, 141, 159, 179, 180, 206, 208, 254, 257, 272, 277, 300, 301.

On Natural Frequencies, Population Waves, Biotic Potential

18, 23, 32, 34, 35, 43, 48, 57, 65, 75, 102, 112, 113, 126, 132, 133, 135, 141, 150, 151, 210, 215, 229, 233, 238, 257, 264, 265, 294, 301.

On Control, Optimization, Sensitivity

3, 32, 50, 86, 92, 118, 119, 121, 122, 141, 144, 147, 172, 203, 219, 236, 242, 246, 257, 259, 261, 276, 280, 291, 292, 293, 294, 301.

Appendix A. Probability Arrays, Array Manipulation

A.1. Definitions

a) Let W, X, Y, and Z be discrete random variables, each capable of assuming the values $\mathbb{0}$, $\mathbb{1}$, $\mathbb{2}$, $\mathbb{3}$, $\mathbb{4}$, ...

b) Let Pr[– –] be the probability that whatever statement is contained in the brackets is true. Thus, $\Pr[W=k]$ is the probability that k is the value assumed by the random variable W.

c) Let Pr[– –;– –] be the conditional probability that the statement on the left of the semicolon is true whenever the statement on the right of it is true. Thus $\Pr[X=j; Y=k]$ is the conditional probability that X has taken on the value j, given that we know Y to have taken on the value k.

d) Let Pr[– –,– –] be the joint probability that two statements are true at the same time. Thus, $\Pr[X=j, Y=k]$ is the joint probability that X has assumed the value j *and*, at the same time, Y has assumed the value k.

e) Let $\mathbf{Pr}[W]$, $\mathbf{Pr}[X]$, $\mathbf{Pr}[Y]$, and $\mathbf{Pr}[Z]$ be arrays of the probabilities associated with each possible value of $W, X, Y,$ and Z respectively. Thus, for example,

$$\mathbf{Pr}[W]=\begin{bmatrix}\Pr[W=\mathbb{0}]\\ \Pr[W=\mathbb{1}]\\ \Pr[W=\mathbb{2}]\\ \vdots\\ \Pr[W=W_m]\end{bmatrix}$$

Column Vector

or

$$\mathbf{Pr}[X]=[\Pr[X=0], \Pr[X=1], \ldots, \Pr[X=X_m]] \tag{A.1}$$

Row Vector

where W_m and X_m are the largest values that can be taken on by W and X respectively. The *vertical array* (on the top) is known as a *column vector*, the *horizontal array* as a *row vector*. Both vectors share the property of having elements (the individual probabilities) the sum of whose values is one. Vectors with this property are called *probability-distribution vectors*, or *probability vectors*.

f) Let $\mathbf{Pr}[W;--]$, $\mathbf{Pr}[X;--]$, $\mathbf{Pr}[Y;--]$ and $\mathbf{Pr}[Z;--]$ be arrays of the conditional probabilities associated with each possible value of W, X, Y, and Z respectively, under the assumption that the statement to the right of the semicolon is true. Thus, for example,

$$\mathbf{Pr}[X;\, Y=k] = \begin{bmatrix} \Pr[X=\mathbb{0};\, Y=k] \\ \Pr[X=\mathbb{1};\, Y=k] \\ \Pr[X=\mathbb{2};\, Y=k] \\ \vdots \\ \Pr[X=X_m;\, Y=k] \end{bmatrix}$$

Column Vector

or

$$\mathbf{Pr}[W; Z=j] = [\Pr[W=\mathbb{0}; Z=j], \Pr[W=\mathbb{1}; Z=j], \ldots, \Pr[W=W_m; Z=j]]. \quad \text{(A.2)}$$

Row Vector

Once again, the element values sum to one (i.e., if all possible values of X and W have been included, then the probability is one that X and W each will take on one of the included values). Therefore, once again we have probability vectors. In this case, they can be called *conditional-probability vectors.*

g) Let $\mathbf{Pr}[W;X]$, $\mathbf{Pr}[Z;Y]$, $\mathbf{Pr}[X;Y]$ be arrays of conditional probability vectors, one for each possible value of the random variable on the right-hand side of the semicolon. By convention, if the probability vectors are horizontal (row vectors), they will be arrayed vertically:

$$\mathbf{Pr}[Y;Z] = \begin{bmatrix} \mathbf{Pr}[Y;Z=\mathbb{0} \\ \mathbf{Pr}[Y;Z=\mathbb{1}] \\ \mathbf{Pr}[Y;Z=\mathbb{2}] \\ \vdots \\ \mathbf{Pr}[Y;Z=Z_m] \end{bmatrix} \quad \text{(A.3)}$$

and if the probability vectors are vertical (column vectors), they will be arrayed horizontally:

$$\mathbf{Pr}[W;X] = [\mathbf{Pr}[W;X=\mathbb{0}], \mathbf{Pr}[W;X=\mathbb{1}], \mathbf{Pr}[W;X=\mathbb{2}], \ldots, \mathbf{Pr}[W;X=X_m]]. \quad \text{(A.4)}$$

Thus we have a two-dimensional array of individual conditional probabilities (the elements of the vectors). Each of the random variables in the argument of the array (e.g., W and X in $\mathbf{Pr}[W;X]$) can be considered as one of the axes of the array. Thus, the coordinates of an element simply become the corresponding values of the arguments.

$$\mathbf{Pr}[Y;Z] =$$

$$\begin{bmatrix} \Pr[Y=\mathbb{0};Z=\mathbb{0}] & \Pr[Y=\mathbb{1};Z=\mathbb{0}] & \Pr[Y=\mathbb{2};Z=\mathbb{0}] & \ldots & \Pr[Y=Y_m;Z=\mathbb{0}] \\ \Pr[Y=\mathbb{0};Z=\mathbb{1}] & \Pr[Y=\mathbb{1};Z=\mathbb{1}] & \Pr[Y=\mathbb{2};Z=\mathbb{1}] & \ldots & \Pr[Y=Y_m;Z=\mathbb{1}] \\ \Pr[Y=\mathbb{0};Z=\mathbb{2}] & \Pr[Y=\mathbb{1};Z=\mathbb{2}] & \Pr[Y=\mathbb{2};Z=\mathbb{2}] & \ldots & \Pr[Y=Y_m;Z=\mathbb{2}] \\ \ldots & \ldots & \ldots & \ldots & \ldots \\ \Pr[Y=\mathbb{0};Z=Z_m] & \Pr[Y=\mathbb{1};Z=Z_m] & \Pr[Y=\mathbb{2};Z=Z_m] & \ldots & \Pr[Y=Y_m;Z=Z_m] \end{bmatrix}$$

(A.5)

or

$$\mathbf{Pr}[W;X]=$$

$$\begin{bmatrix} \Pr[W=\mathbb{0};X=\mathbb{0}] & \Pr[W=\mathbb{0};X=\mathbb{1}] & \Pr[W=\mathbb{0};X=\mathbb{2}] & \dots & \Pr[W=\mathbb{0};X=X_m] \\ \Pr[W=\mathbb{1};X=\mathbb{0}] & \Pr[W=\mathbb{1};X=\mathbb{1}] & \Pr[W=\mathbb{1};X=\mathbb{2}] & \dots & \Pr[W=\mathbb{1};X=X_m] \\ \Pr[W=\mathbb{2};X=\mathbb{0}] & \Pr[W=\mathbb{2};X=\mathbb{1}] & \Pr[W=\mathbb{2};X=\mathbb{2}] & \dots & \Pr[W=\mathbb{2};X=X_m] \\ \dots & \dots & \dots & \dots & \dots \\ \Pr[W=W_m;X=\mathbb{0}] & \Pr[W=W_m;X=\mathbb{1}] & \Pr[W=W_m;X=\mathbb{2}] & \dots & \Pr[W=W_m;X=X_m] \end{bmatrix} \tag{A.6}$$

In A.5, the Z-axis is vertical, the Y-axis is horizontal; in A.6, the W-axis is vertical, the X-axis is horizontal. In A.6, the coordinates of $\Pr[W=k;X=j]$ simply are j (horizontal) and k (vertical); in A.5, the coordinates of $\Pr[Y=i;\ Z=h]$ are i (horizontal), h (vertical). The origin, by convention, is in the upper left-hand corner of the array; and horizontal is measured to the right, vertical is measured down. Such arrays are known as *conditional-probability matrices.*

Generally, conditional probability matrices display an inherent anisotropy. The values of their elements sum to one, in one direction, but not necessarily in the other. Thus, in A.5, the sum of elements taken in the horizontal direction (i.e., the sum of elements in a given row) is one; but taken in the vertical direction (i.e., the sum of elements in a given column) it need not be one. Similarly, in A.6, the sum of elements down a given column is one; but the sum of elements across a given row need not be one. Therefore, we can consider A.5 to be a vertical array of probability row vectors, but we cannot consider it to be a horizontal array of probability column vectors. In A.6, the situation is reversed. In neither case is the situation ambiguous. In other words, given the conditional-probability matrix with element values, one generally would be able immediately to determine whether it was a vertical array of probability row vectors or a horizontal array of probability column vectors.

h) Let $\mathbf{Pr}[W;--,--]$ be the probability vector whose elements are the conditional probabilities for the possible values of the variable on the left, given that the two statements on the right are true. $\mathbf{Pr}[W;X,Y]$ then can be considered an array of conditional probability vectors distributed over the X and Y axes (i.e., a vector $\mathbf{Pr}[W;\ X=i,\ Y=j]$ for every pair of coordinates, i and j). By extension, $\mathbf{Pr}[W;X,Y,Z]$ can be considered an array of conditional-probability vectors distributed over the X, Y and Z axes (i.e., a three-dimensional array of vectors). In each case, the sums of elements taken along the axis of the vector itself (i.e., the W axis) will be one; taken along any other axis it need not be one (and generally will not be one). Therefore, the arrays exhibit inherent anisotropy and are unambiguous with respect to which sets of element form conditional-probability vectors.

i) Let $\mathbf{Pr}[W,X]$, $\mathbf{Pr}[X,Z]$, $\mathbf{Pr}[Y,W]$, etc., be two-dimensional arrays of joint probabilities. Thus, for example

$\mathbf{Pr}[W, Y] =$

$$\begin{bmatrix} \Pr[W=0, Y=0] & \Pr[W=1, Y=0] & \Pr[W=2, Y=0] & \dots & \Pr[W=W_m, Y=0] \\ \Pr[W=0, Y=1] & \Pr[W=1, Y=1] & \Pr[W=2, Y=1] & \dots & \Pr[W=W_m, Y=1] \\ \Pr[W=0, Y=2] & \Pr[W=1, Y=2] & \Pr[W=2, Y=2] & \dots & \Pr[W=W_m, Y=2] \\ \dots & \dots & \dots & \dots & \dots \\ \Pr[W=0, Y=Y_m] & \Pr[W=1, Y=Y_m] & \Pr[W=2, Y=Y_m] & \dots & \Pr[W=W_m, Y=Y_m] \end{bmatrix}. \tag{A.7}$$

By extension, let $\mathbf{Pr}[W, Y, Z]$, $\mathbf{Pr}[W, X, Y, Z]$ be multidimensional arrays of joint probabilities. None of these arrays exhibits the inherent anisotropy of arrays of conditional-probability vectors. For example, in the matrix of A.7 each pair of coordinates $(W=i, Y=j)$ represents a possibility, and all possibilities are represented. Therefore, the sum of *all* elements in the matrix will be one; but, except under extraordinary circumstances, the sum of the elements along any row or any column will be less than one. Similarly, in a complete n-dimensional array of joint probabilities the sum of all elements will be one, and the sum along any single axis generally will be less than one. Clearly, an array of joint probabilities is unambiguously distinct from an array of conditional-probability vectors.

A.2. Manipulation of Arrays

a) Addition. This operation is carried out on an element-by-element basis with arrays containing the same number of elements. For example, the addition of two vectors is carried out as follows:

$$\begin{bmatrix} a_1 \\ b_1 \\ c_1 \\ \vdots \\ m_1 \end{bmatrix} + \begin{bmatrix} a_2 \\ b_2 \\ c_2 \\ \vdots \\ m_2 \end{bmatrix} = \begin{bmatrix} a_1+a_2 \\ b_1+b_2 \\ c_1+c_2 \\ \vdots \\ m_1+m_2 \end{bmatrix} \tag{A.8}$$

and the addition of two matrices is carried out as follows:

$$\begin{bmatrix} a_1 & b_1 & c_1 \\ d_1 & e_1 & f_1 \\ g_1 & h_1 & i_1 \end{bmatrix} + \begin{bmatrix} a_2 & b_2 & c_2 \\ d_2 & e_2 & f_2 \\ g_2 & h_2 & i_2 \end{bmatrix} = \begin{bmatrix} a_1+a_2 & b_1+b_2 & c_1+c_2 \\ d_1+d_2 & e_1+e_2 & f_1+f_2 \\ g_1+g_2 & h_1+h_2 & i_1+i_2 \end{bmatrix}. \tag{A.9}$$

Clearly, the addition of arrays conforms to the commutative and associative laws of addition. If $\boldsymbol{A}$, $\boldsymbol{B}$, and $\boldsymbol{C}$ are arrays (with the same number of elements, then

$$\boldsymbol{A}+\boldsymbol{B}=\boldsymbol{B}+\boldsymbol{A} \tag{A.10}$$

and

$$\boldsymbol{A}+(\boldsymbol{B}+\boldsymbol{C})=(\boldsymbol{A}+\boldsymbol{B})+\boldsymbol{C}. \tag{A.11}$$

Equally clearly, the appropriate registration must be maintained between arrays if the proper elements are to be added to one another. By convention, the ordering of vectors is from left to right or from top down, so alignment for addition is unambiguous as long as the convention is followed. In matrices and arrays of higher dimension, however, the axes must be aligned appropriately.

b) Scalar Multiplication. This operation is carried out with any factor that either is a real number or is a symbol representing a real number. Such a factor is called a *scalar*. Thus, for example, the elements of a probability vector all are scalars. If any array is to be multiplied by a scalar, every element in that array simply is multiplied by the scalar in order to form the product array. Thus, for example,

$$k\begin{bmatrix} a_1 \\ b_1 \\ c_1 \\ \vdots \\ m_1 \end{bmatrix} = \begin{bmatrix} ka_1 \\ kb_1 \\ kc_1 \\ \vdots \\ km_1 \end{bmatrix} \quad \text{and} \quad k\begin{bmatrix} a_1 & b_1 \\ c_1 & d_1 \end{bmatrix} = \begin{bmatrix} ka_1 & kb_1 \\ kc_1 & kd_1 \end{bmatrix}. \tag{A.12}$$

The operation is taken to be commutative,

$$k\boldsymbol{A} = \boldsymbol{A}k \tag{A.13}$$

and it is associative in the following sense:

$$h(k\boldsymbol{A}) = (hk)\boldsymbol{A}. \tag{A.14}$$

c) Inner Product of Two Arrays. This operation is carried out between an array of scalars and a second array of the same dimension, with an element corresponding to each scalar in the first array. However, the elements of the second array may themselves be arrays of any dimension. Scalar multiplication is carried out between the element of the second array and the corresponding scalar of the first array; then all of the scalar products so formed are summed to yield a single, inner product array whose dimensions are precisely those of the elements of the second array. Thus, for example, the inner product of two arrays of scalars is itself a scalar:

$$\begin{aligned} &[a_1, a_2, a_3, \dots, a_m] \cdot [b_1, b_2, b_3, \dots, b_m] \\ &= a_1 b_1 + a_2 b_2 + a_3 b_3 + \cdots + a_m b_m \end{aligned} \tag{A.15}$$

(note that the conventional symbol for inner product is a dot, $\boldsymbol{A} \cdot \boldsymbol{B}$). The inner product of an array of scalars and an array of arrays is an array:

$$\begin{aligned} &[a_1, a_2, a_3, \dots, a_m] \cdot [\boldsymbol{B}_1, \boldsymbol{B}_2, \boldsymbol{B}_3, \dots, \boldsymbol{B}_m] \\ &= a_1 \boldsymbol{B}_1 + a_2 \boldsymbol{B}_2 + a_3 \boldsymbol{B}_3 + \cdots + a_m \boldsymbol{B}_m. \end{aligned} \tag{A.16}$$

The array of scalars may be of any dimension; the array of arrays must be of the same dimension. Clearly, appropriate registration once again is crucial to the success of the operation. Ambiguities certainly may arise when the elements of one of the arrays are themselves arrays. For example, consider the formation of the inner product of a square matrix and a vector:

$$[a_2, b_2, c_2] \cdot \begin{bmatrix} a_1 & b_1 & c_1 \\ d_1 & e_1 & f_1 \\ g_1 & h_1 & i_1 \end{bmatrix} = ? \tag{A.17}$$

do we consider the matrix to be an array of three row vectors or an array of three column vectors? By convention, when scalars are arrayed horizontally to form vectors, vectors are arrayed vertically to form matrices; and vice versa. Therefore, there is a standard, unambiguous interpretation of the situation depicted in A.17. We are dealing with a vertical array of row vectors:

$$
\begin{aligned}
&[a_2, b_2, c_2] \cdot \begin{bmatrix} a_1 & b_1 & c_1 \\ d_1 & e_1 & f_1 \\ g_1 & h_1 & i_1 \end{bmatrix} \\
&= a_2[a_1, b_1, c_1] + b_2[d_1, e_1, f_1] + c_2[g_1, h_1, i_1] \\
&= [(a_2 a_1 + b_2 d_1 + c_2 g_1), (a_2 b_1 + b_2 e_1 + c_2 h_1), (a_2 c_2 + b_2 f_1 + c_2 i_1)].
\end{aligned} \tag{A.18}
$$

Although conventions analogous to this could be extended to arrays of higher dimension, ambiguities with respect to the registration of such arrays can be eliminated rather easily by careful notation with respect to axes. Furthermore, the formation of the inner product in such cases generally is carried out on a digital computer, in which case registration usually is achieved through consistent use of indices.

d) Array Multiplication. The multiplier by convention is taken to be an array of elements distributed over one set of dimensions; and by the process of array multiplication those elements are to be transformed into a set of similar elements (elements of the same dimensionality and size) distributed over another set of dimensions. Thus, the multiplier might be an array of $\mathfrak{G}$-element vectors (vectors comprising $\mathfrak{G}$ scalars each) distributed over the X and Y dimensions; and the process of multiplication might transform them into an array of $\mathfrak{G}$-element vectors distributed over the Z dimension. Thus, in general, one can consider array multiplication as a mapping of arrays from one space into another. Corresponding to each location in the target space there must be an array of scalars, one for each location in the original space. Therefore, the multiplicand conventionally is taken to be an array of elements distributed over the target space, with the elements themselves being arrays of scalars, one representing every location in the original space. The inner product is taken between the element at each location in the target space and the entire array of elements in the original space; and the result becomes the new element at that location in the target space. Thus, if $\boldsymbol{A}$ is the multiplier array and if $\boldsymbol{B}_i$ is the array of scalars making up the element at location i in the target space, then the process of array multiplication leads to a new element, $\boldsymbol{C}_i$, at location i:

$$\boldsymbol{C}_i = \boldsymbol{A} \cdot \boldsymbol{B}_i. \tag{A.19}$$

In the case of two matrices, the multiplier and the multiplicand both are considered to be horizontal arrays of column vectors. Those in the multiplicand have a scalar for each vector in the multiplier. The product is formed as illustrated in the following example

$$\boldsymbol{A} = \begin{bmatrix} a_{11} & a_{21} & a_{31} & a_{41} \\ a_{12} & a_{22} & a_{32} & a_{42} \\ a_{13} & a_{23} & a_{33} & a_{43} \end{bmatrix}$$

$$\boldsymbol{B}=\begin{bmatrix} b_{11} & b_{21} \\ b_{12} & b_{22} \\ b_{13} & b_{23} \\ b_{14} & b_{24} \end{bmatrix}$$

$$\boldsymbol{B}_1=\begin{bmatrix} b_{11} \\ b_{12} \\ b_{13} \\ b_{14} \end{bmatrix} \qquad \boldsymbol{B}_2=\begin{bmatrix} b_{12} \\ b_{22} \\ b_{23} \\ b_{24} \end{bmatrix}$$

$$\boldsymbol{A}_1=\begin{bmatrix} a_{11} \\ a_{12} \\ a_{13} \end{bmatrix} \qquad \boldsymbol{A}_2=\begin{bmatrix} a_{22} \\ a_{22} \\ a_{23} \end{bmatrix} \qquad \boldsymbol{A}_3=\begin{bmatrix} a_{31} \\ a_{32} \\ a_{33} \end{bmatrix} \qquad \boldsymbol{A}_4=\begin{bmatrix} a_{41} \\ a_{42} \\ a_{43} \end{bmatrix} \tag{A.20}$$

$$\boldsymbol{AB}=[\boldsymbol{B}_1\cdot\boldsymbol{A},\boldsymbol{B}_2\cdot\boldsymbol{A}]$$

$$\boldsymbol{B}_1\cdot\boldsymbol{A}=b_{11}\boldsymbol{A}_1+b_{12}\boldsymbol{A}_2+b_{13}\boldsymbol{A}_3+b_{14}\boldsymbol{A}_4 \quad =\begin{bmatrix} b_{11}a_{11}+b_{12}a_{21}+b_{13}a_{31}+b_{14}a_{41} \\ b_{11}a_{12}+b_{12}a_{22}+b_{13}a_{32}+b_{14}a_{42} \\ b_{11}a_{13}+b_{12}a_{23}+b_{13}a_{33}+b_{14}a_{43} \end{bmatrix}$$

$$\boldsymbol{B}_2\cdot\boldsymbol{A}=b_{21}\boldsymbol{A}_1+b_{22}\boldsymbol{A}_2+b_{23}\boldsymbol{A}_3+b_{24}\boldsymbol{A}_4 \quad =\begin{bmatrix} b_{21}a_{11}+b_{22}a_{21}+b_{23}a_{31}+b_{24}a_{41} \\ b_{21}a_{12}+b_{22}a_{22}+b_{23}a_{32}+b_{24}a_{42} \\ b_{21}a_{13}+b_{22}a_{23}+b_{23}a_{33}+b_{24}a_{43} \end{bmatrix}$$

Thus, for two-dimensional arrays registration is prescribed by convention. For multiplication of arrays of higher dimensionality, registration must be achieved through notation (e.g., through consistent use of indices or symbols for axes).

e) Relationship between Array Multiplication and Inner Product Formation. The process of array multiplication obeys the associative law, but not the commutative law. Thus, if we have several arrays, $\boldsymbol{A}$, $\boldsymbol{B}$, $\boldsymbol{C}$, $\boldsymbol{D}$, etc., the following statements are true:

$$\begin{gathered} (\boldsymbol{A}(\boldsymbol{BC}))=(\boldsymbol{AB})\boldsymbol{C}=(\boldsymbol{ABC}) \\ \boldsymbol{A}(\boldsymbol{BC})\boldsymbol{D}=(\boldsymbol{AB})(\boldsymbol{CD})=(\boldsymbol{ABC})\boldsymbol{D}=\boldsymbol{A}(\boldsymbol{BCD}) \end{gathered} \tag{A.21}$$

$$\begin{gathered} \boldsymbol{AB}\neq\boldsymbol{BA} \\ \boldsymbol{ABC}\neq\boldsymbol{CAB}. \end{gathered} \tag{A.22}$$

Conversely, the process of inner-product formation obeys the commutative law, but not the associative law. Thus

$$\begin{gathered} \boldsymbol{A}\cdot(\boldsymbol{B}\cdot\boldsymbol{C})\neq(\boldsymbol{A}\cdot\boldsymbol{B})\cdot\boldsymbol{C} \\ \boldsymbol{A}\cdot(\boldsymbol{B}\cdot\boldsymbol{C})\cdot\boldsymbol{D}\neq(\boldsymbol{A}\cdot\boldsymbol{B})\cdot(\boldsymbol{C}\cdot\boldsymbol{D})\neq(\boldsymbol{A}\cdot\boldsymbol{B}\cdot\boldsymbol{C})\cdot\boldsymbol{D} \\ \boldsymbol{A}\cdot\boldsymbol{B}=\boldsymbol{B}\cdot\boldsymbol{A} \\ \boldsymbol{A}\cdot\boldsymbol{B}\cdot\boldsymbol{C}=\boldsymbol{C}\cdot\boldsymbol{A}\cdot\boldsymbol{B}. \end{gathered} \tag{A.23}$$

The two processes are interchangeable, however; so one may choose between them on the basis of whether the associative property or the commutative property is more convenient. Thus, an ordered sequence of nonassociative inner-product formations may be interchanged for an ordered array of noncommutative multiplications:

$$(\boldsymbol{ABCDEFG})\cdot\boldsymbol{H}=\boldsymbol{A}\cdot(\boldsymbol{B}\cdot(\boldsymbol{C}\cdot(\boldsymbol{D}\cdot(\boldsymbol{E}\cdot(\boldsymbol{F}\cdot(\boldsymbol{G}\cdot\boldsymbol{H})))))). \tag{A.24}$$

Of course there are other ways to generalize array multiplication and inner-product formation. However, with some reflection, the reader will find that the generalizations presented here are quite heuristic. Consider for example, the state projection matrix for a three-state system (see section 1.13):

$$\boldsymbol{P}=\begin{bmatrix} p_{1,1} & p_{1,2} & p_{1,3} \\ p_{2,1} & p_{2,2} & p_{2,3} \\ p_{3,1} & p_{3,2} & p_{3,3} \end{bmatrix}. \tag{A.25}$$

It can be envisioned as comprising three column vectors:

$$\boldsymbol{P}_1=\begin{bmatrix} p_{1,1} \\ p_{2,1} \\ p_{3,1} \end{bmatrix} \qquad \boldsymbol{P}_2=\begin{bmatrix} p_{1,2} \\ p_{2,2} \\ p_{3,2} \end{bmatrix} \qquad \boldsymbol{P}_3=\begin{bmatrix} p_{1,3} \\ p_{2,3} \\ p_{3,3} \end{bmatrix} \tag{A.26}$$

where $\boldsymbol{P}_1$ is the probability distribution that one would expect at $\tau+\mathbb{1}$ if he or she knew that the object was in state U_1 at τ; $\boldsymbol{P}_2$ would be the expected distribution at $\tau+\mathbb{1}$ if the object were in U_2 at τ; and $\boldsymbol{P}_3$ is the expected distribution at $\tau+\mathbb{1}$ if the object were in U_3 at τ. To find the actual expected distribution at $\tau+\mathbb{1}$, one simply takes the sum of these three conditional probability distributions, each weighted by the probability that its condition was fulfilled:

$$\mathbf{Pr}[U(\tau+\mathbb{1})]=\boldsymbol{P}_1\Pr[U_1(\tau)]+\boldsymbol{P}_2\Pr[U_2(\tau)]+\boldsymbol{P}_3\Pr[U_3(\tau)]. \tag{A.27}$$

Thus,

$$\mathbf{Pr}[U(\tau+\mathbb{1})]=\boldsymbol{P}\cdot\mathbf{Pr}[U(\tau)] \tag{A.28}$$

and $\mathbf{Pr}[U(\tau)]$ is taken to be an array of scalars, each being a weighting factor for one of the vectors in $\boldsymbol{P}$.

Next, consider the product of arrays

$$\boldsymbol{P}'=\boldsymbol{PP}. \tag{A.29}$$

It is our intention that $\boldsymbol{P}'$ be a matrix that projects $\mathbf{Pr}[U(\tau)]$ into $\mathbf{Pr}[U(\tau+\mathbb{2})]$. Thus,

$$\mathbf{Pr}[U(\tau+\mathbb{2})]=\boldsymbol{P}'\cdot\mathbf{Pr}[U(\tau)] \tag{A.30}$$

from which it is clear, by extension, that $\boldsymbol{P}'$ should comprise three column vectors, $\boldsymbol{P}_1'$, $\boldsymbol{P}_2'$, and $\boldsymbol{P}_3'$, the expected distributions at $\tau+\mathbb{2}$ if the object were in state U_1, U_2, or U_3 respectively at τ. Therefore, the process of multiplication of $\boldsymbol{P}$ by itself should generate $\boldsymbol{P}_1'$, $\boldsymbol{P}_2'$ and $\boldsymbol{P}_3'$. Now, if the object were in state U_1 at τ, its distribution would have been $\boldsymbol{P}_1$ at $\tau+\mathbb{1}$. Therefore, its distribution at $\tau+\mathbb{2}$ would be $\boldsymbol{P}\cdot\boldsymbol{P}_1$. Similarly, if it had been in U_2 at τ, its distribution at $\tau+\mathbb{2}$ would be $\boldsymbol{P}\cdot\boldsymbol{P}_2$; and if it had been in U_3 at τ, its distribution would be $\boldsymbol{P}\cdot\boldsymbol{P}_3$ at $\tau+\mathbb{2}$.

Therefore, ***PP*** should be an array of the following column vectors

$$\begin{aligned} \boldsymbol{P}_1' &= \boldsymbol{P} \cdot \boldsymbol{P}_1 \\ \boldsymbol{P}_2' &= \boldsymbol{P} \cdot \boldsymbol{P}_2 \\ \boldsymbol{P}_3' &= \boldsymbol{P} \cdot \boldsymbol{P}_3 \end{aligned} \tag{A.31}$$

just as prescribed by the method of array multiplication.

f) Transpose of a Matrix: Considering a matrix to be an array of row vectors, one can form its *transpose* simply by rearranging the elements so that the same vectors appear in column form and in the same order, with the uppermost row vector becoming the leftmost column vector. Thus if the matrix $\boldsymbol{A}$ has the following arrangement of elements:

$$\boldsymbol{A} = \begin{bmatrix} a & b & c & d \\ e & f & g & h \\ i & j & k & l \end{bmatrix} \tag{A.32}$$

then its transpose, $\boldsymbol{A}^t$, has the following arrangement:

$$\boldsymbol{A}^t = \begin{bmatrix} a & e & i \\ b & f & j \\ c & g & k \\ d & h & l \end{bmatrix}. \tag{A.33}$$

A.3. Operations on Probability Arrays

a) Rearrangement of Coordinates. This operation is precisely analogous to the formation of the transpose of a matrix, but it can be carried out in arrays of any dimensionality. If probability arrays and arrays of probability arrays are designated by the notation of section A.1 of this appendix (i.e., in the form $\mathbf{Pr}[Q, R, S, T;\ W, X, Y, Z]$) then any pair of axes (e.g., Q and R) separated by a comma can be exchanged by simple rearrangement of the array.

b) Expanding the Dimensionality of a Conditional-Probability Array. This operation simply involves distribution of an existing array over the values of another axis. In arrays of the form $\mathbf{Pr}[Q, R, S, T;\ W, X, Y, Z]$, joint-probability arrays of the form $\mathbf{Pr}[Q, R, S, T]$ are distributed over the coordinates of W, X, Y, Z. Any axis on the right-hand side of the semicolon can be carried across to the left-hand side to provide a new array, such as $\mathbf{Pr}[Q, R, S, T, W;\ W, X, Y, Z]$. In this new array, joint-probability arrays of the form $\mathbf{Pr}[Q, R, S, T, W]$ are distributed over W, X, Y, Z. Of course the element values of this new array are zero except for the value of W corresponding to the array's location in W, X, Y, Z. In other words,

$$\begin{aligned} &\Pr[Q=a,\ R=b,\ S=c,\ T=d,\ W=e;\ W=e,\ X=f,\ Y=g,\ Z=h] \\ &\quad = \Pr[Q=a,\ R=b,\ S=c,\ T=d;\ W=e,\ X=f,\ Y=g,\ Z=h] \end{aligned} \tag{A.34}$$

but

$$\Pr[Q=a,\ R=b,\ S=c,\ T=d,\ W=e;\ W=i,\ X=f,\ Y=g,\ Z=h]=0.$$

c) Reconditioning an Array. This operation simply transforms the coordinates over which probability vectors or joint-probability arrays are distributed, basically exchanging one set of coordinates for another. It is accomplished directly by array multiplication, as described in section A.2d of this appendix. Thus, for example,

$$\begin{aligned}
&\mathbf{Pr}[W;Z]=\mathbf{Pr}[W;X]\,\mathbf{Pr}[X;Z]\\
&\mathbf{Pr}[W;Z]=\mathbf{Pr}[W;X,Y]\,\mathbf{Pr}[X,Y;Z]\\
&\mathbf{Pr}[Q,R;Y,Z]=\mathbf{Pr}[Q,R;W,X]\,\mathbf{Pr}[W,X;Y,Z] \qquad \text{(A.35)}\\
&\mathbf{Pr}[Q;Z]=\mathbf{Pr}[Q;R]\,\mathbf{Pr}[R;S]\,\mathbf{Pr}[S;T]\,\mathbf{Pr}[T;W]\,\mathbf{Pr}[W;X]\,\mathbf{Pr}[X;Y]\,\mathbf{Pr}[Y;Z]\\
&\mathbf{Pr}[U(\tau);U(\mathbb{0})]=\mathbf{Pr}[U(\tau);U(\tau-\mathbb{1})]\,\mathbf{Pr}[U(\tau-\mathbb{1});U(\tau-\mathbb{2})]\ldots\mathbf{Pr}[U(\mathbb{1});U(\mathbb{0})].
\end{aligned}$$

d) Deconditioning an Array. This operation simply combines a set of probability vectors or joint-probability arrays distributed over a set of coordinates, such as X, Y, Z, into a single, average vector or array. This is accomplished directly by the process of inner-product formation, described in section A.2c. Thus

$$\begin{aligned}
&\mathbf{Pr}[W]=\mathbf{Pr}[W;X]\cdot\mathbf{Pr}[X]\\
&\mathbf{Pr}[W]=\mathbf{Pr}[W;X,Y,Z]\cdot\mathbf{Pr}[X,Y,Z] \qquad \text{(A.36)}\\
&\mathbf{Pr}[Q,R,S,T]=\mathbf{Pr}[Q,R,S,T;W,X,Y,Z]\cdot\mathbf{Pr}[W,X,Y,Z].
\end{aligned}$$

e) Combination of a Conditional-Probability Array and a Probability Vector or Joint-Probability Array to Produce an Expanded Joint-Probability Array. This operation is carried out in two steps. First, elements of the original probability vector or joint-probability array are redistributed into a new space, with twice the dimensionality of the old one. The way that this is done can be illustrated by an example. Let the original joint-probability array be $\mathbf{Pr}[Q,R,S,T]$. The redistributed array becomes $\mathbf{Pr}[(Q,R,S,T),\ Q,R,S,T]$. In other words, at each location in Q,R,S,T we now have a partial joint-probability array of the form $\mathbf{Pr}[Q,R,S,T]$. That array is *partial* because its elements do not sum to one; in fact it has only one nonzero element, namely that corresponding to the location of the array in Q,R,S,T. The array $\mathbf{Pr}[(Q,R,S,T),\ Q,R,S,T]$ is called a diagonal array, because its only nonzero elements occur along a single diagonal in the space or hyperspace Q, R, S, T, Q, R, S, T. Thus, for example, the elements of the vector $[a,b,c,d]$ would be redistributed to form the diagonal matrix

$$\begin{bmatrix} a & 0 & 0 & 0\\ 0 & b & 0 & 0\\ 0 & 0 & c & 0\\ 0 & 0 & 0 & d \end{bmatrix}.$$

The second step of the operation simply is array multiplication of the redistributed array and the conditional-probability array, with the elements of the multiplicand (the redistributed array) taken to be the partial joint probability arrays. Thus, for example,

$$\begin{aligned}&\mathbf{Pr}[X, Y]=\mathbf{Pr}[X; Y]\,\mathbf{Pr}[(Y), Y]\\&\mathbf{Pr}[W, X, Y]=\mathbf{Pr}[W; X, Y]\,\mathbf{Pr}[(X, Y), X, Y].\end{aligned} \tag{A.37}$$

The entire process can be described much more simply on an element-by-element basis. For example

$$\begin{aligned}&\mathbf{Pr}[X, Y=i]=\mathbf{Pr}[X; Y=i]\,\Pr[Y=i]\\&\mathbf{Pr}[W, X=i, Y=j]=\mathbf{Pr}[W; X=i, Y=j]\,\Pr[X=i, Y=j].\end{aligned} \tag{A.38}$$

The formation of a diagonal array provides an excellent example of a task that is difficult to describe in terms of general array operations yet is extremely easy and straightforward to execute. This simply is because one can ignore completely all of the zero elements in the diagonal array when he carries out the operation. Once the operation itself is understood, however, the array notation of equation A.37 provides an excellent shorthand method by which to represent it explicitly. This is true as well of the other operations on probability arrays described in this appendix.

f) Bayes's Rule. The operation corresponding to Bayes's Rule combines a conditional probability array and a probability vector or joint-probability array to produce the inverse of the original conditional-probability array. It is most easily described on an element-by-element basis:

$$\begin{aligned}&\Pr[Q=a,\ R=b,\ S=c,\ T=d;\ X=e,\ Y=f,\ Z=g]\\&\quad=\Pr[X=e,\ Y=f,\ Z=g;\ Q=a,\ R=b,\ S=c,\ T=d]\\&\quad\ \ \Pr[Q=a,\ R=b,\ S=c,\ T=d]/\Pr[X=e,\ Y=f,\ Z=g]\end{aligned} \tag{A.39}$$

where $\Pr[X=e,\ Y=f,\ Z=g]$ is one scalar entry in the array

$$\mathbf{Pr}[X, Y, Z]=\mathbf{Pr}[X,\ Y,\ Z; Q, R, S, T]\cdot\mathbf{Pr}[Q, R, S, T]. \tag{A.40}$$

Thus, from $\mathbf{Pr}[X, Y, Z; Q, R, S, T]$ and $\mathbf{Pr}[Q, R, S, T]$ we can derive $\mathbf{Pr}[Q, R, S, T; X, Y, Z]$.

Appendix B. Bernoulli Trials and the Binomial Distribution

In a population with polygamous mating, a receptive female might wander about the habitat in search of males. If she mates with every male she encounters, and if the males are distributed over the habitat in such a fashion that she is equally likely to encounter any one of them, then we have a situation that conforms to a very simple model in probability theory, namely the Bernoulli Trial. Similarly, a predator might wander about the habitat in search of prey. If it eats every prey that it encounters and if the prey are distributed in such a way that the predator is equally likely to encounter any one of them, then once again we have a situation conforming to the Bernoulli Trial model.

A classic example of Bernoulli Trials is the tossing of identical coins. Consider a set of n coins, identical except for the fact that each bears an identifying number. When the coins are tossed, either in sequence or simultaneously, then each toss represents a Bernoulli Trial, and the set of tosses represents a set of Bernoulli Trials. In order to be classified as such, each toss must

1) produces one of two outcomes
2) with the probability of each outcome being absolutely independent of the outcome of any other trial (toss) in the set
3) and with the probability of each outcome being the same for every trial (toss).

The two possible outcomes for each trial are heads (H) and tails (T). Let p be the probability that a particular toss results in heads and q be the probability that it results in tails. Evidently, since all possible outcomes of the toss have been included,

$$p+q=\mathbb{1}. \tag{B.1}$$

For motivation with respect to the following development, the reader might associate H with encountering a particular prey in a given transition interval, T with failure to encounter that prey in a given transition interval.

Now let us consider the various possible combinations of outcomes over the entire set of trials. Since each trial is independent of the others (i.e., the probability of its outcome is totally independent of the outcome of the other trials), the probability of any combination of outcomes simply is the product of the probabilities of the individual outcomes themselves. Because we have numbered the coins, we can depict each combination of outcomes as an ordered set, as is done in the following example:

Trial Number = Coin Number	1 2 3 4 5 6 7 8 9 10 ... n
Outcome = Result of Toss	H H H T T H T T H T ... H
Probability of Outcome	p p p q q p q q p q ... p
Probability of Combination	$p\times p\times p\times q\times q\times p\times q\times q\times p\times q$ ... $p=p^k q^{n-k}$

Beneath each trial in the set, we have placed the probability of the outcome we have specified. Then, in the fourth row, we have computed the probability of the specified combination of outcomes by taking the product of the individual probabilities, leading to the compact form $p^k q^{n-k}$, where k is the number of heads specified in the combination of outcomes and $n-k$ is the number of tails.

Of course, we have specified only one of many possible combinations of outcomes with k heads and $n-k$ tails. If we consider each of these combinations only once, then each will be distinct and will not include any of the others (i.e., the considered combinations will be mutually exclusive). Therefore, we can determine very easily the probability that whatever combination actually occurs will have exactly k heads and $n-k$ tails. That probability is the sum of the probabilities of the distinct individual combinations with k heads and $n-k$ tails, which is the same thing as the product of the number of such combinations and the probability of any given one of them. Thus, we merely must ask the question "How many distinct ways are there to distribute exactly k heads over n coins?" The answer simply is $\binom{n}{k}$, which is often called "n choose k." Of course, if we have exactly k heads and exactly n coins, we must have $n-k$ tails. It follows that the number of ways to distribute $n-k$ tails over n coins is precisely the same as the number of ways to distribute k heads over n-coins:

$$\binom{n}{k}=\binom{n}{n-k}. \tag{B.2}$$

The terms of the expanded version of n choose k can be interpreted rather easily.

$$\binom{n}{k}=\frac{n!}{(n-k)!\,k!}=\frac{n\cdot(n-1)\cdot(n-2)\cdots(n-k+1)}{k!} \tag{B.3}$$

If we are selecting exactly k coins that will be heads, there are n coins from which to select the first coin, $n-1$ coins left from which to select the second, and so forth, on to $n-k+1$ coins from which to select the kth: the remaining coins will be designated as tails. Thus, the numerator on the right is the total number of ways to make the k selections. However, in making such selections, we will repeat each distinct combination many times. If, in a given combination, the xth, yth and zth coins are designated as heads, then the xth coin could have been selected on any of the k choices (i.e., it could have been selected as the first coin to be designated heads, the second coin to be so designated, the third, and so forth, on to the kth). Once we designate the number of the choice on which the

xth coin was selected to be a head, we are left with $k-1$ choices on which the yth coin could be selected as a head, after which we are left with $k-2$ for the zth coin, and so forth, on through all of the k coins to be selected for a given combination. Thus, each distinct combination appears $k!$ times; so the denominator on the right-hand side of equation B.3 represents the redundancy in our prescribed method of selection.

We now can write the important, general relationship for any set of Bernoulli trials:

$$\Pr[\text{exactly } k \text{ of outcome } H \text{ in } n \text{ trials}] = \binom{n}{k} p^k q^{n-k} \tag{B.4}$$

where p is the probability of H on any given trial; $q=1-p$. This, of course, is the binomial distribution (see section 1.14). It is often written in abbreviated form as $b(k;n,p)$, where

BINOMIAL DISTRIBUTION

$$b(k;n,p) = \binom{n}{k} p^k q^{n-k} \tag{B.5}$$

and k is the number of occurrences of a given outcome out of n Bernoulli trials. Another useful relationship, easily derived from equation B.5, is the probability of at least k occurrences of a given outcome. To determine this probability, we simply consider the possibilities of k occurrences, $k+1$ occurrences, $k+2$ occurrences, and so on, up to the maximum possible number of occurrences, which is n, the number of trials. Since each of these possibilities is distinct, the probability that we seek simply is the sum of their individual probabilities:

$$\Pr\begin{bmatrix}k \text{ or more occurrences of a}\\ \text{given outcome in } n \text{ trials}\end{bmatrix} = \sum_{\eta=k}^{n} b(\eta;n,p). \tag{B.6}$$

If k is comparable in magnitude to n, then the summation in equation B.6 will involve a reasonably small number of terms. Quite often, however, k will be a small number, such as one or two, while n is quite large. In such cases, equation B.6 is inefficient, having a large number of terms in the summation. On the other hand, if we consider *at least* k and *less than* k, then we have covered all possibilities. Therefore, the probability of k or more occurrences plus the probability of less than k occurrences must be equal to one:

$$\Pr\begin{bmatrix}k \text{ or more occurrences of a}\\ \text{given outcome in } n \text{ trials}\end{bmatrix} = 1 - \sum_{\eta=0}^{k-1} b(\eta;n,p). \tag{B.7}$$

This formulation of the probability is much more convenient when k is small. In fact, when k equals one, it reduces to the very simple relationship

$$\Pr\begin{bmatrix}\text{one or more occurrences of a}\\ \text{given outcome in } n \text{ trials}\end{bmatrix} = 1 - q^n.$$

Now let us apply these results to random encounters between members of a population. Suppose that there are n members of the population, that the

Table B.1. Encounters among members of a given population: probabilities for a single transition interval

Probability that a particular member will encounter another particular member	p
Probability that a particular member encounters exactly k of the $n-1$ other members	$b(k; n-1, p)$
Probability that a particular member encounters no other members	q^{n-1}
Probability that a particular member encounters exactly one other member	$(n-1)pq^{n-1}$
Probability that a particular member encounters at least k of the $n-1$ other members	$1-\sum_{\eta=0}^{k-1} b(\eta; n-1, p)$
Probability that a particular member encounters at least one other member	$1-q^{n-1}$

probability is p that a given member will encounter another given member sometime during a given transition interval, that this probability will be the same for all members and independent of any encounters that already have occurred or that occur at the same time. Thus, for a given member, each of the $n-1$ other members represents a Bernoulli trial, with probabilities p of encounter and $q=1-p$ of no encounter. With this simple situation, one very easily can apply the equations of the coin-tossing analogy to generate appropriate probabilities, some of which are listed in Table B.1.

Now we are ready to consider random encounters between members of two different populations or subpopulations. Let the populations be i and j, with n_i and n_j members, respectively; let the probability be p_{ij} that a given member of population i encounters a given member of population j during a given transition interval; and let p_{ij} be the same for every pair of individuals, one from i and one from j. Thus, for a given interval, we can consider each member of population i as representing a Bernoulli trial for each member of population j, or vice versa. Applying the equation from the coin-tossing analogy to this situation, we rather easily can derive the basic relationships listed in Table B.2.

Complex events, such as those involving intraspecific as well as interspecific encounters, can be dealt with by considering each component independently and taking the product of the corresponding component probabilities. Table B.3 contains a few illustrative examples. Here p_i is the probability that a particular member of population i will encounter another particular member of i in a given transition interval.

Probabilities such as these can be used as the basis of modeled interactions among members of one or more species. Of course the consequence of such an interaction may alter the parameters of the system (see section 3.3). Thus, for example, an individual prey may be allowed only one encounter with a predator. Furthermore, in the complex encounters described in Table B.3 interference may occur. Thus, for example, two predators encountering one prey might result in the demise of one prey to the benefit of only one predator. It also might result in a shared meal; or merely in a fight between two predators while the prey wanders safely away from the scene.

Table B.2. Encounters between members of two different populations: probabilities for a single transition interval

Probability that a given member of population i encounters a given member of population j	p_{ij}
Probability that a given member of i encounters exactly zero members of j	$q_{ij}^{n_j}$
Probability that a given member of i encounters exactly one member of j	$n_j p_{ij} q_{ij}^{n_j - 1}$
Probability that a given member of i encounters exactly k members of j	$\binom{n_j}{k} p_{ij}^k q_{ij}^{n_j - k}$
Probability that a given member of i encounters at least one member of j	$1 - q_{ij}^{n_j}$
Probability that a given member of i encounters at least two members of j	$1 - q_{ij}^{n_j} - n_j p_{ij} q_{ij}^{n_j - 1}$
Probability that a given member of i encounters at least k members of j	$1 - \sum_{\eta=0}^{k-1} b(\eta; n_j, p_{ij})$

Table B.3. Encounters among and between members of two different populations: probabilities for a single transition interval

Probability that a given member of i meets no members of i or j	$q_{ij}^{n_j} q_i^{n_i - 1}$
Probability that a given member of i meets one member of j and no members of i	$n_j p_{ij} q_{ij}^{n_j - 1} q_i^{n_i - 1}$
Probability that a given member of i meets exactly one member of i and zero members of j	$(n_i - 1) p_i q_i^{n_i - 2} q_{ij}^{n_j}$
Probability that a given member of i meets one or more members of j, but no other members of i	$[1 - q_{ij}^{n_j}] q_i^{n_i}$
Probability that a given member of i meets exactly k_j members of j and k_i members of i	$\binom{n_j}{k_j} p_{ij}^{k_j} q_{ij}^{n_j - k_j} \binom{n_i - 1}{k_i} p_i^{k_i} q_i^{n_i - k_i - 1}$
Probability that a given member of i meets at least k_j members of j and exactly k_i members of i	$\left[1 - \sum_{\eta=0}^{k_j - 1} b(\eta; n_j, p_{ij})\right] \binom{n_i - 1}{k_i} p_i^{k_i} q_i^{n_i - k_i - 1}$
Probability that a given member of i meets at least k_j members of j and at least k_i members of i	$\left[1 - \sum_{\eta=0}^{k_j - 1} b(\eta; n_j, p_{ij})\right] \left[1 - \sum_{v=0}^{k_i - 1} b(v; n_i - 1, p_i)\right]$
Probability that a given member of i meets exactly k_j members of j and at least k_i members of i	$\binom{n_j}{k_j} p_{ij}^{k_j} q_{ij}^{n_j - k_j} \left[1 - \sum_{\eta=0}^{k_j - 1} b(\eta; n_i - 1, p_i)\right]$

Table B.4. Binomial coefficients (Pascal's triangle)

	k=0	1	2	3	4	5	6	7	8	9	10	11	12	13	14	15	16	17	18	19	20
$\binom{0}{k}$	1																				
$\binom{1}{k}$	1	1																			
$\binom{2}{k}$	1	2	1																		
$\binom{3}{k}$	1	3	3	1																	
$\binom{4}{k}$	1	4	6	4	1																
$\binom{5}{k}$	1	5	10	10	5	1															
$\binom{6}{k}$	1	6	15	20	15	6	1														
$\binom{7}{k}$	1	7	21	35	35	21	7	1													
$\binom{8}{k}$	1	8	28	56	70	56	28	8	1												
$\binom{9}{k}$	1	9	36	84	126	126	84	36	9	1											
$\binom{10}{k}$	1	10	45	120	210	252	210	120	45	10	1										
$\binom{11}{k}$	1	11	55	165	330	462	462	330	165	55	11	1									
$\binom{12}{k}$	1	12	66	220	495	792	924	792	495	220	66	12	1								
$\binom{13}{k}$	1	13	78	286	715	1287	1716	1716	1287	715	286	78	13	1							
$\binom{14}{k}$								3432	3003	2002	1001	364	91	14	1						
$\binom{15}{k}$									6435	5005	3003	1365	455	105	15	1					
$\binom{16}{k}$									12870	11440	8008	4368	1820	560	120	16	1				
$\binom{17}{k}$										24310	19448	12376	6188	2380	680	136	17	1			
$\binom{18}{k}$										48620	43758	31824	18564	8568	3060	816	153	18	1		
$\binom{19}{k}$											92378	75582	50388	27132	11628	3876	969	171	19	1	
$\binom{20}{k}$											184756	167960	125970	77520	38760	15504	4845	1140	190	20	1

Bibliography

1. Andrewartha, H.G.: Introduction to the study of animal populations. Chicago: University of Chicago Press 1961.
2. Andrewartha, H.G., Birch, L.C.: The distribution and abundance of animals. Chicago: University of Chicago Press 1954.
3. Ansari, N. (ed.): Epidemiology and control of schistosomiasis (bilharziasis). Baltimore-London-Tokyo: University Park Press 1973.
4. Anon.: Modern computing methods. New York: Philosophical Library 1958.
5. Anon.: Mechanisms in biological competition. Symp. Soc. exp. Biol. **15** (1961).
6. Arnold, C.Y.: The determination and significance of the base temperature in a linear unit system. Proc. Amer. Soc. hort. Sci. **74**, 430–455 (1959).
7. Arnold, C.Y.: Maximum-minimum temperatures as a basis for computing heat units. Proc. Amer. Soc. hort. Sci. **76**, 682–692 (1960).
8. Asdell, S.A.: Patterns of mammalian reproduction. London: Constable 1964.
9. Bailey, N.T.J.: The mathematical theory of epidemics. London: Griffin 1957.
10. Bailey, N.T.J.: The elements of stochastic processes with applications to the natural sciences. New York-London-Sydney-Toronto: John Wiley & Sons 1964.
11. Bailey, V.A.: The interaction between hosts and parasites. Quart. J. Math. **2**, 68–77 (1931).
12. Bailey, V.A., Nicholson, A.J., Williams, E.J.: Interactions between hosts and parasites when some host individuals are more difficult to find than others. J. theor. Biol. **3**, 1–18 (1962).
13. Bartlett, M.S.: Stochastic population models in ecology and epidemiology. New York-London-Sydney-Toronto: John Wiley & Sons 1960.
14. Bartlett, M.S.: An introduction to stochastic processes. 2. Cambridge: University Press 1966.
15. Baskerville, G.L., Emin, P.: Rapid estimation of heat accumulation from maximum and minimum temperature. Ecology **50**, 514–517 (1969).
16. Beddington, J.R.: Mutual interference between parasites or predators, and its effect on searching efficiency. J. Animal Ecol. **44**, 331–340 (1975).
17. Bent, A.C.: Life histories of North American birds (26 volumes). New York: Dover.
18. Bernardelli, H.: Population waves. J. Burma Res. Soc. **31**, 1–18 (1941).
19. Beverton, R.J.H., Holt, S.J.: On the dynamics of exploited fish populations. Great Brit. Min. Agr. Fish. Food, Fish. Invest. ser. 2, **19**, 1–533 (1957).
20. Bharucha-Reid, A.T.: Elements of the theory of markov processes and their applications. New York-Toronto-London-Sydney: McGraw-Hill 1960.
21. Biological clocks. Cold Spr. Harb. Symp. quant. Biol. **25** (1960).
22. Bogue, D.J.: Principles of demography. New York-London-Sydney-Toronto: Wiley 1969.
23. Borror, D.J., DeLong, D.M.: An introduction to the study of insects. 2. New York-San Francisco-Toronto-London: Holt, Rinehart and Winston 1964.
24. Bossert, W.H.: Simulation of character displacement in animals. Thesis (Dept. of Applied Mathematics), Harvard University, Cambridge 1963.
25. Boughey, A.S.: Ecology of populations. 2. New York-London: Macmillan 1973.
26. Brillouin, L.: Science and information theory. New York-San Francisco-London: Academic Press 1956.
27. Bronk, B.V., Dienes, G.J., Paskin, A.: The stochastic theory of cell proliferation. Biophys. J. **8**, 1353–1398 (1968).
28. Brown, W.L., Wilson, E.O.: Character displacement. Syst. Zool. **5**, 49–64 (1956).
29. Bunge, M.: Phenomenological theories. In: Bunge, M. (ed.), The critical approach to philosophy, pp. 234–254. New York-London: Macmillan 1964.
30. Burgers, J.M.: Causality and anticipation. Science **189**, 194–198 (1975).
31. Bustard, H.R., Tognetti, K.P.: Green sea turtles – a discrete simulation of density-dependent population regulation. Science **163**, 939–941 (1969).

32. Cadzrow, J.A.: Discrete-time systems. Englewood Cliffs, N.J.: Prentice-Hall 1973.
33. Calow, P.: The relationship between fecundity, phenology, and longevity: a systems approach. Amer. Naturalist **107**, 559–574 (1973).
34. Cannon, R.H.: Dynamics of physical systems. New York-Toronto-London-Sydney: McGraw-Hill 1967.
35. Capildeo, R., Haldane, J.B.S.: The mathematics of bird population growth and decline. J. Animal Ecol. **23**, 215–223 (1954).
36. Caswell, H.: A simulation study of a time lag population model. J. theor. Biol. **34**, 419–439 (1972).
37. Caughley, G.: Mortality patterns in mammals. Ecology **47**, 906–918 (1966).
38. Charlesworth, B.: Selection in populations with overlapping Generations. Theor. Population Biol. **1**, 352–370 (1970).
39. Chitty, D.: Population processes in the vole and their relevance to general theory. Canad. J. Zool. **38**, 99–113 (1960).
40. Chitty, D.: What regulates bird populations? Ecology **48**, 698–701 (1967).
41. Christian, J., Davis, D.: Endocrines, behavior and population. Science **146**, 1550–1560 (1964).
42. Chu, K.Y., Massoud, J., Sabbaghian, H.: Effect of months of infection on cercaria incubation periods. Bull. Wld. Hlth. Org. **34**, 135–140 (1966).
43. Churchill, R.V.: Introduction to complex variables and applications. New York-Toronto-London-Sydney: McGraw-Hill 1948.
44. Clark, L.R., Geier, P.W., Hughes, R.D., Morris, R.F.: The ecology of insect populations in theory and practice. London: Methuen 1967.
45. Cody, M.L.: Competition and the structure of bird communities. Princeton: Princeton University Press 1974.
46. Cohen, J.E.: Natural primate troops and a stochastic population model. Amer. Naturalist **103**, 455–477 (1969).
47. Cole, L.C.: The population consequences of life history phenomena. Quart. Rev. Biol. **29**, 103–137 (1954).
48. Cole, L.C.: Some features of random population cycles. J. Wildlife Management **18**, 2–24 (1954).
49. Connell, J.H., Mertz, D.B., Murdoch, W.W.: Readings in ecology and ecological genetics. New York-Evanston-London: Harper & Row 1970.
50. Conway, G.R., Murdie, G.: Population models as a basis for pest control. In: Jeffers, J.N.R. (ed.), Mathematical models in ecology, pp. 195–213. Oxford: Blackwell 1972.
51. Coulman, G.A., Reice, S.R., Tummala, R.L.: Population modeling—a systems approach. Science **175**, 518–521 (1972).
52. Coulson, R.N., Mayyasi, A.M., Foltz, J.L., Pulley, P.E.: Production flow system evaluation of within-tree populations of *Dendroctonus frontalis* Zimm. (Coleoptera: Scolytidae). Environm. Entomol. In Press.
53. Cox, P.R.: Demography. 3. Cambridge: University Press 1959.
54. Crow, J.F., Kimura, M.: An introduction to population genetics theory. New York-Evanston-San Francisco-London: Harper & Row 1970.
55. Cunningham, W.J.: A nonlinear differential-difference equation of growth. Proc. nat. Acad. Sci. (Wash.) **40**, 708–713 (1954).
56. Dale, M.B.: Systems analysis and ecology. Ecology **51**, 2–16 (1970).
57. Davis, D.E.: The existence of cycles. Ecology **38**, 163–164 (1957).
58. Deevey, E.S.: Life tables for natural populations of animals. Quart. Rev. Biol. **22**, 283–314 (1947).
59. Dempster, J.P.: The population dynamics of grasshoppers and locusts. Biol. Rev. **38**, 490–529 (1963).
60. Eddington, A.S.: The nature of the physical world. New York-London: Macmillan 1929.
61. Egerton, F.N.: Studies of animal populations from Lamarck to Darwin. J. Hist. Biol. **1**, 225–259 (1968).
62. Elsasser, W.M.: Quanta and the concept of organismic law. J. theor. Biol. **1**, 27–58 (1961).
63. Elsasser, W.M.: Physical aspects of non-mechanistic biological theory. J. theor. Biol. **3**, 164–191 (1962).
64. Elton, C.S.: The ecology of invasions by animals and plants. London: Methuen 1958.

65. Elton, C., Nicholson, M.: The ten-year cycle in numbers of the lynx in Canada. J. Animal Ecol. **11**, 215–244 (1942).
66. Emlen, J.M.: Ecology: An evolutionary approach. Reading, Mass.-Menlo Park, Calif.-London-Don Mills, Ontario: Addison-Wesley 1973.
67. Engstrom-Heg, V.L.: Predation, competition and environmental variables: some mathematical models. J. theor. Biol. **27**, 175–195 (1970).
68. Errington, P.L.: Factors limiting higher vertebrate populations. Science **124**, 304–307 (1956).
69. Euler, L.: A general investigation into the mortality and multiplication of the human species. (Originally published in 1760.) Theoret. Population Biol. **1**, 307–314 (1970).
70. Fano, R.M.: Transmission of information. New York-London-Sydney-Toronto: John Wiley & Sons 1961.
71. Feller, W.: On the integral equation of renewal theory. Ann. Math. Stat. **12**, 243–267 (1941).
72. Feller, W.: An introduction to probability theory and its applications. 3. New York-London-Sydney-Toronto: John Wiley & Sons 1968.
73. Fogel, L.J.: Biotechnology: concepts and applications. Englewood Cliffs, N.J.: Prentice-Hall 1963.
74. Frank, P.W.: Prediction of population growth form in *Daphnia pulex* cultures. Amer. Naturalist. **94**, 357–372 (1960).
75. Franke, E.K.: A mathematical model of synchronized periodic growth of cell populations. J. theor. Biol. **26**, 373–382 (1970).
76. Fredrickson, A.G.: A mathematical theory of age structure in sexual populations: random mating and monogamous marriage models. Math. Biosci. **10**, 117–143 (1971).
77. Fredrickson, A.G., Jost, J.L., Tsuchiya, H.M., Hsu, P.-H.: Predator-prey interactions between malthusian populations. J. theor. Biol. **38**, 487–526 (1973).
78. Fretwell, S.D.: Populations in a seasonal environment. Princeton: Princeton University Press 1972.
79. Frolov, I.T.: Epistemological problems of modeling of biological systems. In: Philosophical problems of modern biology, pp. 255–278. Praha: Nakládatelství Československé Akademie 1963.
80. Gadgil, M., Bossert, W.H.: Life historical consequences of natural selection. Amer. Naturalist **104**, 1–24 (1970).
81. Garfinkel, D., Sack, R.: Digital computer simulation of an ecological system, based on a modified mass action law. Ecology **45**, 502–507 (1964).
82. Gause, G.F.: The influence of ecological factors on the size of population. Amer. Naturalist **65**, 70–76 (1931).
83. Gause, G.F.: The struggle for existence. Baltimore: Williams and Wilkins 1934.
84. Goel, N.S., Maitra, S.C., Montroll, E.W.: Nonlinear models of interacting populations. New York-San Francisco-London: Academic Press 1971.
85. Goel, N.S., Richter-Dyn, N.: Stochastic models in biology. New York-San Francisco-London: Academic Press 1974.
86. Goffman, W., Warren, K.S.: An application of the Kermack-McKendrick theory to the epidemiology of schistosomiasis. Amer. J. trop. Med. Hyg. **19**, 278–283 (1970).
87. Goldberg, S.: Introduction to difference equations. New York-London-Sydney-Toronto: John Wiley & Sons 1958.
88. Goodman, L.: Population growth of the sexes. Biometrics **9**, 212–225 (1953).
89. Greenspan, D.: Discrete models. Reading, Massachusetts-London-Don Mills, Ontario-Sydney-Tokyo: Addison-Wesley 1973.
90. Greniewski, H.: Cybernetics without mathematics. New York: Pergamon 1960.
91. Griffiths, K.S., Holling, C.S.: A competition submodel for parasites and predators. Canad. Entomologist **101**, 785–818 (1969).
92. Gulland, J.A.: The application of mathematical models to fish populations. In: Le Cren, E.D., Holdgate, M.W. (eds.), The exploitation of natural animal populations, pp. 204–217. Oxford: Blackwell 1962.
93. Hairston, N.G.: Population ecology and epidemiological problems. In: Wolstenholme, G.E.W., O'Connor, M. (eds.), Bilharziasis, pp. 36–62. London: Churchill 1962.
94. Hairston, N.G.: On the mathematical analysis of schistosome populations. Bull. Wld. Hlth. Org. **33**, 45–62 (1965).

95. Hairston, N.G., Smith, F.E., Slobodkin, L.B.: Community structure, population control and competition. Amer. Naturalist **94**, 421–425 (1960).
96. Haldane, J.B.S.: A mathematical theory of natural and artificial selection. Proc. Cambridge phil. Soc. **27**, 137–142 (1931).
97. Harper, J.L.: The regulation of numbers and mass in plant populations. In: Lewontin, R.C. (ed.), Population biology and evolution, pp. 139–158. Syracuse-New York: Syracuse University Press 1967.
98. Harris, T.E.: The theory of branching processes. Berlin-Heidelberg-New York: Springer/Englewood Cliffs, N.J.: Prentice-Hall 1963.
99. Hassell, M.P.: Mutual interference between searching insect parasites. J. Animal Ecol. **40**, 473–486 (1971).
100. Hassell, M.P., May, R.: Aggregation of predators and insect parasites and its effect on stability. J. Animal Ecol. **43**, 567–594 (1974).
101. Heisenberg, W.: Physical principles of the quantum theory. New York: Dover 1930.
102. Hirsch, H.R., Engelberg, J.: Decay of cell synchronization: solutions of the cell growth equation. Bull. Math. Biophys. **28**, 391–409 (1966).
103. Holgate, P.: Population survival and life history phenomena. J. theor. Biol. **14**, 1–10 (1967).
104. Holling, C.S.: The components of predation as revealed by a study of small mammal predation of the european pine sawfly. Canad. Entomologist **91**, 293–332 (1959).
105. Holling, C.S.: Some characteristics of simple types of predation. Canad. Entomologist **91**, 385–398 (1959).
106. Holling, C.S.: Principles of insect predation. Ann. Rev. Entomol. **6**, 163–182 (1961).
107. Holling, C.S.: The strategy of building models of complex ecological systems. In: Watt, K.E.F. (ed.), Systems analysis in ecology, pp. 195–214. New York-San Francisco-London: Academic Press 1966.
108. Holmes, R.L.: Reproduction and environment. Edinburgh: Oliver & Boyd 1968.
109. Hoppensteadt, F., Hyman, J.: Periodic solutions of a logistic difference equation. SIAM Reg. Conf., Univ. of Iowa, 1975.
110. Hubbell, S.P.: Populations and simple food webs as energy filters II: two-species systems. Amer. Naturalist **107**, 122–151 (1973).
111. Huffaker, C.B.: Experimental studies on predation: dispersion factors and predator-prey oscillation. Hilgardia **27**, 343–383 (1958).
112. Hutchinson, G.E.: Circular causal systems in ecology. Ann. N.Y. Acad. Sci. **50**, 221–246 (1948).
113. Hutchinson, G.E.: Notes on oscillatory populations. J. Wildlife Management **18**, 107–109 (1954).
114. Iberall, A.S.: On nature, man, and society: a basis for scientific modeling. Ann. Biomed. Engrg. **3**, 344–385 (1975).
115. Iosifescu, M., Tăutu, P.: Stochastic processes and applications in biology and medicine II (models). Berlin-Heidelberg-New York: Springer 1973.
116. Istock, C.A.: On evolution of complex life cycle phenomena: an ecological perspective. Evolution **21**, 592–605 (1967).
117. Jacquard, A.: The genetic structure of populations. NewYork-Heidelberg-Berlin: Springer 1974.
118. Jaquette, D.L.: A stochastic model for the optimal control of epidemics and pest populations. Math. Biosci. **8**, 343–354 (1970).
119. Jaquette, D.L.: A discrete time population control model. Math. Biosci. **15**, 231–252 (1972).
120. Jordan, P., Webee, G.: Human schistosomiasis. Springfield, Ill.: Thomas 1969.
121. Jury, E.I.: Sampled-data control systems. New York-London-Sydney-Toronto: John Wiley & Sons 1958.
122. Jury, E.I.: The theory and applications of the z transform methods. New York-London-Sydney-Toronto: John Wiley & Sons 1964.
123. Jury, E.I.: Inners approach to some problems of system theory. IEEE Trans. Automatic Contr. **AC-16**, 233–240 (1971).
124. Kalman, R.E.: New developments in systems theory relevant to biology. In: Mesarovic, M.D. (ed.), Systems theory and biology, pp. 222–232. New York-Heidelberg-Berlin: Springer 1968.
125. Karlin, S.: A first course in stochastic processes. New York-San Francisco-London: Academic Press 1969.

126. Keith, L.B.: Wildlife's ten-year cycle. Madison, Wisconsin: Univ. of Wisconsin Press 1963.
127. Kelly, C.D., Rahn, O.: The growth rate of individual bacterial cells. J. Bact. **23**, 147–153 (1932).
128. Kendall, D.G.: On the generalized birth and death process. Ann. Math. Stat. **19**, 1–15 (1948).
129. Kendall, D.G.: On the role of variable generation time in the development of a stochastic birth process. Biometrika **35**, 316–339 (1948).
130. Kendall, D.G.: Stochastic processes and population growth. J. roy. Stat. Soc. B **11**, 230–264 (1949).
131. Kermack, W.O., McKendrick, A.G.: A contribution to the mathematical theory of epidemics. Proc. roy. Soc. A **115**, 700–721 (1927).
132. Keyfitz, N.: The intrinsic rate of natural increase and the dominant latent root of the projection matrix. Population Studies **18**, 293–308 (1964).
133. Keyfitz, N.: Introduction to the mathematics of population. Reading, Mass.: Addison-Wesley 1964.
134. Keyfitz, N.: Reconciliation of population models: matrix, integral equation, and partial fraction. J. roy. Stat. Soc. A **130**, 61–83 (1967).
135. Keyfitz, N.: Population waves. In: Greville, T.N.E. (ed.), Population dynamics, pp. 1–38. New York-San Francisco-London: Academic Press 1972.
136. Kilmer, W.L.: On some realistic constraints in prey-predator mathematics. J. theor. Biol. **36**, 9–22 (1972).
137. Kimura, M., Ohta, T.: Theoretical aspects of population genetics. Princeton: Princeton University Press 1971.
138. Klomp, H.: The influence of climate and weather on the mean density level, the fluctuations and the regulation of animal populations. Arch. Neerl. Zool. **15**, 68–109 (1962).
139. Kojima, K.-I. (ed.): Mathematical topics in population genetics. New York-Heidelberg-Berlin: Springer 1970.
140. Krebs, C.J.: Ecology – the experimental analysis of distribution and abundance. New York-Evanston-San Francisco-London: Harper & Row 1972.
141. Kuo, B.C.: Linear networks and systems. New York-San Francisco-Toronto-London-Sydney: McGraw-Hill 1967.
142. Lack, D.: The natural regulation of animal numbers. Oxford: The Clarendon Press 1954.
143. Lack, D.: Ecological adaptations for breeding in birds. London: Methuen 1967.
144. Larkin, P.A., Hourston, A.S.: A model for simulation of the population biology of pacific salmon. J. Fish. Res. Bd. Can. **21**, 1245–1265 (1964).
145. Laws, R.M.: Aspects of reproduction in the African elephant, *Loxodonta africana*. J. Reprod. Fertil., Suppl. **6**, 193–217 (1969).
146. Le Cren, E.D.: The efficiency of reproduction and recruitment in freshwater fishes. In: Le Cren, E.D., Holdgate, M.W. (eds.), The Exploitation of natural animal populations, pp. 283–296. Oxford: Blackwell 1962.
147. Lee, K.-L., Lewis, E.R.: Delay time models of population dynamics with application to schistosomiasis control. IEEE Trans. Biomed. Engrg. **23**, 225–233 (1976).
148. Lefkovitch, L.P.: The study of population growth in organisms grouped by stages. Biometrics **21**, 1–18 (1965).
149. Lefkovitch, L.P.: A population growth model incorporating delayed responses. Bull. Math. Biophys. **28**, 219–233 (1966).
150. Leslie, P.H.: On the use of matrices in certain population mathematics. Biometrika **33**, 183–212 (1945).
151. Leslie, P.H.: Some further notes on the use of matrices in population mathematics. Biometrika **35**, 213–245 (1948).
152. Levin, S.A.: Community equilibria and stability, and an extension of the competitive exclusion principle. Amer. Naturalist **104**, 413–423 (1970).
153. Levin, S.A.: A mathematical analysis of the genetic feedback mechanism. Amer. Naturalist **106**, 145–164 (1972).
154. Levin, S.A., May, R.M.: A note on difference-delay equations. Theoret. Pop. Biol. **9**, 178–187 (1976).
155. Levins, R.: Strategy of model building in population biology. Amer. Sci. **54**, 421–431 (1966).

156. Levins, R.: Evolution in changing environments. Princeton, New Jersey: Princeton University Press 1968.
157. Lewis, E.G.: On the generation and growth of a population. Sankhya **6**, 93–96 (1942).
158. Lewis, E.R.: Delay-line models of population growth. Ecology **53**, 797–807 (1972).
159. Lewis, E.R.: Applications of discrete and continuous network theory to linear population models. Ecology **57**, 33–47 (1976).
160. Lewontin, R.C.: Selection in and of populations. In: More, J.A. (ed.), Ideas in modern biology. Garden City, New York: Doubleday 1965.
161. Lewontin, R.C. (ed.): Population biology and evolution. Syracuse-New York: Syracuse University Press 1968.
162. Lewontin, R.C., Birch, L.C.: Hybridization as a source of variation for adaptation to new environments. Evolution **20**, 315–336 (1966).
163. Li, T.-Y., Yorke, J.A.: Period three implies chaos. Amer. Math. Monthly **82**, 985–992 (1975).
164. Licht, P.: Physiology of reptilian reproductive cycles. In: Hoar, W.S., Bern, H.A. (eds.), Sixth Intern. Symp. Comp. Endocrinol. (J. gen. comp. Endocr. Suppl. 3) 1972.
165. Lofts, B.: Animal photoperiodism. London: Arnold 1970.
166. Lotka, A.J.: The stability of the normal age distribution. Proc. nat. Acad. Sci. (Wash.) **8**, 339–345 (1922).
167. Lotka, A.J.: Growth of mixed populations. J. Wash. Acad. Sci. **22**, 461–469 (1932).
168. Lotka, A.J.: Elements of mathematical biology. New York: Dover 1956.
169. MacArthur, R.H.: Geographical ecology. New York-Evanston-San Francisco-London: Harper & Row 1972.
170. MacArthur, R.H., Connell, J.H.: The biology of populations. New York-London-Sydney-Toronto: John Wiley & Sons 1966.
171. MacArthur, R.H., Wilson, E.O.: The theory of island biogeography. Princeton: Princeton University Press 1967.
172. Macdonald, G.: The dynamics of helminth infections, with special reference to schistosomiasis. Trans. roy. Soc. trop. Med. Hyg. **59**, 489–506 (1965).
173. Macfadyen, A.: Animal ecology. 2. London-Victoria, Australia-Johannesburg-New York-Toronto: Pitman 1963.
174. Maki, D.P., Thompson, M.: Mathematical models and applications. Englewood Cliffs, New Jersey: Prentice-Hall 1973.
175. Malthus, T.R.: An essay on the principle of population (Originally published in 1798). A Norton critical review (Appleman, P., ed.). New York: Norton 1976.
176. Margalef, R.: Simplified physical models of populations of organisms. Mem. Real. Acad. Cien. Art., Barc. **34**, 1–66 (1962).
177. Marshall, A.J. (ed.): Biology and comparative physiology of birds. New York-San Francisco-London: Academic Press 1961.
178. Marshall, N.B.: Life of fishes. London: Weidenfeld and Nicholson 1965.
179. Mason, S.J.: Feedback theory: Further properties of signal flow graphs. Proc. Inst. Radio Engrs. **44**, 920–926 (1956).
180. Mason, S.J., Zimmerman, H.J.: Electronic circuits, signals, and systems. New York-London-Sydney-Toronto: John Wiley & Sons 1960.
181. May, R.M.: Limit cycles in predator-prey communities. Science **177**, 900–902 (1972).
182. May, R.M.: Stability and complexity in model ecosystems. Princeton: Princeton University Press 1973.
183. May, R.M.: Time-delay versus stability in population models with two and three trophic levels. Ecology **54**, 315–325 (1973).
184. May, R., Oster, G.: Bifurcations and dynamic complexity in simple ecological models. Amer. Naturalist **110**, 573–599 (1976).
185. Maynard Smith, J.: Mathematical ideas in biology. Cambridge: University Press 1971.
186. Maynard Smith, J.: Models in ecology. Cambridge: University Press 1974.
187. Mazanov, A.: A multi-stage population Model. J. theor. Biol. **39**, 581–587 (1973).
188. McLaren, I.A. (ed.): Natural regulation of animal populations. New York: Atherton Press 1971.
189. Mesarovic, M.D.: Systems theory and biology—view of a theoretician. In: Mesarovic, M.D. (ed.), Systems theory and biology, pp. 59–87. New York-Heidelberg-Berlin: Springer 1968.

190. Milne, A.: On a theory of natural control of insect population. J. theor. Biol. **3**, 19–50 (1962).
191. Morisita, M.: The fitting of the logistic equation to the rate of increase of population density. Res. Pop. Ecol. **7**, 52–55 (1965).
192. Mostofi, F.K. (ed.): Bilharziasis. Berlin-Heidelberg-New York: Springer 1967.
193. Murdie, G., Hassell, M.P.: Food distribution, searching success and predator-prey models. In: Bartlett, M.S., Hiorns, R.W. (eds.), The mathematical theory of the dynamics of biological populations, pp. 88–101. New York-San Francisco-London: Academic Press 1973.
194. Murdoch, D.C.: Linear algebra for undergraduates. New York-London-Sydney-Toronto: John Wiley & Sons 1957.
195. Murdoch, W.W.: Population stability and life history phenomena. Amer. Naturalist **100**, 5–11 (1966).
196. Murphy, G.I.: Patterns in life history and the environment. Amer. Naturalist **102**, 390–404 (1968).
197. Nasell, I., Hirsch, W.M.: Mathematical models of some parasitic diseases involving an intermediate host. New York: Courant Institute 1971.
198. Neyman, J., Scott, E.L.: Stochastic models of population dynamics. Science **130**, 303–308 (1959).
199. Nicholson, A.J.: The balance of animal populations. J. Animal Ecol. **2**, 132–178 (1933).
200. Nicholson, A.J.: Compensatory reactions of populations to stress, and their evolutionary significance. Aust. J. Zool. **2**, 1–8 (1954).
201. Nicholson, A.J.: An outline of the dynamics of animal populations. Aust. J. Zool. **2**, 9–65 (1954).
202. Nicholson, A.F., Bailey, A.V.: The balance of animal populations. Proc. Zool. Soc. London **3**, 551–598 (1935).
203. Nikol'skii, G.V.: Some adaptations to the regulation of population density in fish species with different types of stock structure. In: Le Cren, E.D., Holdgate, M.W. (eds.), The exploitation of natural animal populations, pp. 265–282. Oxford: Blackwell 1962.
204. Nikolsky, A.V.: The ecology of fishes. New York-San Francisco-London: Academic Press 1963.
205. Odum, E.P.: Fundamentals of ecology. 3. Philadelphia-London-Toronto: Saunders 1971.
206. Odum, H.T.: Environment, power and society. New York-London-Sydney-Toronto: John Wiley & Sons 1971.
207. Olsen, O.W.: Animal parasites—their life cycles and ecology. 3. Baltimore-London-Tokyo: University Park Press 1974.
208. Ore, O.: Number theory and it history. New York-Toronto-London: McGraw-Hill 1948.
209. Orians, G.H.: On the evolution of mating systems in birds and mammals. Amer. Natur. **103**, 589–603 (1963).
210. Oster, G., Takahashi, Y.: Models for age-specific interactions in a periodic environment. Ecol. Monogr. **44**, 488–501 (1974).
211. Paloheimo, J.: On the theory of search. Biometrika **58**, 61–75 (1971).
212. Park, T.: Beetles, competition, and populations. Science **138**, 1369–1375 (1962).
213. Park, T., Leslie, P.H., Mertz, D.B.: Genetic strains and competition in populations of *Tribolium*. Physiol. Zool. **37**, 97–162 (1964).
214. Parkes, A.S. (ed.): Marshall's physiology of reproduction. London: Longman's, Green (vol. 1, part 1) 1956; (vol. 1, part 2) 1960; (vol. 2) 1952.
215. Parlett, B.: Ergodic properties of populations I: The one sex model. Theoret. Pop. Biol. **1**, 191–207 (1970).
216. Parzen, E.: Modern probability theory and its applications. New York-London-Sydney-Toronto: John Wiley & Sons 1960.
217. Parzen, E.: Stochastic processes. San Francisco: Holden-Day 1962.
218. Patten, B.C. (ed.): Systems analysis and simulation in ecology. New York-San Francisco-London: Academic Press vol. 1 (1971); vol. 2 (1972); vol. 3 (1975); vol. 4 (1976).
219. Paulik, G.J., Greenough, J.W., Jr.: Management analysis for a salmon resource system. In: Watt, K.E.F. (ed.), Systems analysis in ecology, pp. 215–252. New York-San Francisco-London: Academic Press 1966.
220. Pearl, R.: The growth of populations. Quart. Rev. Biol. **2**, 532–548 (1927).

221. Pearl, R., Reed, L.J.: On the rate of growth of the population of the United States since 1790 and its mathematical representation. Proc. nat. Acad. Sci. (Wash.) **6**, 275–288 (1920).
222. Pearl, R., Reed, L.J., Kish, J.F.: The logistic curve and the census count of 1940. Science **92**, 486–488 (1940).
223. Pennycuick, C.J., Compton, R.M., Beckingham, L.: A computer model for simulating the growth of a population or of two interacting populations. J. theor. Biol. **18**, 316–329 (1968).
224. Pennycuick, L.: A computer model of the Oxford great tit population. J. theor. Biol. **22**, 381–400 (1969).
225. Perrins, C.M.: Population fluctuations and clutch size in the great tit, *Parus major* L. J. Animal Ecol. **34**, 601–647 (1965).
226. Perrins, C., Wynne-Edwards, V.C.: Survival of young swifts in relation to brood size. Nature (Lond.) **201**, 1147–1149 (1964).
227. Perry, J.S., Rowlands, I.W.: The ovarian cycle in vertebrates. In: Zuckerman, S. (ed.), The ovary. New York-San Francisco-London: Academic Press 1962.
228. Pianka, E.R.: On r- and K-selection. Amer. Naturalist **104**, 592–597 (1970).
229. Pielou, E.C.: An introduction to mathematical ecology. New York-London-Sydney-Toronto: Wiley-Interscience 1969.
230. Pimentel, D.: Animal population regulation by the genetic feedback mechanism. Amer. Naturalist **95**, 65–79 (1961).
231. Pimentel, D.: Natural population regulation and interspecies evolution. Proc. 16th Intern. Zool. Congr. **3**, 329–336 (1963).
232. Pimentel, D.: Population regulation and genetic feedback. Science **159**, 1432–1437 (1968).
233. Pitelka, F.A.: Some aspects of population structure in the short-term cycle of the brown lemming in Northern Alaska. Cold Spr. Harbor Symp. quant. Biol. **22**, 237–251 (1957).
234. Pitelka, F.A., Tomich, P.Q., Triechel, G.W.: Ecological relations of jaegers and owls as lemming predators near Barrow, Alaska. Ecol. Monogr. **25**, 85–117 (1955).
235. Platt, J.R.: Strong inference. Science **146**, 347–353 (1964).
236. Polak, E.: Computational methods in optimization: a unified approach. New York-San Francisco-London: Academic Press 1971.
237. Pollard, J.H.: On the use of the direct matrix product in analysing certain stochastic population models. Biometrika **53**, 397–415 (1966).
238. Pollard, J.H.: Mathematical models for the growth of human populations. Cambridge: University Press 1973.
239. Popper, K.R.: Conjectures and refutations. London: Routledge and Kegan Paul 1963.
240. Population Studies: Animal ecology and demography. Cold Spr. Harb. Symp. quant. Biol. **22** (1957).
241. Rafferty, J.A.: Mathematical models in biological theory. Amer. Scientist **38**, 549–567 (1950).
242. Raines, G.E., Bloom, S.G., McKee, P.A., Bell, J.C.: Mathematical simulation of sea otter population dynamics, Amchitka Island, Alaska. Bioscience **21**, 686–691 (1971).
243. Randolph, A.D., Larson, M.A.: Theory of particulate processes. New York-San Francisco-London 1971.
244. Rashevsky, N.: Mathematical biophysics: physico-mathematical foundations of biology. 2. Chicago: University of Chicago Press 1948.
245. Rhodes, E.C.: Population mathematics. J. roy. Statist. Soc. **103**, 68–89, 218–245, 362–387 (1940).
246. Ricker, W.E.: Stock and recruitment. J. Fish. Res. Bd Canada **11**, 559–623 (1954).
247. Rogers, D.J.: Random search and insect population models. J. Animal Ecol. **41**, 369–383 (1972).
248. Rogers, D.J., Hassell, M.P.: General models for insect parasite and predator searching behavior: interference. J. Animal Ecol. **43**, 239–253 (1974).
249. Rosen, R.: Dynamical system theory in biology. New York-London-Sydney-Toronto: John Wiley & Sons 1970.
250. Rosenblueth, A., Wiener, N.: The role of models in science. Phil. of Sci. **12**, 316–322 (1945).
251. Rosenzweig, M.L.: Why the prey curve has a hump. Amer. Naturalist **103**, 81–87 (1969).
252. Rosenzweig, M., MacArthur, R.H.: Graphical representation and stability conditions of predator-prey interaction. Amer. Naturalist **97**, 209–223 (1963).

253. Sadleir, R.M.F.S.: The reproduction of vertebrates. New York-San Francisco-London: Academic Press 1973.
254. Salvadori, M.G., Baron, M.L.: Numerical methods in engineering. Englewood Cliffs, N.J.: Prentice-Hall 1961.
255. Schaffner, K.F.: Antireductionism and molecular biology. Science **157**, 644–647 (1967).
256. Schoener, T.W.: The evolution of bill size differences among sympatric congeneric species of birds. Evolution **19**, 189–213 (1965).
257. Schwartz, R.J., Friedland, B.: Linear systems. New York-Toronto-London-Sydney: McGraw Hill 1965.
258. Shannon, C.E., Weaver, W.: The mathematical theory of communication. Urbana, Ill.: University of Illinois Press 1949.
259. Shoemaker, C.: Applications of mathematical control theory to agricultural pest management (Thesis, Dept. of Mathematics). University of Southern California, Los Angeles 1971.
260. Shryock, H.S., Siegel, J.S.: The materials and methods of demography, vols. I and II. Washington: U.S. Government Printing Office 1973.
261. Silliman, R.P.: Analog computer models of fish populations. Fishery Bull. **66**, 31–46 (1966).
262. Sinko, J.W., Streifer, W.: A new model for age-size Structure of a population. Ecology **48**, 910–918 (1967).
263. Sinko, J.W., Streifer, W.: A model for populations reproducing by fission. Ecology **52**, 330–335 (1971).
264. Slobodkin, L.B.: An algebra of population growth. Ecology **34**, 513–519 (1953).
265. Slobodkin, L.B.: Cycles in animal populations. Amer. Scientist **42**, 658–660 (1954).
266. Slobodkin, L.B.: Growth and regulation of animal populations. New York-Chicago-San Francisco-Toronto-London: Holt, Rinehart and Winston 1961.
267. Smale, S., Williams, R.F.: The qualitative analysis of a difference equation of population growth. J. Math. Biol. **3**, 1–4 (1976).
268. Smith, J.D.: Introduction to animal parasitology. Springfield, Illinois: Thomas 1962.
269. Solodovnokov, V.V.: Introduction to the statistical dynamics of automatic control systems. New York: Dover 1960.
270. Sommerfeld, A.: Thermodynamics and statistical mechanics. New York-San Francisco-London: Academic Press 1956.
271. Southwick, C.: Population characteristics of house mice living in English corn ricks: density relationships. Proc. Zool. Soc. (London) **131**, 163–175 (1958).
272. Steiglitz, K.: An introduction to discrete systems. New York-London-Sydney-Toronto: John Wiley & Sons 1974.
273. Streifer, W.: Realistic models in population biology. In: Macfadyen, A. (ed.), Advances in ecological research (vol. 8), pp. 199–266. New York-San Francisco-London: Academic Press 1975.
274. Takahashi, F.: Reproduction curve with two equilibrium points: a consideration on the fluctuation of insect population. Res. Pop. Ecol. **6**, 28–36 (1964).
275. Tanner, J.T.: Effects of population density on growth rates of animal populations. Ecology **47**, 733–745 (1966).
276. Taylor, H.M.: Some models in epidemic control. Math. Biosci. **3**, 383–398 (1968).
277. Tewarson, R.P.: Sparse matrices. New York-San Francisco-London: Academic 1973.
278. Thom, R.: Structural stability and morphogenesis: an outline of a general theory of models. Reading, Mass.-London-Don Mills, Ontario-Sydney-Tokyo: Addison Wesley 1974.
279. Tognetti, K.P., Mazanov, A.: A two-stage population model. Math. Biosci. **8**, 371–378 (1970).
280. Tomović, R., Vukobratović, M.: General sensitivity theory. New York-London-Amsterdam: Elsevier 1972.
281. Trucco, E.: Mathematical models for cellular systems: the von Foerster equation. Bull. Math. Biophys. **27**, 449–471 (1965).
282. Utida, S.: Cyclic fluctuations of population density intrinsic to the host-parasite system. Ecology **38**, 442–449 (1957).
283. Tienhoven, A. van: Reproductive physiology of vertebrates. Philadelphia-London-Toronto: Saunders 1968.
284. Verhulst, P.F.: Notice sur la loi que la population suit dans son accroissement. Corr. math. et phys. **10**, 113–121 (1839).

285. Volterra, V.: Fluctuations in the abundance of a species considered mathematically. Nature (Lond.) **118**, 558–560 (1926).
286. Volterra, V.: Variations and fluctuations of the number of individuals in animal species living together. In: Chapman, R.N., Animal ecology. New York-Toronto-London-Sydney: McGraw-Hill 1931.
287. Foerster, H. von: Some remarks on changing populations. In: Stohlman, F.S. (ed.), The kinetics of cellular proliferation. New York: Grune and Stratton 1959.
288. Wang, J.Y.: A critique of the heat unit approach to plant response studies. Ecology **41**, 785–790 (1960).
289. Wangersky, P.J., Cunningham, W.J.: On time lags in equations of growth. Proc. nat. Acad. Sci. (Wash.) **42**, 699–702 (1956).
290. Watt, K.E.F.: A mathematical model for the effect of attacked and attacking species on the number attacked. Canad. Entomol. **91**, 129–144 (1959).
291. Watt, K.E.F.: Mathematical models for use in insect pest control. Canad. Entomol., Suppl. **19**, 5–62 (1961).
292. Watt, K.E.F.: The use of mathematics and computers to determine optimal strategies and tactics for a given insect control problem. Canad. Entomol. **96**, 202–220 (1964).
293. Watt, K.E.F. (ed.): Systems analysis in ecology. New York-San Franciso-London: Academic Press 1966.
294. Watt, K.E.F.: Ecology and resource management. New York-Toronto-London-Sydney: McGraw-Hill 1968.
295. Weaver, W.: Science and complexity. Amer. Scientist. **36**, 536–544 (1948).
296. Wilson, E.O., Bossert, W.H.: A primer of population biology. Stamford, Connecticut: Sinauer Associates 1971.
297. Winsor, C.P.: The Gompertz curve as a growth curve. Proc. nat. Acad. Sci. (Wash.) **18**, 1–8 (1932).
298. Wynne-Edwards, V.C.: Animal dispersion in relation to social behavior. New York: Hafner 1962.
299. Wynne-Edwards, V.C.: Self-regulating systems in populations of animals. Science **147**, 1543–1548 (1965).
300. Young, D.M., Gregory, R.T.: A survey of numerical mathematics. Reading, Mass.: Addison-Wesley 1972.
301. Zadeh, L.A., Desoer, C.A.: Linear system theory. New York-Toronto-London-Sydney: McGraw-Hill 1963.

Subject Index

Lecture Notes in Biomathematics

This series aims to report new developments in biomathematics research and teaching – quickly, informally and at a high level.

Volume 1: P. Waltman
Deterministic Threshold Models in the Theory of Epidemics

Volume 2:
Mathematical Problems in Biology
Victoria Conference
Editor: P. van den Driessche

Volume 3: D. Ludwig
Stochastic Population Theories
Notes by M. Levandowsky

Volume 4:
Physics and Mathematics of the Nervous System
Proceedings of a Summer School organized by the International Centre for Theoretical Physics, Trieste, and the Institute for Information Sciences, University of Tübingen, held at Trieste, August 21-31, 1973
Editors: M. Conrad, W. Güttinger, M. Dal Cin

Volume 5:
Mathematical Analysis of Decision Problems in Ecology
Proceedings of the NATO Conference held in Istanbul, Turkey, July 9-13, 1973
Editors: A. Charnes. W. R. Lynn

Volume 6: H. T. Banks
Modeling and Control in the Biomedical Sciences

Volume 7: M. C. Mackey
Ion Transport trough Biological Membranes
In Integrated Theoretical Approach

Volume 8: C. Delisi
Antigen Antibody Interactions

Volume 9: N. Dubin
A Stochastic Model for Immunological Feedback in Carcinogenesis: Analysis and Approximations

Volume 10: J. J. Tyson
The Belousov-Zhabotinskii Reaction

Volume 11:
Mathematical Models in Medicine
Workshop, Mainz, March 1976
Editors: J. Berger, W. Bühler, R. Repges, P. Tautu

Volume 12: A. V. Holden,
Models of the Stochastic Activity of Neurones

Volume 13:
Mathematical Models in Biological Discovery
Editors: D.L. Solomon, C.F. Walter

Volume 14: L. M. Ricciardi
Diffusion Processes and Related Topics in Biology
Notes taken by C. E. Smith

Volume 15: T. Nagylaki
Selection in One- and Two-Locus Systems

Volume 16: G. Sampath, S.K. Srinivasan
Stochastic Models for Spike Trains of Single Neurons

Volume 17: T. Maruyama
Stochastic Problems in Population Genetics

Volume 18:
Mathematics and the Life Sciences
Selected Lectures, Canadian Mathematical Congress, August 1975
Editor: D. E. Matthews

Springer-Verlag
Berlin
Heidelberg
New York